Design of Structural Timber
to Eurocode 5

William M. C. McKenzie BSc PhD CPhys MInstP CEng
Teaching Fellow, Napier University, Edinburgh

Binsheng Zhang BEng MSc PhD
Lecturer, Napier University, Edinburgh

palgrave
macmillan

First published 2007 by
PALGRAVE MACMILLAN
Houndmills, Basingstoke, Hampshire RG21 6XS and
175 Fifth Avenue, New York, N.Y. 10010
Companies and representatives throughout the world

PALGRAVE MACMILLAN is the global academic imprint of the Palgrave Macmillan division of St. Martin's Press, LLC and of Palgrave Macmillan Ltd. Macmillan® is a registered trademark in the United States, United Kingdom and other countries. Palgrave is a registered trademark in the European Union and other countries.

ISBN 978–0230–00777–2

This book is printed on paper suitable for recycling and made from fully managed and sustained forest sources. Logging, pulping and manufacturing processes are expected to conform to the environmental regulations of the country of origin.

A catalogue record for this book is available from the British Library.

10   9   8   7   6   5   4   3   2   1
16   15   14   13   12   11   10   09   08   07

Printed and bound in Great Britain by
Cromwell Press Ltd, Trowbridge, Wiltshire

Contents

Preface

Rationale

Existing design textbooks for undergraduate engineering students neglect, to a large extent, the importance of timber as a structural building material. As a consequence, relatively few textbooks provide information on the design of timber structures. *Design of Structural Timber to Eurocode 5* has therefore been written to:

♦ provide a comprehensive source of information on practical timber design,
♦ introduce the nature and inherent characteristics of timber given in relation to the requirements of the structural Eurocodes,
♦ provide detailed guidance on the use of Eurocodes EC and EC1 with respect to structural loading,
♦ provide detailed guidance on the use of EC5 with respect to structural timber design.

The book's content ranges from an introduction to timber as a material to the design of realistic structures including and beyond that usually considered essential for undergraduate study.

Readership

Design of Structural Timber is written primarily for undergraduate: civil, structural, building engineers and architects. Chapters 2 and 3 provide detailed information and fully worked examples relating to the assessment of structural loading, i.e. dead, imposed, snow and wind loading.

Whilst Chapter 4 provides a summary of elastic analysis techniques frequently used to determine load effects, students will normally be expected to have covered these techniques in their courses on structural theory and analysis.

The book will also provide an invaluable reference source for practising engineers in many building, civil and architectural design offices.

Worked Examples

The design of structures/elements is explained and illustrated using numerous detailed, relevant and practical worked examples. These design examples are presented in a format typical of that used in design office practice in order to encourage students to adopt a methodical and rational approach when preparing structural calculations.

Design Codes

It is essential when undertaking structural design to make frequent reference to the appropriate design codes. Students are encouraged to do this whilst using this text. It is assumed that readers will have access to either "*Structural Eurocodes. Guide to the Structural Eurocodes for students of structural design*" or the complete versions of the necessary codes.

William M.C. McKenzie
Binsheng Zhang

x

To Lena and Ling, Caroline, Karen and Gordon.

Acknowledgements

The authors wish to thank Mr. Peter Steer and Dr. Bill Chan for their valuable technical advice. We aslo wish to thank Caroline and Lena for their endless suppport, encouragement and patient proof-reading.

Permission to reproduce extracts from BS EN 1995-1-1:2004, BS EN 14081-1:2005, BS EN 363:2003 and BS 5268-3:2006 is granted by BSI. British Standards can be obtained from BSI Customer Services, 389 Chiswick High Road, London W4 4AL. Tel: +44 (0)20 8996 9001. email: cservices@bsi-global.com

Cover photo: Scottish Parliament Debating Chamber © Scottish Parliamentary copyright material is reproduced with the permission of the Queen's Printer for Scotland on behalf of the Scottish Parliamentary Corporate Body.

1. Structural Timber

Objective: to introduce the inherent botanical and structural characteristics of timber in addition to the classification and philosophies used in the design process.

1.1 Introduction

The use of timber as a structural material probably dates back to primitive times when man used fallen trees to bridge streams and gain access to hunting grounds or new pastures. It is only during the latter half of the twentieth century that detailed knowledge regarding the physical properties and behaviour of timber has been developed on a scientific basis and subsequently used in design. The inherent variability of a material such as timber, which is unique in its structure and mode of growth, results in characteristics and properties which are distinct and more complex than those of other common structural materials such as concrete, steel and brickwork. Some of the characteristics which influence design and are specific to timber are:

♦ the moisture content,
♦ the difference in strength when loads are applied parallel and perpendicular to the grain direction,
♦ the duration of the application of the load,
♦ the method adopted for strength grading of the timber.

As a live growing material, every identified tree has a name based on botanical distinction, for example '*Pinus sylvestris*' is commonly known as 'Scots pine.' The botanical names have a Latin origin, the first part indicating the genus, e.g. '*Pinus*', and the second part indicating the species, e.g. '*sylvestris.*'

A general botanical classification of trees identifies two groups: endogenous and exogenous.

♦ **Endogenous:** This type includes trees such as palms and bamboos which are inward growing and are generally found in the tropics; they are not considered in this text.

♦ **Exogenous:** This type is outward growing and includes all of the commercial timbers used for construction in the UK. There is a sub-division of exogenous trees into two main groups which are familiar to most designers; they are softwoods and hardwoods. These timbers are considered in this text and in the Eurocode BS 1995-1-1:2004 'Design of Timber Structures.' Softwoods such as pine, Douglas fir and spruce supply the bulk of the world's commercial timber.

Hardwoods including timbers such as iroko, teak, keruing and greenheart comprise the vast majority of species throughout the world.

This type of classification is of little value to a structural designer and consequently design codes adopt a classification based on stress grading. Stress grading is discussed in Section 1.4. The growth of a tree depends on the ability of the cells to perform a number of functions, primarily conduction, storage and mechanical support. The stem (or trunk) conducts essential mineral salts and moisture from the roots to the leaves, stores food materials and provides rigidity to enable the tree to compete with surrounding vegetation for air and sunlight. Chemical processes, which are essential for growth, occur in the branches, twigs and leaves in the crown of the tree. A typical cross-section of a tree trunk is shown in Figure 1.1.

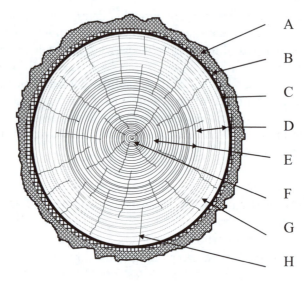

Figure 1.1   Typical cross-section of tree trunk

- **A Outer bark:** dry dead tissue and providing general protection against external elements.
- **B Inner bark:** moist, soft material which conducts food synthesised by the leaves to all growing parts of the tree.
- **C Cambium layer:** a microscopic layer on the inside of the inner bark. Cell division within this layer is responsible for growth in the thickness of the tree.
- **D Sapwood:** the younger growth containing living cells which store foods and conduct moisture from the roots to the leaves. This zone of wood is normally lighter in colour than the inner heartwood and varies in thickness from approximately 30 mm in Western red cedar and Douglas fir to thicknesses in excess of 75 mm in maples, white ash and some pines.
- **E Heartwood:** the inner layers in which the sapwood has become inert, with the cells dying and the remaining food stores undergoing chemical change to produce substances such as tannin, a colourless amorphous mass. The main

function of the inactive heartwood tissue is to provide mechanical rigidity to the tree.

- ◆ **F Pith:** the core of the tree around which the first wood growth takes place.
- ◆ **G Growth rings:** changes occurring in timber grown in climates having distinct seasonal changes. These rings are the result of cycles of growth and rest, where changes occur in the tissues formed between the beginning and the end of the growing season. In general, tissues formed at the beginning of a season conduct moisture, whilst toward the end of a season the demand for moisture diminishes and the conduction vessels become smaller with proportionately thicker walls. A consequence of this is that springwood tends to be light, porous and relatively weak and summerwood stronger and more dense.
- ◆ **H Rays:** narrow bands of tissue running radially across the growth rings. Their purpose is to store and conduct food to the various layers between the pith and the bark.

1.2 Moisture Content

Unlike most structural materials, the behaviour of timber is significantly influenced by the existence and variation of its moisture content. The moisture content, as determined by oven drying of a test piece, is defined in Clause 1.5.2.7 of BS EN 1995-1-1 as '*The mass of water in wood expressed as a proportion of its oven-dry mass*', i.e.

$$w = 100(m_1 - m_2)/m_2$$

where:
w is the percentage moisture content,
m_1 is the mass of the test piece before drying (in g),
m_2 is the mass of the test piece after drying (in g).

Moisture contained in 'green' timber is held both within the cells (free water) and the cell walls (bound water). The condition where all free water has been removed but the cell walls are still saturated is known as the 'fibre saturation point' (FSP). At levels of moisture above the FSP, most physical and mechanical properties remain constant. Variations in moisture content below the FSP cause considerable changes to properties such as weight, strength, elasticity, shrinkage and durability. The controlled drying of timber is known as seasoning. There are two methods generally used:

- ◆ **Air seasoning** in which the timber is stacked and layered with air-space in open sided sheds to promote natural drying. This method is relatively inexpensive with very little loss in the quality of timber if carried out correctly. It has the disadvantage that the timber and the space it occupies are unavailable for long periods. In addition, only a limited control is possible by varying the spaces between the layers and/or by using mobile slatted sides to the sheds.
- ◆ **Kiln drying** in which timber is dried out in a heated, ventilated and humidified oven. This requires specialist equipment and is more expensive in terms of energy input. The technique does offer a more controlled environment in which to achieve the required reduction in moisture content and is much quicker.

The anisotropic nature of timber and differential drying out caused by uneven exposure to drying agents e.g. wind, sun or applied heat can result in a number of defects such as *twisting*, *cupping*, *bowing* and *cracking* as shown in Figure 1.2.

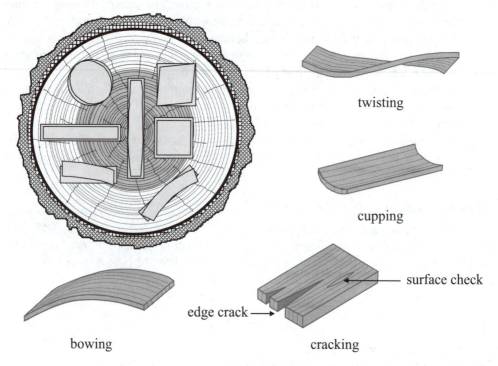

twisting

cupping

surface check

edge crack ⟶

bowing   cracking

Figure 1.2   Distortions due to differential directional shrinkage

1.3 Defects in Timber

In addition to the defects indicated in Figure 1.2 there are a number of naturally occurring defects in timber. The most common and familiar of such defects is a ***knot*** (see Figure 1.3). Normal branch growth originates near the pith of a tree and consequently its base develops new layers of wood each season which develop with the trunk. The cells of the new wood grow into the lower parts of the branches, maintaining a flow of moisture to the leaves. The portion of a branch which is enclosed within the main trunk constitutes a *live* or *intergrown* knot and has a firm connection with surrounding wood.

When lower branches in forest trees die and drop off as a result of being deprived of sunlight, the dead stubs become overgrown with new wood but have no connection to it. This results in dead or enclosed knots which are often loose and, when cut, fall out.

The presence of knots is often accompanied by a decrease in the physical properties of timber such as the tensile and compressive strength. The reduction in strength is primarily due to the distortion of the grain passing around the knots and the large angle between the grain of the knot and the piece of timber in which it is present.

During the seasoning of timber, checks often develop around the location of knots.

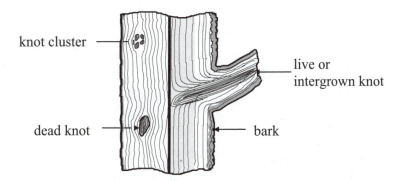

Figure 1.3   Defects due to knots

In a mill when timber is converted from a trunk into suitable commercial sizes (see Figure 1.4).

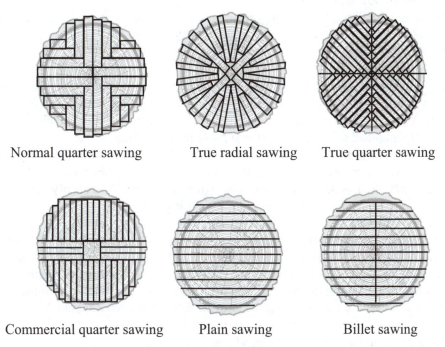

Normal quarter sawing   True radial sawing   True quarter sawing

Commercial quarter sawing   Plain sawing   Billet sawing

Figure 1.4 Typical sawing patterns

A *wane* can occur when part of the bark or rounded periphery of the trunk is present in a cut length as shown in Figure 1.5. The effect of a wane is to reduce the cross-sectional area with a resultant reduction in strength. A *shake* is produced when fibres separate along the grain: this normally occurs between the growth rings, as shown in Figure 1.5. The effect of a shake in the cross-section is to reduce the shear strength of beams; it does not significantly affect the strength of axially loaded members.

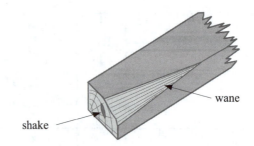

Figure 1.5   shake   wane

1.4 Classification of Timber

The efficient and economic use of any structural material requires a knowledge of its physical characteristics and properties such that a representative mathematical model can be adopted to predict behaviour in qualitative terms. Manufactured materials such as steel and concrete, in which quality can be tightly controlled and monitored during production, readily satisfy this. As a natural product, timber is subject to a wide range of variation in quality which cannot be controlled.

   Uniformity and reliability in the quality of timber as a structural material is achieved by a process of selection based on established grading systems, i.e. appearance grading and strength grading:

♦ appearance grading is frequently used by architects to reflect the warm, attractive features of the material such as the surface grain pattern, the presence of knots, colour, etc. In such circumstances the timber is left exposed and remains visible after completion and may be either structural or non-structural,

♦ all structural (load-bearing) timber must be strength graded according to criteria which reflect its strength and stiffness. In some cases timber may be graded according to both appearance and strength. Strength grading is normally carried out either visually or mechanically, using purpose built grading machines.

Every piece of strength graded timber should be marked clearly and indelibly with the following information in accordance with the BS EN 14081-1/2/3/4:2005 'Strength Graded Structural Timber – Rectangular Sections' This standard gives the requirements for machine strength graded timber and grading machines and visually graded timber. The marking should be in accordance with Section 7 of BS EN 14081-1 and include:

♦ the name or identifying mark of the producer,
♦ species code (Table 4 for species combination or Table B1 for single species),
♦ number of the European Standard used (e.g. BS EN 14081-1),
♦ identification of documentation providing essential information not given,
♦ grade and strength class (e.g. GS, SS, C16, C24),
♦ identification of the notified body,
♦ timber condition during grading if appropriate (i.e. Dry Graded).

A typical grading mark/stamp indicating the minimum information for untreated machine graded timber is illustrated in Figure 1.6.

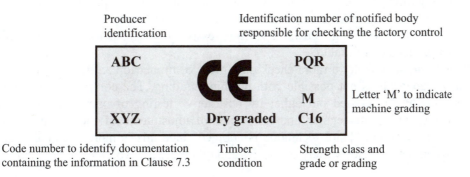

Figure 1.6

1.4.1 Visual Strength Grading

As implied by the name, this method of grading is based on the physical observation of strength-reducing defects such as knots, rate of growth, cracks, wane, bowing, etc. Since the technique is based on the experience and judgement of the grader it is inherently subjective. In addition, important properties such as density, which has a significant influence on stiffness, and strength are not considered. Numerous grading rules and specifications have been developed throughout Europe, Canada and the USA during the last fifty years. Visual grading is governed by the requirements of BS 4978 : 2007 – '*Visual strength grading of softwood*' 'and the Eurocode BS EN 14081-1/2: 2005 – '*Strength graded structural timber with rectangular cross section*'.

Visual defects considered when assessing timber strength include: location and extent of knots, slope of grain, rate of growth, fissures, wane, distortions such as bowing, springing, twisting, cupping, resin and bark pockets, and insect damage. Two strength grades are specified: General Structural (GS) and Special Structural (SS), the latter being the higher quality material. Timber which contains abnormal defects e.g. compression wood (see Section 1.5), insect damage such as worm holes, or fungal decay (not sapstain), or which is likely to impair the serviceability of the pieces, is excluded from the grades.

1.4.2 Machine Strength Grading

The requirements for machine strength grading are specified in BS EN 14081-1/2: 2005 – '*Strength graded structural timber with rectangular cross section*'. Timber is classified into:

♦ twelve classes of poplar and coniferous species ranging from the weakest grade C14 to the highest grade C50,
♦ six classes for deciduous species ranging from the weakest grade D30 to the highest grade D70.

In each case the number following either the 'C' or the 'D' represents the characteristic bending strength of the timber.

The inherently subjective nature of visual strength grading results in a lower yield of higher strength classes than would otherwise be achieved. Machine strength grading is generally carried out by conducting bending tests on planks of timber which are fed continuously through a grading machine. The results of such tests produce a value for the modulus of elasticity. The correlation between the modulus of elasticity and strength properties such as bending, tensile and compressive strength can be used to define a particular grade/class of timber.

Visual grading enables a rapid check on a piece of timber to confirm, or otherwise, an assigned grade; this is not possible with machine grading. The control and reliability of machine grading is carried out either by destructive testing of output samples (output controlled systems), or regular and strict control/adjustments of the grading machine settings. In Europe the latter technique is adopted since this is more economic than the former when using a wide variety of species and relatively low volumes of production.

1.5 Material Properties

The strength of timber is due to certain types of cells (called tracheids in softwoods and fibres in hardwoods) which make up the many minute hollow cells of which timber is composed. These cells are roughly polygonal in cross-section and the dimension along the grain is many times larger than across it. The principal constituents of the cells are cellulose and lignin. Individual cell walls comprise four layers, one of which is more significant with respect to strength than the others. This layer contains chains of cellulose which run nearly parallel to the main axis of the cell. The structure of the cell enhances the strength of the timber in the grain direction.

Density, which is expressed as mass per unit volume, is one of the principal properties affecting strength. The heaviest species, i.e. those with most wood substance, have thick cell walls and small cell cavities. They also have the highest densities and consequently are the strongest species. Numerous properties in addition to strength, e.g. shrinkage, stiffness and hardness, increase with increasing density.

When timber is seasoned the cell contents dry out leaving only cell walls. Shrinkage occurs during the drying process as absorbed moisture begins to leave the cell walls. The cell walls become thinner as they draw closer together. However the length of the cell layers is only marginally affected. A consequence of this is that as shrinkage occurs the width and thickness change but the length remains the same. The degree to which shrinkage occurs is dependent upon its initial moisture content value and the value at which it stabilises in service. A number of defects such as bowing, cupping, twisting and surface checks are a direct result of shrinkage.

Since timber is hygroscopic, and can absorb moisture whilst in service, it can also swell until it reaches an equilibrium moisture content.
Anisotropy is a characteristic of timber because of the long fibrous nature of the cells and their common orientation, the variation from early to late wood, and the differences between sapwood and hardwood. The elastic modulus of a fibre in a direction *along* its axis is considerably greater than that *across* it, resulting in the strength and elasticity of timber parallel to the grain being much higher than in the radial and tangential directions.
The slope of the grain can have an important effect on the strength of a timber member.
Typically a reduction of 4% in strength can result from a slope of 1 in 25, increasing to an 11% loss for slopes of 1 in 15.

The strength of timber is also affected by the ratio of growth as indicated by the width of the annual growth rings. For most timbers the number of growth rings to produce the optimum strength is approximately in the range of 6–15 per 25 mm measured radially. Timber which has grown either much more quickly or much more slowly than that required for the optimum growth rate is likely to be weaker.

In timber from a tree which has grown with a pronounced lean, wood from the compression side (compression wood) is characterised by much greater shrinkage than normal. In softwood planks containing compression wood, bowing is likely to develop in the course of seasoning and the bending strength will be low. In hardwoods, the tension wood has abnormally high longitudinal shrinkage and although stronger in tension is much weaker in longitudinal compression than normal wood.

Like many materials, e.g. concrete, the stress–strain relationship demonstrated by timber under load is linear for low stress values. For all species the strains for a given load increase with moisture content. A consequence of this is that the strain in a beam under constant load will increase in a damp environment and decrease as it dries out again.

Timber demonstrates viscoelastic behaviour (creep) as high stress levels induce increasing strains with increasing time. The magnitude of long-term strains increases with higher moisture content. In structures where deflection is important, the duration of the loading must be considered. This is reflected in BS EN 1995-1-1:2004 by the use of modifying factor applied to the characteristic strength. The factor (k_{mod}) is dependent on the type of loading, e.g. long-term, medium-term, short- and very short-term and the moisture content (see Section 4.2 of Chapter 4).

The cellular structure of timber results in a material which is a poor conductor of heat. The air trapped within its cells greatly improves its insulating properties. Heavier timbers having smaller cell cavities are better conductors of heat than lighter timbers. Timber does expand when heated but this effect is more than compensated for by the shrinkage caused by loss of moisture.

The fire resistance of timber generally compares favourably with other structural materials and is often better than most. Steel is subject to loss of strength, distortion, expansion and collapse, whilst concrete may spall and crack.

Whilst small timber sections may ignite easily and support combustion until reduced to ash, this is not the case with large structural sections. At temperatures above 250°C material at the exposed surface decomposes, producing inflammable gases and charcoal. These gases, when ignited, heat the timber to a greater depth and the fire continues. The charcoal produced during the fire is a poor conductor and will eventually provide an insulating layer between the flame and the unburned timber. If there is sufficient heat, charcoal will continue to char and smoulder at a very slow rate, particularly in large timber sections with a low surface to mass ratio.

Fire authorities usually consider that a normal timber door will prevent the spread of fire to an adjoining room for about 30 minutes. The spread of fire is then often due to flames and hot gases permeating between the door and its frame or through cracks between door panels and styles produced by shrinkage (see BS EN 1995-1-2:2004 'Design of timber structures – Part 1-2: General – Structural fire design').

The durability of timbers to resist the effects of weathering, chemical or fungal attack varies considerably from one species to another. In general the heartwood is more durable to fungal decay than the sapwood. This is due to the presence of organic compounds within the cell walls and cavities which are toxic to fungi and insects. Provided timber is kept dry, or is continuously immersed in fresh water, decay will generally not be a problem. Where timber is used in seawater, particularly in harbours, there is always a risk of severe damage due to attack by molluscs. The pressure impregnation of timber with suitable preservatives will normally be sufficient to prevent damage due to fungal, insect or mollusc attack.

1.6 Preservative Treatments

A number of chemical treatments are available to prevent the degeneration of timber due to fungi, insect or mollusc attack. The extent to which a structural member is susceptible to attack is dependent on several factors including the species and the environmental conditions. The treatments can be classified into three types:

- ♦ tar oil preservatives,
- ♦ organic solvents,
- ♦ water borne preservatives.

Tar oil preservatives such as creosote are restricted in their use to a limited type of external members (railway sleepers, telegraph poles, fences, etc.) because of their eco-toxicological characteristics.

Organic solvent preservatives are widely used and have the advantage of being readily absorbed by the timber, even when applied using simple techniques such as brushing, spraying or dipping,

Water borne solvents, e.g. chromated copper arsenate, are the most widely used preservatives and are normally introduced into the timber under pressure. Preservative treatments should be selected in accordance with BS EN 351-1:1996 'Durability of wood and wood-based products – Preservative treated solid wood. Part 1: Classification of preservative penetration and retention' and BS EN 460:1994 'Durability of wood and wood-based products – Natural durability of solid wood – Guide of the durability requirements for wood to be used in hazard classes.'

2. Design Philosophies and Eurocodes

Objective: *to introduce design philosophies and Eurocodes as used in the design process.*

2.1 Design Philosophies

The successful completion of any structural design project is dependent on many variables. There are, however, a number of fundamental objectives, which must be incorporated in any design philosophy to provide a structure and which, throughout its intended lifespan:

(i)   will possess an acceptable margin of safety against collapse whilst in use,
(ii) is serviceable and will perform its intended purpose whilst in use,
(iii) is sufficiently robust such that damage to an extent disproportionate to the original cause will not occur,
(iv) is economic to construct and
(v)   is economic to maintain.

Historically, structural design was carried out on the basis of intuition, trial and error, and experience which enabled empirical *design rules*, generally relating to structure/member proportions, to be established. These rules were used to minimise structural failures and consequently introduced a *margin-of-safety* against collapse. In the latter half of the 19th century the introduction of modern materials and the development of mathematical modelling techniques led to the introduction of a design philosophy which incorporated the concept of *factor-of-safety* based on known material strength, e.g. ultimate tensile stress; this is known as **permissible stress design**. During the 20th century two further design philosophies were developed and are referred to as **load-factor** design and **limit-state** design; each of the three philosophies is discussed separately in Sections 2.2 to 2.4.

2.2 Permissible Stress Design

When using permissible stress design, the margin of safety is introduced by considering structural behaviour under working/service load conditions and comparing the stresses under these conditions with permissible values. The permissible values are obtained by dividing the failure stresses by an appropriate factor of safety. The applied stresses are determined using elastic analysis techniques, i.e.

$$stress\ induced\ by\ working\ loads \leq \frac{failure\ stress}{factor\ of\ safety}$$

2.3 Load Factor Design

When using load factor design, the margin of safety is introduced by considering structural behaviour at collapse load conditions. The ultimate capacities of sections based on yield strength (e.g. axial, bending moment and shear force capacities) are compared with the design effects induced by the ultimate loads. The ultimate loads are determined by multiplying the working/service loads by a factor of safety. Plastic methods of analysis are used to determine section capacities and design load effects. Despite being acceptable, this method has never been widely used.

$$\begin{array}{ccc} \textit{Ultimate design load effects due to} & & \textit{Ultimate capacity based on the} \\ \textit{(working loads} \times \textit{factor of safety)} & \leq & \textit{failure stress of the material} \end{array}$$

2.4 Limit State Design

The limit state design philosophy, which was formulated for reinforced concrete design in Russia during the 1930s, achieves the objectives set out in Section 2.1 by considering two 'types' of limit state under which a structure may become unfit for its intended purpose. They are:

(i) the *Ultimate Limit State* in which the structure, or some part of it, is unsafe for its intended purpose, e.g. compressive, tensile, shear or flexural failure or instability leading to partial or total collapse and

(ii) the *Serviceability Limit State* in which a condition, e.g. deflection, vibration or cracking, occurs to an extent, which is unacceptable to the owner, occupier, client, etc.

The basis of the approach is statistical and lies in assessing the probability of reaching a given limit state and deciding upon an acceptable level of that probability for design purposes. The method in most codes is based on the use of *characteristic values* and *partial safety factors*.

Partial Safety Factors: The use of partial safety factors, which are applied separately to individual parameters, enables the degree of risk for each one to be varied. This reflects the differing degrees of control which are possible in the manufacturing process of building structural materials/units (e.g. steel, concrete, timber, mortar and individual bricks) and construction processes such as steel fabrication, in-situ/pre-cast concrete, or building in masonry.

Characteristic Values: The use of characteristic values enables the statistical variability of various parameters such as material strength, different load types, etc., to be incorporated in an assessment of the acceptable probability that the value of the parameter will be exceeded during the life of a structure. The term 'characteristic' in current design codes normally refers to a value of such magnitude that statistically for loads, there is a 5% probability of it being exceeded, whilst for strengths, there is a 5% probability of the actual strength being less.

In the design process the characteristic loads are multiplied by the partial safety factors to obtain the design values of design effects such as axial or flexural stress, and the design strengths are obtained by dividing the characteristic strengths by appropriate partial safety factors for materials. To ensure an adequate margin of safety the following must be satisfied:

$$\textbf{Design strength} \quad \geq \quad \textbf{Design load effects}$$

e.g. $\qquad \dfrac{f_k}{\gamma_m} \geq [(\text{stress due to } G_k \times \gamma_{f,dead}) + (\text{stress due to } Q_k \times \gamma_{f,imposed}) + \ldots\ldots]$

where:

f_k	is the characteristic compressive strength,
γ_m	is the partial safety factor for materials,
G_k	is the characteristic dead load,
Q_k	is the characteristic imposed load,
$\gamma_{f,dead}$	is the partial safety factor for dead loads,
$\gamma_{f,imposed}$	is the partial safety factor for imposed loads.

The limit state philosophy can be expressed with reference to frequency distribution curves for design strengths and design effects, as shown in Figure 2.1.

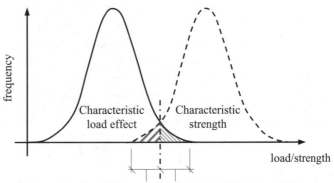

Figure 2.1   5% probability of strength being below   5% probability of exceeding the
the characteristic strength value   characteristic load effect

The shaded area represents the probability of failure, i.e. the level of design load effect, which can be expected to be exceeded by 5%, and the level of design strength which 5% of samples can be expected to fall below. The point of intersection of these two distribution curves represents the ultimate limit state, i.e. the design strength equals the design load effects.

The partial safety factors represent the uncertainty in the characteristic values. The lack of detailed statistical data on all of the parameters considered in design and the complexity of the statistical analysis has resulted in the use of a more subjective assessment of the values of partial safety factors than is mathematically consistent with the philosophy.

2.5 Eurocodes

The European Standards Organisation, CEN, is the umbrella organisation under which a set of common structural design standards (EC1, EC2, EC3, etc.) are being developed. The Structural Eurocodes are the result of attempts to eliminate barriers to trade throughout the European Union. Separate codes exist for each structural material, as indicated in Table 2.1. The basis of design and loading considerations are included in EC1.

Each country publishes its own European Standards (EN - EuroNorm), e.g. in the UK the British Standards Institution (BSI) issues documents (which are based on the Eurocodes developed under CEN), with the designation BS EN.

Prior to publication of the final EN versions, structural Eurocodes are issued as Pre-standards (ENV – Euronorm Vornrom) which can be used as an alternative to existing national rules. In the UK the BSI has used the designation DD ENV; the pre-standards are equivalent to the traditional 'Draft for development' Documents.

CEN Number	Ref.	Title of Eurocode
EN 1990	**EC**	**Eurocode: Basis of structural design**
EN 1991	**EC1**	**Eurocode 1: Actions on structures**
EN 1992	EC2	Eurocode 2: Design of concrete structures
EN 1993	EC3	Eurocode 3: Design of steel structures
EN 1994	EC4	Eurocode 4: Design of composite steel and concrete structures
EN 1995	**EC5**	**Eurocode 5: Design of timber structures**
EN 1996	EC6	Eurocode 6: Design of masonry structures
EN 1997	EC7	Eurocode 7: Geotechnical design
EN 1998	EC8	Eurocode 8: Design of structures for earthquake resistance
EN 1999	EC9	Eurocode 9: Design of aluminium structures

Table 2.1

Each Eurocode generally consists of a number of parts, e.g.

BS EN 1991-1-1:2002   Eurocode 1: Actions on structures -
General actions - Densities, self-weight, imposed loads on buildings

BS EN 1991-1-2:2003   Eurocode 1: Actions on structures –
General actions -
Actions on structures exposed to fire

BS EN 1991-1-3:2003   Eurocode 1: Actions on structures -
General actions - Snow loads

BS EN 1991-1-4:2002   Eurocode 1: Actions on structures -
General actions - Wind actions

BS EN 1991-2:2003   Eurocode 1: Actions on structures - Traffic loads on bridges

BS EN 1991-3:2006   Eurocode 1: Actions on structures -
Actions induced by cranes and machines

After approval of an EN by CEN the Standard reaches the **Date of Availability** (DAV). During the two years following this date, National Calibration is expected to be carried out in each country to determine the **Nationally Determined Parameters** (NDPs). The NDPs represent the safety factors to be adopted when using the EN, and setting their values remains the prerogative for each individual country.

Following this will be a **Co-Existence Period** for a maximum of three years during which each country will withdraw all of the existing National Codes which have a similar scope to the EC being introduced, (this is scheduled to be completed by each country on or before, 1st April 2010).

The development of the Eurcodes is intended to facilitate both 'everyday design' requirements and more complex, innovative design, without being too onerous for either.

2.5.1 National Annex

Each country which issues a European Standard also issues a **National Annex** currently published in the UK as a 'NAD' – National Application Document, for use with the EN. There is no legal requirement for a country to produce a National Annex; however, it is recommended to do so.

The harmonisation of structural design practice must accommodate the wide ranging and varied geographical/geophysical conditions which exist throughout central Europe.

The purpose of the National Annex is to provide information to designers relating to product standards for materials, partial safety factors and any additional rules and/or supplementary information specific to design within that country.

(**Note:** The contents of the core document of an EC cannot be changed to include the NDPs being adopted: these should be provided if required in a separate document.)

2.5.2 Normative and Informative Text

There are two types of information contained within the core documents: **normative** and **informative**. Normative Annexes have the same status as the main body of the text whilst Informative Annexes provide additional information. The Annexes generally contain more detailed material or material which is used less frequently.

2.5.3 Non-Contradictory Complimentary Information (NCCI)

This information may be given in the National Annex and contains additional information or guidance which does not contradict the Eurocode.

2.5.4 Terminology, Symbols and Conventions

The terminology, symbols and conventions used in **Eurocode 5** differ from those used by current British Standards. The code indicates **Principles** which are general statements and definitions which must be satisfied and **Rules** which are recommended design procedures which comply with the Principles. The Rules can be substituted by alternative procedures provided that these can be shown to be in accordance with the Principles,
e.g. in Section 3 of BS EN 1990:2002,

'Clause 3.1(1)P A distinction shall be made between ultimate limit states and serviceability limit states.'

e.g. in Section 3 of BS EN 1990:2002,
'Clause 3.1(2) Verification of one of the two categories of limit states may be omitted provided that sufficient information is available to prove that it is satisfied by the other.'

2.5.4.1 Decimal Point

Standard ISO practice has been adopted in representing a decimal point by a comma, i.e. $5,3 \equiv 5.3$. In the UK, to avoid confusion, engineers should avoid using a comma to indicate 'thousands' (as is common practice) if they do not adopt Standard ISO practice.

Initially a Pre-standard 'ENV Eurocode 1: Basis of design and actions on structures' was produced as the primary document upon which all other codes would be developed. This was subsequently changed resulting in:

'BS EN 1990:2002   Eurocode: Basis of structural design'   and

'BS EN 1991:2002   Eurocode 1: Actions on structures' (see Chapter 3)

2.6 BS EN 1990:2002 (Eurocode - Basis of Structural Design)

As indicated in Clause 1.1(1) of BS EN 1990:

'EN 1990 establishes Principles and requirements for the safety, serviceability and durability of structures, describes the basis for their design and verification and gives guidelines for related aspects of structural reliability.'

The 'basic requirements' of Eurocode are that all structures are required to have adequate:

- ♦ **Resistance** (strength)   Clause 2.1(2)P,
- ♦ **Serviceability**   Clause 2.1(2)P,
- ♦ **Durability**   Clause 2.1(2)P and Clause 2.4,
- ♦ **Fire resistance**   Clause 2.1(3)P   and
- ♦ **Robustness**   Clause 2.1(4)P.

Compliance with the Principles and Rules set out in the Eurocodes should ensure that these criteria are satisfied. It is presumed that design and construction in accordance with the Eurocodes is carried out/supervised by appropriately qualified and experienced personnel.

There are four categories of design situations defined in the code, they are:

- (i)   **Persistent:** conditions of normal use,
- (ii) **Transient:** temporary conditions such as during construction or repair,
- (iii) **Accidental:** exceptional conditions, e.g. fire, impact loading, explosion, etc.,
- (iv) **Seismic:** due to earthquake activity.

The applied forces on a structure are referred to as either '**direct**' or '**indirect**' actions such as those caused by e.g. shrinkage or settlement; actions are classified using two methods in the code, i.e.

In Clause 4.1.1.(1)P in terms of their variation in time:
♦ **Permanent,**
♦ **Variable**   or
♦ **Accidental**,

In Clause 4.1.1(4)P by their:
♦ **origin – direct or indirect,**
♦ **spatial variation - fixed or free**   or
♦ **nature and/or structural response - static or dynamic.**

The principal variables and their symbols, as used in Eurocode are summarized in Sections 2.6.1.1 to 2.6.1.4 and Table 2.2 and Table 2.3.

2.6.1 F: Action

A force (load) applied to a structure or an imposed deformation (indirect action), such as temperature effects or settlement, i.e.

G: permanent action such as dead loads due to self-weight, e.g.
Characteristic value of a permanent action = G_k
Design value of a permanent action = G_d
Favourable design value of a permanent action = $G_{d,inf}$
Unfavourable design value of a permanent action = $G_{d,sup}$

Q: variable actions such as imposed, wind or snow loads,
Characteristic value of a variable action = Q_k
Design value of a variable action = Q_d

A: accidental actions such as explosions, fire or vehicle impact.

E: *effect of actions* on static equilibrium or of gross displacements, etc., e.g.
Design effect of a destabilising action = $E_{d,dst}$
Design effect of a stabilising action = $E_{d,stb}$

Note: $E_{d,dst} \leq E_{d,stb}$

2.6.2 R: Design Resistance of Structural Elements

Design axial resistance = N_{Rd}
Design shear resistance = V_{Rd}
Design moment resistance = M_{Rd}

2.6.3 S: Design Value of Actions

Factored values of externally applied loads or load effects such as axial load, shear force, bending moment, etc., e.g.

Design axial load $\qquad N_{Sd} \leq N_{Rd}$
Design shear force $\qquad V_{Sd} \leq V_{Rd}$
Design bending moment $\quad M_{Sd} \leq M_{Rd}$

2.6.4 X: Material Property

Physical properties such as tension, compression, shear and bending strength, modulus of elasticity, etc., e.g.

Characteristic compressive strength $= X_k = f_k$

Design compressive strength $= X_d = \dfrac{X_k}{\gamma_M} = f_d$

Variables:			
A	area	*k*	coefficient or factor (used with a subscript)
F	force	*l*	span; contact length
E	modulus of elasticity	*m*	mass
G	shear modulus	*r*	radius
I	second moment of area	*s*	spacing
L	length	*t*	thickness
M	bending moment	*u*	deformation
N	axial force	*w*	deflection
R	capacity (resistance)	*v*	unit impulse velocity response
V	shear force; volume	α	angle
W	elastic section modulus	β	angle
X	strength property	γ	angle; partial factor
a	distance	λ	slenderness ratio
b	width	ρ	density
d	diameter	σ	normal stress
f	strength	τ	shear stress
h	depth; distance	ψ	combination factor
i *	notch inclination	ζ	modal damping ratio
* Although not given in the code *i* is also used for the radius of gyration			

Table 2.2

Subscripts:			
c	compression	q or Q	variable action
cr	critical	Rd	design capacity
d	design value	rel	relative
ef	effective	Rk	characteristic capacity
Ed	design value	sup	superior; upper
f	flange	ser	serviceability
g	permanent action	sys	system
inf	inferior; lower	t	tension
inst	instantaneous	tor	torsional
k	characteristic	u	ultimate
l	low; lower	v	shear
ls	load sharing	w	web
m	material or bending	y	direction of the major principal axis
max	maximum	z	direction of the minor principal axis
mean	mean value	α	angle between force/stress and grain direction
min	minimum	0	parallel to the grain direction
mod	modification	90	perpendicular to the grain direction
nom	nominal	0,05	fifth percentile value

Table 2.3

Multiple subscripts are used to denote variables; e.g.
$\sigma_{m,y,d}$ is the design (d), bending (m) stress (σ) about the principal y-y axis (y),
$f_{m,y,d}$ is the corresponding design bending strength.

2.6.5 Limit State Design

The limit states are states beyond which a structure can no longer satisfy the design performance requirements (see Section 2.4). The two classes of limit state are:

- ◆ *Ultimate limit states:* These include failures such as full or partial collapse due to e.g. rupture of materials, excessive deformations, loss of equilibrium or development of mechanisms. Limit states of this type present a direct risk to the safety of individuals.

- ◆ *Serviceability limit states:* Whilst not resulting in a direct risk to the safety of people, serviceability limit states still render the structure unsuitable for its intended purpose. They include failures such as excessive deformation resulting in unacceptable appearance or non-structural damage, loss of durability or excessive vibration causing discomfort to the occupants.

The limit states are quantified in terms of design values for actions, material properties and geometric characteristics in any particular design. Essentially the following conditions must be satisfied:

Ultimate limit state: **Rupture** $S_d \le R_d$

where:
S_d is the design value of the effects of the actions imposed on the structure/structural elements,
R_d is the design resistance of the structure/structural elements to the imposed actions.

Similarly: **Stability** $S_{d,dst} \le R_{d,stb}$

where:
$S_{d,dst}$ is the design value of the destabilising effects of the actions imposed on the structure (including self-weight where appropriate).
$S_{d,stb}$ is the design value of the stabilising effects of the actions imposed on the structure (including self-weight where appropriate).

The Ultimate Limit States are considered in four categories in Clause 6.4.1 of Eurocode. They are:

  (i) **EQU:** relating to the static equilibrium of a structure or any part of it which is considered as a rigid body,
  (ii) **STR:** relating to internal failure or excessive deformation of a structure or structural member,
  (iii) **GEO:** relating to failure or excessive deformation of the ground,
  (iv) **FAT:** relating to fatigue failure of structural members.

Serviceability limit state: **Serviceability** $S_d \le C_d$
where:
S_d is the design value of the effects of the actions imposed on the structure/structural elements,
C_d is a prescribed value, e.g. a limit of vibration.

2.6.6 Combination Values
Combination values in Eurocodes relate to the values of design actions which are to be considered acting simultaneously. The combination value makes an allowance for the reduced probability that the full value of all variable actions will occur together.

When a number of variable actions are possible each action should be considered in turn as a leading action (i.e. at its full value) and combined with the other actions considered as accompanying variables. The accompanying action is always considered as the combination value.

Three different types of combination are given in Section 6.4.3 of Eurocode. They are:

- ♦ Combinations of actions for persistent or transient design situations (fundamental combinations),
- ♦ Combinations of actions for accidental design situations,
- ♦ Combinations of actions for seismic design situations.

In each case equations are given in the code from which design values of load can be determined.

2.6.7 Representative Values

The representative value of an action is the value used for the verification of a limit state. A representative value may be the characteristic value (F_k), a nominal value or an accompanying value (ψF_k) in a combination as indicated in Clause 1.5.3.20 of EC.

2.6.8 Design Values

The term *design* is used for factored loading and member resistance.

Design loading $\quad F_d =$ partial safety factor $(\gamma_f) \times$ characteristic value (F_k)

e.g. $\qquad\qquad G_d = \gamma_G G_k$

where:

γ_G is the partial safety factor for permanent actions,

G_k is the characteristic value of permanent actions.

Note: $G_{d,sup}$ $(= \gamma_{G,sup} G_{k,sup}$ or $\gamma_{G,sup} G_k)$ represents the 'upper' design value of a permanent action,

$G_{d,inf}$ $(= \gamma_{G,inf} G_{k,inf}$ or $\gamma_{G,inf} G_k)$ represents the 'lower' design value of a permanent action.

Design resistance $\quad (R_d) = \dfrac{\text{material characteristic strength } (X_k)}{\text{material partial safety factor } (\gamma_m)}$

e.g.   Design bending strength about the y-y axis $= f_{m,y,d} = \dfrac{f_{m,y,k}}{\gamma_m}$

The design values of the actions vary depending upon the limit state being considered. All of the possible load cases should be considered as follows:

Ultimate Limit States:

When considering persistent and transient design situations two options are given in Clause 6.4.3.2:

(i) $\quad F_d = \displaystyle\sum_{j\geq1}\gamma_{G,j}G_{k,j} \; "+" \; \gamma_p P \; "+" \; \gamma_{Q,1}Q_{k,1} \; "+" \; \sum_{i>1}\gamma_{Q,i}\psi_{0,i}Q_{k,i}$

(Equation (6.10) from EC)

Alternatively, the following can be used,

(ii) the less favourable of:

$$F_d = \sum_{j\geq1}\gamma_{G,j}G_{k,j} \ "+" \ \gamma_p P \ "+" \ \gamma_{Q,1}\psi_{0,1}Q_{k,1} \ "+" \ \sum_{i>1}\gamma_{Q,i}\psi_{0,i}Q_{k,i}$$

(Equation (6.10(a)) from EC)

$$F_d = \sum_{j\geq1}\xi_j\gamma_{G,j}G_{k,j} \ "+" \ \gamma_p P \ "+" \ \gamma_{Q,1}Q_{k,1} \ "+" \ \sum_{i>1}\gamma_{Q,i}\psi_{0,i}Q_{k,i}$$

(Equation (6.10(b)) from EC)

When considering accidental design situations: Clause 6.4.3.3

$$F_d = \sum_{j\geq1}G_{k,j} \ "+" \ P \ "+" \ A_d \ "+" \ \left(\psi_{1,1} \ or \ \psi_{2,1}\right)Q_{k,1} \ "+" \ \sum_{i>1}\psi_{2,i}Q_{k,i}$$

(Equation (6.11(b)) from EC)

When considering seismic design situations: Clause 6.4.3.4

$$F_d = \sum_{j\geq1}G_{k,j} \ "+" \ P \ "+" \ A_{Ed} \ "+" \ \sum_{i\geq1}\psi_{2,i}Q_{k,i}$$

(Equation (6.12(b)) from EC)

Note: In the case of the latter two, partial safety factors are not included, i.e. these events generally occur when a structure is in use.

Serviceability Limit States:
Three possible combinations of actions are considered in Clause 6.5.3 with respect to serviceability. They are:
(i) the *characteristic combination* when considering the function of a structure and damage to structural/non-structural elements and irreversible limit states:

$$F_d = \sum_{j\geq1}G_{k,j} \ "+" P \ "+" \ Q_{k,1} \ "+" \ \sum_{i>1}\psi_{0,i}Q_{k,i}$$

(Equation (6.14(b)) from EC)

(ii) the *frequent combination* when considering the comfort of the occupants, the function of machinery, avoiding ponding of water, i.e. reversible limit states:

$$F_d = \sum_{j\geq1}G_{k,j} \ "+" P \ "+" \ \psi_{1,1}Q_{k,1} \ "+" \ \sum_{i>1}\psi_{2,i}Q_{k,i}$$

(Equation (6.15(b)) from EC)

(i) the *quasi-permanent combination* when considering the long-term effects, e.g. creep effects, and the appearance of a structure:

$$F_d = \sum_{j\geq1}G_{k,j} \ "+" P \ "+" \ \sum_{i\geq1}\psi_{2,i}Q_{k,i}$$

(Equation (6.16(b)) from EC)

where:
"+" implies "to be combined with",
Σ implies "the combined effect of",
F_d is the design value of the action,
P is the representative value of a prestressing action,

$G_{k,j}$ is the characteristic values of permanent actions,

$Q_{k,1}$ is characteristic value of the *leading* variable action,

$Q_{k,i}$ is the characteristic value of the *accompanying* variable actions.

$\gamma_{G,j}$ is the safety factor for permanent actions,

$\gamma_{Q,1}$ is the partial safety factor for the *leading* variable action,

$\gamma_{Q,i}$ is the partial safety factor for the *accompanying* variable actions,

$\psi_{0,i}$ is the combination factor which is applied to the characteristic value Q_k of an action not being considered as $Q_{k,1}$,

$\psi_{1,i}$ is the factor for frequent the value of a variable action,

$\psi_{2,i}$ is the factor for the quasi-permanent value of a variable action.

The values for these factors can be found in the Eurocode and the National Annex to the code as indicated in Table 2.4

Factor	EC Table	UK NA	Comments
ψ_0	A1.1	NA.A1.1	Related to various imposed load categories, snow loads, wind loads and temperature (non-fire) effects. See NA.
ψ_1	A1.1	NA.A1.1	
ψ_2	A1.1	NA.A1.1	
$\gamma_{Gj,sup}$	A1.2(A/B)	NA.A1.2(A/B)	Relates to permanent, leading variable and accompanying variable actions. See NA. - Clauses NA.2.2.3.1/ NA.2.2.3.2 for EQU (Set A) and STR/GEO (Set B) respectively.
$\gamma_{Gj,inf}$	A1.2(A/B)	NA.A1.2(A/B)	
$\gamma_{Q,1}$	A1.2(A/B)	NA.A1.2(A/B)	
$\gamma_{Q,i}$	A1.2(A/B)	NA.A1.2(A/B)	
ξ	A1.2(A/B)	NA.A1.2(A/B)	
$\psi_{1,1}$	A1.3	NA.A1.3	Relates to accidental and seismic combinations of actions. See NA. Clause NA.2.2.5
$\psi_{2,1}$	A1.3	NA.A1.3	
$\psi_{2,i}$	A1.3	NA.A1.3	

Table 2.4

2.6.9 Partial Safety Factors

The Eurocode provides indicative values for various safety factors: these may be changed using Nationally Determined Parameters within the National Annex document to reflect the levels of safety required by the appropriate authority of the national government. In the UK this is the British Standards Institution.

2.6.10 Conventions

The difference in conventions most likely to cause confusion with UK engineers is the change in the symbols used to designate the major and minor axes of a cross-section. Traditionally in the UK the **y-y axis** has represented the minor axis; in Eurocodes this represents the **MAJOR axis** and the minor axis is represented by the **z-z** axis. The **x-x axis** defines the **LONGITUDINAL axis**. All three axes are shown in Figure 2.2.

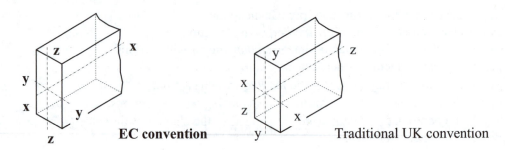

Figure 2.2

A summary of the abbreviations used in relation to Eurocodes is given in Table 2.5.

Abbreviation	Meaning
CEN	European Standards Organisation
DAV	Date of Availability
DD ENV	UK version of Pre-standard (BSI)
EC	Eurocode produced by CEN
EN	European Standard based on Eurocode and issued by member countries
ENV	Pre-standard of Eurocode issued by member countries
National Annex (NAD)	National Application Document issued by member countries (BSI)
NSB	National Standards Bodies
NCCI	Non-contradictory complimentary information
NDP	Nationally Determined Parameter (contained with National Annexes)

Table 2.5

2.7 Example 2.1: Stability

A timber beam with a cantilever section, is simply supported on two masonry walls as shown in Figure 2.3. The beam supports permanent and variable actions as indicated which can exist independently on spans AB and BC. Using the design data given and the ultimate limit state of EQU, check the stability of the beam.

Design Data:

Permanent characteristic uniformly distributed load (g_k) 2,0 kN/m
Variable characteristic uniformly distributed load - action 1 ($q_{k,1}$) 1,0 kN/m
Variable characteristic concentrated load - action 2 ($Q_{k,2}$) 1,5 kN

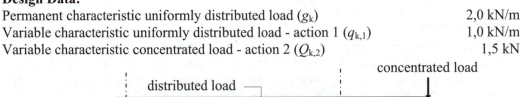

Figure 2.3

Solution:
Stability (EQU) $E_{d,dst} \leq E_{d,stb}$ (Equation (6.7) of EC)
where:
$E_{d,dst}$ is the design value of the effect of destabilising actions,
$E_{d,stb}$ is the design value of the effect of stabilising actions.

The stability of the beam is dependent on the vertical reaction (V_A) at support A ≥ 0 i.e. upwards. Overturning of the beam will occur when $V_A < 0$. The most unfavourable combinations of actions, using Equation (6.10), must be considered such that the minimum value of V_A is determined.

(Equation (6.10) from EC 1990)

$$F_d = \sum_{j \geq 1} \gamma_{G,j} G_{k,j} \ "+" \ \gamma_{Q,1} Q_{k,1} \ "+" \ \sum_{i>1} \gamma_{Q,i} \psi_{0,i} Q_{k,i}$$

The values of $\gamma_{G,inf}$, $\gamma_{G,sup}$, $\gamma_{Q,1}$, $\gamma_{Q,2}$, and ψ_0 can be obtained from the UK National Annex in NA.A1.2(A) and Tables NA.A1.1 as follows:

Table NA.A1.1: $\psi_0 = 0,7$
Table NA.A1.2(A): $\gamma_{G,inf} = 0,9$ $\gamma_{G,sup} = 1,1$ $\gamma_{Q,1} = 1,5$ $\gamma_{Q,2} = 1,5$

or alternatively, when the verification of static equilibrium also involves the resistance of structural members:

Table NA.A1.2(A): $\gamma_{G,inf} = 1,15$ $\gamma_{G,sup} = 1,35$ $\gamma_{Q,1} = \gamma_{Q,2} = 1,5$ (where unfavourable)
$\gamma_{Q,1} = \gamma_{Q,2} = 0$ (where favourable)

The arrangement of loading in the combination to be considered is the addition of the permanent and variable actions as indicated in (i) to (iv):

(i) unfavourable permanent action on span from A to B,
(ii) favourable permanent action on span from B to C,
(iii) variable action 1 on span from B to C,
(iv) variable action 2 on span from B to C.

Each variable action is considered as the leading variable in turn. Four cases are considered below for the calculations using the recommended and alternative values as indicated above.

Case 1: assume the distributed load to be the leading variable.

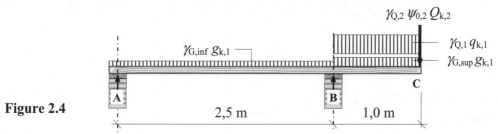

Figure 2.4

$\gamma_{G,inf}\, g_{k,1} = (0,9 \times 2,0) = 1,8$ kN/m

$\gamma_{G,sup}\, g_{k,1} = (1,1 \times 2,0) = 2,2$ kN/m

$\gamma_{Q,1}\, q_{k,1} = (1,5 \times 1,0) = 1,5$ kN/m

$\gamma_{Q,2}\, \psi_{0,2}\, Q_{k,2} = (1,5 \times 0,7 \times 1,5) = 1,58$ kN

+ve $\uparrow \Sigma F_y = 0 \quad V_A - (1,8 \times 2,5) - (2,2 \times 1,0) - (1,5 \times 1,0) - 1,58 + V_B = 0$

$$V_A + V_B = 9,78 \text{ kN}$$

+ve $\circlearrowright \Sigma M_B = 0 \quad (V_A \times 2,5) - (1,8 \times 2,5 \times 1,25) + [(2,2 + 1,5) \times 1,0 \times 0,5] + (1,58 \times 1,0)$

$$= 0$$

$$\therefore V_A = \textbf{0,88 kN} \uparrow \quad \text{and} \quad V_B = \textbf{8,9 kN} \uparrow$$

Case 2: assume the concentrated load to be the leading variable.

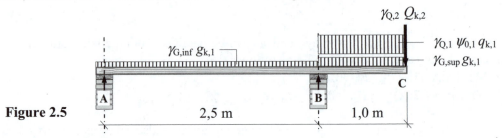

Figure 2.5

$\gamma_{G,inf}\, g_{k,1} = (0,9 \times 2,0) = 1,8$ kN/m

$\gamma_{G,sup}\, g_{k,1} = (1,1 \times 2,0) = 2,2$ kN/m

$\gamma_{Q,1}\, \psi_{0,1}\, q_{k,1} = (1,5 \times 0,7 \times 1,0) = 1,05$ kN/m

$\gamma_{Q,2}\, Q_{k,2} = (1,5 \times 1,5) = 2,25$ kN

+ve $\uparrow \Sigma F_y = 0 \quad V_A - (1,8 \times 2,5) - (2,2 \times 1,0) - (1,5 \times 0,7 \times 1,0) - 2,25 + V_B = 0$

$$V_A + V_B = 10,0 \text{ kN}$$

+ve $\circlearrowright \Sigma M_B = 0 \quad (V_A \times 2,5) - (1,8 \times 2,5 \times 1,25) + [(2,2 + 1,05) \times 1,0 \times 0,5] + (2,25 \times 1,0)$

$$= 0$$

$$\therefore V_A = \textbf{0,70 kN} \uparrow \quad \text{and} \quad V_B = \textbf{9,30 kN} \uparrow$$

Considering the alternative values:

Case 3: assume the distributed load to be the leading variable.

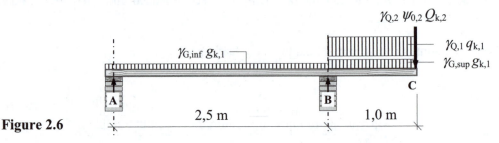

Figure 2.6

$\gamma_{G,inf}\, g_{k,1} = (1,15 \times 2,0) = 2,3$ kN/m

$\gamma_{G,sup}\, g_{k,1} = (1,35 \times 2,0) = 2,7$ kN/m

$\gamma_{Q,1}\, q_{k,1} = (1,5 \times 1,0) = 1,5 \text{ kN/m}$
$\gamma_{Q,2}\, \psi_{0,2}\, Q_{k,2} = (1,5 \times 0,7 \times 1,5) = 1,58 \text{ kN}$

+ve $\uparrow \Sigma F_y = 0$ $V_A - (2,3 \times 2,5) - (2,7 \times 1,0) - (1,5 \times 1,0) - 1,58 + V_B = 0$
$$V_A + V_B = 11,53 \text{ kN}$$
+ve $\curvearrowright \Sigma M_B = 0$ $(V_A \times 2,5) - (2,3 \times 2,5 \times 1,25) + [(2,7 + 1,5) \times 1,0 \times 0,5] + (1,58 \times 1,0)$
$$= 0$$
$$\therefore V_A = 1,4 \text{ kN} \uparrow \quad \text{and} \quad V_B = 10,13 \text{ kN} \uparrow$$

Case 4: assume the concentrated load to be the leading variable.

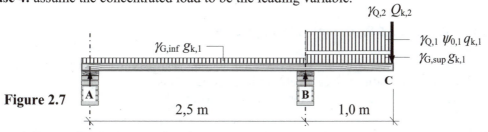

Figure 2.7

$\gamma_{G,\inf}\, g_{k,1} = (1,15 \times 2,0) = 2,3 \text{ kN/m}$
$\gamma_{G,\sup}\, g_{k,1} = (1,35 \times 2,0) = 2,7 \text{ kN/m}$
$\gamma_{Q,1}\, \psi_{0,1}\, q_{k,1} = (1,5 \times 0,7 \times 1,0) = 1,05 \text{ kN/m}$
$\gamma_{Q,2}\, Q_{k,2} = (1,5 \times 1,5) = 2,25 \text{ kN}$

+ve $\uparrow \Sigma F_y = 0$ $V_A - (2,3 \times 2,5) - (2,7 \times 1,0) - (1,05 \times 1,0) - 2,25 + V_B = 0$
$$V_A + V_B = 11,75 \text{ kN}$$
+ve $\curvearrowright \Sigma M_B = 0$ $(V_A \times 2,5) - (2,3 \times 2,5 \times 1,25) + [(2,7 + 1,05) \times 1,0 \times 0,5] + (2,25 \times 1,0)$
$$= 0$$
$$\therefore V_A = 1,23 \text{ kN} \uparrow \quad \text{and} \quad V_B = 10,52 \text{ kN} \uparrow$$

The values obtained considering each of the equations and leading variables are given in Table 2.6

Summary of vertical reactions at A					
Case No.	Equation	Leading variable	$\gamma_{G,\inf}$	$\gamma_{G,\sup}$	Vertical reaction at A (kN)
1	(6.10)	distributed load	0,9	1,1	+ 0,88
2		concentrated load			**+ 0,70**
3		distributed load	1,15	1,35	+ 1,40
4		concentrated load			+ 1,23

Table 2.6

Since all values are > zero the beam is stable - (Case 2 is the most critical case).

2.8 Example 2.2: Design Bending Moment

Considering the beam and shown in Figure 2.3. The beam supports permanent and variable actions as indicated which can exist independently on spans AB and BC. Using the design data given and the ultimate limit state of STR, determine the design bending moment in span AB.

Design Data:

Permanent characteristic uniformly distributed load (g_k)   2,0 kN/m
Variable characteristic uniformly distributed load - action 1 ($q_{k,1}$)   1,0 kN/m
Variable characteristic concentrated load - (action 2) ($Q_{k,2}$)   1,5 kN

Rupture (STR) $E_d \leq R_d$ (Equation 6.8 of EC 1990)

where:

E_d is the design value of the effects of the actions such as internal force, moment or vector representing several internal forces or moments,

R_d is the design value of the corresponding resistance.

The design bending moment should be calculated in accordance with Clause 6.4.3.2 and either Equation 6.10 as above or the less favourable of Equations 6.10(a) and 6.10(b):

$$F_d = \sum_{j\geq1} \gamma_{G,j} G_{k,j} \; ''+'' \; \gamma_{Q,1} \psi_{0,1} Q_{k,1} \; ''+'' \; \sum_{i>1} \gamma_{Q,i} \psi_{0,1} Q_{k,i}$$

(Equation (6.10(a)) from EC)

$$F_d = \sum_{j\geq1} \xi_j \gamma_{G,j} G_{k,j} \; ''+'' \; \gamma_{Q,1} Q_{k,1} \; ''+'' \; \sum_{i>1} \gamma_{Q,i} \psi_{0,i} Q_{k,i}$$

(Equation (6.10(b)) from EC)

The UK National Annex indicates the following values:

Table NA.A1.1: $\psi_0 = 0,7$

Clause NA. 2.2.3.2 and Table NA.A1.2(B):

For Equation (6.10) $\gamma_{G,inf} = 1,0$ $\gamma_{G,sup} = 1,35$ $\gamma_{Q,1} = \gamma_{Q,2} = 1,5$

For Equation (6.10(a)) $\gamma_{G,inf} = 1,0$ $\gamma_{G,sup} = 1,35$ $\gamma_{Q,1} = \gamma_{Q,2} = 1,5$

For Equation (6.10(b)) $\gamma_{G,inf} = 1,0$ $\gamma_{G,sup} = 1,35$ $\gamma_{Q,1} = \gamma_{Q,2} = 1,5$ $\xi = 0,925$

The arrangement of loading in the combination to be considered to determine the maximum bending moment in span AB is the addition of the permanent and variable actions as indicated in (i) to (iv):

(i)   favourable permanent action on span from A to B ,
(ii)   unfavourable permanent action on span from B to C,
(iii)   variable action 1 on span from A to B,
(iv)   no variable action 2.

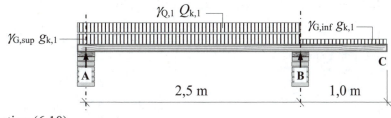

Figure 2.8

Case 1: Using Equation (6.10)

$$F_d = \sum_{j \geq 1} \gamma_{G,j} G_{k,j} \ "+" \ \gamma_{Q,1} Q_{k,1} \ "+" \ \sum_{i>1} \gamma_{Q,i} \psi_{0,i} Q_{k,i}$$

For Equation (6.10) $\gamma_{G,inf} = 1{,}0$ $\gamma_{G,sup} = 1{,}35$ $\gamma_{Q,1} = \gamma_{Q,2} = 1{,}5$

$\gamma_{G,inf}\, g_{k,1} = (1{,}0 \times 2{,}0) = 2{,}0$ kN/m
$\gamma_{G,sup}\, g_{k,1} = (1{,}35 \times 2{,}0) = 2{,}7$ kN/m
$\gamma_{Q,1}\, q_{k,1} = (1{,}5 \times 1{,}0) = 1{,}5$ kN/m
$\gamma_{Q,2}\, \psi_{0,2}\, Q_{k,2} = $ zero

+ve $\uparrow \Sigma F_y = 0$ $V_A - (2{,}7 \times 2{,}5) - (1{,}5 \times 2{,}5) - (2{,}0 \times 1{,}0) + V_B = 0$
$\hspace{10cm} V_A + V_B = 12{,}5$ kN

+ve $\circlearrowright \Sigma M_B = 0$ $(V_A \times 2{,}5) - (2{,}7 \times 2{,}5 \times 1{,}25) - (1{,}5 \times 2{,}5 \times 1{,}25) + (2{,}0 \times 1{,}0 \times 0{,}5) = 0$

$$\therefore \ V_A = \textbf{4{,}85 kN} \uparrow \quad \text{and} \quad V_B = \textbf{7{,}65 kN} \uparrow$$

4,85 kN

2,0 kN

$x = (4{,}85/4{,}2) = 1{,}15$ m

Shear force diagram

5,65 kN

Figure 2.9

Design bending moment = shaded area of the shear force diagram
$\hspace{5cm} = (0{,}5 \times 1{,}15 \times 4{,}85) = 2{,}79$ kNm

Case 2: Using Equations (6.10(a)) and (6.10(b))
Equation (6.10(a))

$$F_d = \sum_{j \geq 1} \gamma_{G,j} G_{k,j} \ "+" \ \gamma_{Q,1} \psi_{0,1} Q_{k,1} \ "+" \ \sum_{i>1} \gamma_{Q,i} \psi_{0,i} Q_{k,i}$$

For Equation (6.10(a)) $\gamma_{G,inf} = 1{,}0$ $\gamma_{G,sup} = 1{,}35$ $\gamma_{Q,1} = \gamma_{Q,2} = 1{,}5$ $\psi_0 = 0{,}7$
$\gamma_{G,inf}\, g_{k,1} = (1{,}0 \times 2{,}0) = 2{,}0$ kN/m
$\gamma_{G,sup}\, g_{k,1} = (1{,}35 \times 2{,}0) = 2{,}7$ kN/m
$\gamma_{Q,1}\, \psi_{0,1}\, q_{k,1} = (1{,}5 \times 0{,}7 \times 1{,}0) = 1{,}05$ kN/m
$\gamma_{Q,2}\, \psi_{0,2}\, Q_{k,2} = $ zero
+ve $\uparrow \Sigma F_y = 0$ $V_A - (2{,}7 \times 2{,}5) - (1{,}05 \times 2{,}5) - (2{,}0 \times 1{,}0) + V_B = 0$
$\hspace{9cm} V_A + V_B = 11{,}38$ kN

+ve $\circlearrowleft$ $\Sigma M_B = 0$ $(V_A \times 2,5) - (2,7 \times 2,5 \times 1,25) - (1,05 \times 2,5 \times 1,25) + (2,0 \times 1,0 \times 0,5) = 0$

$$\therefore V_A = \mathbf{4,29 \ kN} \uparrow \quad \text{and} \quad V_B = \mathbf{7,09 \ kN} \uparrow$$

4,29 kN

2,0 kN

$x = (4,29/3,75) = 1,14 \text{ m}$

Shear force diagram   5,09 kN

Figure 2.10

Design bending moment = shaded area of the shear force diagram
$$= (0,5 \times 1,14 \times 4,29) = 2,45 \text{ kNm}$$

Equation (6.10(b))
$$F_d = \sum_{j \geq 1} \xi_j \gamma_{G,j} G_{k,j} \ '' + '' \ \gamma_{Q,1} Q_{k,1} \ '' + '' \ \sum_{i > 1} \gamma_{Q,i} \psi_{0,i} Q_{k,i}$$

For Equation (6.10(b)) $\gamma_{G,inf} = 1,0$ $\gamma_{G,sup} = 1,35$ $\gamma_{Q,1} = \gamma_{Q,2} = 1,5$ $\xi = 0,925$ $\psi_0 = 0,7$

$\xi_1 \gamma_{G,inf} \ g_{k,1} = (0,925 \times 1,0 \times 2,0) = 1,85 \text{ kN/m}$
$\xi_1 \gamma_{G,sup} \ g_{k,1} = (0,925 \times 1,35 \times 2,0) = 2,5 \text{ kN/m}$
$\gamma_{Q,1} \ q_{k,1} = (1,5 \times 1,0) = 1,5 \text{ kN/m}$
$\gamma_{Q,2} \ \psi_{0,2} \ Q_{k,2} = \text{zero}$

+ve $\uparrow$ $\Sigma F_y = 0$ $V_A - (2,5 \times 2,5) - (1,5 \times 2,5) - (1,85 \times 1,0) + V_B = 0$
$$V_A + V_B = 11,85 \text{ kN}$$
+ve $\circlearrowleft$ $\Sigma M_B = 0$ $(V_A \times 2,5) - (2,5 \times 2,5 \times 1,25) - (1,5 \times 2,5 \times 1,25) + (1,85 \times 1,0 \times 0,5)$
$$= 0$$

$$\therefore V_A = \mathbf{4,63 \ kN} \uparrow \quad \text{and} \quad V_B = \mathbf{7,22 \ kN} \uparrow$$

4,63 kN

1,85 kN

$x = (4,63/4,0) = 1,16 \text{ m}$

5,37 kN

Shear force diagram

Figure 2.11

Design bending moment   = shaded area of the shear force diagram
$$= (0{,}5 \times 1{,}16 \times 4{,}63) = 2{,}69 \text{ kNm}$$

The values obtained considering each of the Equations and Leading variables are given in Table 2.7

Summary of maximum design bending moment in span AB			
Case No.	Equation	Leading variable	Design bending moment (kNm)
1	(6.10)		**+ 2,79**
2	(6.10(a)) (6.10(b))	no variable action 2	+ 2,45 **+ 2,69**

Table 2.7

Using Equation (6.10) the design bending moment in span AB = + 2,79kNm.
Using Equations (6.10(a))/(6.10(b)) the design bending moment in span AB = + 2,69 kNm

2.9 Example 2.3: Design Bending Moment over Support B

Considering the beam and shown in Figure 2.3. The beam supports permanent and variable actions as indicated which can exist independently on spans AB and BC. Using the design data given and the ultimate limit state of STR, determine the design bending moment over support B.

Design Data:
Permanent characteristic uniformly distributed load (g_k)   2,0 kN/m
Variable characteristic uniformly distributed load - action 1 ($q_{k,1}$)   1,0 kN/m
Variable characteristic concentrated load - (action 2) ($Q_{k,2}$)   1,5 kN

Rupture (STR) $E_d \le E_{Rd}$ (Equation (6.8) of EC 1990)
The arrangement of loading in the combination to be considered is the addition of the permanent and variable actions as indicated in (i) to (iv):

(i)   favourable permanent action on span from A to B,
(ii)   unfavourable permanent action on span from B to C,
(iii)   variable action 1 on span from B to C,
(iv)   variable action 2 on span from B to C.

Each variable action is considered as the leading variable in turn. Four cases are considered below for the calculations using the Equation (6.10) and Equations (6.10(a)) and (6.10(b)) as indicated above.

Note: The value of the bending moment over the support at B is independent of the loading which is applied to span AB.

Case 1: Using Equations (6.10) assuming the uniformly distributed load to be the leading variable action.

$$F_d = \sum_{j \geq 1} \gamma_{G,j} G_{k,j} \ "+" \ \gamma_{Q,1} Q_{k,1} \ "+" \ \sum_{i>1} \gamma_{Q,i} \psi_{0,i} Q_{k,i}$$

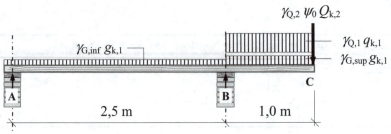

Figure 2.12

Table NA.A1.1: $\psi_0 = 0,7$

Clause NA.2.2.3.2 and Table NA.A1.2(B):

For Equation (6.10) $\gamma_{G,inf} = 1,0$ $\gamma_{G,sup} = 1,35$ $\gamma_{Q,1} = \gamma_{Q,2} = 1,5$

$\gamma_{G,inf} \, g_{k,1} = (1,0 \times 2,0) = 2,0 \ \text{kN/m}$
$\gamma_{G,sup} \, g_{k,1} = (1,35 \times 2,0) = 2,7 \ \text{kN/m}$
$\gamma_{Q,1} \, q_{k,1} = (1,5 \times 1,0) = 1,5 \ \text{kN/m}$
$\gamma_{Q,2} \, \psi_{0,2} \, Q_{k,2} = (1,5 \times 0,7 \times 1,5) = 1,58 \ \text{kN}$

Design bending moment $= - \, [(2,7 \times 1,0 \times 0,5) + (1,5 \times 1,0 \times 0,5) + (1,58 \times 1,0)]$
$\qquad\qquad\qquad\qquad = - \, 3,68 \ \text{kNm}$

Case 2: Using Equations (6.10) assuming the concentrated load to be the leading variable action.

$$F_d = \sum_{j \geq 1} \gamma_{G,j} G_{k,j} \ "+" \ \gamma_{Q,1} Q_{k,1} \ "+" \ \sum_{i>1} \gamma_{Q,i} \psi_{0,i} Q_{k,i}$$

Figure 2.13

Table NA.A1.1: $\psi_0 = 0,7$
Clause NA.2.2.3.2 and Table NA.A1.2(B):

For Equation (6.10) $\gamma_{G,inf} = 1,0$ $\gamma_{G,sup} = 1,35$ $\gamma_{Q,1} = \gamma_{Q,2} = 1,5$

$\gamma_{G,inf}\, g_{k,1} = (1,0 \times 2,0) = 2,0$ kN/m
$\gamma_{G,sup}\, g_{k,1} = (1,35 \times 2,0) = 2,7$ kN/m
$\gamma_{Q,1}\, \psi_{0,1}\, q_{k,1} = (1,5 \times 0,7 \times 1,0) = 1,05$ kN/m
$\gamma_{Q,2}\, Q_{k,2} = (1,5 \times 1,5) = 2,25$ kN

Design bending moment $= - [(2,7 \times 1,0 \times 0,5) + (1,05 \times 1,0 \times 0,5) + (2,25 \times 1,0)]$
$= - 4,13$ kNm

Case 3: Using Equations (6.10(a)) and (6.10(b)) assuming the uniformly distributed load as the leading variable action.

Equation (6.10(a)): $F_d = \displaystyle\sum_{j\geq1} \gamma_{G,j} G_{k,j}\ "+"\ \gamma_{Q,1}\psi_{0,1}Q_{k,1}\ "+"\ \sum_{i>1} \gamma_{Q,i}\psi_{0,i}Q_{k,i}$

For Equation (6.10(a)) $\gamma_{G,inf} = 1,0$ $\gamma_{G,sup} = 1,35$ $\gamma_{Q,1} = \gamma_{Q,2} = 1,5$ $\psi_0 = 0,7$

$\gamma_{G,inf}\, g_{k,1} = (1,0 \times 2,0) = 2,0$ kN/m
$\gamma_{G,sup}\, g_{k,1} = (1,35 \times 2,0) = 2,7$ kN/m
$\gamma_{Q,1}\, \psi_{0,1}\, q_{k,1} = (1,5 \times 0,7 \times 1,0) = 1,05$ kN/m
$\gamma_{Q,2}\, \psi_{0,2}\, Q_{k,2} = (1,5 \times 0,7 \times 1,5) = 1,58$ kN

Design bending moment $= - [(2,7 \times 1,0 \times 0,5) + (1,05 \times 1,0 \times 0,5) + (1,58 \times 1,0)]$
$= - 3,46$ kNm

Equation (6.10(b))

$$F_d = \sum_{j\geq1} \xi_j\gamma_{G,j} G_{k,j}\ "+"\ \gamma_{Q,1}Q_{k,1}\ "+"\ \sum_{i>1} \gamma_{Q,i}\psi_{0,i}Q_{k,i}$$

For Equation (6.10(b)) $\gamma_{G,inf} = 1,0$ $\gamma_{G,sup} = 1,35$ $\gamma_{Q,1} = \gamma_{Q,2} = 1,5$ $\xi = 0,925$ $\psi_0 = 0,7$
$\xi_1\gamma_{G,inf}\, g_{k,1} = (0,925 \times 1,0 \times 2,0) = 1,85$ kN/m
$\xi_1\gamma_{G,sup}\, g_{k,1} = (0,925 \times 1,35 \times 2,0) = 2,5$ kN/m
$\gamma_{Q,1}\, q_{k,1} = (1,5 \times 1,0) = 1,5$ kN/m
$\gamma_{Q,2}\, \psi_{0,2}\, Q_{k,2} = (1,5 \times 0,7 \times 1,5) = 1,58$ kN

Design bending moment $= - [(2,5 \times 1,0 \times 0,5) + (1,5 \times 1,0 \times 0,5) + (1,58 \times 1,0)]$
$= - 3,58$ kNm

Case 4: Using Equations (6.10(a)) and (6.10(b)) assuming the concentrated load as the leading variable action.

Equation (6.10(a)): $F_d = \sum_{j \geq 1} \gamma_{G,j} G_{k,j} \; "+" \; \gamma_{Q,1} \psi_{0,1} Q_{k,1} \; "+" \; \sum_{i>1} \gamma_{Q,i} \psi_{0,i} Q_{k,i}$

For Equation (6.10(a)) $\gamma_{G,inf} = 1,0$ $\gamma_{G,sup} = 1,35$ $\gamma_{Q,1} = \gamma_{Q,2} = 1,5$ $\psi_0 = 0,7$

$\gamma_{G,inf} \, g_{k,1} = (1,0 \times 2,0) = 2,0 \text{ kN/m}$

$\gamma_{G,sup} \, g_{k,1} = (1,35 \times 2,0) = 2,7 \text{ kN/m}$

$\gamma_{Q,1} \, \psi_{0,1} \, q_{k,1} = (1,5 \times 0,7 \times 1,0) = 1,05 \text{ kN/m}$

$\gamma_{Q,2} \, \psi_{0,2} \, Q_{k,2} = (1,5 \times 0,7 \times 1,5) = 1,58 \text{ kN}$

Design bending moment $= - [(2,7 \times 1,0 \times 0,5) + (1,05 \times 1,0 \times 0,5) + (1,58 \times 1,0)]$
$ = - 3,46 \text{ kNm}$

Equation (6.10(b)): $F_d = \sum_{j \geq 1} \xi_j \gamma_{G,j} G_{k,j} \; "+" \; \gamma_{Q,1} Q_{k,1} \; "+" \; \sum_{i>1} \gamma_{Q,i} \psi_{0,i} Q_{k,i}$

For Equation (6.10(b)) $\gamma_{G,inf} = 1,0$ $\gamma_{G,sup} = 1,35$ $\gamma_{Q,1} = \gamma_{Q,2} = 1,5$ $\xi = 0,925$ $\psi_0 = 0,7$

$\xi_1 \gamma_{G,inf} \, g_{k,1} = (0,925 \times 1,0 \times 2,0) = 1,85 \text{ kN/m}$

$\xi_1 \gamma_{G,sup} \, g_{k,1} = (0,925 \times 1,35 \times 2,0) = 2,5 \text{ kN/m}$

$\gamma_{Q,1} \, \psi_{0,1} \, q_{k,1} = (1,5 \times 0,7 \times 1,0) = 1,05 \text{ kN/m}$

$\gamma_{Q,2} \, Q_{k,2} = (1,5 \times 1,5) = 2,25 \text{ kN}$

Design bending moment $= - [(2,5 \times 1,0 \times 0,5) + (1,05 \times 1,0 \times 0,5) + (2,25 \times 1,0)]$
$ = - 4,03 \text{ kNm}$

The values obtained considering each of the equations and leading variables are given in Table 2.8.

Summary of maximum design bending moments over support B			
Case No.	Equation	Leading variable	Design bending moment (kNm)
1	(6.10)	distributed load	− 3,68
2		concentrated load	**− 4,13**
3	(6.10(a)) (6.10(b))	distributed load	− 3,46 − 3,58
4	(6.10(a)) (6.10(b))	concentrated load	− 3,46 **− 4,03**

Table 2.8

Using Equation (6.10) the design bending moment over support B $= - 4,13 \text{ kNm}$.
Using Equations (6.10(a)) / (6.10(b)) the design bending over support B $= - 4,03 \text{ kNm}$.

2.10 Example 2.4: Maximum Design Vertical Reaction at Support B

Considering the beam and shown in Figure 2.3. The beam supports permanent and variable actions as indicated which can exist independently on spans AB and BC. Using the design data given and the ultimate limit state STR, determine the maximum design vertical reaction at support B.

Design Data:

Permanent characteristic uniformly distributed load (g_k)	2,0 kN/m
Variable characteristic uniformly distributed load - action 1 ($q_{k,1}$)	1,0 kN/m
Variable characteristic concentrated load - (action 2) ($Q_{k,2}$)	1,5 kN

Rupture (STR/GEO) $E_d \leq E_{Rd}$ (Equation (6.8) of EC 1990)

The arrangement of loading in the combination to be considered is the addition of the permanent and variable actions as indicated in (i) to (iv):

(i)   unfavourable permanent action on span from A to B ,
(ii)   unfavourable permanent action on span from B to C,
(iii) variable action 1 on spans A to B and B to C,
(iv)   variable action 2 on spans A to B and B to C.

Each variable action is considered as the leading variable in turn. Four cases are considered below for the calculations using the recommended and alternative values as indicated above.

Case 1: Using Equations (6.10) assuming the uniformly distributed load to be the leading variable action.

$$F_d = \sum_{j \geq 1} \gamma_{G,j} G_{k,j} \; "+" \; \gamma_{Q,1} Q_{k,1} \; "+" \; \sum_{i>1} \gamma_{Q,i} \psi_{0,i} Q_{k,i}$$

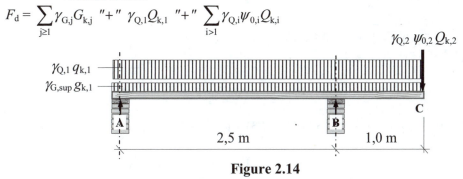

Figure 2.14

Table NA.A1.1: $\psi_0 = 0,7$

Clause NA.2.2.3.2 and Table NA.A1.2(B):

For Equation (6.10) $\gamma_{G,inf} = 1,0$ $\gamma_{G,sup} = 1,35$ $\gamma_{Q,1} = \gamma_{Q,2} = 1,5$

$\gamma_{G,inf} \, g_{k,1} = (1,0 \times 2,0) = 2,0$ kN/m
$\gamma_{G,sup} \, g_{k,1} = (1,35 \times 2,0) = 2,7$ kN/m

$\gamma_{Q,1}\, q_{k,1} = (1,5 \times 1,0) = 1,5 \text{ kN/m}$
$\gamma_{Q,2}\, \psi_{0,2}\, Q_{k,2} = (1,5 \times 0,7 \times 1,5) = 1,58 \text{ kN}$

+ve $\uparrow$ $\Sigma F_y = 0$ $V_A - (2,7 \times 3,5) - (1,5 \times 3,5) - 1,58 + V_B = 0$ $V_A + V_B = 16,28 \text{ kN}$
+ve $\curvearrowright$ $\Sigma M_A = 0$ $- (V_B \times 2,5) + (2,7 \times 3,5 \times 1,75) + (1,5 \times 3,5 \times 1,75) + (1,58 \times 3,5) = 0$

$\therefore$ $V_B = 12.50 \text{ kN} \uparrow$ and $V_A = 3,78 \text{ kN} \uparrow$

Case 2: Using Equations (6.10) assuming the concentrated load to be the leading variable.

$$F_d = \sum_{j\geq1}\gamma_{G,j}G_{k,j} \ '' + '' \ \gamma_{Q,1}Q_{k,1} \ '' + '' \ \sum_{i>1}\gamma_{Q,i}\psi_{0,i}Q_{k,i}$$

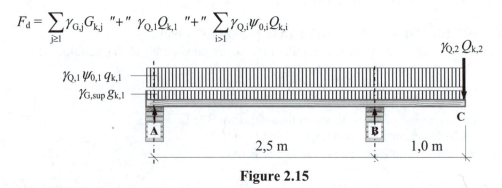

Figure 2.15

Table NA.A1.1: $\psi_0 = 0,7$

Clause NA.2.2.3.2 and Table NA.A1.2(B):
For Equation (6.10) $\gamma_{G,inf} = 1,0$ $\gamma_{G,sup} = 1,35$ $\gamma_{Q,1} = \gamma_{Q,2} = 1,5$
$\gamma_{G,inf}\, g_{k,1} = (1,0 \times 2,0) \ = 2,0 \text{ kN/m}$
$\gamma_{G,sup}\, g_{k,1} = (1,35 \times 2,0) = 2,7 \text{ kN/m}$
$\gamma_{Q,1}\, \psi_{0,1}\, q_{k,1} = (1,5 \times 0,7 \times 1,0) = 1,05 \text{ kN/m}$
$\gamma_{Q,2}\, Q_{k,2} = (1,5 \times 1,5) = 2,25 \text{ kN}$

+ve $\uparrow$ $\Sigma F_y = 0$ $V_A - (2,7 \times 3,5) - (1,05 \times 3,5) - 2,25 + V_B = 0$ $V_A + V_B = 15,38 \text{ kN}$

+ve $\curvearrowright$ $\Sigma M_A = 0$ $- (V_B \times 2,5) + (2,7 \times 3,5 \times 1,75) + (1,05 \times 3,5 \times 1,75) + (2,25 \times 3,5) = 0$

$\therefore$ $V_B = 12,34 \text{ kN} \uparrow$ and $V_A = 3,04 \text{ kN} \uparrow$

Case 3: Using Equations (6.10(a)) and (6.10(b)) assuming the uniformly distributed load
          as the leading variable.

Equation (6.10(a))

$$F_d = \sum_{j\geq1}\gamma_{G,j}G_{k,j} \ '' + '' \ \gamma_{Q,1}\psi_{0,1}Q_{k,1} \ '' + '' \ \sum_{i>1}\gamma_{Q,i}\psi_{0,i}Q_{k,i}$$

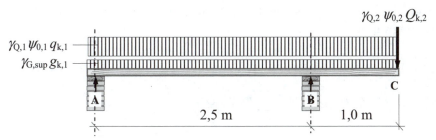

Figure 2.16

Table NA.A1.1: $\quad \psi_0 = 0,7$
Clause NA.2.2.3.2 and Table NA.A1.2(B):
For Equation (6.10(a)) $\quad \gamma_{G,inf} = 1,0 \quad \gamma_{G,sup} = 1,35 \quad \gamma_{Q,1} = \gamma_{Q,2} = 1,5$

$\gamma_{G,inf}\, g_{k,1} = (1,0 \times 2,0) = 2,0$ kN/m
$\gamma_{G,sup}\, g_{k,1} = (1,35 \times 2,0) = 2,7$ kN/m
$\gamma_{Q,1} \psi_{0,1}\, q_{k,1} = (1,5 \times 0,7 \times 1,0) = 1,05$ kN/m
$\gamma_{Q,2} \psi_{0,2}\, Q_{k,2} = (1,5 \times 0,7 \times 1,5) = 1,58$ kN

+ve $\uparrow$ $\Sigma F_y = 0 \quad V_A - (2,7 \times 3,5) - (1,05 \times 3,5) - 1,58 + V_B = 0 \qquad V_A + V_B = 14,71$ kN

+ve $\circlearrowright$ $\Sigma M_A = 0 \quad -(V_B \times 2,5) + (2,7 \times 3,5 \times 1,75) + (1,05 \times 3,5 \times 1,75) + (1,58 \times 3,5) = 0$

$$\therefore\ V_B = 11,40 \text{ kN} \uparrow \ \text{ and } \ V_A = 3,31 \text{ kN} \uparrow$$

Equation (6.10(b))

$$F_d = \sum_{j\geq 1} \xi_j \gamma_{G,j} G_{k,j} \ "+" \ \gamma_{Q,1} Q_{k,1} \ "+" \ \sum_{i>1} \gamma_{Q,i} \psi_{0,i} Q_{k,i}$$

For Equation (6.10(b)) $\quad \gamma_{G,inf} = 1,0 \quad \gamma_{G,sup} = 1,35 \quad \gamma_{Q,1} = \gamma_{Q,2} = 1,5 \quad \xi = 0,925 \quad \psi_0 = 0,7$

$\xi_1 \gamma_{G,inf}\, g_{k,1} = (0,925 \times 1,0 \times 2,0) = 1,85$ kN/m
$\xi_1 \gamma_{G,sup}\, g_{k,1} = (0,925 \times 1,35 \times 2,0) = 2,5$ kN/m
$\gamma_{Q,1}\, q_{k,1} = (1,5 \times 1,0) = 1,5$ kN/m
$\gamma_{Q,2} \psi_{0,2}\, Q_{k,2} = (1,5 \times 0,7 \times 1,5) = 1,58$ kN

+ve $\uparrow$ $\Sigma F_y = 0 \quad V_A - (2,5 \times 3,5) - (1,5 \times 3,5) - 1,58 + V_B = 0 \qquad V_A + V_B = 15,58$ kN

+ve $\circlearrowright$ $\Sigma M_A = 0 \quad -(V_B \times 2,5) + (2,5 \times 3,5 \times 1,75) + (1,5 \times 3,5 \times 1,75) + (1,58 \times 3,5) = 0$

$$\therefore\ V_B = 12,01 \text{ kN} \uparrow \ \text{ and } \ V_A = 3,57 \text{ kN} \uparrow$$

Case 4: Using Equations (6.10(a)) and (6.10(b)) assuming the concentrated load as the leading variable.

Equation (6.10(a))

$$F_d = \sum_{j\geq1} \gamma_{G,j} G_{k,j} \ "+" \ \gamma_{Q,1}\psi_{0,1}Q_{k,1} \ "+" \ \sum_{i>1}\gamma_{Q,i}\psi_{0,i}Q_{k,i}$$

For Equation (6.10(a)) $\gamma_{G,inf} = 1,0$ $\gamma_{G,sup} = 1,35$ $\gamma_{Q,1} = \gamma_{Q,2} = 1,5$ $\psi_0 = 0,7$

$\gamma_{G,inf}\, g_{k,1} = (1,0 \times 2,0) = 2,0$ kN/m
$\gamma_{G,sup}\, g_{k,1} = (1,35 \times 2,0) = 2,7$ kN/m
$\gamma_{Q,1}\, \psi_{0,1}\, q_{k,1} = (1,5 \times 0,7 \times 1,0) = 1,05$ kN/m
$\gamma_{Q,2}\, \psi_{0,2}\, Q_{k,2} = (1,5 \times 0,7 \times 1,5) = 1,58$ kN

+ve $\uparrow$ $\Sigma F_y = 0$ $V_A - (2,7 \times 3,5) - (1,05 \times 3,5) - 1,58 + V_B = 0$ $V_A + V_B = 14,71$ kN

+ve $\searpoon$ $\Sigma M_A = 0$ $- (V_B \times 2,5) + (2,7 \times 3,5 \times 1,75) + (1,05 \times 3,5 \times 1,75) + (1,58 \times 3,5) = 0$

$$\therefore \ V_B = 11,40 \text{ kN} \uparrow \ \text{ and } \ V_A = 3,31 \text{ kN} \uparrow$$

Equation (6.10(b)):

$$F_d = \sum_{j\geq1} \xi_j\gamma_{G,j} G_{k,j} \ "+" \ \gamma_{Q,1}Q_{k,1} \ "+" \ \sum_{i>1}\gamma_{Q,i}\psi_{0,i}Q_{k,i}$$

For Equation (6.10(b)) $\gamma_{G,inf} = 1,0$ $\gamma_{G,sup} = 1,35$ $\gamma_{Q,1} = \gamma_{Q,2} = 1,5$ $\xi = 0,925$ $\psi_0 = 0,7$

$\xi_1\gamma_{G,inf}\, g_{k,1} = (0,925 \times 1,0 \times 2,0) = 1,85$ kN/m
$\xi_1\gamma_{G,sup}\, g_{k,1} = (0,925 \times 1,35 \times 2,0) = 2,5$ kN/m
$\gamma_{Q,1}\, \psi_{0,1}\, q_{k,1} = (1,5 \times 0,7 \times 1,0) = 1,05$ kN/m
$\gamma_{Q,2}\, Q_{k,2} = (1,5 \times 1,5) = 2,25$ kN

+ve $\uparrow$ $\Sigma F_y = 0$ $V_A - (2,5 \times 3,5) - (1,05 \times 3,5) - 2,25 + V_B = 0$ $V_A + V_B = 14,68$ kN

+ve $\searrow$ $\Sigma M_A = 0$ $- (V_B \times 2,5) + (2,5 \times 3,5 \times 1,75) + (1,05 \times 3,5 \times 1,75) + (2,25 \times 3,5) = 0$

$$\therefore \ V_B = 11,85 \text{ kN} \uparrow \ \text{ and } \ V_A = 2,83 \text{ kN} \uparrow$$

The values obtained considering each of the equations and leading variables are given in Table 2.9.

Summary of maximum design vertical reactions at B			
Case No.	Equation	Leading variable	Design vertical reaction (kN)
1	(6.10)	distributed load	**12,5**
2		concentrated load	12,34
3	(6.10(a)) (6.10(b))	distributed load	11,40 **12,01**
4	(6.10(a)) (6.10(b))	concentrated load	11,40 11,85

Table 2.9

Using Equation (6.10) the design vertical reaction at support B = + 12,5 kN.
Using Equations (6.10(a))/(6.10(b)) the design vertical reaction at support B = + 12,01 kN.

3. Design Loading

Objective: *to introduce design loading and Eurocode 1 for dead, imposed, snow and wind loading.*

3.1 Design Loading

All structures are subjected to loading from various sources. The main categories of loading are: dead, imposed and wind loads. In some circumstances there may be other loading types which should be considered, such as settlement, fatigue, temperature effects, dynamic loading, or impact effects (e.g. when designing bridge decks, crane-gantry girders or maritime structures). In the majority of cases, design considering combinations of dead, imposed, snow and wind loads is the most appropriate.

The definitions of 'actions on structures', as given in 'BS EN 1991 Eurocode 1 – Actions on structures' and to be used in conjunction with 'BS EN 1990:2002 Eurocode – Basis of structural design', are considered in four parts as shown in Table 3.1.

Part 1	BS EN 1991-1-1:2002	General actions – Densities, self-weight, imposed loads for buildings
	BS EN 1991-1-2:2002	General actions – Actions on structures exposed to fire
	BS EN 1991-1-3:2003	General actions – Snow loads
	BS EN 1991-1-4:2005	General actions – Wind actions
	BS EN 1991-1-5:2003	General actions – Thermal actions
	BS EN 1991-1-6:2005	General actions – Actions during execution
	BS EN 1991-1-7:2006	General actions – Accidental actions
Part 2	BS EN 1991-2:2003	General actions – Traffic loads on bridges
Part 3	BS EN 1991-3:2006	General actions – Actions induced by cranes and machinery
Part 4	BS EN 1991- 4:2006	General actions – Silos and tanks

Table 3.1

In the majority of design situations, Part 1 of Eurocode 1 and its associated National Annexes will provide sufficient information relating to structural actions. In this text further details relating to sub-sections 1, 3 and 4 of Part 1 are given in Sections 3.7.1 to 3.7.10 respectively. In all cases it is necessary to consider the distribution of loading throughout a structure. This frequently involves the lateral distribution on floor systems as described in Section 3.2.

3.2 Floor Load Distribution

The Principles and Rules given in Eurocode and Eurocode 1 enable various load types to be evaluated and hence to produce a system of equivalent static forces which can be used in the analysis and design of a structure. The application of the load types indicated in Table 3.1, to structural beams and frames results in axial loads, shear forces, bending moments and deformations being induced in the floor/roof slabs, beams, columns and other structural elements which comprise a structure. The primary objective of structural analysis is to determine the distribution of internal moments and forces throughout a structure such that they are in equilibrium with the applied design loads.

As indicated in Chapter 4, there are a number of manual mathematical models (in addition to computer-based models) which can be used to idealise structural behaviour. These methods include: two-dimensional and three-dimensional elastic behaviour, elastic behaviour considering a redistribution of moments, plastic behaviour and non-linear behaviour. Detailed explanations of these techniques can be found in the numerous structural analysis text books which are available.

In braced structures (see Chapter 10, Section 10.1.2) where floor slabs and beams are considered to be simply supported, vertical loads give rise to three basic types of beam loading condition:

(i)   uniformly distributed line loads,
(ii) triangular and trapezoidal loads, and
(iii) concentrated point loads.

These load types are illustrated in Examples 3.1 to 3.4 (self-weights have been ignored).

3.2.1 Example 3.1: Load Distribution – One-way Spanning Slabs

Consider the floor plan shown in Figure 3.1(a) where two one-way spanning slabs are supported on three beams AB, CD and EF. Both slabs are assumed to be carrying a uniformly distributed design load of 5,0 kN/m^2.

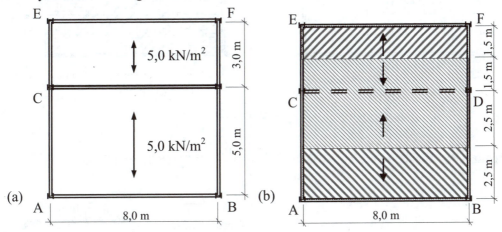

Figure 3.1

Both slabs have continuous contact with the top flanges of their supporting beams and span in the directions indicated. The floor area supported by each beam is indicated in Figure 3.1(b).

Beam AB: Total load = (floor area supported × magnitude of distributed load/m²)
$$= (2,5 \times 8,0) \times (5,0) = 100,0 \text{ kN}$$

Beam CD: Total load $= (4,0 \times 8,0) \times (5,0) = 160,0$ kN

Beam EF: Total load $= (1,5 \times 8,0) \times (5,0) = 60,0$ kN

Check: Total load on both slabs $= (8,0 \times 8,0 \times 5,0) = 320,0$ kN

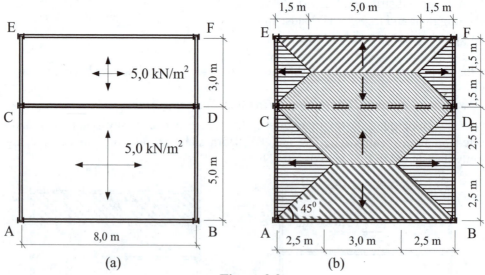

Figure 3.1(c)

3.2.2 Example 3.2: Load Distribution – Two-way Spanning Slabs

Consider the same floor plan as in Example 3.1 but now with the floor slabs two-way spanning, as shown in Figure 3.2(a).

Since both slabs are two-way spanning, their loads are distributed to supporting beams on all four sides assuming a 45° dispersion as indicated in Figure 3.2(b).

Figure 3.2

Beam AB: Load due to slab ACDB $= \left(\dfrac{8,0 + 3,0}{2} \times 2,5 \right) \times (5,0) = 68,75$ kN

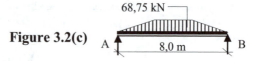

Figure 3.2(c)

Beam EF: Load due to slab CEFD $= \left(\dfrac{8,0 + 5,0}{2} \times 1,5 \right) \times (5,0) = 48,75$ kN

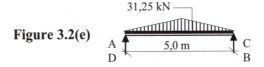

Figure 3.2(d)

Beams AC and BD: Load due to slab ACDB $= \left(\dfrac{5,0}{2} \times 2,5 \right) \times (5,0) = 31,25$ kN

Figure 3.2(e)

Beams CE and DF: Load due to slab CEFD $= \left(\dfrac{3,0}{2} \times 1,5 \right) \times (5,0) = 11,25$ kN

Figure 3.2(f)

The loading on beam CD can be considered to be the addition of two separate loads, i.e.

Load due to slab ACDB = 68,75 kN   (as for beam AB)
Load due to slab CEFD = 48,75 kN   (as for beam EF)

Note: Both loads are trapezoidal, but they are different.

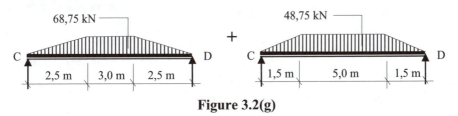

Figure 3.2(g)

Check: Total load on all beams $= 2 \times (68,75 + 48,75 + 31,25 + 11,25) = 320$ kN

3.2.3 Example 3.3: Load Distribution – Secondary Beams

Consider the same floor plan as in Example 3.2 with the addition of a secondary beam GH spanning between beams AB and CD as shown in Figure 3.3(a). The load carried by this new beam imposes a concentrated load at the mid-span points G and H respectively.

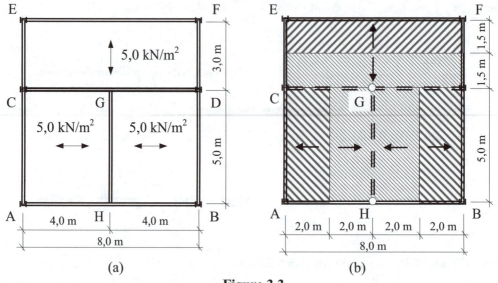

Figure 3.3

Beam EF:　Total load $= (1{,}5 \times 8{,}0) \times (5{,}0) = 60{,}0$ kN

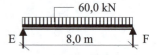

Beam GH:　Total load $= (4{,}0 \times 5{,}0) \times (5{,}0) = 100{,}0$ kN

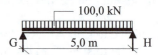

Beams AC and BD:
　　　　Total load $= (2{,}0 \times 5{,}0) \times (5{,}0) = 50{,}0$ kN

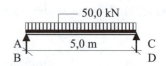

Figure 3.3(c)

Beam AB:　Total load = End reaction from beam GH $= 50{,}0$ kN

Figure 3.3(d)

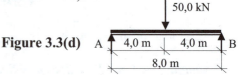

Beam CD:
The loading on beam CD can be considered to be the addition of two separate loads, i.e.

Load due to slab CEFD $= 60{,}0$ kN　　(as for beam EF)
Load due to beam GH $= 50{,}0$ kN　　(as for beam AB)

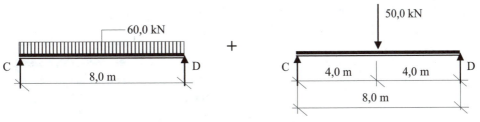

Figure 3.3(e)

3.2.4 Example 3.4: Combined One-way Slabs, Two-way Slabs and Beams

Considering the floor plan shown in Figure 3.4(a), with the one-way and two-way spanning slabs indicated, determine the type and magnitude of the loading on each of the supporting beams.

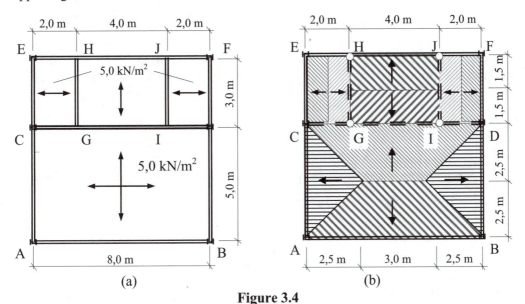

(a)   (b)

Figure 3.4

The loads on beams AB, AC and BD are the same as in Example 3.2.

Beams CE, DF, GH and IJ:
Total load = $(3,0 \times 1,0) \times (5,0) = 15,0$ kN

Figure 3.4(c)

Beam EF: The loads on EF are due to the end reactions from beams GH and IJ and a distributed load from GHJI.
End reaction from beam GH = 7,5 kN
End reaction from beam JI = 7,5 kN

Figure 3.4(d)

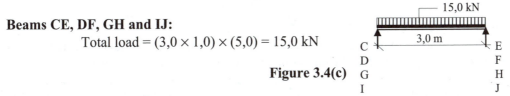

Beam EF: Load from slab GHJI $= [(4,0 \times 1,5) \times (5,0)]$
$= 30,0$ kN

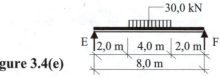

Figure 3.4(e)

Total loads on beam EF due to beams GH, JI and slab GHJI:

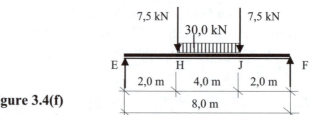

Figure 3.4(f)

Beam CD: The loads on CD are due to the end reactions from beams GH and IJ, a distributed load from GHIJ and a trapezoidal load from slab ABCD as in member AB of Example 3.2.

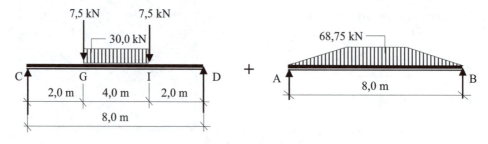

Figure 3.4(g)

3.3 BS EN 1991 (Eurocode 1 – General actions on structures)

The UK National Annex has a number of NDPs relating to categories of both residential and industrial building use and imposed loading, as indicated in Table 3.2.

3.4 Dead Loads: BS EN 1991-1-1:2002

Dead loads are loads which are due to the effects of gravity, i.e. the self-weight of all permanent construction such as beams, columns, floors, walls, roofs and finishes.

If the position of permanent partition walls is known, their weight can be assessed and included in the dead load. In speculative developments, internal partitions are regarded as imposed loading.

The nominal density of construction materials, and nominal density and angles of repose for stored materials, are given in an 'informative' annex – Annex A of EC1-1-1. The values are described as 'nominal values' since they do not include an underlying statistical basis in their determination, (i.e. they are not mean or characteristic values).

As indicated in Clause 2.1, self-weight should be classified as a permanent fixed action in accordance with EC 1990. In circumstances where the self-weight is time-dependent, the 'upper' and 'lower' (superior and inferior - see EC 1990) values should be used to induce the most critical design effect.

NDP values	BS EN 1991-1-1:2002	UK National Annex
Categories for residential, social, commercial and administrative areas	Table 6.1	Table NA.2
Imposed loads on floors, balconies and stairs in buildings	Table 6.2	Table NA.3
Reduction factor for imposed loads for floors and accessible roofs - α_A	Equation (6.1)	Equation (NA.1)[1]
Reduction factor for imposed loads from several storeys - α_n	Equation (6.2)	Equation (NA.2)[1]
Categories for storage and industrial areas	Table 6.3	Table NA.4
Imposed loads on floors due to storage	Table 6.4	Table NA.5
Imposed loads on garages and vehicle traffic areas	Table 6.8	Table NA.6
Imposed loads on roofs	Table 6.10	Table NA.7
Horizontal loads on parapets and partition walls acting as barriers	Table 6.12	Table NA.8

Notes:

1 – Loads that have been specifically determined from knowledge of the proposed use of the structure do not qualify for reduction.

2 – On roofs, imposed loads and snow loads or wind actions should not be applied together simultaneously, (Clause 3.3.2. of BS EN 1991-1-1:2002).

3 – Provided that the structural system permits lateral distribution of load, the self-weight of moveable partitions may be taken into account assuming a uniformly distributed load which should be added to the imposed load obtained from Table 6.2 or Table NA.1 in the UK (Clause 6.3.1.2(8) of BS EN 1991-1-1:2002).

Table 3.2

3.5 Imposed Loads: BS EN 1991-1-1:2002

Imposed loads are loads which are due to variable effects such as the movement of people, furniture, equipment and traffic. The values adopted are based on observation and measurement and are inherently less accurate than the assessment of dead loads.

In Clause 6.3.1.2(8) relating to the values of actions, the self-weight of internal moveable partitions (e.g. in speculative developments), should be considered as an additional uniformly distributed imposed load. The magnitude of the assumed distributed load is dependent on the weight/unit length of the partitions as shown in Table 3.3, (see Clause 6.3.1.2(8) and (9)). The values given in Table 3.3 may be used provided that the floor system is capable of lateral distribution of the loads.

Equivalent uniformly distributed loads for moveable partitions	
Weight of moveable partition/ m length w (kN/m)	Assumed uniformly distributed load (q_k kN/m^2)
$w \leq 1,0$	0,5
$1,0 < w \leq 2,0$	0,8
$2,0 < w \leq 3,0$	1,2
$w > 3,0$	The design should take account of: (i) the locations and directions of the partitions, (ii) the structural form of the floors.

Table 3.3

Imposed loads are normally considered as quasi-static actions (i.e. their dynamic characteristics are represented in a static analysis assuming an equivalent static action). In situations where resonance effects may be significant e.g. dancing movements on a suspended timber floor, a dynamic analysis is more appropriate. There are four classes of imposed load in building structures as follows:

1 residential, social, commercial and administrative areas – Tables 6.1 and 6.2 of EC1-1-1, (Tables NA.2 and NA.3 of the UK National Annex),
2 storage and industrial areas – Table 6.3 and Table 6.4 of EC1-1-1, (Tables NA..4 and NA..5 of the UK National Annex),
3 garages and vehicle traffic areas – Tables 6.7 and 6.8 of EC1-1-1, (Tables NA..6 of the UK National Annex),
4 roofs – Table 6.9 and Table 6.10 of EC1-1-1, (Tables NA..7 of the UK National Annex).

The tables referred to above identify individual categories within each of the first three classes and define the magnitude of uniformly distributed and concentrated point loads which are recommended for the design of floors, ceilings and their supporting elements.

The following categories are considered:

A Domestic and residential activities,
B Offices areas,
C Areas where people may congregate,
D Shopping areas,
E Storage and industrial areas (e.g. warehouses),
F Garages and vehicle and traffic areas (gross vehicle weight $\leq$ 30 kN),
G Garages and vehicle and traffic areas (30 kN $<$ gross vehicle weight $\leq$ 160 kN).

Most floor systems are capable of lateral distribution of loading and the recommended concentrated load need not be considered. In situations where lateral distribution is not possible, the effects of the concentrated loads should be considered with the load applied at locations which will induce the most adverse effect, e.g. maximum bending moment, shear and deflection. In addition, local effects such as crushing and punching should be considered where appropriate.

In multi-storey structures it is very unlikely that all floors will be required to carry the full imposed load at the same time. Statistically it is acceptable to reduce the total floor loads carried by a supporting member by varying amounts depending on the number of floors or floor area carried. This is reflected in Clause 6.2.1(4) for floors, beams and roofs using a reduction factor α_A and in Clause 6.2.2(2) for columns and walls using a reduction factor α_n.

For floors, beams and roofs:
Clause 6.2.1(4): *"Imposed loads from a single category may be reduced according to the areas supported by the appropriate member, by a reduction factor α_A according to 6.3.1.2(10)"*.

Clause 6.3.1.2(10) – EC1-1-1
The recommended value for the reduction factor α_A is determined as follows:
For categories A, B, E

$$\alpha_A = \frac{5}{7}\psi_0 + \frac{A_0}{A} \leq 1{,}0$$

For categories C and D

$$0{,}6 \leq \alpha_A = \frac{5}{7}\psi_0 + \frac{A_0}{A} \leq 1{,}0$$

(Equation (6.1) in EC1-1-1)

where:
ψ_0 is the factor given in Table NA.A1.1 of the UK National Annex,
$A_0 = 10{,}0 \text{ m}^2$,
A is the loaded area.

Clause NA.2.5 – UK National Annex
The recommended value for the reduction factor α_A is determined as follows:
$\alpha_A = 1{,}0 - A/1000 \geq 0{,}75$ (Equation (NA.1) in the UK National Annex)
where A is the supported area in m^2.

For columns and walls:
Clause 6.2.2(2): *"Where imposed loads from several storeys act on columns and walls, the total imposed loads may be reduced by a factor α_n according to 6.3.1.2(11) and 3.3.1(2)P."*

Clause 6.3.1.2(11) – EC1-1-1
The recommended value for the reduction factor α_n is determined as follows:

For categories A, B, C and D

$$\alpha_n = \frac{2+(n-2)\psi_0}{n} \qquad \text{(Equation (6.2) in EC1-1-1)}$$

where:
n is the number of storeys (> 2) above the loaded structural elements from the same category,
ψ_0 is the factor given in Table A1.1 of the UK National Annex.

Clause NA.2.6 – UK National Annex
The recommended value for the reduction factor α_A is determined as follows:
For categories A, B, C and D

$$\left.\begin{array}{ll}
\alpha_n = 1{,}0 - n/10 & \text{for } 1 \leq n \leq 5 \\
\alpha_n = 0{,}6 & \text{for } 5 < n \leq 10 \\
\alpha_n = 0{,}5 & \text{for } n > 10
\end{array}\right\} \text{(Equation (NA.2) in the UK National Annex)}$$

where n is the number of storeys with loads qualifying for reduction.

If $\alpha_A < \alpha_n$ load reductions based on Equation (NA.1) may be used.
Note: reductions given by Equation (NA.1) cannot be used in combination with those determined from Equation (NA.2).

3.5.1 Example 3.5: BS EN 1991-1-1:2002 - Dead and Imposed Loads

The floor plan of an industrial building is shown in Figure 3.5. Using the characteristic permanent and variable actions given, determine:

(i)   the design loads carried by beams B1 and B2,

(ii) the maximum shear force in each case, and

(iii) the maximum bending moment in each case.

The solution is presented using Equation (6.10) of BS EN 1990:2002. The reader should evaluate a solution using the alternative, i.e. using Equations (6.10(a)) and (6.10(b)).

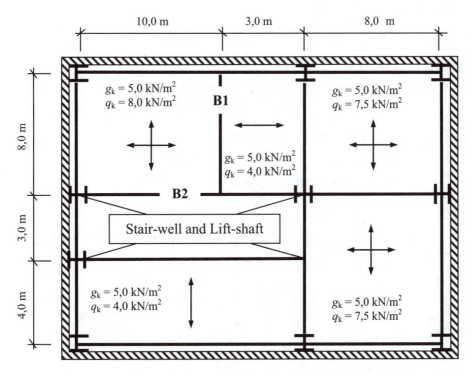

Figure 3.5

Solution
(i) Design Loads:
Beam B1
The load on beam B1 is equal to a triangular load from the two-way spanning slab combined with a uniformly distributed load from the one-way spanning slab.

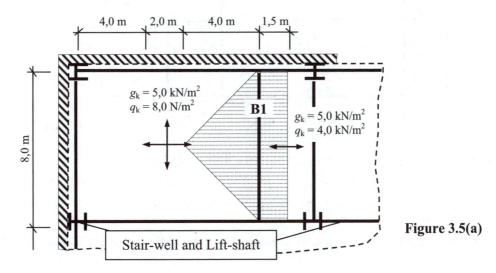

Figure 3.5(a)

BS EN 1990:2002

Equation (6.10) $F_d = \sum_{j\geq 1} \gamma_{G,j} G_{k,j}\ "+"\ \gamma_{Q,1} Q_{k,1}\ "+"\ \sum_{i>1} \gamma_{Q,i} \psi_{0,i} Q_{k,i}$

UK National Annex to BS EN 1990:2002

Table NA.A1.2(B) For permanent actions: Unfavourable $\gamma_{G,sup} = 1{,}35$
                   For variable actions: Leading action $\gamma_{Q,1} = 1{,}5$

Triangular area $A_1 = (0{,}5 \times 8{,}0 \times 4{,}0) = 16{,}0\ \text{m}^2$

                       $F_d = \left(\gamma_{G,1} g_{k,1} + \gamma_{Q,1} q_{k,1}\right) = [(1{,}35 \times 5{,}0) + (1{,}5 \times 8{,}0)] = 18{,}75\ \text{kN/m}^2$

                       $W_1 = (18{,}75 \times 16{,}0) = 300{,}0\ \text{kN}$

Rectangular area $A_2 = (1{,}5 \times 8{,}0) = 12{,}0\ \text{m}^2$

                       $F_d = \left(\gamma_{G,1} g_{k,1} + \gamma_{Q,1} q_{k,1}\right) = [(1{,}35 \times 5{,}0) + (1{,}5 \times 4{,}0)] = 12{,}75\ \text{kN/m}^2$

                       $W_2 = (12{,}75 \times 12{,}0) = 153{,}0\ \text{kN}$

Total design load $= (W_1 + W_2) = (300 + 153{,}0) = 453{,}0$ **kN**

Beam B2:

The load on beam B2 is equal to a trapezoidal load from the two-way spanning slab combined with a point load from beam B1.

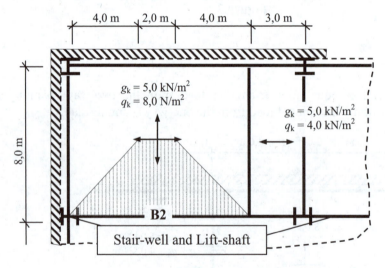

Figure 3.5(b)

Trapezoidal area $A_3 = [0{,}5 \times (2{,}0 + 10{,}0) \times 4{,}0] = 24{,}0\ \text{m}^2$

                       $F_d = \left(\gamma_{G,1} g_{k,1} + \gamma_{Q,1} q_{k,1}\right) = [(1{,}35 \times 5{,}0) + (1{,}5 \times 8{,}0)] = 18{,}75\ \text{kN/m}^2$

                       $W_3 = (18{,}75 \times 24{,}0) = 450{,}0$ **kN**

Point load due to end reaction of Beam 1 $W_4 = (W_1 + W_2)/2 = (0{,}5 \times 453{,}0) = 226{,}5$ **kN**

(ii) Maximum Shear Force:

Beam B1

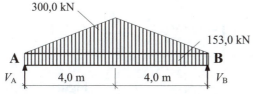

Figure 3.5(c)

+ve ↑ $\Sigma F_y = 0$ $V_A - 300{,}0 - 153{,}0 + V_B = 0$ ∴ $V_A + V_B = 453{,}0$ kN

+ve ↻ $\Sigma M_A = 0$ + $(300{,}0 \times 4{,}0) + 153{,}0 \times 4{,}0) - (V_B \times 8{,}0) = 0$

∴ $V_B = $ **226,5 kN** ↑ and $V_A = $ **226,5 kN** ↑

$F_{v,max}$ = **maximum end reaction** = 453,0/2 = **226,5 kN**

Beam B2

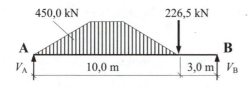

Figure 3.5(d)

+ve ↑ $\Sigma F_y = 0$ $V_A - 450{,}0 - 226{,}5 + V_B = 0$ ∴ $V_A + V_B = 676{,}5$ kN

+ve ↻ $\Sigma M_A = 0$ + $(450{,}02 \times 5{,}0) + (226{,}5 \times 10{,}0) - (V_B \times 13{,}0) = 0$

∴ $V_B = $ **347,31 kN** ↑ and $V_A = $ **329,19 kN** ↑

$F_{v,max}$ = **maximum end reaction = 347,31 kN**

(iii) Maximum Bending Moment:
Beam B1

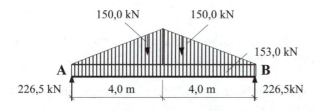

Figure 3.5(e)

The maximum bending moment occurs at the mid-span.
Design bending moment = $[(226{,}5 \times 4{,}0) - (76{,}5 \times 2{,}0) - (150{,}0 \times 4{,}0/3{,}0)]$
                    $M = $ **553,0 kNm**

Beam B2
The maximum value of the trapezoidal load = (2,0 × 450,0)/(10,0 + 2,0) = 75,0 kN/m

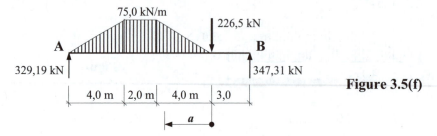

Figure 3.5(f)

Assume the point of zero shear occurs at a point between 6,0 m and 10,0 m from the support A.

Considering the sum of the forces from the right-hand side:

$$+ 347,31 - 226,5 - (0,5 \times a)[(75,0 \times a)/4,0] = 0 \qquad \therefore a^2 = 12,886$$

The point of zero shear occurs at $a = 3,59$ m, which is 6,41 m from the support A, i.e. between the assumed values.

Considering the bending moment due to the forces to the right-hand side of the position of zero shear:

$$M = (347,31 \times 6,59) - (226,5 \times 3,59) - (0,5 \times 3,59) \times [(75,0 \times 3,59)/4,0] \times (3,59/3,0)$$
$$M = 1331,05 \text{ kNm}$$

3.6 Snow Loads: BS EN 1991-1-3:2003

The actions on structural roofs due to snow accumulations can be derived for altitudes less than or equal to 1500 m as indicated in Clause 5.2 on the basis of:

   (a)   undrifted snow:
         (i.e. the assumed uniformly distributed snow load on the roof, affected only by the shape of the roof before any redistribution of snow due to other climatic actions),

   (b)   drifted snow:
         (i.e. the assumed snow distribution resulting from snow having been moved from one location to another location on a roof, e.g. by the action of wind).

For altitudes greater than 1500 m specialist advice should be sought regarding snow loads. The main features which influence the design snow load on a roof are:

    ♦ the characteristic value of the snow load on the ground,
    ♦ the shape of the roof (e.g. mono-pitch, duo-pitch, multi-pitch),
    ♦ the thermal properties (e.g. the thermal transmittance characteristics),

- the surface roughness,
- the proximity of nearby buildings (e.g. adjacent taller constructions),
- the existence of projections, obstructions, etc. (e.g. local effects due to snow boards),
- the surrounding terrain (e.g. sheltered, normal or windswept topography),
- the local meteorological climate (e.g. conditions characteristic of geographical location).

These features are incorporated in the derivation of the snow loads (s) on the roof by multiplying the characteristic snow load on the ground (s_k), by three factors as follows:

Clause 5.2(3)P – EC1-1-3
For persistent/transient design situations:

$s = \mu_i\, C_e\, C_t\, s_k$ (Equation (5.1) in EC1-1-3)

For design situations where exceptional **snow load** is considered as an accidental action:

$s = \mu_i\, C_e\, C_t\, s_{Ad}$ (Equation (5.2) in EC1-1-3)

where
$s_{Ad} = C_{esl}\, s_k$ (Equation (4.1) in EC1-1-3)
C_{esl} = the coefficient for exceptional snow loads. The recommended value in EC1-1-3 and in the UK National Annex is equal to 2,0.

For design situations where exceptional **snow drifts** are considered as an accidental action and where the conditions indicated in Annex B apply, (i.e. snow drifts for multi-span roofs, roofs abutting and close to taller construction works or where projections, obstructions or parapets exist):

$s = \mu_i\, s_k$ (Equation (5.3) in EC1-1-3)

where
μ_i is the snow load coefficient,
C_e is the exposure coefficient,
C_t is the thermal coefficient,
s_k is the characteristic value of the snow on the ground for a given location. This value can be found from snow maps given in the National Annex for any given country. The snow maps in the UK National Annex provide values to be used for the ground snow load in kN/m^2 at 100 m above mean sea level (a.m.s.l) in six different zones throughout the country. This value can be modified for a particular site altitude (m) a.m.s.l. using the expression NA.1 given in Clause NA.2.8.

The authors believe there is a typing error in this expression in the UK Annex to determine

s_k. The second term should read $+\left(\dfrac{A-100}{525}\right)$ and not $+\left(\dfrac{A+100}{525}\right)$ as given.

This is consistent with the values given in Figure NA.1 for 100 m a.m.s.l, with the exception of zone1, which should read 0,3, and not 0,25 as given; i.e.

Equation (NA.1) $s_k = \left[0,15 + \left(0,1Z + 0,05\right)\right] + \left(\dfrac{A-100}{525}\right)$

Figure NA.1   Zone number $Z = 1$
                    Site altitude $A = 100,0$ m

$$s_k = \left[0,15 + \left(\left(0,1\times1,0\right) + 0,05\right)\right] + \left(\dfrac{100,0 - 100}{525}\right) = 0,3 \text{ kN/m}^2$$

3.6.1 Example 3.6: BS EN 1991-1-3:2003 -Snow Load on Mono-pitched Roof

A mono-pitched, three-storey timber framed house, is shown in Figure 3.6. Using the design data given, determine the overall snow load on the roof in accordance with BS EN 1991-1-3 when considering the undrifted load arrangement only.

Design data:
Location   Stirling - Scotland
Altitude   40 m above mean sea level
Thermal transmittance of roof < 1 W/m^{2}K
Similar buildings are planned for the surrounding area.

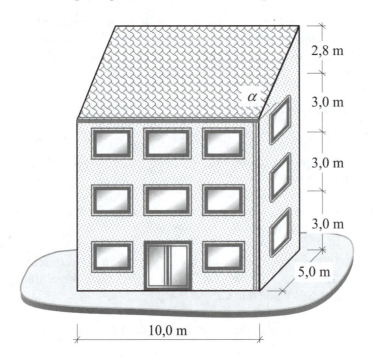

2,8 m

3,0 m

3,0 m

3,0 m

5,0 m

10,0 m

Figure 3.6

Solution:
BS EN 1991-1-3:(2003)
Clause 5.2(3)P   For persistent/transient design situations: $s = \mu_i\, C_e\, C_t\, s_k$

Figure 5.2   Roof slope for monopitch roof $\alpha = \tan^{-1}(2{,}8/5{,}0) = 29{,}25°$

National Annex to BS EN 1991-1-3:(2003)
Clause NA.2.8   Characteristic ground snow load:

Equation (NA.1) $s_k = \left[0{,}15 + \left(0{,}1Z + 0{,}05\right)\right] + \left(\dfrac{A-100}{525}\right)$

Figure NA.1   Zone number $Z = 4$
                      Site altitude $A = 40{,}0$ m

Equation (NA.1) $s_k = \left[0{,}15 + \left(\left(0{,}1 \times 4{,}0\right) + 0{,}05\right)\right] + \left(\dfrac{40{,}0-100}{525}\right) = 0{,}486$ kN/m^2

Clause 5.2(7)/Table 5.1   Assuming sheltered topography
                           Exposure coefficient $C_e = 1{,}2$

Clause 5.2(8)   Thermal transmittance of roof < 1 W/m^{2}K
                      Thermal coefficient $C_t = 1{,}0$

Figure 5.1/Table 5.2   Snow load shape coefficient for $0° \leq \alpha \leq 30°$ $\mu_1 = 0{,}8$

Equation (5.1) $s = \mu_i\, C_e\, C_t\, s_k = (0{,}8 \times 1{,}2 \times 1{,}0 \times 0{,}486) = 0{,}467$ kN/m^2

The roof area for the snow load $A_{snow} =$ plan area $= (5{,}0 \times 10{,}0) = 50{,}0$ m^2

Overall snow load on the roof $F_{snow} = (s \times A_{snow}) = (0{,}467 \times 50{,}0) = 23{,}35$ kN

3.6.2 Example 3.7: BS EN 1991-1-3:2003 - Snow Load on Duo-pitched Roof

An asymmetric duo-pitched, two-storey timber framed building for a student dormitory is shown in Figure 3.7. Using the design data given, determine the overall snow load on the roof in accordance with BS EN 1991-1-3, when considering both the undrifted and drifted load arrangements.

Design data:
Location   South of Glasgow - Scotland
Altitude   50 m above mean sea level
Thermal transmittance of roof < 1 W/m^{2}K
Normal topography conditions exist.

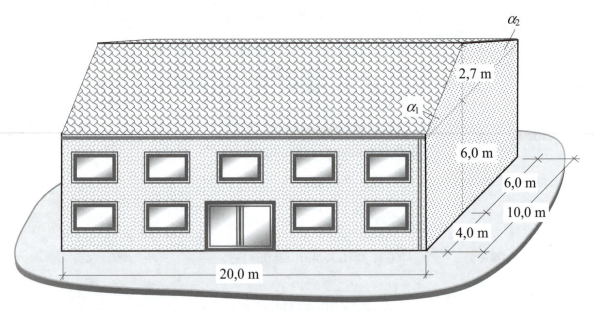

Figure 3.7

Solution:
Three load arrangements for undrifted and drifted snow are given in Figure 5.3 of BS EN 1991-1-3:(2003). The UK National Annex specifies alternative arrangements in the case of drifted snow loads for duo-pitched roofs, i.e. see Clause NA.2 and Table NA.1.
Load arrangements:

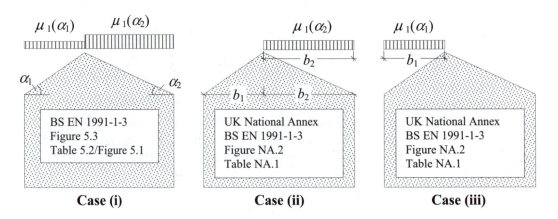

Figure 3.7(a)

BS EN 1991-1-3:(2003)

Clause 5.2(3)P   For persistent/transient design situations: $s = \mu_i\, C_e\, C_t\, s_k$
Figure 5.3   Left-hand roof slope $\alpha_1 = \tan^{-1}(2,7/4,0) = 34,02°$
                         Right-hand roof slope $\alpha_2 = \tan^{-1}(2,7/6,0) = 24,23°$

National Annex to BS EN 1991-1-3:(2003)

Clause NA.2.8   Characteristic ground snow load:

Equation (NA.1) $s_k = \left[0{,}15 + \left(0{,}1Z + 0{,}05\right)\right] + \left(\dfrac{A - 100}{525}\right)$

Figure NA.1   Zone number $Z = 3$
                         Site altitude $A = 50{,}0$ m

Equation (NA.1) $s_k = \left[0{,}15 + \left((0{,}1 \times 3{,}0) + 0{,}05\right)\right] + \left(\dfrac{50{,}0 - 100}{525}\right) = 0{,}405$ kN/m^2

Clause 5.2(7)/Table 5.1   Normal topography
                                    Exposure coefficient $C_e = 1{,}0$

Clause 5.2(8)   Thermal transmittance of roof < 1 W/m^{2}K
                                    Thermal coefficient $C_t = 1{,}0$

Case (i):

Figure 5.1/Table 5.2 $\alpha_1 = 34{,}02°$
                                    Snow load shape coefficient for $30° \le \alpha \le 60°$
                                    $\mu_{1,(\alpha_1)} = 0{,}8(60 - \alpha_1)/30 = 0{,}8(60 - 34{,}02)/30 = 0{,}693$

Figure 5.1/Table 5.2 $\alpha_2 = 24{,}23°$
                                    Snow load shape coefficient for $0° \le \alpha \le 30°$
                                    $\mu_{1,(\alpha_2)} = 0{,}8$

Equation (5.1) $s_1 = \mu_1\, C_e\, C_t\, s_k = (0{,}693 \times 1{,}0 \times 1{,}0 \times 0{,}405) = 0{,}281$ kN/m^2
                                    $s_2 = \mu_2\, C_e\, C_t\, s_k = (0{,}8 \times 1{,}0 \times 1{,}0 \times 0{,}405) = 0{,}324$ kN/m^2

The left-hand roof slope area for the snow load $A_{snow,1} = (4{,}0 \times 20{,}0) = 80{,}0$ m^2
                                          $F_{snow} = (s \times A_{snow}) = (0{,}281 \times 80{,}0) = 22{,}48$ kN
The right-hand roof slope area for the snow load $A_{snow,2} = (6{,}0 \times 20{,}0) = 120{,}0$ m^2
                                       $F_{snow} = (s \times A_{snow}) = (0{,}324 \times 12{,}0) = 38{,}88$ kN

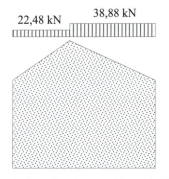

Figure 3.7(b)

The reader should complete the calculation for Cases (ii) and (iii) using Figure NA.2 and Table NA.1 from the UK National Annex.

3.7 Wind Loads: BS EN 1991-1-4:2005

Environmental loading such as wind loading is clearly variable and its source is outwith human control. In most structures the dynamic effects of wind loading are small, and static methods of analysis are adopted. The nature of such loading dictates that a statistical approach is the most appropriate in order to quantify the magnitudes and directions of the related design loads. The main features which influence the wind loading imposed on a structure are:

- geographical location   – Edinburgh, London, Paris, Berlin, ...
- terrain roughness   – town, open country, coastal areas, ...
- orography   – exposed hill top, escarpment, ...
- altitude   – height above mean sea level
- building shape   – square, rectangular, cruciform, irregular, ...
- roof pitch   – shallow, steep, mono-pitch, duo-pitch, multi-bay,…
- building dimensions
- wind speed and direction
- wind gust peak factor.

Guidance on the determination of wind actions on land-based structures, parts of a structure or elements attached to a structure (e.g. cladding, fixings, etc), is given in BS EN 1991-1-4. There are a considerable number of Nationally Determined Parameters, (NDPs), applicable to this part of the code.

Generally the procedures, values and recommendations are intended for structures which are not susceptible to dynamic excitation by virtue of their structural properties, e.g. mass, stiffness, natural response frequencies or structural form such as slender suspended bridge decks or long span cable stayed roofs. Structures of this type will normally require more complex mathamatical modelling and/or wind tunnel testing.

The information given in the following text is based on the '*Draft*' version of the UK National Annex for BS EN 1991-1-4 (July 2006), which provides alternative procedures, values and recommendations to satisfy the Principles defined in the code.

The classification of '*terrain roughness*' is given in terms of five categories (0 to IV) in Annex A of the code ranging from coastal sea areas to urban city areas. In the UK National Annex the categories have been reduced to three:

1 Sea terrain   – (corresponding to terrain roughness category 0),
2 Country terrain – (corresponding to terrain roughness categories I and II) and
3 Town terrain– (corresponding to terrain roughness categories III and IV).

A roughness factor $c_r(z)$ which allows for the variability of the mean wind velocity at a site, z m above ground level, and the roughness of the terrain upwind of the structure can be determined using Equations (4.4) and (4.5) in the code. These expressions **do not apply** when using the UK National Annex. In the Annex the roughness factor is determined as follows:

For sea terrain: use Figure NA3 assuming that the distance upwind from the shoreline is equal to 0,1 km,

For country terrain: use Figure NA3,

For town terrain: use Figure NA3 multiplied by a correction factor obtained from Figure NA4.

An orography factor c_o allows for the effects of significant orography on wind velocities. In the UK National Annex significant orography is defined in Figure NA2. In such cases the value can be determined using the procedure given in A.3 of the code. In most cases, structures will not be sited in areas of significant orography and a value of 1,0 can be assumed as indicated in Clause NA 2.13 of the UK National Annex and Clause 4.3.3(2) of the code.

The wind forces applied to a complete structure or structural component can be evaluated using either:

(a) force coefficients, or
(b) surface pressures.

Method (a) using force coefficients is described in relation to:

(i)	signboards	Clause 7.4.3
(ii)	structural elements with rectangular section	Clause 7.6
(iii)	structural elements with sharp edged sections	Clause 7.7
(iv)	structural elements with regular polygonal section	Clause 7.8
(v)	circular cylinders	Clause 7.9.2
(vi)	vertical cyliners in a row arrangement	Clause 7.9.3
(vii)	spheres	Clause 7.10
(viii)	lattice structures and scaffolding	Clause 7.11
(ix)	flags	Clause 7.12
(x)	bridges	Section 8

The use of force coefficients is not illustrated in this text. Method (b) is used in Examples 3.8 to 3.10.

3.7.1 Surface Pressures: (Clause 5.2(1) and Clause 5.2(2))

The wind pressure acting on a surface can be determined considering one of two cases:

1. External pressure
$$w_e = q_p(z_e) \times c_{pe}$$ (Equation (5.1) in EC1-1- 4)

2. Internal pressure
$$w_i = q_p(z_i) \times c_{pi}$$ (Equation (5.2) in EC1-1- 4)

where:

$q_p(z_e)$ is the peak velocity pressure on the exernal sufaces

$q_p(z_i)$ is the peak velocity pressure on the internal surfaces

z_e is the reference height for the external pressure

z_i is the reference height for the internal pressure

c_{pe} is the pressure coefficient for external pressure   } see Section 7.0 of EC-1-1- 4

c_{pi} is the pressure coefficient for internal pressure

The sign convention adopted for surface pressures is shown in Figures 3.8(a) and 3.8(b), i.e. pressure directed towards a surface is regarded as positive and pressure directed away from a surface is regarded as negative.

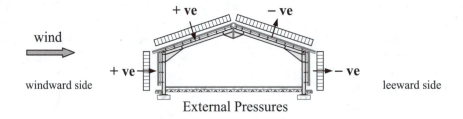

Figure 3.8(a)

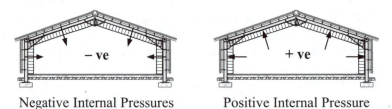

Figure 3.8(b)

3.7.2 Peak Velocity Pressure - $q_p(z)$: (Clause 4.5(1))

The peak velocity pressure at height 'z' in accordance with the code using Equation (4.8):

$$q_p(z) = \frac{\left[1 + 7I_v(z)\right] \times \rho \times v_m^2(z)}{2}$$
(Equation (4.8) in EC1-1- 4)

$$= c_e(z) \times q_b$$

This **does not apply** in the UK. In the UK in accordance with the National Annex:

$$q_p(z) = \frac{\left[1 + 3I_v(z)\right] \times \rho \times v_m^2(z)}{2}$$
(Equation (NA3) in the UK National Annex)

$$= c_e(z) \times q_b$$

where:

q_b is the basic velocity pressure given by $\dfrac{\rho v_b^2}{2}$ (Equation (4.10) in EC1-1- 4)

v_b is the basic wind velocity (see Section 3.7.3 of this text),
v_m is the mean wind velocity (see Section 3.7.4 of this text),

$I_v(z)$ is the turbulence intensity at height (z):

$$
\begin{aligned}
&= \frac{\sigma_v}{v_m(z)} = \frac{k_I}{c_o(z) \times ln(z/z_0)} \quad && \text{for } z_{min} \leq z \leq z_{max} \\
&= I_v(z_{min}) && \text{for } z \leq z_{min}
\end{aligned}
\left.\rule{0pt}{40pt}\right\} \text{Equation (4.7) of EC 1-1- 4}
$$

ρ is the air density. The recommended value in Note 2 of Clause 4.5(1) of the code is equal to 1,25 kg/m³. In the UK National Annex, Clause NA 2.18, a value of 1,226 kg/m³ is given,

z_{min} is the minimum height defined in Table 4.1 relating to the terrain category,

z_{max} is to be taken as 200 m (see Clause 4.3.2),

z_0 is the roughness length defined in Table 4.1 relating to the terrain category,

$c_e(z)$ is the exposure factor $= \dfrac{q_p(z)}{q_b}$ (Equation (4.9) in EC1-1- 4)

This is also given in Figure 4.2 of the code as a function of terrain categories 0 to IV and height (z) above the terrain, assuming flat terrain (i.e. $c_o(z) = 1,0$) and $k_I = 1,0$

In the UK National Annex, $c_e(z)$ can be found directly using Figure NA7 for sites in Country terrain. In Town terrain this value should be modified by multiplying by the exposure correction factor for Town terrain, obtained from NA8.

When the orography is significant the value of $q_p(z)$ should be determined in accordance with the expression NA4 of the UK National Annex.

3.7.3 Basic Wind Velocity: (Clause 4.2(2)P)
The basic wind velocity can be determined from:

$$v_b = c_{dir} \times c_{season} \times v_{b,0} \qquad \text{(Equation (4.1) in EC1-1- 4)}$$

where:
c_{dir} is the directional factor. The recommended value of 1,0 is conservative for all directions. The UK National Annex gives alternative values in Table NA1 for various angles measured in a clockwise direction from due North.
c_{season} is the season factor. The recommended value is 1,0. The UK National Annex gives alternative values in Table NA2 for each month and various time periods.

$v_{b,0}$ is the fundamental value of the basic wind velocity normally given in map form. In the UK National Annex the value of $v_{b,0}$ is obtained from Equation (NA1):

$v_{b,0} = v_{b,map} \times c_{alt}$ (Equation (NA1) in the UK National Annex)

where $v_{b,map}$ is the fundamental value of the basic wind velocity before the altitude correction has been applied; obtained from Figure NA1. The altitude correction factor, c_{alt}, is given by Equations (NA2a) and (NA2b) as follows:

$$c_{alt} = 1 + 0,001 \times A \qquad\qquad\quad \textbf{for } z \leq 10 \text{ m} \qquad \text{Equation (NA2a)}$$
$$c_{alt} = 1 + 0,001 \times A \times (10/z)^{0,2} \quad \textbf{for } z > 10 \text{ m} \qquad \text{Equation (NA2b)}$$

A is the altitude of the site in metres above mean sea level,
z is either z_s as defined in Figure 6.1 of the code or z_e, the height of the part above ground as defined in Figure 7.4 of the code.
Equation (NA2a) always gives a conservative value and can be used for any site altitude.

3.7.4 Mean Wind Velocity: (Clause 4.3.1(1))

The mean wind velocity can be determined from:

$v_m(z) = c_r(z) \times c_o(z) \times v_b$ (Equation (4.1) in EC1-1- 4)

where:
$c_r(z)$ is the roughness factor,
$c_o(z)$ is the orography factor, $\Big\}$ see Section 3.7.4
$v_b,$ is the basic wind velocity.

3.7.5 External Pressure Coefficients: (Clause 7.2)

The aerodynamic characteristics of a building are allowed for by the inclusion of external pressure coefficients. Values for these are given in relation to the various surfaces on which the wind acts. The surfaces are defined for vertical walls as A, B, C, D and E in Figure 7.5 of the code and similarly as F, G, H, I, J, K, L, M and N for a variety of roof shapes in Figures 7.6 to 7.9 of the code.

When considering rectangular plan buildings, Table 7.1 in BS EN 1991-1-4 and Tables NA4a and NA4b in the UK National Annex provide values corresponding to different surfaces (zones) and h/d ratios where 'h' is the full height of the building and 'd' is the dimension parallel to the wind direction as shown in Figure 3.9.

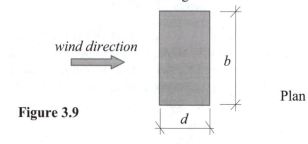

Figure 3.9

The surface of a building on which it is assumed that the wind pressures are acting may be considered to comprise one or several discrete areas in accordance with Clause 7.2.2 and Figure 7.4 of the code. The top level of each area considered is the reference height 'z_e' as shown in Figure 3.10.

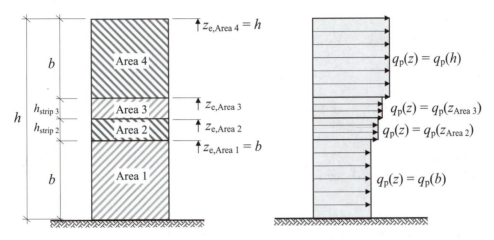

Velocity pressure profile over the height of the building for windward walls.

Figure 3.10

The values for the external pressure coefficients given in the Tables in BS EN 1991-1-4 are for loaded areas = 1,0 m² ($c_{pe,1}$) and loaded areas ≥ 10,0 m² ($c_{pe,10}$). These are intended for use in asessing wind loads for localised areas and overall structures respectively.

For loaded areas between 1,0 m² and 10,0 m², the code gives a logorithmic equation from which c_{pe} can be determined. This **does not** apply in the UK.

The UK National Annex indicates in Clause NA 2.26, that the $c_{pe,1}$ values should be applied to loaded areas ≤ 1,0 m² and that the $c_{pe,10}$ values apply to loaded areas > 1,0 m².

Note: The reference height 'z_e' for the leeward wall and the sidewalls should be taken as the full height of the building as indicated in the Note in Clause 7.2.2(1) of the code and Clause NA 2.26 of the UK National Annex.

Coefficients are also given for a variety of other structural situations, e.g. canopy roofs, free-standing walls and parapets etc., and reference should also be made to the UK National Annex.

3.7.6 Internal Pressure Coefficients - c_{pi}: (Clause 7.2.9)

The internal pressure coefficients depend on the size and distribution of openings in a building envelope. A number of rules are given in Clause 7.2.9 from which values can be derived. Where it is not possible, or not considered justified, to calculate values based on the area of openings in a building c_{pi} should be taken as the most onerous of + 0,2 and − 0,3.

3.7.7 Wind Forces using Surface Pressures: (Clause 5.3(3))

The wind force on a structure can be determined by evaluating the vectorial summation of the forces due to external and internal pressures and friction forces using Equations (5.5), (5.6) and (5.7) as follows:

External forces:

$$F_{w,e} = c_s c_d \sum_{\text{surfaces}} w_e \times A_{ref} \qquad \text{(Equation (5.5) in EC1-4)}$$

Internal forces:

$$F_{w,i} = c_s c_d \sum_{\text{surfaces}} w_i \times A_{ref} \qquad \text{(Equation (5.6) in EC1-4)}$$

Friction forces:

$$F_{fr} = c_{fr} \times q_p(z_e) \times A_{fr} \qquad \text{(Equation (5.7) in EC1-4)}$$

where:
$c_s c_d$ is the structural factor,
w_e is the the external pressure on the individual surface at height z_e,
w_i is the the internal pressure on the individual surface at height z_i,
A_{ref} is the reference area of the individual surface,
c_{fr} is the friction coefficient,
A_{fr} is the area of the external surface parallel to the wind.

As indicated in Clause 5.3(4):
"*The effects of wind friction can be ignored when the total area of all surfaces parallel with (or at a small angle to) the wind is equal to or less than 4 times the total area of all external surfaces perpendicular to the wind (windward or leeward)*"; friction forces are not considered further in this text.

The structural factor '$c_s c_d$' has two components, 'c_s', the size effect factor and 'c_d', the dynamic factor. The determination of '$c_s c_d$' is given in Section 6 of BS EN 1991-1-4 or alternatively in the UK National Annex. (In the UK National Annex, 'c_s' and 'c_d' can be evaluated seperately using Table NA3 and Figure NA9). In most typical framed buildings '$c_s c_d$' can be taken as 1,0.
The application of BS EN 1990 and BS EN 1991 is illustrated in Examples 3.8 to 3.10.

3.7.8 Example 3.8: BS EN 1991-1-4:2005 - Wind Load on Storage Silo

A closed top storage silo, as shown in Figure 3.11, is situated in an industrial development in the suburbs of Edinburgh. Using the design data given and considering the wind to be acting in the direction indicated, determine the overall horizontal wind loading on the windward and leeward faces of the structure.

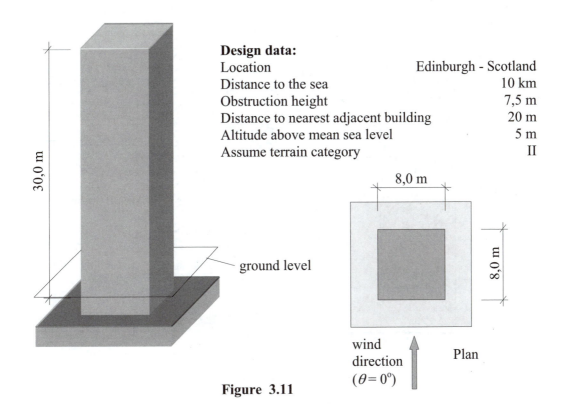

Design data:

Location	Edinburgh - Scotland
Distance to the sea	10 km
Obstruction height	7,5 m
Distance to nearest adjacent building	20 m
Altitude above mean sea level	5 m
Assume terrain category	II

Figure 3.11

Solution:

BS EN 1991-1-4: 2005

Clause 7.2.2   A building whose height 'h' is greater than or equal to '$2b$', the crosswind dimension (see Figure 7.5 of the code) may be considered to be in multiple parts as indicated in Figure 7.4 of the code.

Height of the silo h = 30,0 m,   Crosswind dimension b = 8,0 m
$h/b = (30,0/8,0) = 3,75 > 2$

Figure 7.4   Consider the windward surface to be divided into four parts A, B, C and D as indicated in Figure 3.20.

Reference heights for each part A, B, C and D of the surface are shown in Figure 3.11(a).

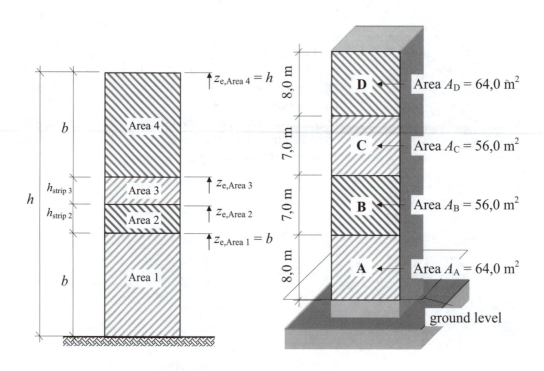

Figure. 3.11(a)

Clause 7.2.2(1)   Reference heights:
                   For Part A $z_e = 8{,}0$ m
                   For Part B $z_e = 15{,}0$ m
                   For Part C $z_e = 22{,}0$ m
                   For Part D $z_e = 30{,}0$ m

UK NA.
Clause NA 2.19   The reduction factor indicated in Clause 7.2.2(3) of the code may be applied to the summation of the loads on all windward and leeward surfaces.

Clause 7.2.2(3)   For buildings with $h/d < 1$ the resulting force may be multiplied by 0,85. For $h/d \geq 5$ a value of 1 should be used.

                   For intermediate values of h/d, linear interpolation may be applied. (**Note:** d is the depth of the building measured in the direction of the wind as indicated in Figure 7.5 of the code.)

                   In this case $d = 8{,}0$ m and $h/d = (30{,}0/8{,}0) = 3{,}75$

                   The reduction factor $= 0{,}85 + \dfrac{(3{,}75-1)(1-0{,}85)}{(5-1)} = 0{,}953$

Clause 5.3(3)   The silo is closed and there is no need to consider any internal pressure caused by the wind.

Equation (5.5) $F_{w,e} = c_s c_d \times \left(\displaystyle\sum_{\text{surfaces}} w_{e,\text{front}} \times A_{\text{ref}} \right)$

$$F_{w,e} = 0{,}953 \times c_s c_d \times \left(\sum_{\text{surfaces}} w_{e,\text{front}} \times A_{\text{ref}} - \sum_{\text{surfaces}} w_{e,\text{rear}} \times A_{\text{ref}} \right)$$

$$= 0{,}953 \times c_s c_d \times \left(\sum_{\text{surfaces}} F_{w,e,\text{front}} - \sum_{\text{surfaces}} F_{w,e,\text{rear}} \right)$$

UK NA.
Clause NA 2.7   The values of the season factor c_s and the directional factor c_d can be taken as 1,0

BS EN 1991-1-4
Clause 5.2(1)   The external pressures acting on the windward and leeward surfaces are given by:

Equation (5.1) $w_{e,\text{front}} = q_{p,\text{front}}(z_e) \times c_{pe,\text{front}}$ and $w_{e,\text{rear}} = q_{p,\text{rear}}(z_e) \times c_{pe,\text{rear}}$

Clause 7.2.2(1)   The external pressure for each of the surface areas on the windward face should be determined in relation to the corresponding reference heights. **On the leeward surface, the value corresponding to the full height of the building (i.e. Part D) should be used as indicated in the Note to Clause 7.2.2(1).**

UK NA.
Clause NA 2.11   Since terrain is category II, this is reclassified in the UK National Annex as Country terrain.

Equation (NA3)   Peak velocity pressure $q_p(z) = \dfrac{\left[1 + 3I_v(z)\right]^2 \times \rho \times v_m^2(z)}{2} = c_e(z) \times q_b$

When the orography is not significant (see Figure NA2), then $q_p(z)$ can be calculated by using $q_p(z) = c_e(z) \times q_b$

where $c_e(z) = c_{e,\text{flat}}(z)$ from Figure NA7.

Figure NA7   The value of $c_{e,\text{flat}}(z)$ is dependent on the reference height z_e, the displacement height h_{dis} and the distance upwind to the shore line.

BS EN 1991-1-4
Figure 7.4   Reference heights:
Zone A: $z_e = 8{,}0$ m   Zone B: $z_e = 15{,}0$ m
Zone C: $z_e = 22{,}0$ m   Zone D: $z_e = 30{,}0$ m

Annex A
A.5   The horizontal distance to the nearest adjacent building $x = 20{,}0$ m
The obstruction height $h_{ave} = 7{,}5$ m
$(2 \times h_{ave}) = 15{,}0$ m and $(6 \times h_{ave}) = 45{,}0$ m

Equation (A.15)   The displacement height is the lesser of:

$(1,2 \times h_{ave}) - (0,2 \times x) = (1,2 \times 7,5) - (0,2 \times 20,0) = 5,0$ m   and

$(0,6 \times h) = (0,6 \times 30,0) = 18,0$ m

$\therefore \ h_{dis} = 5,0$ m

Figure NA7   The values of $c_e(z)$ for the windward and leeward surfaces are summarized in Table 3.4 below corresponding to 10 km from the sea.

Parts	z_e (m)	h_{dis} (m)	$(z - h_{dis})$ (m)	$c_{e,front}(z_e)$	$c_{e,rear}(z_e)$
A	8,0		3,0	1,77	
B	15,0	5,0	10,0	2,66	3,40
C	22,0		17,0	3,05	
D	30,0		25,0	3,40	

Table 3.4

Equation (4.10)   Basic velocity pressure $q_b = \dfrac{\rho v_b^2}{2}$

Equation (4.1)   Basic wind velocity $v_b = c_{dir} \times c_{season} \times v_{b,0}$

UK NA.

Clause NA 2.4   Fundamental basic wind velocity $v_{b,0} = v_{b,map} \times c_{alt}$

Figure NA1   For Edinburgh $v_{b,map} = 24,5$ m/s

Clause NA 2.5   Equation (NA2a) always gives a conservative value and can be used for any site altitude.

$c_{alt} = \ 1 + 0,001 \times A$ for $z \leq 10$ m   Equation (NA2a)

$c_{alt} = \ 1 + 0,001 \times A \times (10/z)^{0,2}$ for $z > 10$ m   Equation (NA2b)

z is either z_s as defined in Figure 6.1 of the code or z_e, the height of the part above ground as defined in Figure 7.4.

BS EN 1991-1-4

Figure 6.1 $z_s = (0,6 \times h) \geq z_{min}$

Table 4.1   For Terrain category III $z_{min} = 2,0$ m

$z_s = (0,6 \times 30,0) = 18,0$ m

$\geq z_{min}$

Equation (NA2b) $c_{alt} = 1 + 0,001 \times A \times (10/z)^{0,2} = 1 + [0,001 \times 5,0 \times (10/18,0)^{0,2}]$

$= 1,004$

Clause NA 2.4   Fundamental basic wind velocity $v_{b,0} = v_{b,map} \times c_{alt}$

$v_{b,0} = (24,5 \times 1,004) = 24,61$ m/s

Equation (4.1)   Basic wind velocity $v_b = (c_{dir} \times c_{season} \times v_{b,0}) = (1,0 \times 1,0 \times 24,61)$
$= 24,61$ m/s

UK NA.
Clause NA 2.18   The air density in the UK $\rho = 1,226$ kg/m^3
BS EN 1991-1-4

Equation (4.10)   Basic velocity pressure $q_b = \dfrac{\rho v_b^2}{2} = \dfrac{1,226 \times 24,61^2}{2 \times 1000} = 0,371$ kN/m^2

Equation (NA3)   Peak velocity pressure $q_p(z)$ values for the windward and leeward surfaces are summarized in Table 3.5 below.

Parts	z_e	q_b (kN/m^2)	$c_{e,front}(z_e)$	$q_{p,front}$ (kN/m^2)	$c_{e,rear}(z_e)$	$q_{p,rear}$(kN/m^2)
A	8		1,77	0,657		
B	15	0,371	2,66	0,987	3,40	1,261
C	22		3,05	1,132		
D	30		3,40	1,261		

Table 3.5

BS EN 1991-1-4
Clause 5.2(1)   The external pressures acting on the windward and leeward surfaces are given by:

Equation (5.1) $w_{e,front} = q_{p,front}(z_e) \times c_{pe,front}$ and $w_{e,rear} = q_{p,rear}(z_e) \times c_{pe,rear}$
UK NA.
Clause NA 2.25   The external pressure coefficients to be used in accordance with the National Annex are:
For loaded areas $\leq 1,0$ m^2 use $c_{pe,1}$ $\Big\}$ Table NA4a of UK NA
For loaded areas $> 1,0$ m^2 use $c_{pe,10}$

Table NA4a   Zone D is the windward face and zone E is the leeward face.
$h/b = 30,0/8,0 = 3,75$
$d/b = 8,0/8,0 = 1,0$
For the windward surface (D) $c_{pe,10,front} = +0,8$
For the leeward surface (E) use interpolation between $-0,4$ and $-0,5$
$c_{pe,10,rear} = -0,4 + [-(0,5-0,4) \times (3,75-2,0)/(5,0-2,0)] = -0,458$

Parts	$q_{p,front}$ (kN/m^2)	$c_{pe,10,front}$	$w_{e,front}$ (kN/m^2)	$q_{p,rear}$ (kN/m^2)	$c_{pe,10,rear}$	$w_{e,rear}$ (kN/m^2)
A	0,657		+ 0,526			
B	0,987	+ 0,8	+ 0,790	1,261	− 0,458	− 0,578
C	1,132		+ 0,906			
D	1,261		+ 1,009			

Table 3.6

The distribution of the horizontal wind load on the building is given by:

$$F_{w,e} = 0,953 \times c_s c_d \times \left(\sum_{\text{surfaces}} w_{e,\text{front}} \times A_{\text{ref}} - \sum_{\text{surfaces}} w_{e,\text{rear}} \times A_{\text{ref}} \right)$$

$$= 0,953 \times c_s c_d \times \left(\sum_{\text{surfaces}} F_{w,e,\text{front}} - \sum_{\text{surfaces}} F_{w,e,\text{rear}} \right)$$

The corresponding values are summarized in Table 3.7 and indicated in Figure 3.11(b).

Parts	A_{ref} (m²)	$w_{e,\text{front}}$ (kN/m²)	$F_{w,e,\text{front}}$* (kN)	$w_{e,\text{rear}}$ (kN/m²)	$F_{w,e,\text{rear}}$* (kN)	$F_{w,e,\text{total}}$* (kN)
A	64,0	0,526	32,082		− 35,253	67,335
B	56,0	0,790	42,161	− 0,578	− 30,847	73,008
C	56,0	0,906	48,351		− 30,847	79,198
D	64,0	1,009	61,541		− 35,253	96,794
Total	240,0		184,135		132,200	316,335

* These values include the 0,953 factor. This only applies when considering oveall structural stabiity, (see Clause 7.2.2(3)).

Table 3.7

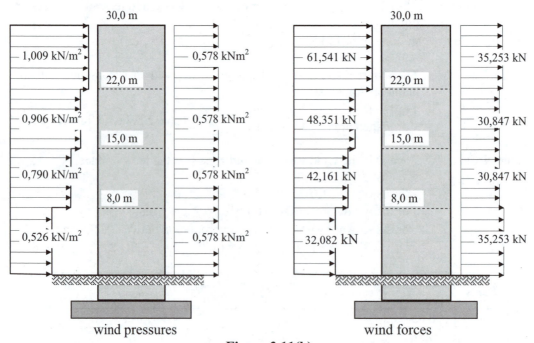

Figure 3.11(b)

The overall external wind force on the building $F_{w,e} = 316,335$ kN

3.7.9 Example 3.9: BS EN 1991-1-4:2005 - Wind Load on Building with Mono-pitch Roof

A sports complex is constructed using masonry external and internal walls with principal dimensions as shown in Figure 3.12. Using the data given, determine the surface loads which act on the walls and the roof for the wind direction indicated. (Do not consider internal pressures and suctions in this example; see Example 3.10).

Design data:

Location	Aberdeen - Scotland
Distance to the sea	5 km
Obstruction height	10,5 m
Distance to nearest adjacent building	30 m
Altitude above mean sea level	6 m
Assume terrain category	II

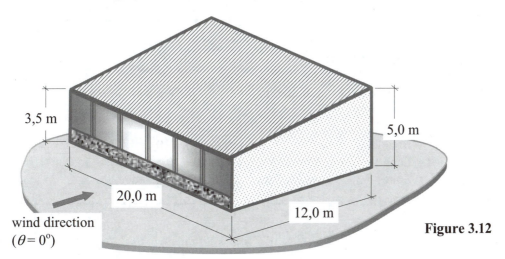

3,5 m

5,0 m

20,0 m

12,0 m

wind direction
($\theta = 0°$)

Figure 3.12

Solution:
BS EN 1991-1-4: 2005

Clause 7.2.2   A building whose height 'h' is less than or equal to 'b', the crosswind dimension (see Figure 7.5 of the code) should be considered to be one part (i.e. Part A) as indicated in Figure 7.4 of the code.

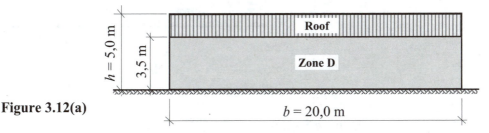

$h = 5,0$ m

3,5 m

Roof

Zone D

Figure 3.12(a)

$b = 20,0$ m

The building surfaces are considered as zones A to E for the vertical faces and F to H for the roof as indicated in Figure 7.5 and Figure 7.7 of the code.

Figure 7.5   Key for zones in vertical walls:

e is defined as the smaller of b or $2h$, i.e. 20,0 m or $(2 \times 5,0) = 10,0$ m
$\therefore e = 10,0$ m
The dimension in the wind direction $d = 12,0$ m
For an elevation with $e < d$ zone widths A, B and C are as follows:
Zone A $= (e/5) = (10,0/5,0) = 2,0$ m
Zone B $= (4e/5) = (4 \times 10,0/5,0) = 8,0$ m
Zone C $= (d - e) = (12,0 - 10,0) = 2,0$ m

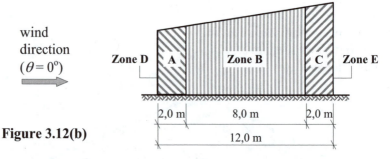

Figure 3.12(b)

Figure 7.7   Key for zones in monopitch roofs for the wind angle $\theta = 0°$.

Length of Zone F $= (e/4) = (10,0/4,0) = 2,5$ m
Breadth of Zone F and G $= (e/10) = (10,0/10) = 1,0$ m
Length of Zone G $= (b - 0,5e) = [20,0 - (0,5 \times 10,0)] = 15,0$ m
Zones F and G are used for local effects at the edges.

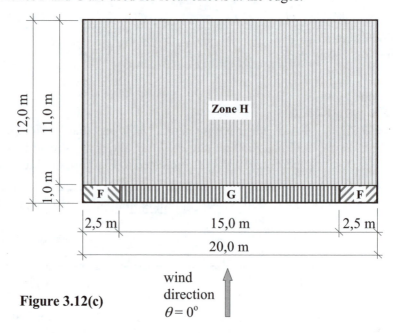

Figure 3.12(c)

UK NA.

Clause NA 2.19   The reduction factor indicated in Clause 7.2.2(3) of the code may be applied to the summation of the loads on all windward and leeward surfaces.

Clause 7.2.2(3)   For buildings with $h/d < 1$ the resulting force may be multiplied by 0,85. For $h/d \geq 5$ a value of 1 should be used.
For intermediate values of h/d, linear interpolation may be applied. (**Note:** d is the depth of the building measured in the direction of the wind as indicated in Figure 7.5 of the code.)

In this case $d = 12,0$ m and $h/d = (5,0/12,0) = 0,42 < 1$
The reduction factor = 0,85

Clause 5.3(3)   Assume the building has no major opening and there is no need to consider any internal pressure caused by the wind.

Equation (5.5)

$$F_{w,e} = c_s c_d \times \left(\sum_{surfaces} w_{e,front} \times A_{ref} \right)$$

$$F_{w,e} = 0,85 \times c_s c_d \times \left(\sum_{surfaces} w_{e,front} \times A_{ref} - \sum_{surfaces} w_{e,rear} \times A_{ref} \right)$$

$$= 0,85 \times c_s c_d \times \left(\sum_{surfaces} F_{w,e,front} - \sum_{surfaces} F_{w,e,rear} \right)$$

UK NA.

Clause NA 2.6/7   The values of the season factor c_s and the directional factor c_d can be taken as 1,0 (or in Clause 6.2.(1) of the code for $h < 15$ m $c_s = c_d = 1,0$).

BS EN 1991-1-4

Clause 5.2(1)   The external pressures acting on the windward and leeward surfaces are given by:

Equation (5.1) $w_{e,front} = q_{p,front}(z_e) \times c_{pe,front}$ and $w_{e,rear} = q_{p,rear}(z_e) \times c_{pe,rear}$

UK NA.

Clause NA 2.11   Since terrain is category II this is reclassified in the UK National Annex as Country terrain.

Equation (NA3)   Peak velocity pressure $q_p(z) = \dfrac{\left[1 + 3I_v(z)\right]^2 \times \rho \times v_m^2(z)}{2} = c_e(z) \times q_b$

When the orography is not significant (see Figure NA2), then $q_p(z)$ can be calculated by using:

$q_p(z) = c_e(z) \times q_b$
where $c_e(z) = c_{e,flat}(z)$ from Figure NA7.

Figure NA7   The value of $c_{e,flat}(z)$ is dependent on the reference height z_e, the
                     displacement height h_{dis} and the distance upwind to the shore line.

BS EN 1991-1-4
Figure 7.4   Reference heights:
                     Windward vertical surface Zone D: $z_e = 5{,}0$ m
                     Leeward vertical surface Zone E: $z_e = 5{,}0$ m

Annex A   The horizontal distance to the nearest adjacent building $x = 30{,}0$ m
A.5   The obstruction height $h_{ave} = 10{,}5$ m
                     $(2 \times h_{ave}) = 21{,}0$ m and $(6 \times h_{ave}) = 63{,}0$ m

Equation (A.15)   The displacement height is the lesser of:
                     $(1{,}2 \times h_{ave}) - (0{,}2 \times x) = (1{,}2 \times 10{,}5) - (0{,}2 \times 30{,}0) = 6{,}6$ m and
                     $(0{,}6 \times h) = (0{,}6 \times 5{,}0) = 3{,}0$ m
                     $\therefore\ h_{dis} = 3{,}0$ m

Figure NA7   The values of $C_e(z)$ for the windward and leeward surfaces are
                     summarized in Table 3.8 below corresponding to 5 km from the sea.

Zones	z_e (m)	h_{dis} (m)	$(z - h_{dis})$ (m)	$c_{e,front}(z_e)$	$c_{e,rear}(z_e)$
D and E	5,0	3,0	2,0	1,55	1,55

Table 3.8

Equation (4.10)   Basic velocity pressure $q_b = \dfrac{\rho v_b^2}{2}$

Equation (4.1)   Basic wind velocity $v_b = c_{dir} \times c_{season} \times v_{b,0}$

UK NA.
Clause NA 2.4   Fundamental basic wind velocity $v_{b,0} = v_{b,map} \times c_{alt}$
Figure NA1   For Aberdeen $v_{b,map} = 25{,}7$ m/s

Clause NA 2.5   Equation (NA2a) always gives a conservative value and can be used
                     any for site altitude.
                     $c_{alt} = 1 + 0{,}001 \times A$ for $z \le 10$ m   Equation (NA2a)
                     $c_{alt} = 1 + 0{,}001 \times A \times (10/z)^{0,2}$ for $z > 10$ m   Equation (NA2b)
                     z is either z_s as defined in Figure 6.1 of the code or z_e, the height of the
                     part above ground as defined in Figure 7.4.

BS EN 1991-1-4
Figure 6.1 $z_s = (0{,}6 \times h) \ge z_{min}$
Table 4.1   For Terrain category II $z_{min} = 2{,}0$ m
                     $z_s = (0{,}6 \times 5{,}0) = 3{,}0$ m
                     $\ge z_{min}$

Equation (NA2a) $c_{alt} = 1 + (0{,}001 \times A) = 1 + (0{,}001 \times 6{,}0) = 1{,}006$

Clause NA 2.4   Fundamental basic wind velocity $v_{b,0} = v_{b,map} \times c_{alt}$
$v_{b,0} = (25,7 \times 1,006) = 25,85$ m/s

Equation (4.1)   Basic wind velocity $v_b = (c_{dir} \times c_{season} \times v_{b,0}) = (1,0 \times 1,0 \times 25,85)$
$= 25,85$ m/s

UK NA.
Clause NA 2.18   The air density in the UK $\rho = 1,226$ kg/m^3

BS EN 1991-1-4

Equation (4.10)   Basic velocity pressure $q_b = \dfrac{\rho v_b^2}{2} = \dfrac{1,226 \times 25,85^2}{2 \times 1000} = 0,410$ kN/m^2

Equation (NA3)   Peak velocity pressure $q_p(z)$ values for the windward and leeward surfaces are summarized in Table 3.9 below.

Zones	z_e	q_b (kN/m^2)	$c_{e,front}(z_e)$	$q_{p,front}$ (kN/m^2)	$c_{e,rear}(z_e)$	$q_{p,rear}$(kN/m^2)
D and E	5,0	0,410	1,55	0,636	1,55	0,636

Table 3.9

BS EN 1991-1-4
Clause 5.2(1)   The external pressures acting on the windward and leeward surfaces are given by:

Equation (5.1) $w_{e,front} = q_{p,front}(z_e) \times c_{pe,front}$ and $w_{e,rear} = q_{p,rear}(z_e) \times c_{pe,rear}$

UK NA.
Clause 2.25   The external pressure coefficients to be used in accordance with the National Annex are:
For loaded areas $\leq 1,0$ m^2 use $c_{pe,1}$ ⎫
For loaded areas $> 1,0$ m^2 use $c_{pe,10}$ ⎬   Table NA4a of UK NA

Table NA4a   Zone D is the windward face and zone E is the leeward face.
$h/b = 5,0/20,0 = 0,25$
$d/b = 12,0/20,0 = 0,6$
For the windward surface (D) $c_{pe,10,front} = + 0,6$
For the leeward surface (E) use interpolation between $-0,4$ and $-0,5$
$c_{pe,10,rear} = -0,4 + [-(0,5 - 0,4) \times (0,6 - 0,25)/(1,0 - 0,25)] = -0,45$

Zone	$q_{p,front}$ (kN/m^2)	$c_{pe,10,front}$	$w_{e,front}$ (kN/m^2)	$q_{p,rear}$ (kN/m^2)	$c_{pe,10,rear}$	$w_{e,rear}$ (kN/m^2)
D and E	0,636	+ 0,6	+ 0,382	0,636	− 0,45	− 0,286

Table 3.10

The distribution of the horizontal wind load on the building is given by:

$$F_{w,e} = 0,85 \times c_s c_d \times \left(\sum_{surfaces} w_{e,front} \times A_{ref} - \sum_{surfaces} w_{e,rear} \times A_{ref} \right)$$

$$= 0,85 \times c_s c_d \times \left(\sum_{surfaces} F_{w,e,front} - \sum_{surfaces} F_{w,e,rear} \right)$$

The corresponding values are summarized in Table 3.11 and indicated on Figure 3.29.

$$A_{ref,windward} = (3,5 \times 20,0) = 70,0 \text{ m}^2$$
$$A_{ref,leedward} = (5,0 \times 20,0) = 100,0 \text{ m}^2$$

Zone	$A_{windward}$ (m²)	$w_{e,front}$ (kN/m²)	$F_{w,e,front}$ * (kN)	$A_{leeward}$ (m²)	$w_{e,rear}$ (kN/m²)	$F_{w,e,rear}$ * (kN)	$F_{w,e,total}$ * (kN)
D and E	70,0	+ 0,382	+ 22,729	100,0	− 0,286	− 24,310	47,039
* These values include the 0,85 factor. This only applies when considering overall structural stability, (see Clause 7.2.2(3)).							

Table 3.11

The wind loads on surfaces A, B and C (i.e. the side walls)
Clause 7.2.2(2)
Figure 7.5   Key for zones in vertical walls: $e = 10,0$ m

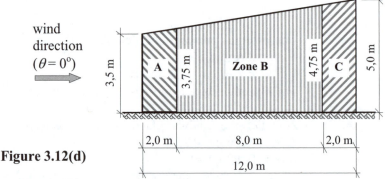

Figure 3.12(d)

Peak velocity pressure $q_p(z) = 0,636$ kN/m²

UK NA.
Clause NA 2.27   External pressure coefficients
                      In NOTE 8(a) of this Clause:
                      "where the gap between the buildings is < e/4 or > e, the isolated coefficient values should be used."

                      Distance to nearest adjacent building 30 m $> e$ ∴ building is isolated.
                      Use Table NA4b to determine the $c_{pe,10}$ values for zones A, B and C.

BS EN 1991-1-4
Clause 5.2(1)   The external pressures acting on zones A, B and C are given by:

Equation (5.1) $w_e = q_p(z_e) \times c_{pe}$
                    $w_{e,A} = q_p(z_e) \times c_{pe,A}$; $w_{e,B} = q_p(z_e) \times c_{pe,B}$; $w_{e,C} = q_p(z_e) \times c_{pe,C}$

UK NA.
Table NA4b   Exposure case: Isolated

Zone	q_p (kN/m²)	$c_{pe,10}$	w_e (kN/m²)
A		− 1,3	− 0,827
B	0,636	− 0,8	− 0,509
C		− 0,5	− 0,318

Table 3.12

The distribution of the horizontal wind load on the building is given by:

$$F_{w,e} = c_s c_d \times \left(\sum_{surfaces} w_{e,front} \times A_{ref} - \sum_{surfaces} w_{e,rear} \times A_{ref} \right)$$

$$= c_s c_d \times \left(\sum_{surfaces} F_{w,e,front} - \sum_{surfaces} F_{w,e,rear} \right)$$

Note: The reduction factor only applies to the windward and leeward faces.

The corresponding values are summarized in Table 3.13 and indicated on Figure 3.12(e).

$$A_{ref,A} = 0,5 \times (3,5 + 3,75) \times 2,0 = 7,25 \text{ m}^2$$
$$A_{ref,B} = 0,5 \times (3,75 + 4,75) \times 8,0 = 34,0 \text{ m}^2$$
$$A_{ref,C} = 0,5 \times (4,75 + 5,0) \times 2,0 = 9,75 \text{ m}^2$$

Zone	A_{ref} (m²)	w_e (kN/m²)	$F_{w,e}$ (kN)
A	7,25	− 0,827	− 5,996
B	34,0	− 0,509	− 17,306
C	9,75	− 0,318	− 3,101

Table 3.13

The wind loads on roof surfaces F, G and H (i.e. the monopitch roof)

Figure 7.7(a) The pitch angle $\alpha = \tan^{-1} \left(\dfrac{1,5}{12,0} \right) = 7,13°$

Figure 7.7(b)

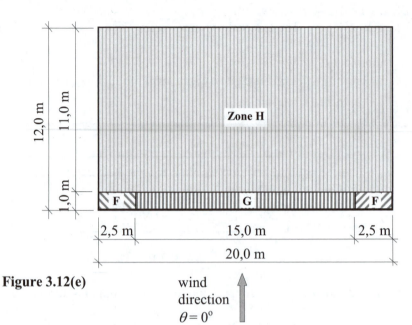

Figure 3.12(e)   wind
                                     direction
                                     $\theta = 0°$

UK NA.
Clause 2.25

The external pressure coefficients to be used in accordance with the
National Annex are:

For loaded areas $\leq 1,0$ m² use $c_{pe,1}$
For loaded areas $> 1,0$ m² use $c_{pe,10}$ $\Bigg\}$ Table 7.3a of BS EN 1991-1-4

All loaded areas $\geq (2,5 \times 1,0) = 2,5$ m² $> 1,0$ m² $\therefore$ use $c_{pe,10}$ values.

BS EN 1991-1-4
Table 7.3a

	Zone for wind direction $\theta = 0°$					
Pitch Angle α	**F**		**G**		**H**	
	$c_{pe,10}$	$c_{pe,1}$	$c_{pe,10}$	$c_{pe,1}$	$c_{pe,10}$	$c_{pe,1}$
5°	$-1,7$	$-2,5$	$-1,2$	$-2,0$	$-0,6$	$-1,2$
	$+0,0$		$+0,0$		$+0,0$	
15°	$-0,9$	$-2,0$	$-0,8$	$-1,5$	$-0,3$	
	$+0,2$		$+0,2$		$+0,2$	

Table 3.14

Use interpolation to determine the values for $\alpha = 7,13°$

Zone F: $c_{pe,10} = -1,7 + [(1,7 - 0,9) \times (7,13 - 5,0)/(15 - 5)]$ $= -1,53$
$= +0,0 + [(0,2 - 0,0) \times (7,13 - 5,0)/(15 - 5)]$ $= +0,04$
Zone G: $c_{pe,10} = -1,2 + [(1,2 - 0,8) \times (7,13 - 5,0)/(15 - 5)]$ $= -1,12$
$=$ as for zone F $= +0,04$
Zone H: $c_{pe,10} = -0,6 + [(0,6 - 0,3) \times (7,13 - 5,0)/(15 - 5)]$ $= -0,54$
$=$ as for zone F $= +0,04$

Clause 5.2(1)   The external pressures acting on the roof are given by:

Equation (5.1) $w_e = q_p(z_e) \times c_{pe}$

$w_{e,F} = q_p(z_e) \times c_{pe,F}$; $w_{e,G} = q_p(z_e) \times c_{pe,G}$; $w_{e,H} = q_p(z_e) \times c_{pe,H}$

UK NA.   Exposure case: Isolated

Table NA4b

Zone	q_p (kN/m²)	$c_{pe,10}$	w_e (kN/m²)	$c_{pe,10}$	w_e (kN/m²)
F		− 1,53	− 0,973		
G	0,636	− 1,12	− 0,712	+ 0,04	+ 0,025
H		− 0,54	− 0,343		

Table 3.15

The distribution of the horizontal wind load on the building is given by:

$$F_{w,e} = c_s c_d \times \left(\sum_{surfaces} w_{e,front} \times A_{ref} - \sum_{surfaces} w_{e,rear} \times A_{ref} \right)$$

$$= c_s c_d \times \left(\sum_{surfaces} F_{w,e,front} - \sum_{surfaces} F_{w,e,rear} \right)$$

Note: The reduction factor only applies to the vertical windward and leeward faces.

The corresponding values are summarized in Table 3.16 and indicated on Figure 3.29.

$A_{ref,F} = (2,5 \times 1,0)/\text{Cos } 7,13° = 2,52 \text{ m}^2$
$A_{ref,G} = (15,0 \times 1,0)/\text{Cos } 7,13° = 15,12 \text{ m}^2$
$A_{ref,H} = (20,0 \times 11,0)/\text{Cos } 7,13° = 221,71 \text{ m}^2$

Zone	A_{ref} (m²)	w_e (kN/m²)	$F_{w,e}$ (kN)	w_e (kN/m²)	$F_{w,e}$ (kN)
F	2,52	− 0,973	− 2,452		+ 0,063
G	15,12	− 0,712	− 10,765	+ 0,025	+ 0,378
H	221,71	− 0,343	− 76,047		+ 5,543

Table 3.16

Zones F and G occupy less than 10% of the total area and are normally used when designing for local effects where high local suction can occur. When calculating the load on the entire structure including roofs and walls, then the value for zone H (termed as H*) here for single roof area) should be adopted as shown below.

$A_{ref,H*} = (20,0 \times 12,0)/\text{Cos } 7,13° = 241,87 \text{ m}^2$
$F_{w,e,H*} = - (241,87 \times 0,343) = - 82,961 \text{ kN};$ or $F_{w,e,H*} = + (241,87 \times 0,025) = + 6,047 \text{ kN}$

(i) External forces: wind pressures on the roof

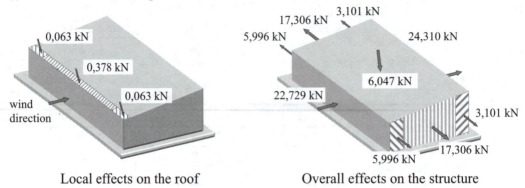

Local effects on the roof   Overall effects on the structure

(ii) External forces: wind suctions on the roof

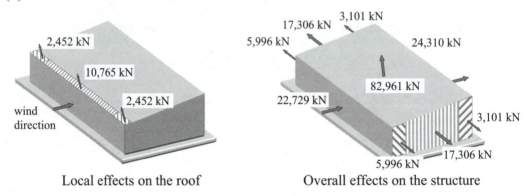

Local effects on the roof   Overall effects on the structure

Figure 3.12(f)

3.7.10 Example 3.10: BS EN 1991-1-4:2005 - Wind Load on Building with Duo-pitched Roof

An timber-framed industrial warehouse is to be constructed with treated timber and clad with external masonry walls as shown in Figure 3.13. Using the data given, determine the wind loads which act on the walls and the roof of the building for the two orthogonal wind directions A and B as indicated. Assume that the building does not have any dominant faces when considering internal pressures and suctions.

Design data:

Location	Open country near Preston
Distance to the sea	7 km
Obstruction height	7,5 m
Distance to nearest adjacent building	50 m
Altitude above mean sea level	15 m
Assume terrain category	I

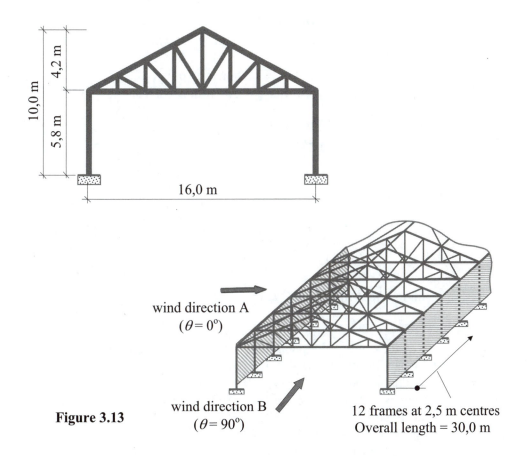

Figure 3.13

wind direction A
$(\theta = 0^\circ)$

wind direction B
$(\theta = 90^\circ)$

12 frames at 2,5 m centres
Overall length = 30,0 m

Solution:
Consider wind direction A: $\theta = 0^\circ$
BS EN 1991-1-4: 2005

Clause 7.2.2   A building whose height 'h' is less than or equal to 'b', the crosswind dimension (see Figure 7.5 of the code) should be considered to be one part (i.e. Part A) as indicated in Figure 7.4 of the code.

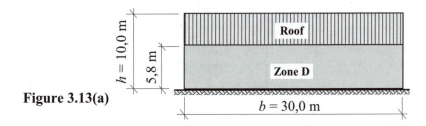

Figure 3.13(a)

$h = 10,0$ m

5,8 m

Roof

Zone D

$b = 30,0$ m

The building surfaces are considered as zones A to E for the vertical faces and F to J for the roof as indicated in Figure 7.5 and Figure 7.8 of the code.

Figure 7.5   Key for zones in vertical walls:

e is defined as the smaller of b or $2h$, i.e. 30,0 m or $(2 \times 10,0) = 20,0$ m

∴ $e = 20,0$ m

The dimension in the wind direction $d = 16,0$ m

For an elevation with $d \leq e \leq 5d$, widths for zones A and B are as follows:

Zone A = $(e/5) = (20,0/5,0) = 4,0$ m

Zone B = $(d - e/5) = (16,0 - 4,0) = 12,0$ m

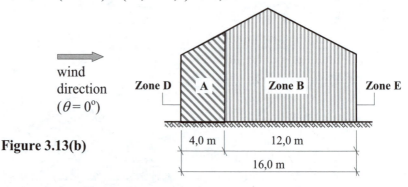

wind
direction
$(\theta = 0^\circ)$

Figure 3.13(b)

Figure 7.8   Key for zones in duopitch roofs for the wind angle $\theta = 0^\circ$.

Length of Zone F = $(e/4) = (20,0/4,0) = 5,0$ m

Width of Zone F and G = $(e/10) = (20,0/10) = 2,0$ m

Length of Zone G = $(b - 0,5e) = [30,0 - (0,5 \times 20,0)] = 20,0$ m

Length of Zone J = $b = 30,0$ m

Width of Zone J = $(e/10) = (20,0/10) = 2,0$ m

Zones F, G and J are used for local effects at the edges.

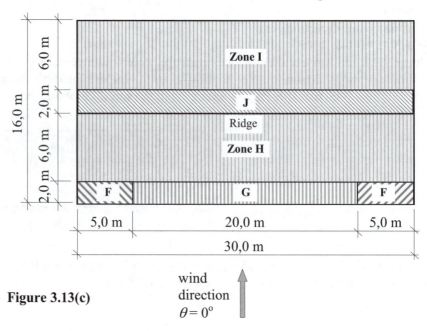

Figure 3.13(c)

wind
direction
$\theta = 0^\circ$

UK NA.

Clause NA 2.19   The reduction factor indicated in Clause 7.2.2(3) of the code may be applied to the summation of the loads on all windward and leeward surfaces when considering overall structural loads.
It is not included in the following calculations to determine the forces on each of the individual faces of the structure.

External forces: Equation (5.5) $F_{w,e} = c_s c_d \times \left(\sum_{\text{surfaces}} w_e \times A_{ref} \right)$

$$F_{w,e} = c_s c_d \times \left(\sum_{\text{surfaces}} w_{e,\text{front}} \times A_{ref} - \sum_{\text{surfaces}} w_{e,\text{rear}} \times A_{ref} \right)$$

$$= c_s c_d \times \left(\sum_{\text{surfaces}} F_{w,e,\text{front}} - \sum_{\text{surfaces}} F_{w,e,\text{rear}} \right)$$

Internal forces: Equation (5.6) $F_{w,i} = \sum_{\text{surfaces}} w_i \times A_{ref}$

Friction forces: Equation (5.7) $F_{fr} = c_{fr} \times q_p(z_e) \times A_{fr}$

Friction forces on the walls and roof surfaces should be considered as defined in Clause 5.3(3), Figure 7.22 and the friction coefficients c_{fr}, given in Table 7.10. The forces are applied on the part of the external surfaces parallel to the wind, located beyond a distance from the upwind eaves or corners, equal to the smallest value of $2b$ or $4h$. The reference height z_e should be taken equal to the structure height above ground or building height h.

$2b = (2 \times 30) = 60,0$ m, $4h = (4 \times 10) = 40$ m $\therefore$ Distance to reference area $= 40$ m
Since the distance to the reference area $> d = 16$ m, friction forces need not be considered.

In addition, as indicated in Clause 5.3(4), "*the effects of wind friction on a surface can be disregarded when the total area of all surfaces parallel with (or at a small angle to) the wind is equal to or less than 4 times the total area of all external surfaces perpendicular to the wind (windward and leeward)*".

UK NA.

Clause NA 2.6/7   The values of the season factor c_s and the directional factor c_d can be taken as 1,0 (or in Clause 6.2.(1) of the code for $h < 15$ m $c_s = c_d = 1,0$)

BS EN 1991-1-4

Clause 5.2(1)   The external pressures acting on the windward and leeward surfaces are given by:

Equation (5.1) $w_{e,\text{front}} = q_{p,\text{front}}(z_e) \times c_{pe,\text{front}}$ and $w_{e,\text{rear}} = q_{p,\text{rear}}(z_e) \times c_{pe,\text{rear}}$

UK NA.

Clause NA 2.11   Since terrain is category I this is reclassified in the UK National Annex as Country terrain.

Equation (NA3)   Peak velocity pressure $q_p(z) = \dfrac{\left[1+3I_v(z)\right]^2 \times \rho \times v_m^2(z)}{2} = c_e(z) \times q_b$

When the orography is not significant (see Figure NA2), then $q_p(z)$ can be calculated by using:

$q_p(z) = c_e(z) \times q_b$

where $c_e(z) = c_{e,flat}(z)$ from Figure NA7.

Figure NA7   The value of $c_{e,flat}(z)$ is dependent on the reference height z_e, the displacement height h_{dis} and the distance upwind to the shore line.

BS EN 1991-1-4

Figure 7.4   Reference heights:

Windward vertical surface Zone D: $z_e = 10{,}0$ m

Leeward vertical surface Zone E: $z_e = 10{,}0$ m

Annex A

A.5   The horizontal distance to the nearest adjacent building $x = 50{,}0$ m

The obstruction height $h_{ave} = 7{,}5$ m

$(2 \times h_{ave}) = 15{,}0$ m and $(6 \times h_{ave}) = 45{,}0$ m

Equation A.15   The displacement height:

Since $x \geq 6\, h_{ave}$ $h_{dis} = 0{,}0$ m

Figure NA7   The values of $c_e(z)$ for the windward and leeward surfaces are summarized in Table 3.17 below corresponding to 7 km from the sea.

Zone	z_e (m)	h_{dis} (m)	$(z - h_{dis})$ (m)	$c_{e,front}(z_e)$	$c_{e,rear}(z_e)$
D and E	10,0	0,0	10,0	2,68	2,68

Table 3.17

Equation (4.10)   Basic velocity pressure $q_b = \dfrac{\rho v_b^2}{2}$

Equation (4.1)   Basic wind velocity $v_b = c_{dir} \times c_{season} \times v_{b,0}$

UK NA.

Clause NA 2.4   Fundamental basic wind velocity $v_{b,0} = v_{b,map} \times c_{alt}$

Figure NA1   For Preston $v_{b,map} = 23{,}2$ m/s

Clause NA 2.5   Equation (NA2a) always gives a conservative value and can be used for any site altitude.

$c_{alt} = 1 + 0{,}001 \times A$ for $z \leq 10$ m   Equation (NA2a)

$c_{alt} = 1 + 0{,}001 \times A \times (10/z)^{0{,}2}$ for $z > 10$ m   Equation (NA2b)

The value of z is either z_s as defined in Figure 6.1 of the code or z_e, the height of the part above ground as defined in Figure 7.4.

BS EN 1991-1-4
Figure 6.1 $\quad\quad\quad z_s = (0,6 \times h) \geq z_{min}$
Table 4.1 $\quad\quad\quad$ For Terrain category I $\quad z_{min} = 1,0$ m
$\quad\quad\quad\quad\quad\quad\quad z_s = (0,6 \times 10,0) = 6,0$ m $\geq z_{min}$

Equation (NA2a) $\quad c_{alt} = 1 + (0,001 \times A) = 1 + (0,001 \times 15,0) = 1,015$
Clause NA 2.4 $\quad\quad$ Fundamental basic wind velocity $v_{b,0} = v_{b,map} \times c_{alt}$
$\quad\quad\quad\quad\quad\quad\quad v_{b,0} = (23,2 \times 1,015) = 23,55$ m/s

Equation (4.1) $\quad\quad$ Basic wind velocity $v_b = (c_{dir} \times c_{season} \times v_{b,0}) = (1,0 \times 1,0 \times 23,55)$
$\quad\quad\quad\quad\quad\quad\quad\quad\quad\quad\quad = 23,55$ m/s

UK NA.
Clause NA 2.18 $\quad$ The air density in the UK $\rho = 1,226$ kg/m^3

BS EN 1991-1-4

Equation (4.10) $\quad$ Basic velocity pressure $q_b = \dfrac{\rho v_b^2}{2} = \dfrac{1,226 \times 23,55^2}{2 \times 1000} = 0,340$ kN/m^2

Equation (NA3) $\quad$ Peak velocity pressure $q_p(z)$ values for the windward and leeward surfaces are summarized in Table 3.18 below.

Zone	z_e	q_b (kN/m^2)	$c_{e,front}(z_e)$	$q_{p,front}$ (kN/m^2)	$c_{e,rear}(z_e)$	$q_{p,rear}$ (kN/m^2)
D and E	10,0	0,340	2,68	0,911	2,68	0,911

Table 3.18

BS EN 1991-1-4
Clause 5.2(1) $\quad\quad$ The external pressures acting on the windward and leeward surfaces are given by:

Equation (5.1) $\quad\quad w_{e,front} = q_{p,front}(z_e) \times c_{pe,front}$ $\quad$ and $\quad$ $w_{e,rear} = q_{p,rear}(z_e) \times c_{pe,rear}$

UK NA.
Clause 2.25 $\quad\quad$ The external pressure coefficients to be used in accordance with the National Annex are:
$\quad\quad\quad\quad\quad\quad$ For loaded areas $\leq 1,0$ m^2 use $c_{pe,1}$
$\quad\quad\quad\quad\quad\quad$ For loaded areas $> 1,0$ m^2 use $c_{pe,10}$ $\quad$ Table NA4a of UK NA
Table NA4a $\quad\quad$ Zone D is the windward face and zone E is the leeward face.
$\quad\quad\quad\quad\quad\quad h/b = 10,0/30,0 = 0,33$
$\quad\quad\quad\quad\quad\quad d/b = 16,0/30,0 = 0,53$

The required values can be obtained using interpolation for h/b between 0,25 and 1,0 and d/b between 0,25 and 1,0 as shown in Table 3.19.

For the windward surface (D): (**Note:** Values are the same for $d/b \leq 0,25$ and $d/b = 1,0$) Use interpolation between $+ 0,6$ and $+ 0,7$ for $h/b = 0,33$

$$c_{pe,10,front} = + 0,60 + [(0,7 - 0,6) \times (0,33 - 0,25)/(1,0 - 0,25)] = + 0,61$$

For the leeward surface (E): (**Note:** Values are the same for $h/b \leq 0,25$ and $h/b = 1,0$) Use interpolation between $- 0,5$ and $- 0,4$ for $d/b = 0,53$

$$c_{pe,10,rear} = - 0,5 - [- (0,5 - 0,4) \times (0,53 - 0,25)/(1,0 - 0,25)] = - 0,46$$

h/b	d/b					
	$\leq 0,25$		1,0		$d/b = 0,53$	
	D	E	D	E	D	E
$\leq 0,25$	+ 0,6	− 0,5	+ 0,6	− 0,4	+ 0,6	− 0,46
1,0	+ 0,7	− 0,5	+ 0,7	− 0,4	+ 0,7	− 0,46
				$h/b = 0,33$	+ 0,61	− 0,46

Table 3.19

The external pressure acting on the windward and leeward faces are given in Table 3.20.

Zone	$q_{p,front}$ (kN/m^2)	$c_{pe,10,front}$	$w_{e,front}$ (kN/m^2)	$q_{p,rear}$ (kN/m^2)	$c_{pe,10,rear}$	$w_{e,rear}$ (kN/m^2)
D and E	0,911	+ 0,61	+ 0,556	0,911	− 0,46	− 0,419

Table 3.20

The distribution of the horizontal wind load on the building is given by:

$$F_{w,e} = c_s c_d \times \left(\sum_{surfaces} w_{e,front} \times A_{ref} - \sum_{surfaces} w_{e,rear} \times A_{ref} \right)$$

$$= c_s c_d \times \left(\sum_{surfaces} F_{w,e,front} - \sum_{surfaces} F_{w,e,rear} \right)$$

The corresponding values are summarized in Table 3.21 and indicated on Figure 3.13(d) and Figure 3.13(g).

$$A_{ref,windward} = A_{ref,leedward} = (5,8 \times 30,0) = 174,0 \text{ m}^2$$

Zone	$A_{windward}$ (m^2)	$w_{e,front}$ (kN/m^2)	$F_{w,e,front}$ (kN)	$A_{leeward}$ (m^2)	$w_{e,rear}$ (kN/m^2)	$F_{w,e,rear}$ (kN)	$F_{w,e,total}$ (kN)
D and E	174,0	+ 0,556	+ 96,744	174,0	− 0,419	− 72,906	169,65

Table 3.21

The wind loads on surfaces A and B (i.e. the side walls)
Clause 7.2.2(2)
Figure 7.5   Key for zones in vertical walls: $e = 20,0$ m

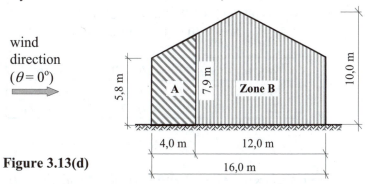

Figure 3.13(d)

Peak velocity pressure $q_p(z) = 0,911$ kN/m²

UK NA.
Clause NA 2.27 External pressure coefficients
            In NOTE 8(a) of this Clause:
            "where the gap between the buildings is < e/4 or > e, the isolated coefficient values should be used."

            Distance to nearest adjacent building 50 m $> e$ ∴ building is isolated.
            Use Table NA4b to determine the $c_{pe,10}$ values for zones A, B and C.

BS EN 1991-1-4
Clause 5.2(1)   The external pressures acting on zones A and B are given by:
Equation (5.1) $w_e = q_p(z_e) \times c_{pe}$
            $w_{e,A} = q_p(z_e) \times c_{pe,A}$; $w_{e,B} = q_p(z_e) \times c_{pe,B}$; $w_{e,C} = q_p(z_e) \times c_{pe,C}$

UK NA.
Table NA4b   Exposure case: Isolated

Zone	q_p (kN/m²)	$c_{pe,10}$	w_e (kN/m²)
A	0,911	− 1,3	− 1,184
B		− 0,8	− 0,729

Table 3.22

The distribution of the horizontal wind load on the building is given by:

$$F_{w,e} = c_s c_d \times \left(\sum_{\text{surfaces}} w_{e,\text{front}} \times A_{\text{ref}} - \sum_{\text{surfaces}} w_{e,\text{rear}} \times A_{\text{ref}} \right)$$

$$= c_s c_d \times \left(\sum_{\text{surfaces}} F_{w,e,\text{front}} - \sum_{\text{surfaces}} F_{w,e,\text{rear}} \right)$$

The corresponding values are summarized in Table 3.23 and indicated on Figure 3.36 and Figure 3.37.

$$A_{\text{ref,A}} = 0,5 \times (5,8 + 7,9) \times 4,0 = 27,4 \text{ m}^2$$
$$A_{\text{ref,B}} = [(5,8 \times 16,0) + (0,5 \times 16,0 \times 4,2)] - 27,4 = 99,0 \text{ m}^2$$

Zone	A_{ref} (m^2)	w_e (kN/m^2)	$F_{w,e}$ (kN)
A	27,4	− 1,184	− 32,442
B	99,0	− 0,729	− 72,171

Table 3.23

The wind loads on surfaces F, G, H, J and I (i.e. the duopitch roof)

Figure 7.7(a) The pitch angle $\alpha = \tan^{-1}\left(\dfrac{4,2}{8,0}\right) = 27,7^\circ$

Figure 7.7(b)

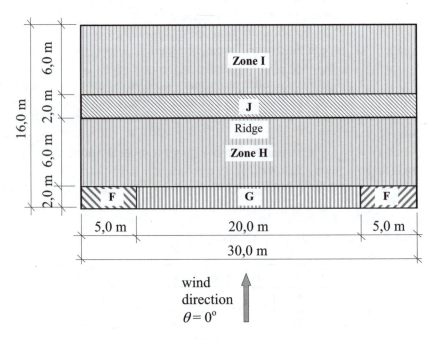

Figure 3.13(e)

UK NA.

Clause 2.25   The external pressure coefficients to be used in accordance with the National Annex are:

For loaded areas $\leq 1,0$ m^2 use $c_{pe,1}$
For loaded areas $> 1,0$ m^2 use $c_{pe,10}$ Table 7.3a of BS EN 1991-1- 4

All loaded areas $\geq (5,0 \times 2,0) = 10,0$ m$^2 > 1,0$ m^2 $\therefore$ use $c_{pe,10}$ values.

BS EN 1991-1-4: Table 7.4(a)

Pitch Angle α	Zone for wind direction $\theta = 0°$				
	F	G	H	I	J
	$c_{pe,10}$	$c_{pe,10}$	$c_{pe,10}$	$c_{pe,10}$	$c_{pe,10}$
15°	$-0,9$	$-0,8$	$-0,3$	$-0,4$	$-1,0$
	$+0,2$	$+0,2$	$+0,2$	$+0,0$	$+0,0$
30°	$-0,5$	$-0,5$	$-0,2$	$-0,4$	$-0,5$
	$+0,7$	$+0,7$	$+0,4$	$+0,0$	$+0,0$

Table 3.24

Use interpolation to determine the values for $\alpha = 7,13°$
Use interpolation to determine the values for $\alpha = 27,7°$

Zone F: $c_{pe,10} = -0,9 + [(0,9 - 0,5) \times (27,7 - 15,0)/(30 - 15)] = -0,56$
$= +0,2 + [(0,7 - 0,2) \times (27,7 - 15,0)/(30 - 15)] = +0,62$

Zone G: $c_{pe,10} = -0,8 + [(0,8 - 0,5) \times (27,7 - 15,0)/(30 - 15)] = -0,55$
$=$ as for zone F $\qquad = +0,62$

Zone H: $c_{pe,10} = -0,3 + [(0,3 - 0,2) \times (27,7 - 15,0)/(30 - 15)] = -0,22$
$= +0,2 + [(0,4 - 0,2) \times (27,7 - 15,0)/(30 - 15)] = +0,37$

Zone I: $c_{pe,10} = -0,4 + [(0,4 - 0,4) \times (27,7 - 15,0)/(30 - 15)] = -0,4$
$= +0,0$

Zone J: $c_{pe,10} = -1,0 + [(1,0 - 0,5) \times (27,7 - 15,0)/(30 - 15)] = -0,58$
$= +0,0$

Clause 5.2(1)   The external pressures acting on the roof are given by:
Equation (5.1) $w_e = q_p(z_e) \times c_{pe}$
$w_{e,F} = q_p(z_e) \times c_{pe,F}$; $w_{e,G} = q_p(z_e) \times c_{pe,G}$; $w_{e,H} = q_p(z_e) \times c_{pe,H}$, etc.

UK NA: Table NA4b - Exposure case: Isolated

Zone	q_p (kN/m^2)	$c_{pe,10}$	w_e (kN/m^2)	$c_{pe,10}$	w_e (kN/m^2)
F		$-0,56$	$-0,510$	$+0,62$	$+0,565$
G		$-0,55$	$-0,501$		$+0,565$
H	0,911	$-0,22$	$-0,200$	$+0,37$	$+0,337$
I		$-0,40$	$-0,364$	$+0,0$	$+0,0$
J		$-0,58$	$-0,528$		$+0,0$

Table 3.25

The distribution of the horizontal wind load on the building is given by:

$$F_{w,e} = c_s c_d \times \left(\sum_{\text{surfaces}} w_{e,\text{front}} \times A_{\text{ref}} - \sum_{\text{surfaces}} w_{e,\text{rear}} \times A_{\text{ref}} \right)$$

$$= c_s c_d \times \left(\sum_{\text{surfaces}} F_{w,e,\text{front}} - \sum_{\text{surfaces}} F_{w,e,\text{rear}} \right)$$

The corresponding values are summarized in Table 3.26 and indicated on Figure 3.13(f) and Figure 3.13(g).

$$A_{\text{ref,F}} = (5,0 \times 2,0)/\text{Cos } 27,7° = 11,29 \text{ m}^2$$
$$A_{\text{ref,G}} = (20,0 \times 2,0)/\text{Cos } 27,7° = 45,18 \text{ m}^2$$
$$A_{\text{ref,H}} = (30,0 \times 6,0)/\text{Cos } 27,7° = 203,30 \text{ m}^2$$
$$A_{\text{ref,I}} = (30,0 \times 6,0)/\text{Cos } 27,7° = 203,30 \text{ m}^2$$
$$A_{\text{ref,J}} = (30,0 \times 2,0)/\text{Cos } 27,7° = 67,77 \text{ m}^2$$

Zone	A_{ref} (m²)	w_e (kN/m²)	$F_{w,e}$ (kN)	w_e (kN/m²)	$F_{w,e}$ (kN)
F	11,29	− 0,510	− 5,758	+ 0,565	+ 6,379
G	45,18	− 0,501	− 22,635	+ 0,565	+ 25,527
H	203,30	− 0,200	− 40,600	+ 0,337	+ 68,512
I	203,30	− 0,364	− 74,001	0,0	0,0
J	67,77	− 0,528	− 35,783	0,0	0,0

Table 3.26

Zones F, G and J occupy less than 25% of the total area and are normally used when designing for local effects where high local suction can occur. When calculating the load on the entire structure including roofs and walls, then the value for zones H and I (termed as H* and I* here for single roof area) should be adopted as shown below.

$$A_{\text{ref,H*}} = (30,0 \times 8,0)/\text{Cos } 27,7° = 271,07 \text{ m}^2$$
$$F_{w,e,\text{H*}} = - (271,07 \times 0,200) = - 54,214 \text{ kN}; \quad \text{or} \quad F_{w,e,\text{H*}} = + (271,07 \times 0,337) = + 91,351 \text{ kN}$$

$$A_{\text{ref,I*}} = (30,0 \times 8,0)/\text{Cos } 27,7° = 271,07 \text{ m}^2$$
$$F_{w,e,\text{I*}} = - (271,07 \times 0,364) = - 98,669 \text{ kN}; \quad \text{or} \quad F_{w,e,\text{I*}} = + (271,07 \times 0,0) = + 0,0$$

Internal pressure coefficients:
Clause 7.2.9(6) specifies that for buildings without a dominant face, the internal pressure coefficient c_{pi} should be determined from Figure 7.13, and is a function of the ratio of the height and the depth of the building, h/d, and the opening ratio μ for each wind direction θ. NOTE 2 in Clause 7.2.9(6) further specifies that where it is not possible, or not considered justified, to estimate μ for a particular case, c_{pi} should be taken as the more onerous of + 0,2 (pressure) and − 0,3 (suction). In this example it is assumed that the warehouse has

no dominant faces and it is not possible to estimate μ and consequently the internal pressure coefficients c_{pi} can be taken either $+ 0,2$ or $- 0,3$ whichever gives a larger net pressure coefficient across the walls.

Equation (5.6) $\qquad F_{w,i} = \sum_{\text{surfaces}} w_i \times A_{ref} = \sum_{\text{surfaces}} \left(q_p \times c_{pi} \times A_{ref} \right)$

The forces due to internal pressures and suctions are summarized in Table 3.27.

Zone	A_{ref} (m²)	q_p (kN/m²)	c_{pi} (pressure)	$F_{w,i}$ (kN)	c_{pi} (suction)	$F_{w,i}$ (kN)
A	27,4			+ 4,992		− 7,488
B	99,0			+ 18,038		− 27,057
D	174,0			+ 31,703		− 47,554
E	174,0			+ 31,703		− 47,554
F	11,29	0,911	+ 0,2	+ 2,057	− 0,3	− 3,086
G	45,18			+ 8,232		− 12,348
H	203,30			+ 37,041		− 55,562
I	203,30			+ 37,041		− 55,562
J	67,77			+ 12,348		− 18,522

Table 3.27

The forces on the surfaces due to internal pressures and suctions for wind angle $\theta = 0°$ are indicated in Figures 3.13(h) to Figure 3.13(i).

(i) Forces on surfaces due to external wind pressure on the roof:

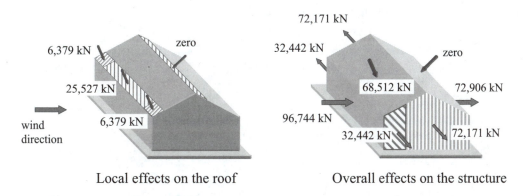

Local effects on the roof   Overall effects on the structure

Figure 3.13(f)

(ii) Forces on surfaces due to external wind suction on the roof:

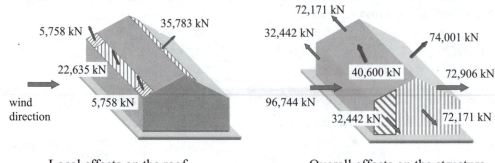

Local effects on the roof   Overall effects on the structure

Figure 3.13(g)

(iii) Forces on surfaces due to internal pressure:

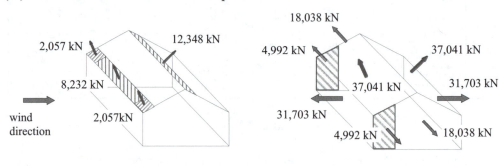

Local effects on the roof   Overall effects on the structure

Figure 3.13(h)

(iv) Forces on surfaces due to internal suction:

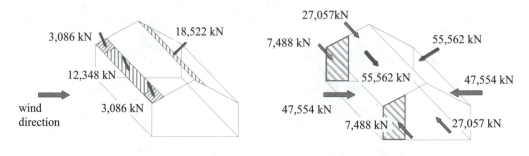

Local effects on the roof   Overall effects on the structure

Figure 3.13(i)

Consider wind direction B: $\theta = 90°$
BS EN 1991-1-4: 2005

Clause 7.2.2   A building whose height 'h' is less than or equal to 'b', the crosswind dimension (see Figure 7.5 of the code) should be considered to be one part (i.e. Part A) as indicated in Figure 7.4 of the code.

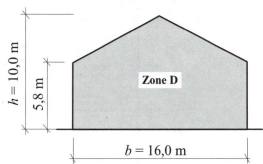

Figure 3.13(j)

The building surfaces are considered as zones A to E for the vertical faces and F to I for the roof as indicated in Figure 7.5 and Figure 7.8 of the code.

Figure 7.5   Key for zones in vertical walls:
e is defined as the smaller of b or $2h$, i.e. 16,0 m or $(2 \times 10,0) = 20,0$ m
$\therefore e = 16,0$ m
The dimension in the wind direction $d = 30,0$ m
For an elevation with $e \le d$ zone widths for A, B and C are as follows:
Zone A $= (e/5) = (16,0/5,0) = 3,2$ m
Zone B $= (4e/5) = (4 \times 16,0)/5 = 12,8$ m
Zone C $= (d - e) = (30,0 - 16,0) = 14,0$ m

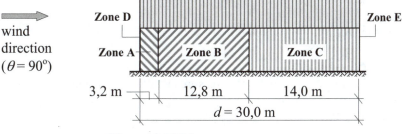

Figure 3.13(k)

Figure 7.8   Key for zones in duopitch roofs for the wind angle $\theta = 90°$.
Length of Zone F $= (e/4) = (16,0/4,0) = 4,0$ m
Width of Zone F and G $= (e/10) = (16,0/10) = 1,6$ m
Length of Zone G $= (0,5b - 0,25e) = [8,0 - (0,25 \times 16,0)] = 4,0$ m
Length of Zone H $= (e/2 - e/10) = (8,0 - 1,6) = 6,4$ m
Width of Zone I $= (d - e/2) = (30,0 - 8,0) = 22,0$ m
Zones F and G are used for local effects at the edges.

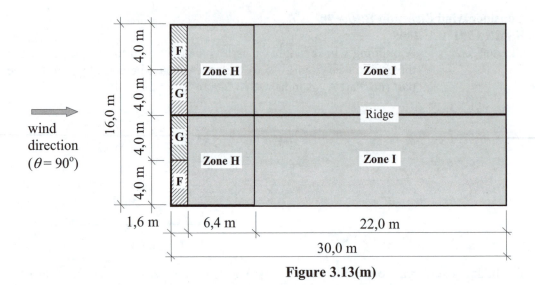

Figure 3.13(m)

Friction forces on the walls and roof surfaces should be considered as defined in Clause 5.3(3), Figure 7.22 and the friction coefficients c_{fr}, given in Table 7.10. The forces are applied on the part of the external surfaces parallel to the wind, located beyond a distance from the upwind eaves or corners, equal to the smallest value of $2b$ or $4h$. The reference height z_e should be taken equal to the structure height above ground or building height h.

$2b = (2 \times 16) = 32,0$ m, $4h = (4 \times 10) = 40$ m $\therefore$ distance to reference area $= 32$ m
Since the distance to the reference area $> d = 30$ m friction forces need not be considered.

UK NA.
Equation (NA3)   Peak velocity pressure $q_p(z) = 0,911$ kN/m^2 (as before)

BS EN 1991-1-4
Clause 5.2(1)   The external pressures acting on the windward and leeward surfaces are given by:
Equation (5.1) $w_{e,front} = q_{p,front}(z_e) \times c_{pe,front}$ and $w_{e,rear} = q_{p,rear}(z_e) \times c_{pe,rear}$

UK NA.
Clause 2.25   The external pressure coefficients to be used in accordance with the National Annex are:
For loaded areas $\leq 1,0$ m^2 use $c_{pe,1}$ ⎫
For loaded areas $> 1,0$ m^2 use $c_{pe,10}$ ⎬ Table NA4a of UK NA

Table NA4a   Zone D is the windward face and zone E is the leeward face.
$h/b = 10,0/16,0 = 0,63$
$d/b = 30,0/16,0 = 1,88$

The required values can be obtained using interpolation for h/b between 0,25 and 1,0 and d/b between 1,0 and 2,0 as shown in Table 3.28

For the windward surface (D): (**Note:** Values are the same for $d/b = 1,0$ and $d/b = 2,0$)
Use interpolation between $+ 0,6$ and $+ 0,7$ for $h/b = 0,63$
$c_{pe,10,ront} = + 0,60 + [(0,7 - 0,6) \times (0,63 - 0,25)/(1,0 - 0,25)] = + 0,65$
For the leeward surface (E): (**Note:** Values are the same for $h/b \leq 0,25$ and $h/b = 1,0$)
Use interpolation between $- 0,4$ and $- 0,3$ for $d/b = 1,88$
$c_{pe,10,rear} = - 0,4 - [- (0,4 - 0,3) \times (1,88 - 1,0)/(2,0 - 1,0)] = - 0,31$

h/b	d/b					
	1,0		**2,0**		**$d/b = 1,88$**	
	D	**E**	**D**	**E**	**D**	**E**
$\leq 0,25$	$+ 0,6$	$- 0,4$	$+ 0,6$	$- 0,3$	$+ 0,6$	$- 0,46$
1,0	$+ 0,7$	$- 0,4$	$+ 0,7$	$- 0,3$	$+ 0,7$	$- 0,46$
				$h/b = 0,63$	$+ 0,65$	$- 0,31$

Table 3.28

The external pressure acting on the windward and leeward faces are given in Table 3.29

Zone	$q_{p,front}$ (kN/m²)	$c_{pe,10,front}$	$w_{e,front}$ (kN/m²)	$q_{p,rear}$ (kN/m²)	$c_{pe,10,rear}$	$w_{e,rear}$ (kN/m²)
D and E	0,911	$+ 0,65$	$+ 0,592$	0,911	$- 0,31$	$- 0,282$

Table 3.29

The distribution of the horizontal wind load on the building is given by:

$$F_{w,e} = c_s c_d \times \left(\sum_{\text{surfaces}} w_{e,front} \times A_{ref} - \sum_{\text{surfaces}} w_{e,rear} \times A_{ref} \right)$$

$$= c_s c_d \times \left(\sum_{\text{surfaces}} F_{w,e,front} - \sum_{\text{surfaces}} F_{w,e,rear} \right)$$

The corresponding values are summarized in Table 3.30 and indicated on Figure 3.45.
$$A_{ref,windward} = A_{ref,leedward} = [(5,8 \times 16,0) + (0,5 \times 16,0 \times 4,2)] = 126,4 \text{ m}^2$$

Zone	$A_{windward}$ (m²)	$w_{e,front}$ (kN/m²)	$F_{w,e,front}$ (kN)	$A_{leeward}$ (m²)	$w_{e,rear}$ (kN/m²)	$F_{w,e,rear}$ (kN)	$F_{w,e,total}$ (kN)
D and E	126,40	$+ 0,592$	$+ 74,829$	126,40	$- 0,282$	$- 35,645$	110,474

Table 3.30

The wind loads on zones A, B and C (i.e. the side walls)
Clause 7.2.2(2)
Figure 7.5   Key for zones in vertical walls: $e = 16,0$ m

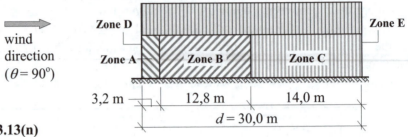

Figure 3.13(n)

Peak velocity pressure $q_p(z) = 0,911$ kN/m^2

UK NA.
Clause NA 2.27 External pressure coefficients
              In NOTE 8(a) of this Clause:
              "where the gap between the buildings is < e/4 or > e, the isolated coefficient values should be used"
              Distance to nearest adjacent building 50 m $> e$ ∴ building is isolated.
              Use Table NA4b to determine the $c_{pe,10}$ values for zones A, B and C.

BS EN 1991-1-4
Clause 5.2(1)   The external pressures acting on zones A, B and C are given by:
Equation (5.1) $w_e = q_p(z_e) \times c_{pe}$
                   $w_{e,A} = q_p(z_e) \times c_{pe,A}$; $w_{e,B} = q_p(z_e) \times c_{pe,B}$; $w_{e,C} = q_p(z_e) \times c_{pe,C}$

UK NA.
Table NA4b   Exposure case: Isolated

Zone	q_p (kN/m^2)	$c_{pe,10}$	w_e (kN/m^2)
A		$-1,3$	$-1,184$
B	0,911	$-0,8$	$-0,729$
C		$-0,5$	$-0,456$

Table 3.31

The distribution of the horizontal wind load on the building is given by:

$$F_{w,e} = c_s c_d \times \left(\sum_{\text{surfaces}} w_{e,\text{front}} \times A_{\text{ref}} - \sum_{\text{surfaces}} w_{e,\text{rear}} \times A_{\text{ref}} \right)$$

$$= c_s c_d \times \left(\sum_{\text{surfaces}} F_{w,e,\text{front}} - \sum_{\text{surfaces}} F_{w,e,\text{rear}} \right)$$

The corresponding values are summarized in Table 3.32 and indicated on Figure 3.45.

$$A_{ref,A} = (5,8 \times 3,2) = 18,56 \text{ m}^2$$
$$A_{ref,B} = (5,8 \times 12,8) = 74,24 \text{ m}^2$$
$$A_{ref,B} = (5,8 \times 14,0) = 81,20 \text{ m}^2$$

Zone	A_{ref} (m²)	w_e (kN/m²)	$F_{w,e}$ (kN)
A	18,56	− 1,184	− 21,975
B	74,24	− 0,729	− 54,121
C	81,20	− 0,456	− 37,027

Table 3.32

The wind loads on surfaces F, G, H and I (i.e. the duopitch roof)

Figure 7.7(a) The pitch angle $\alpha = \tan^{-1}\left(\dfrac{4,2}{8,0}\right) = 27,7°$

Figure 7.7(b)

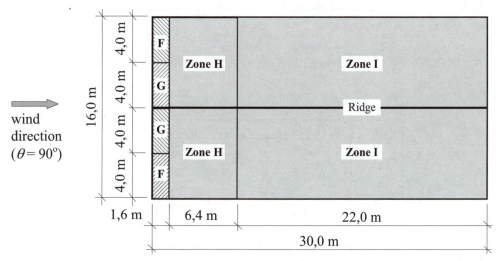

Figure 3.13(p)

UK NA.
Clause 2.25   The external pressure coefficients to be used in accordance with the National Annex are:

For loaded areas ≤ 1,0 m² use $c_{pe,1}$
For loaded areas > 1,0 m² use $c_{pe,10}$ } Table 7.3a of BS EN 1991-1- 4

All loaded areas ≥ (4,0 × 1,6) = 6,4 m² > 1,0 m² ∴ use $c_{pe,10}$ values.

BS EN 1991-1-4: Table 7.4(b)

Pitch Angle α	Zone for wind direction $\theta = 90°$			
	F	G	H	I
	$c_{pe,10}$	$c_{pe,10}$	$c_{pe,10}$	$c_{pe,10}$
15°	– 1,3	– 1,3	– 0,6	– 0,5
30°	– 1,1	– 1,4	– 0,8	– 0,5

Table 3.33

Use interpolation to determine the values for $\alpha = 27,7°$

Zone F: $c_{pe,10} = -1,3 + [(1,3 - 1,1) \times (27,7 - 15,0)/(30 - 15)] = -1,13$
Zone G: $c_{pe,10} = -1,3 + [(1,3 - 1,4) \times (27,7 - 15,0)/(30 - 15)] = -1,38$
Zone H: $c_{pe,10} = -0,6 + [(0,6 - 0,8) \times (27,7 - 15,0)/(30 - 15)] = -0,77$
Zone I: $c_{pe,10} = -0,5$

Clause 5.2(1)   The external pressures acting on the roof are given by:
Equation (5.1) $w_e = q_p(z_e) \times c_{pe}$
$w_{e,F} = q_p(z_e) \times c_{pe,F}$; $w_{e,G} = q_p(z_e) \times c_{pe,G}$; $w_{e,H} = q_p(z_e) \times c_{pe,H}$
$w_{e,I} = q_p(z_e) \times c_{pe,I}$;

UK NA.   Exposure case: Isolated
Table NA4b

Zone	q_p (kN/m^2)	$c_{pe,10}$	w_e (kN/m^2)
F		– 1,13	– 1,029
G	0,911	– 1,38	– 1,257
H		– 0,77	– 0,701
I		– 0,50	– 0,456

Table 3.34

The distribution of the horizontal wind load on the building is given by:

$$F_{w,e} = c_s c_d \times \left(\sum_{\text{surfaces}} w_{e,\text{front}} \times A_{\text{ref}} - \sum_{\text{surfaces}} w_{e,\text{rear}} \times A_{\text{ref}} \right)$$

$$= c_s c_d \times \left(\sum_{\text{surfaces}} F_{w,e,\text{front}} - \sum_{\text{surfaces}} F_{w,e,\text{rear}} \right)$$

The corresponding values are summarized in Table 3.35 and indicated on Figure 3.13(q).

$$A_{\text{ref,F}} = A_{\text{ref,G}} = (4,0 \times 1,6)/\text{Cos } 27,7° = 7,23 \text{ m}^2$$

$A_{ref,H} = (8,0 \times 6,4)/ \text{Cos } 27,7^\circ = 57,83 \text{ m}^2$
$A_{ref,I} = (22,0 \times 8,0)/ \text{Cos } 27,7^\circ = 198,78 \text{ m}^2$

Zone	A_{ref} (m²)	w_e (kN/m²)	$F_{w,e}$ (kN)
F	7,23	− 1,029	− 7,44
G	7,23	− 1,257	− 9,09
H	57,83	− 0,701	− 40,54
I	198,78	− 0,456	− 90,65

Table 3.35

Zones F, and G occupy less than 10% of the total area and are normally used when designing for local effects where high local suction can occur. When calculating the load on the entire structure including roofs and walls, then the value for zone H (termed as H* here for single roof area) should be adopted as shown below.

$A_{ref,H*} = (8,0 \times 8,0)/\text{Cos } 27,7^\circ = 72,28 \text{ m}^2$
$F_{w,e,H*} = - (72,284 \times 0,701) = - 50,67 \text{ kN}$

Internal pressure coefficients:

Equation (5.6) $F_{w,i} = \sum_{surfaces} w_i \times A_{ref} = \sum_{surfaces} \left(q_p \times c_{pi} \times A_{ref} \right)$

The forces due to internal pressures and suctions are summarized in Table 3.36.

Zone	A_{ref} (m²)	q_p (kN/m²)	c_{pi}	$F_{w,i}$ (kN)	c_{pi}	$F_{w,i}$ (kN)
A	18,56			+ 3,382		− 5,072
B	74,24			+ 13,527		− 20,290
C	81,20			+ 14,795		− 22,192
D	126,4			+ 23,030		− 34,545
E	126,4	0,911	+ 0,2	+ 23,030	− 0,3	− 34,545
F	7,23			+ 1,317		− 1,976
G	7,23			+ 1,317		− 1,976
H	57,83			+ 10,537		− 15,805
I	198,78			+ 36,218		− 54,327

Table 3.36

All of the forces on the surfaces due internal pressures and suctions for wind angle $\theta = 90^\circ$ are indicated in Figure 3.13(r) and Figure 3.13(s).

(i) Forces on surfaces due wind suctions on the roof:

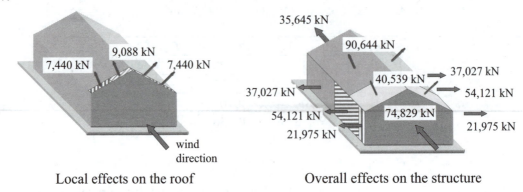

Local effects on the roof   Overall effects on the structure

Figure 3.13(q)

(iii) Forces on surfaces due to internal pressure:

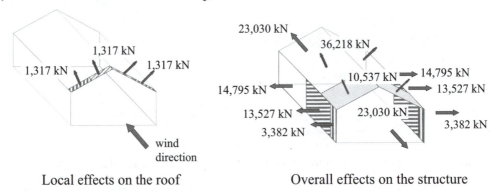

Local effects on the roof   Overall effects on the structure

Figure 3.13(r)

(iv) Forces on surfaces due to internal suction:

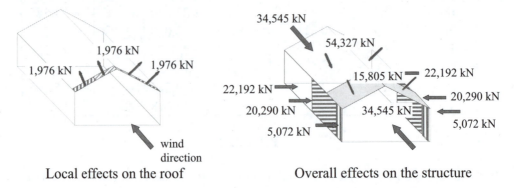

Local effects on the roof   Overall effects on the structure

Figure 3.13(s)

There are a number of possible combinations to consider for the wind loads, e.g.

(i)   wind direction $\theta = 0°$ combined with internal pressure,
(ii) wind direction $\theta = 0°$ combined with internal suction,
(iii) wind direction $\theta = 90°$ combined with internal pressure,
(iv) wind direction $\theta = 90°$ combined with internal suction.

Consider combination (ii): (ignoring local effects, i.e. zones F, G and J)
Wind direction $\theta = 0°$

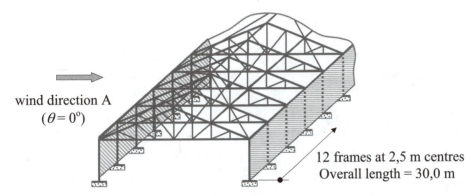

wind direction A
$(\theta = 0°)$

12 frames at 2,5 m centres
Overall length = 30,0 m

Figure 3.13(t)

As indicated in Note 1 of Table 7.4a in BS EN 1991-1-4:2005, the pressure changes rapidly between positive and negative values on the windward roof surface around a pitch angle of $\alpha = -5°$ to $45°$. In such roofs, four cases should be considered where the largest or smallest values of all areas F, G and H are combined with the largest or smallest values in areas I and J. No mixing of positive and negative values is allowed on the same face.

In this example consider the following combination of pressure coefficients:
External pressures coefficients for zone D - (see Table 3.20) $c_{pe,D} = +0,61$
External pressures coefficients for zone E - (see Table 3.20) $c_{pe,E} = -0,46$
External pressures coefficients for zone H - (see Table 3.25) $c_{pe,H} = +0,37$
External pressures coefficients for zone I -   (see Table 3.25) $c_{pe,I} = -0,40$
External pressures coefficients for zone A - (see Table 3.22) $c_{pe,A} = -1,3$
External pressures coefficients for zone B - (see Table 3.22) $c_{pe,B} = -0,80$
Internal pressure coefficient - (see Table 3.27) $c_{pi} = -0,30$

The combined wind pressure coefficients and the corresponding wind pressures on the walls and roof for the data given are shown in the Table 3.37 and Figure 3.13(u).

Pressure due to internal suction $w_i = (q_p \times c_{pi}) = -(0,911 \times 0,3) = -0,273 \text{ kN/m}^2$

Zone	c_{pe}	w_e (kN/m²)	c_{pe}	w_i (kN/m²)	$w = w_e - w_i$ (kN/m²)
A	− 1,3	−1,184			− 0,911
B	− 0,8	− 0,729			− 0,456
D	+ 0,61	+ 0,556	− 0,3	− 0,273	+ 0,829
E	− 0,46	− 0,419			− 0,146
H	+ 0,37	+ 0,337			+ 0,610
I	− 0,40	− 0,364			− 0,091

Table 3.37

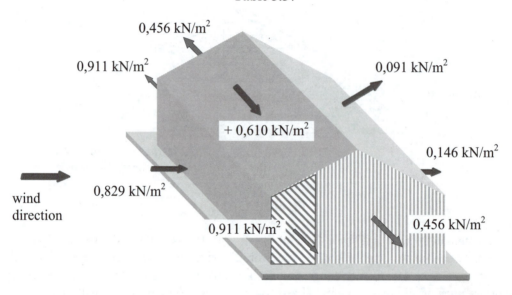

Figure 3.13(u)

Consider the wind forces on a typical individual frame within the structure:

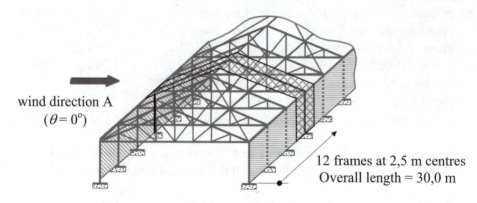

Figure 3.13(v)

The reference areas for the four surfaces of the typical frame are:

$A_{\text{ref,windward wall}} = A_{\text{ref,leeward wall}} = (5{,}8 \times 2{,}5) = 14{,}5 \text{ m}^2$
$A_{\text{ref,windward roof}} = A_{\text{ref,leeward roof}} = (8{,}0 \times 2{,}5)/\cos 27{,}7° = 22{,}59 \text{ m}^2$

The wind forces on the considered surfaces are calculated as:

$F_{\text{w,windward wall}} = (w_{\text{windward wall}} \times A_{\text{ref,windward wall}}) = + (0{,}829 \times 14{,}5) = +12{,}02 \text{ kN}$

(pressure)

$F_{\text{w,leeward wall}} = (w_{\text{leeward wall}} \times A_{\text{ref,leeward wall}}) = - (0{,}146 \times 14{,}5) = -2{,}12 \text{ kN}$

(suction)

$F_{\text{w,windward roof}} = (w_{\text{windward roof}} \times A_{\text{ref,windward roof}}) = + (0{,}610 \times 22{,}59) = +13{,}78 \text{ kN}$

(pressure)

$F_{\text{w,leeward roof}} = w_{\text{leeward roof}} \times A_{\text{ref,leeward wall}} = - (0{,}091 \times 22{,}59) = -2{,}06 \text{ kN}$

(suction)

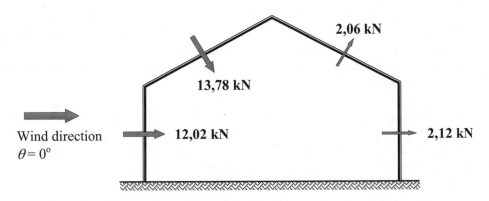

Figure 3.13(x)

Here only a typical internal frame is considered to indicate how to calculate the wind load on each surface. When designing such a structure, all wind load cases should be considered, using appropriate partial load factors, in combination with permanent loads, imposed loads, snow loads, etc., to determine the critical design load case.

4. Structural Analysis Techniques

Objective: to provide a résumé of the elastic methods of structural analysis most commonly used when undertaking structural design.

4.1 *Analysis Techniques*

BS EN 1995-1-1:2004 is based on the concepts of Limit State Design and as indicated in Clause 2.2.2, mathematical modelling of the behaviour of timber elements and structures may be based on either first or second order linear elastic analysis; allowing for creep effects where appropriate.

In general, the verification of strength requirements are based on the assumption of a linear stress-strain relationship until failure. In addition, in the case of members or parts of members which are subject to compression, elastic-plastic behaviour may be used. The *Principles* on which analyses should be carried out are defined in Section 5 of the code: 'Basis of structural analysis.'

The laws of structural mechanics are those well established in recognised 'elastic theory', i.e.

♦ The material is *homogeneous,* which implies that its constituent parts have the same physical properties throughout its entire volume. This assumption is clearly violated in the case of timber. The constituent fibres are large in relation to the mass of which they form a part when compared with a material such as steel. In the case of steel there are millions of very small crystals per cubic centimetre, randomly distributed and of similar quality creating an amorphous mass. In addition, the presence of defects such as knots, shakes, etc., as described in Section 1.3 of Chapter 1 represent the inclusion of elements with differing physical properties.

♦ The material is *isotropic,* which implies that the elastic properties are the same in all directions. The main constituent of timber is cellulose, which occurs as long chain molecules. The chemical/electrical forces binding the molecules together in these chains are much stronger than those which hold the chains to each other. A consequence of this is that considerable differences in elastic properties occur according to the grain orientation. Timber exhibits a marked degree of anisotropic behaviour. The design of timber is generally based on an assumption of orthotropic behaviour with three principal axes of symmetry: the longitudinal axis (parallel to the grain), the radial axis and the tangential axis, as shown in Figure 4.1. The properties relating to the tangential and radial directions are often treated together and regarded as properties perpendicular to the grain.

♦ The material obeys *Hooke's Law,* i.e. when subjected to an external force system the deformations induced will be directly proportional to the magnitude of the applied force. A typical stress/strain curve for a small wood specimen (with as few variations or defects as possible and loaded for a short-term duration) exhibits linearity prior to failure when loaded in tension or compression as indicated in Figures 4.2(a) and 4.2(b).

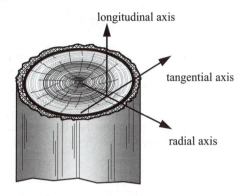

Figure 4.1

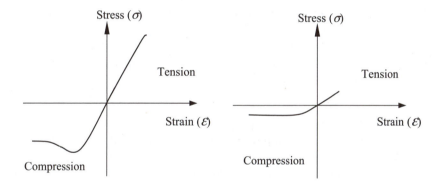

(a) load parallel to the grain   (b) load perpendicular to the grain

Figure 4.2

In Figure 4.2(a) the value of tensile strength is greater than the compressive strength. In both compression and tension linear behaviour occurs. In the case of compression, ductility is present before failure occurs, whilst in tension a brittle, sudden failure occurs. These characteristics are reflected in the interaction behaviour of timber elements designed to resist combined bending and axial stresses as discussed in Chapter 7.0.

♦ The material is *elastic*, which implies that it will recover completely from any deformation after the removal of load. Elastic behaviour is generally observed in timber subject to compression up to the limit of proportionality. The elastic properties of timber in tension are more sensitive to the level of moisture content. Permanent strain occurs at very low stress levels in timber which contains a high percentage of moisture.

♦ The *modulus of elasticity* is the same in tension and compression. This assumption is reasonable for both compression and tension. The value is much lower when the load is applied perpendicular to the grain than when it is applied parallel to the grain, as shown in Figure 4.2. The values for various modulii of elasticity of timber are given in several codes, e.g. BS EN 338:2003(E) for solid timber of various strength classes and BS EN 12369-1:2001 for OSB, particleboards and fibreboards. In the case of solid timber these are:

$E_{0,mean}$ – the mean modulus of elasticity parallel to the grain,
$E_{0,05}$ – the 5% modulus of elasticity parallel to the grain,
$E_{90,mean}$ – the mean modulus of elasticity perpendicular to the grain.

♦ *Plane sections remain plane* during deformation. During bending this assumption is violated and is reflected in a non-linear bending stress diagram throughout cross-sections subject to a moment.

The behaviour and properties of timber do not satisfy the basic assumptions used in simple elastic theory. These deficiencies are accommodated in the design processes by the introduction of numerous modification factors and a factor of safety which are applied to produce a design strength which is then compared with the design stress induced by the applied load system and calculated using elastic theory.

Extensive research and development during the latter half of the twentieth century has resulted in more representative mathematical models of timber behaviour as reflected in the equations and requirements of Eurocode 5.

4.2 *Modification Factors*

The inherently variable nature of timber and its effects on structural material properties such as stress/strain characteristics, elasticity and creep has resulted in numerous modification factors which are used in addition to partial safety factors to convert *characteristic* strengths (see Section 2.4 of Chapter 2) to design strengths for design purposes. In general, when designing to satisfy strength requirements (e.g. axial, bending or shear strength) the following relationship must be satisfied:

Design stress ≤ Design strength

The design stresses are calculated using elastic theory and the design strengths are determined from the code using the appropriate values relating to the characteristic strength multiplied by the modification factors which are relevant to the stress condition being considered.

There are many symbols defined in Section 1.6 of the code in the form of 'Latin upper case letters', 'Latin lower case letters' and 'Greek lower case letters.' Generally strengths are indicated by the letter 'f' and stress by the symbol 'σ'.

As mentioned previously, the design strength is evaluated by multiplying the *characteristic* strength for a particular strength class by the appropriate modification factors and dividing the result by the appropriate partial safety factor, e.g.

$$f_{m,y,d} = (f_{m,y,k} \times k_{mod} \times k_h \times k_{m,\alpha} \times k_{sys})/\gamma_M$$

where:

$f_{m,y,k}$ relates to the *characteristic bending strength* of the timber about the *y-y* axis,

k_{mod} relates to the *service and load-duration classes* of the timber,

k_h relates to the *depth* of the section being considered,

$k_{m,\alpha}$ relates to the *bending strength* for single tapered beams,

k_{sys} relates to the existence of lateral load distribution systems enabling *load sharing*,

γ_M relates to the material property.

In each case a definition and the method of evaluating a coefficient is given in the code. In this text a table is given in each chapter which identifies each coefficient and reference clause number for the coefficients which are pertinent to the structural elements being considered. Three of the most frequently used coefficients are k_{mod}, k_{def} and k_{sys}:

k_{mod}:   The value of k_{mod} is governed by the material, e.g. solid timber, plywood, OSB, etc., the average moisture content likely to be attained in service conditions and the load duration.

These variables are allowed for in the code by identifying a '*load-duration class*', a '*service class*' and the '*material type*' for a particular element being designed and obtaining the k_{mod} value from Table 3.1

The Load-duration Classes as given in Clause 2.3.1.2, Tables 2.1 and 2.2 are:

Load-duration Class	Order of accumulated duration of characteristic load
Permanent: (e.g. self-weight)	more than 10 years
Long-term: (e.g. storage, water tank)	6 months to 10 years
Medium-term: (e.g. imposed floor load, snow)	1 week to 6 months
Short-term: (e.g. snow, wind)	less than one week
Instantaneous: (e.g. wind, accidental load)	

Table 4.3

The Service Classes as given in Clause 2.3.1.3 are:

Service Class 1:   This is characterized by a moisture content in the materials corresponding to a temperature of 20°C and the relative humidity of the surrounding air only exceeding 65% for a few weeks per year. In such conditions most timber will attain an average moisture content not exceeding 12%.

Service Class 2:   This is characterized by a moisture content in the materials corresponding to a temperature of 20°C and the relative humidity of the surrounding air only exceeding 85% for a few weeks per year. In such conditions most timber will attain an average moisture content not exceeding 20%.

Service Class 3:   This is characterized, due to climatic conditions, by moisture contents higher than service class 2.

In cases where more than one load-duration class is included in a particular load combination, the k_{mod} value relating to the action corresponding to the shortest duration should be used (see Clause 3.1.3(2)).

> **k_{def}:** The *deformation factor*, k_{def}, is related to the creep characteristics of the material, e.g. solid timber, plywood, OSB, etc., and Service Class as defined above. It is used to modify the stiffness parameters (see Equations 2.7 to 2.12 of EC5):
> E_{mean} – the mean value of modulus of elasticity,
> G_{mean} – the mean value of shear modulus,
> K_{ser} – the slip modulus.

The use of k_{def} applies to both ultimate and serviceability limit states criteria, e.g. when calculating structural deformations in structures containing elements having different creep characteristics, and in ultimate limit state criteria where the distribution of element forces and moments is affected by the stiffness distribution in the structure.

In the case of connections comprising timber elements with the *same* creep characteristics, the value of k_{def} should be doubled as indicated in Clause 2.3.2.2(3).

In the case of connections comprising timber elements with the *different* creep characteristics, the value of k_{def} should be doubled as indicated in Clause 2.3.2.2(4),

i.e. $k_{def} = 2 \times \sqrt{k_{def,1}\, k_{def,2}}$ (Equation 2.13 in EC5)

Values of k_{def} are given in Table 3.2 of EC5.

> **k_{sys}:** When designing structures in which members are connected by structural elements which provide lateral distribution of load (i.e. load-sharing) the characteristic strength properties can be enhanced by multiplying by k_{sys} as indicated in Clause 6.6. Typical elements, which provide lateral distribution of load, are purlins, binders, boarding, battens etc.
>
> In general the value of the *system strength factor*, k_{sys}, is 1,1, however, in the case of glued laminated members a linearly increasing enhancement is permitted depending on the number of loaded laminations. The maximum value is also dependent on the method of connection as indicated in Figure 6.12 of the code.

4.3 *Résumé of Analysis Techniques*

The following résumé gives a brief summary of the most common manual techniques adopted to determine the forces induced in the members of statically determinate structures. There are numerous structural analysis books available which give comprehensive detailed explanations of these and other more advanced techniques.

4.4 *Method of Sections for Pin-jointed Frames*

The *method of sections* involves the application of the three equations of static equilibrium to two-dimensional plane frames. The sign convention adopted to indicate ties (i.e. tension members) and struts (i.e. compression members) in frames is as shown in Figure 4.3.

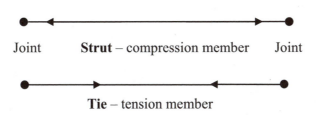

Tie – tension member

Figure 4.3

The method involves considering an imaginary section line which cuts the frame under consideration into two parts A and B as shown in Figure 4.6.

Since only three independent equations of equilibrium are available any section taken through a frame must not include more than three members for which the internal force is unknown.

Consideration of the equilibrium of the resulting force system enables the magnitude and sense (i.e. compression or tension) of the forces in the cut members to be determined.

4.4.1 Example 4.1: Pin-jointed Truss

A pin-jointed truss supported by a pinned support at A and a roller support at G carries three loads at joints C, D and E as shown in Figure 4.4. Determine the magnitude and sense of the forces induced in members X, Y and Z as indicated.

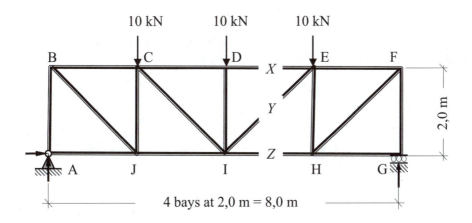

Figure 4.4

Step 1: Evaluate the support reactions. It is not necessary to know any information regarding the frame members at this stage other than dimensions as shown, in Figure 4.5, since only **externally** applied loads and reactions are involved.

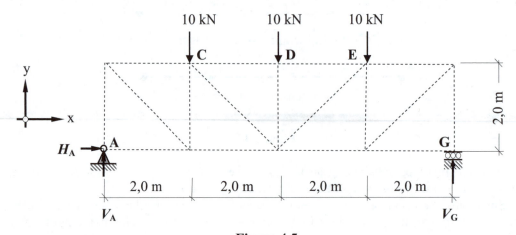

Figure 4.5

Apply the three equations of static equilibrium to the force system:

+ve ↑ $\Sigma F_y = 0$ $V_A - (10 + 10 + 10) + V_G = 0$ $V_A + V_G = 30$ kN
+ve → $\Sigma F_x = 0$ $\therefore H_A = 0$
+ve ↘ $\Sigma M_A = 0$ $(10 \times 2,0) + (10 \times 4,0) + (10 \times 6,0) - (V_G \times 8,0) = 0$

$$\therefore V_G = 15 \text{ kN} \uparrow$$
$$\text{Hence} \quad V_A = 15 \text{ kN} \uparrow$$

Step 2: Select a section through which the frame can be considered to be cut and using the same three equations of equilibrium, determine the magnitude and sense of the unknown forces (i.e. the internal forces in the cut members).

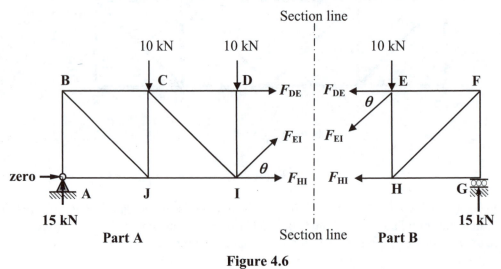

Figure 4.6

It is convenient to **assume** all unknown forces to be tensile and hence at the cut section their direction and lines of action are considered to be pointing away from the joints (refer to Figure 4.6). If the answer results in a negative force this means that the assumption of a tie was incorrect and the member is actually in compression, i.e. a strut.

The application of the equations of equilibrium to either part of the cut frame will enable the forces X (F_{DE}), Y (F_{EI}) and Z (F_{HI}) to be evaluated.

Note: The section considered must not cut through more than three members with unknown internal forces since only three equations of equilibrium are applicable.

Consider Part A:

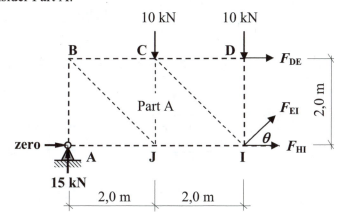

Figure 4.7

Note: $Sin\ \theta = \dfrac{2}{2\sqrt{2}} = 0,707,$ $Cos\ \theta = \dfrac{2}{2\sqrt{2}} = 0,707,$

+ve $\uparrow$ $\Sigma F_y = 0$ $+15,0 - 10,0 - 10,0 + F_{EI}\ Sin\theta = 0$

$$F_{EI} = +\ \frac{5.0}{Sin\theta} = +\ \textbf{7,07 kN}$$

Member EI is a tie

+ve $\rightarrow$ $\Sigma F_x = 0$ $+F_{DE} + F_{HI} + F_{EI}\ Cos\theta\quad = 0$

+ve $\searrow$ $\Sigma M_I = 0$ $+(15,0 \times 4,0) - (10,0 \times 2,0) + (F_{DE} \times 2,0) = 0$
$\qquad\qquad F_{DE} = -\ \textbf{20,0 kN}$

Member DE is a strut

Hence $F_{HI} = -\ F_{DE} - F_{EI}\ Cos\theta = -(-20,0) - (7,07 \times Cos\theta) = +\ \textbf{15,0 kN}$

Member HI is a tie

These answers can be confirmed by considering Part B of the structure and applying the equations as above.

4.5   *Method of Joint Resolution for Pin-jointed Frames*

Considering the same frame using *joint resolution* highlights the advantage of the method of sections when only a few member forces are required.

In this technique (which can be considered as a special case of the method of sections), sections are taken which isolate each individual joint in turn in the frame, e.g.

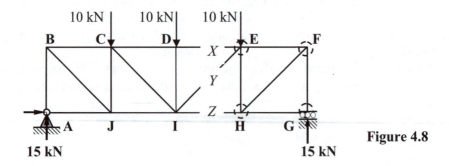

Figure 4.8

15 kN   15 kN

In Figure 4.8 four sections are shown, each of which isolates a joint in the structure as indicated in Figure 4.9.

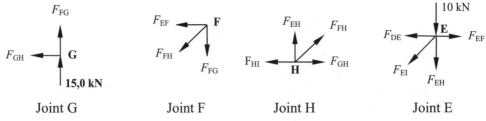

Joint G   Joint F   Joint H   Joint E

Figure 4.9

Since in each case the forces are coincident, the moment equation is of no value, hence only two independent equations are available. It is necessary when considering the equilibrium of each joint to do so in a sequence which ensures that there are no more than two unknown member forces in the joint under consideration. This can be carried out until all member forces in the structure have been determined.
Consider Joint G:

$+ve \uparrow \Sigma F_y = 0 \qquad + 15,0 + F_{FG} = 0$

$$F_{FG} = - 15,0 \text{ kN}$$

$+ve \rightarrow \Sigma F_x = 0 \quad - F_{GH} = 0 \qquad F_{GH} = 0$

Member GH is a zero member
Member FG is a strut

Consider Joint F: substitute for calculated values, i.e. F_{FG} (direction of force is into the joint)

$+ve \uparrow \Sigma F_y = 0 \qquad + 15,0 - F_{FH} \cos\theta = 0$

$$F_{FH} = + 15,0 / 0,707$$
$$F_{FH} = + 21,21 \text{ kN}$$

$+ve \rightarrow \Sigma F_x = 0 \quad - F_{EF} - F_{FH} \sin\theta = 0$

$$F_{EF} = - 21,21 \times 0,707$$
$$F_{EF} = - 15,0 \text{ kN}$$

Member FH is a tie
Member EF is a strut

Consider Joint H: substitute for calculated values, i.e. F_{GH} and F_{FH}

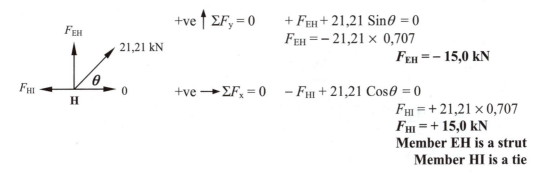

$$+ve \uparrow \Sigma F_y = 0 \qquad + F_{EH} + 21,21 \, Sin\theta = 0$$
$$F_{EH} = -21,21 \times 0,707$$
$$\boxed{F_{EH} = -15,0 \text{ kN}}$$

$$+ve \longrightarrow \Sigma F_x = 0 \qquad - F_{HI} + 21,21 \, Cos\theta = 0$$
$$F_{HI} = +21,21 \times 0,707$$
$$\mathbf{F_{HI} = +15,0 \text{ kN}}$$
Member EH is a strut
Member HI is a tie

Consider Joint E: substitute for calculated values, i.e. F_{EF} and F_{EH}

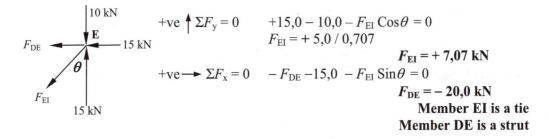

$$+ve \uparrow \Sigma F_y = 0 \qquad +15,0 - 10,0 - F_{EI} \, Cos\theta = 0$$
$$F_{EI} = +5,0 / 0,707$$
$$\boxed{F_{EI} = +7,07 \text{ kN}}$$

$$+ve \longrightarrow \Sigma F_x = 0 \qquad - F_{DE} - 15,0 - F_{EI} \, Sin\theta = 0$$
$$\mathbf{F_{DE} = -20,0 \text{ kN}}$$
Member EI is a tie
Member DE is a strut

4.6 *Unit Load Method to Determine the Deflection of Pin-jointed Frames*

The Unit Load Method of analysis is based on the principles of strain energy and Castigliano's 1st Theorem (71). When structures deflect under load the ***work-done*** by the displacement of the applied loads is stored in the members of the structure in the form of ***strain energy***.

4.6.1 *Strain Energy (Axial Load Effects)*

Consider an axially loaded structural member of length 'L', cross-sectional area 'A', and of material with modulus of elasticity 'E' as shown in Figure 4.10(a)

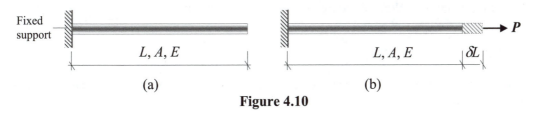

(a)   (b)

Figure 4.10

When an axial load 'P' is applied as indicated, the member will increase in length by 'δL' as shown in Figure 4.10(b). Assuming linear elastic behaviour $\delta L \propto P$, this relationship is represented graphically in Figure 4.11.

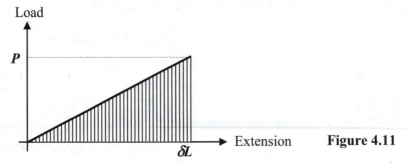

Figure 4.11

The work-done by the externally applied load 'P' is equal to:

(average value of the force × distance through which the force moves in its line of action)

i.e.   Work-done $= \left(\dfrac{P}{2} \times \delta L \right)$

For linearly elastic materials the relationship between the applied axial load and the change in length is:

$$\delta L = \frac{PL}{AE}$$

∴   Work-done $= \left(\dfrac{P}{2} \times \delta L \right) = \left(\dfrac{P}{2} \times \dfrac{PL}{AE} \right) = \dfrac{P^2 L}{2AE}$

This work-done by the externally applied load is equal to the 'energy' stored by the member when it changes length and is known as the strain energy, usually given the symbol 'U'. It is this energy which causes structural members to return to their original length when an applied load system is removed. (**Note:** assuming that the strains are within the elastic limits of the material.)

∴   Strain energy = Work-done by the applied load system

$$U = \frac{P^2 L}{2AE}$$

(**Note:** the principles of strain energy also apply to members subject to shear, bending, torsion, etc.).

4.6.2 Castigliano's 1st Theorem

Castigliano's 1st Theorem relating to strain energy and structural deformation can be expressed as follows:

'If the total strain energy in a structure is partially differentiated with respect to an applied load the result is equal to the displacement of that load in its line of action.'

In mathematical terms this is:

$$\Delta = \frac{\partial U}{\partial W}$$

where:
U is the total strain energy of the structure due to the applied load system,
W is the force acting at the point where the displacement is required,
Δ is the linear displacement in the direction of the line of action of W.

This form of the theorem is very useful in obtaining the deflection at joints in pin-jointed structures. Consider the pin-jointed frame shown in Figure 4.12 in which it is required to determine the vertical deflection of joint B.

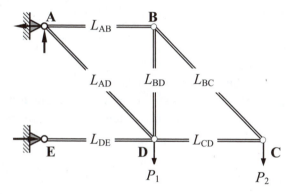

Figure 4.12

Step 1:
The member forces induced by the applied load system are calculated, in this case referred to as the '*P*' forces, as shown in Figure 4.13.

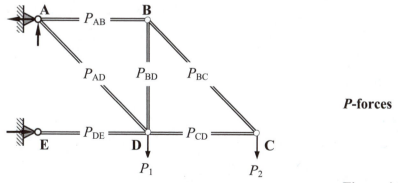

P-forces

Figure 4.13

Step 2:
The applied load system is removed from the structure and an imaginary Unit load is applied at the joint and in the direction of the required deflection, i.e. a vertical load equal to 1,0 at joint B. The resulting member forces due to the Unit load are calculated and referred to as the '*u*' forces, as shown in Figure 4.14.

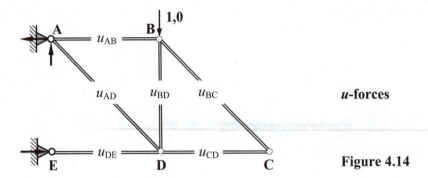

u-forces

Figure 4.14

If both the Step 1 and the Step 2 load systems are considered to act simultaneously, then by superposition the total force in each member is given by:

$$Q = (P + \beta u)$$

where:
P is the force due to the applied load system,
u is the force due to the applied imaginary Unit load applied at B,
β is a multiplying factor to reflect the value of the load applied at B, (since the unit load is an imaginary force the value of β = zero is used here as a mathematical convenience.)

The total strain energy in the structure is equal to the sum of the energy stored in all the members:

$$U = \sum \frac{Q^2 L}{2AE}$$

Using Castigliano's 1st Theorem the deflection of joint B is given by:

$$\Delta = \frac{\partial U}{\partial W}$$

$$\therefore \Delta_B = \frac{\partial U}{\partial \beta} = \frac{\partial U}{\partial Q} \times \frac{\partial Q}{\partial \beta}$$

and

$$\frac{\partial U}{\partial Q} = \sum \frac{QL}{AE}; \quad \frac{\partial Q}{\partial \beta} = u$$

$$\therefore \Delta_B = \frac{\partial U}{\partial \beta} = \frac{\partial U}{\partial Q} \times \frac{\partial Q}{\partial \beta} = \sum \frac{QL}{AE} \times u = \sum \frac{(P + \beta u)L}{AE} \times u$$

Since β = zero the vertical deflection at B (Δ_B) is given by:

$$\Delta_B = \sum \frac{PL}{AE} u$$

i.e. the deflection at any joint in a pin-jointed frame can be determined from:

$$\delta = \sum \frac{PL}{AE} u$$

where:

δ is the displacement of the point of application of any load, along the line of action of that load,

P is the force in a member due to the externally applied loading system,

u is the force in a member due to a **unit load** acting at the position of, and in the direction of the desired displacement,

L/A is the ratio of the length to the cross-sectional area of the members,

E is the modulus of elasticity of the material for each member (i.e. Young's modulus).

4.6.3 Example 4.2: Deflection of a Pin-jointed Truss

A pin-jointed truss ABCD is shown in Figure 4.15 in which both a vertical and a horizontal load are applied at joint B as indicated. Determine the magnitude and direction of the resultant deflection at joint B and the vertical deflection at joint D.

Assume the cross-sectional area of all members is equal to *A* and all members are made from the same material, i.e. have the same modulus of elasticity *E*

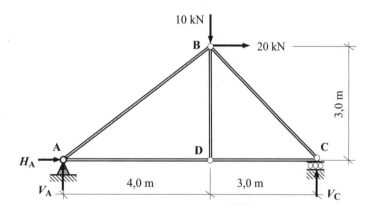

Figure 4.15

Step 1: Evaluate the member forces. The reader should follow the procedure given in Example 4.1 to determine the following results:

Horizontal component of reaction at support A $H_A = -20{,}0$ kN

Vertical component of reaction at support A $V_A = -4{,}29$ kN

Vertical component of reaction at support C $V_C = +14{,}29$ kN

Use the *method of sections* or *joint resolution* as indicated in Sections 4.2 and 4.3 respectively to determine the magnitude and sense of the unknown member forces (i.e. the *P* forces).

The reader should complete this calculation to determine the member forces as indicated in Figure 4.16.

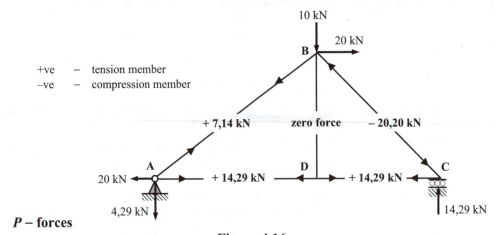

P – forces

Figure 4.16

Step 2: To determine the vertical deflection at joint B, remove the externally applied load system and apply a **unit load only** in a vertical direction at joint B as shown in Figure 4.17. Use the *method of sections* or *joint resolution* as before to determine the magnitude and sense of the unknown member forces (i.e. the *u* forces).

The reader should complete this calculation to determine the member forces as indicated in Figure 4.17.

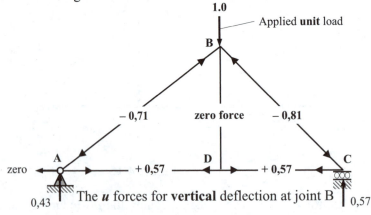

Figure 4.17

The vertical deflection $\delta_{V,B} = \sum \dfrac{PL}{AE} u$

This is better calculated in tabular form as shown in Table 4.4.

Member	Length (L)	Cross-section (A)	Modulus (E)	P forces (kN)	u forces	PL × u (kNm)
AB	5,0 m	A	E	+ 7,14	– 0,71	– 25,35
BC	4,24 m	A	E	– 20,20	– 0,81	+ 69,37
AD	4,0 m	A	E	+ 14,29	+ 0,57	+ 32,58
CD	3,0 m	A	E	+ 14,29	+ 0,57	+ 24,44
BD	3,0 m	A	E	0,0	0,0	0,0
					Σ	+ 101,04

Table 4.4

The +ve sign indicates that the deflection is in the same direction as the applied unit load.

Hence the vertical deflection $\delta_{V,B} = \sum \dfrac{PL}{AE} u = +(101{,}04/AE)$ ↓

Note: Where the members have different cross-sectional areas and/or modulii of elasticity each entry in the last column of the table should be based on $(PL \times u)/AE$ and not only $(PL \times u)$.

A similar calculation can be carried out to determine the horizontal deflection at joint B. The reader should complete this calculation to determine the member forces as indicated in Figure 4.18.

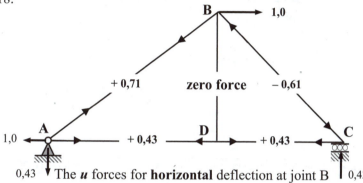

Figure 4.18

$0{,}43$ ↓ The **u** forces for **horizontal** deflection at joint B ↓ $0{,}43$

The horizontal deflection $\delta_{H,B} = \sum \dfrac{PL}{AE} u$

Member	Length (*L*)	Cross-section (*A*)	Modulus (*E*)	*P* forces (kN)	*u* forces	*PL* × *u* (kNm)
AB	5,0 m	A	E	+ 7,14	+ 0,71	+ 25,35
BC	4,24 m	A	E	− 20,20	− 0,61	+ 52,25
AD	4,0 m	A	E	+ 14,29	+ 0,43	+ 24,58
CD	3,0 m	A	E	+ 14,29	+ 0,43	+ 18,43
BD	3,0 m	A	E	0,0	0,0	0,0
					Σ	+ 121,61

Table 4.5

Hence the horizontal deflection $\delta_{H,B} = \sum \dfrac{PL}{AE} u = +(120{,}61/AE)$ →

The resultant deflection at joint B can be determined from the horizontal and vertical components evaluated above, i.e.

$R = \sqrt{\left(101{,}04^2 + 120{,}61^2\right)}/AE = 157{,}34/AE$

$\theta = \tan^{-1}(120{,}61/101{,}04) = 50{,}05°$

A similar calculation can be carried out to determine the vertical deflection at joint D. The reader should complete this calculation to determine the member forces as indicated in Figure 4.19.

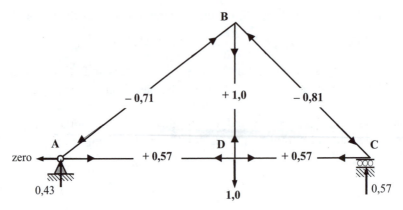

The member *u* forces for **vertical** deflection at joint D

Figure 4.19

The vertical deflection $\delta_{V,D} = \sum \dfrac{PL}{AE} u$

Member	Length (*L*)	Cross-section (*A*)	Modulus (*E*)	*P* forces (kN)	*u* forces	*PL × u* (kNm)
AB	5,0 m	*A*	*E*	+ 7,14	− 0,71	− 25,35
BC	4,24 m	*A*	*E*	− 20,20	− 0,81	+ 69,37
AD	4,0 m	*A*	*E*	+ 14,29	+ 0,57	+ 32,58
CD	3,0 m	*A*	*E*	+ 14,29	+ 0,57	+ 24,44
BD	3,0 m	*A*	*E*	0,0	+1,0	0,0
					Σ	+ 101,04

Table 4.6

Hence the vertical deflection $\delta_{V,D} = \sum \dfrac{PL}{AE} u = + (101{,}04/AE)$ ↓

4.7 *Shear Force and Bending Moment*

Two parameters which are fundamentally important to the design of beams are ***shear force*** and ***bending moment***. These quantities are the result of internal forces acting on the material of a beam in response to an externally applied load system.

4.7.1 *Example 4.3: Beam with Point Loads*

Consider a simply supported beam as shown in Figure 4.20 carrying a series of secondary beams each imposing a point load of 4 kN.

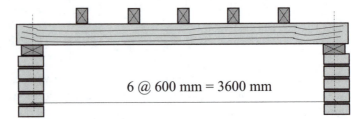

Figure 4.20

6 @ 600 mm = 3600 mm

This structure can be represented as a line diagram as shown in Figure 4.21:

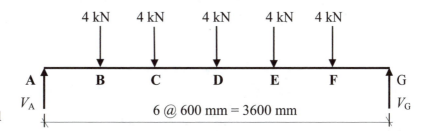

Figure 4.21

Since the externally applied force system is in equilibrium, the three equations of static equilibrium must be satisfied, i.e.

+ve ↑ $\Sigma F_y = 0$ The sum of the vertical forces must equal zero.

+ve ↴ $\Sigma M = 0$ The sum of the moments of all forces about *any* point on the plane of the forces must equal zero.

+ve → $\Sigma F_x = 0$ The sum of the horizontal forces must equal zero.

The assumed positive directions are as indicated. In this particular problem there are no externally applied horizontal forces and consequently the third equation is not required.
(Note: It is still necessary to provide horizontal restraint to a structure since it can be subject to a variety of load cases, some of which may have a horizontal component.)
Consider the vertical equilibrium of the beam:

+ve ↑ $\Sigma F_y = 0$

$+ V_A - (5 \times 4,0) + V_G = 0$ ∴ $V_A + V_G = 20$ kN   Equation (1)
Consider the rotational equilibrium of the beam:

+ve ↴ $\Sigma M_A = 0$

Note: The sum of the moments is taken about one end of the beam (end A) for convenience. Since one of the forces (V_A) passes through this point it does not produce a moment about A and hence does not appear in the equation. It should be recognised that the sum of the moments could have been considered about *any* known point in the same plane.

$+ (4,0 \times 0,6) + (4,0 \times 1,2) + (4,0 \times 1,8) + (4,0 \times 2,4) + (4,0 \times 3,0) - (V_G \times 3,6) = 0$

∴ $V_G = 10$ kN

Substituting for V_G into Equation (1) gives   ∴ $V_A = 10$ kN

This calculation was carried out considering only the externally applied forces, i.e.

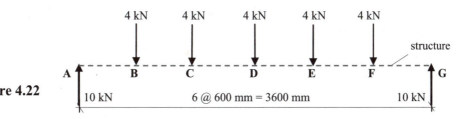

Figure 4.22

The structure itself was ignored, however the applied loads are transferred to the end supports through the material fibres of the beam. Consider the beam to be cut at section X–X producing two sections each of which is in equilibrium as shown in Figure 4.23.

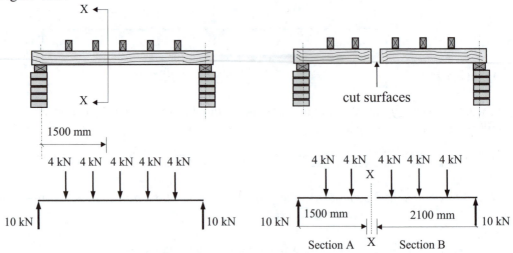

Figure 4.23

Clearly if the two sections are in equilibrium there must be internal forces acting on the cut surfaces to maintain this; these forces are known as the *shear force* and the *bending moment*, and are illustrated in Figure 4.24

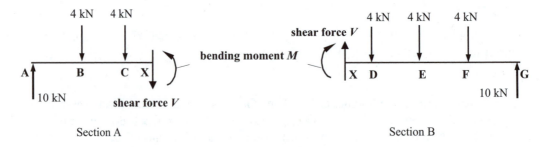

Figure 4.24

The force V and moment M are equal and opposite on each surface. The magnitude and direction of V and M can be determined by considering two equations of static equilibrium for either of the cut sections; both will give the same answer.

Consider the left-hand section with the 'assumed' directions of the internal forces V and M as shown in Figure 4.25.

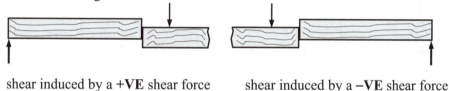

$$+\text{ve} \uparrow \Sigma F_{y} = 0$$
$$+ 10 - 4{,}0 - 4{,}0 - V = 0 \qquad \therefore V = 2{,}0 \text{ kN}$$

$$+\text{ve} \curvearrowright \Sigma M_{A} = 0$$
$$+ (4{,}0 \times 0{,}6) + (4{,}0 \times 1{,}2) - (V \times 1{,}5) - M = 0$$
$$\therefore M = 10{,}2 \text{ kNm}$$

Figure 4.25

4.7.2 Shear Force Diagrams

In a statically determinate beam, the numerical value of the shear force can be obtained by evaluating the algebraic sum of the vertical forces to one side of the section being considered. The convention adopted in this text to indicate positive and negative shear forces is shown in Figure 4.26.

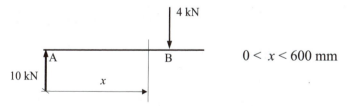

shear induced by a **+VE** shear force   shear induced by a **−VE** shear force

Figure 4.26

The calculation carried out to determine the shear force can be repeated at various locations along a beam and the values obtained plotted as a graph; this graph is known as the *shear force diagram*. The shear force diagram indicates the variation of the shear force along a structural member.

Consider any section of the beam between A and B:

$$0 < x < 600 \text{ mm}$$

Note: The value immediately under the point load at the cut section is not being considered.

The shear force at any position $x = \Sigma$ vertical forces to one side
$$= + 10{,}0 \text{ kN}$$

This value is a constant for all values of x between zero and 600 mm, the graph will therefore be a horizontal line equal to 10,0 kN. This force produces a +ve shear effect, i.e.

+ve shear effect

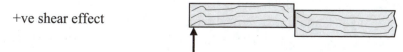

Consider any section of the beam between B and C:

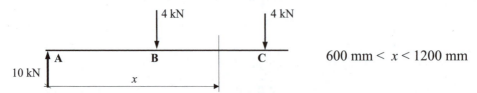

600 mm < x < 1200 mm

The shear force at any position x = Σ vertical force to one side
$$= + 10,0 - 4,0 = 6,0 \text{ kN}$$

This value is a constant for all values of x between 600 mm and 1200 mm. The graph will therefore be a horizontal line equal to 6,0 kN. This force produces a +ve effect shear effect.

Similarly for any section between C and D:

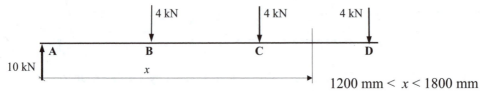

1200 mm < x < 1800 mm

The shear force at any position x = Σ vertical forces to one side
$$= + 10,0 - 4,0 - 4,0 = 2,0 \text{ kN}$$

Consider any section of the beam between D and E:

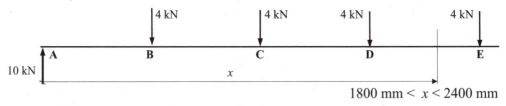

1800 mm < x < 2400 mm

The shear force at any position x = Σ vertical forces to one side
$$= + 10,0 - 4,0 - 4,0 - 4,0 = - 2,0 \text{ kN}$$

In this case the shear force is negative: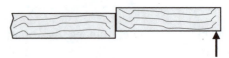

Similarly between E and F   2400 mm < x < 3000 mm
The shear force at any position x = Σ vertical forces to one side
$$= + 10,0 - 4,0 - 4,0 - 4,0 - 4,0 = - 6,0 \text{ kN}$$

and

between F and G   3000 mm < x < 3600 mm
The shear force at any position x = Σ vertical forces to one side
$$= + 10,0 - 4,0 - 4,0 - 4,0 - 4,0 - 4,0 = - 10,0 \text{ kN}$$

In each of the cases above the value has not been considered at the point of application of the load.

Consider the location of the applied load at B shown in Figure 4.27.

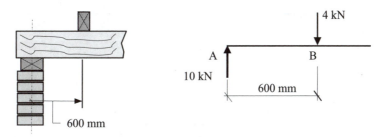

Figure 4.27

The 4,0 kN is not instantly transferred through the beam fibres at B but instead over the width of the actual secondary beam. The change in value of the shear force between $x < 600$ mm and $x > 600$ mm occurs over this width, as shown in Figure 4.28.

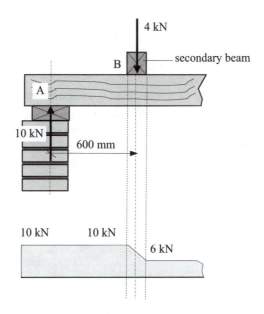

Figure 4.28

The width of the secondary beam is insignificant when compared with the overall span, and the shear force is assumed to change instantly at this point, producing a vertical line on the shear force diagram as shown in Figure 4.29.

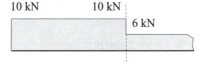

Figure 4.29

The full shear force diagram can therefore be drawn as shown in Figure 4.30.

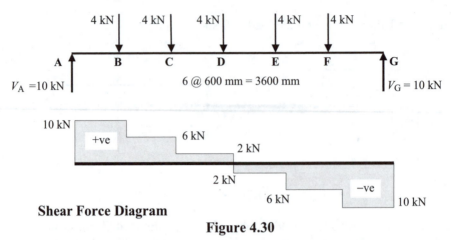

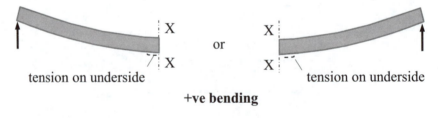

Shear Force Diagram

Figure 4.30

The same result can be obtained by considering sections from the right-hand side of the beam.

4.7.3 Bending Moment Diagrams

In a statically determinate beam the numerical value of the bending moment (i.e. moments caused by forces which tend to bend the beam) can be obtained by evaluating the algebraic sum of the moments of the forces to one side of a section. In the same manner as with shear forces either the left-hand or the right-hand side of the beam can be considered. The convention adopted in this text to indicate positive and negative bending moments is shown in Figures 4.31(a) and (b).

Bending inducing **tension on the underside** of a beam is considered **positive**.

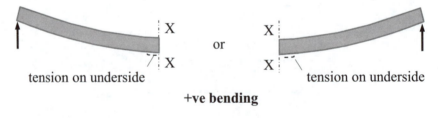

Figure 4.31 (a)

Bending inducing **tension on the top** of a beam is considered **negative**.

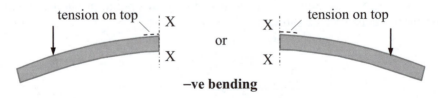

Figure 4.31 (b)

Note: *Clockwise/anti-clockwise moments do **not** define +ve or −ve **bending** moments. The sign of the bending moment is governed by the location of the tension surface at the point being considered.*

As with shear forces, the calculation for bending moments can be carried out at various locations along a beam and the values plotted on a graph. This graph is known as the **bending moment diagram.** The bending moment diagram indicates the variation in the bending moment along a structural member.

Consider sections between A and B of the beam as before:

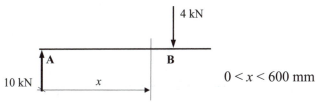

$$0 < x < 600 \text{ mm}$$

In this case when $x = 600$ mm the 4,0 kN load passes through the section being considered and does not produce a bending moment, and can therefore be ignored.

Bending moment = Σ algebraic sum of the moments of the forces to one side of a section.

$$= \Sigma \text{ (Force} \times \text{lever arm)}$$
$$M_x = 10,0 \times x = 10,0 \ x \text{ kNm}$$

Unlike the shear force, this expression is not a constant and depends on the value of 'x' which varies between the limits given. This is a linear expression which should be reflected in the calculated values of the bending moment.

$x = 0$	$M_x = 10,0 \times 0 = \text{zero}$
$x = 200$ mm	$M_x = 10,0 \times 0,2 = 2,0$ kNm
$x = 400$ mm	$M_x = 10,0 \times 0,4 = 4,0$ kNm
$x = 600$ mm	$M_x = 10,0 \times 0,6 = 6,0$ kNm

Clearly the bending moment increases linearly from zero at the simply supported end to a value of 6,0 kNm at point B.

Consider sections between B and C of the beam:

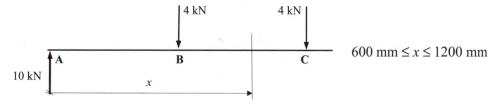

$$600 \text{ mm} \leq x \leq 1200 \text{ mm}$$

Bending moment = Σ algebraic sum of the moments of the forces to 'one' side of a section

$$M_x = + (10,0 \times x) - [4,0 \times (x - 0,6)]$$

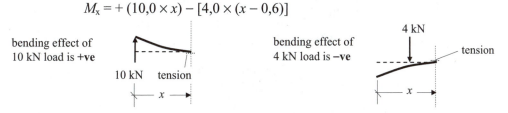

bending effect of
10 kN load is **+ve**

10 kN   tension

bending effect of
4 kN load is **−ve**

4 kN

tension

$x = 800$ mm $M_x = + (10,0 \times 0,8) - (4,0 \times 0,2) = 7,2$ kNm
$x = 1000$ mm $M_x = + (10,0 \times 1,0) - (4,0 \times 0,4) = 8,4$ kNm
$x = 1200$ mm $M_x = + (10,0 \times 1,2) - (4,0 \times 0,6) = 9,6$ kNm

As before the bending moment increases linearly, i.e. from 7,2 kNm at $x = 800$ mm to a value of 9,6 kNm at point C.

Since the variation is linear it is only necessary to evaluate the magnitude and sign of the bending moment at locations where the slope of the line changes, i.e. each of the point load locations.

Consider point D:

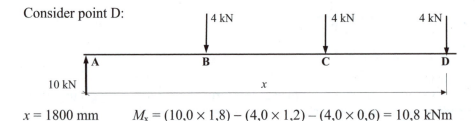

$x = 1800$ mm $M_x = (10,0 \times 1,8) - (4,0 \times 1,2) - (4,0 \times 0,6) = 10,8$ kNm

Consider point E:

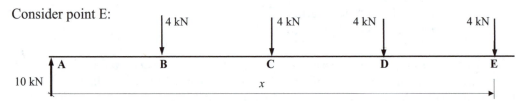

$x = 2400$ mm $M_x = (10,0 \times 2.4) - (4,0 \times 1,8) - (4,0 \times 1,2) - (4,0 \times 0,6) = 9,6$ kNm

Similarly at point F:
$x = 3000$ mm $M_x = (10,0 \times 3,0) - (4,0 \times 2,4) - (4,0 \times 1,8) - (4,0 \times 1,2) - (4,0 \times 0,6)$
                           $= 6,0$ kNm

The full bending moment diagram can therefore be drawn as shown in Figure 4.32.

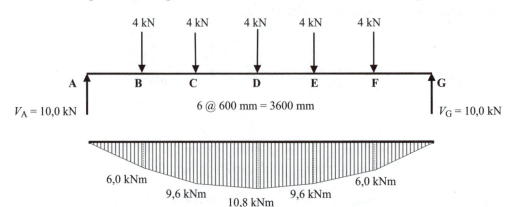

Bending Moment Diagram
Figure 4.32

The same result can be obtained by considering sections from the right-hand side of the beam. The value of the bending moment at any location can also be determined by evaluating the area under the shear force diagram.

Consider point B:

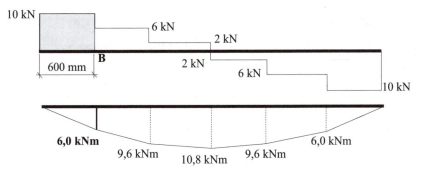

Bending moment at B = shaded area on the shear force diagram
$M_B = (10,0 \times 0,6) = 6,0$ kNm as before

Consider a section at a distance of $x = 900$ mm along the beam between D and E:

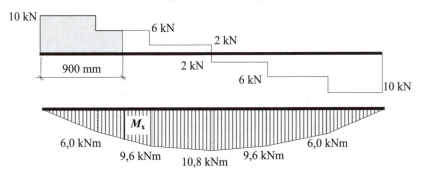

Bending moment at x = shaded area on the shear force diagram

$M_x = (10,0 \times 0,6) + (6,0 \times 0,3) = 7,8$ kNm as before

Consider a section at a distance of $x = 2100$ mm along the beam between D and E:

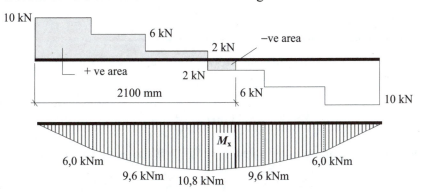

Bending moment at x = shaded area on the shear force diagram

$M_x = (10,0 \times 0,6) + (6,0 \times 0,6) + (2,0 \times 0,6) - (2,0 \times 0,3)$
$\quad = 10,2$ kNm
(**Note:** A maximum bending moment occurs at the same position as a zero shear force.)

4.7.4 Example 4.4: Beam with a Uniformly Distributed Load (UDL)

Consider a simply-supported beam carrying a uniformly distributed load of 5,0 kN/m, as shown in Figure 4.33.

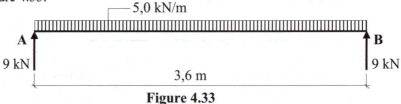

Figure 4.33

The shear force at any section a distance x from the support at A is given by:
V_x = algebraic sum of the vertical forces

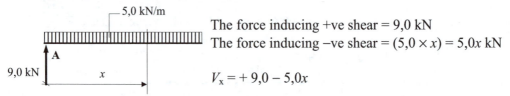

The force inducing +ve shear = 9,0 kN
The force inducing −ve shear = $(5,0 \times x)$ = $5,0x$ kN

$V_x = + 9,0 - 5,0x$

This is a linear equation in which V_x decreases as x increases. The points of interest are at the supports where the maximum shear forces occur, and at the locations where the maximum bending moment occurs, i.e. the point of zero shear.

$V_x = 0$ when $+ 9,0 - 5,0x = 0$ $\therefore x = 1,8$ m

Any intermediate value can be found by substituting the appropriate value of 'x' in the equation for the shear force, e.g.
$x = 600$ mm $V_x = + 9,0 - (5,0 \times 0,6) = + 6,0$ kN
$x = 2100$ mm $V_x = + 9,0 - (5,0 \times 2,1) = - 1,5$ kN
The shear force can be drawn as shown in Figure 4.34.

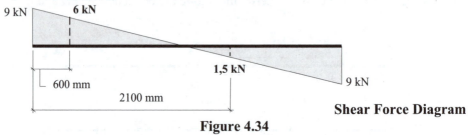

Shear Force Diagram

Figure 4.34

The bending moment can be determined as before, **either** using an equation or evaluating the area under the shear force diagram.

Using an equation:

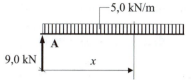

Bending moment at x: $M_x = + (9,0 \times x) - [(5,0 \times x) \times (x/2)] = (9,0x - 2,5x^2)$

In this case the equation is *not* linear, and the bending moment diagram will therefore be *curved.*

Consider several values:

$x = 0$ $M_x = $ zero

$x = 600$ mm $M_x = + (9,0 \times 0,6) - (2,5 \times 0,6^2) = 4,5$ kNm

$x = 1800$ mm $M_x = + (9,0 \times 1,8) - (2,5 \times 1,8^2) = 8,1$ kNm

$x = 2100$ mm $M_x = + (9,0 \times 2.1) - (2,5 \times 2,1^2) = 7,88$ kNm

Using the shear force diagram:

$x = 600$ mm

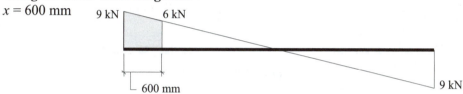

$M_x = $ shaded area $= + [0,5 \times (9,0 + 6,0) \times 0,6] = 4,5$ kNm

$x = 1800$ mm

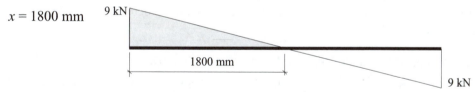

$M_x = $ shaded area $= + [0,5 \times 9,0 \times 1,8] = 8,1$ kNm

$x = 2100$ mm

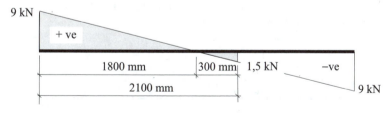

$M_x = $ shaded area $= + [8,1 - (0,5 \times 0,3 \times 1,5)] = 7,88$ kNm

The bending moment diagram is shown in Figure 4.35.

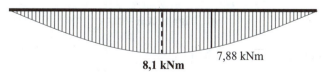

Bending Moment Diagram
Figure 4.35

The UDL loading is a 'standard' load case which occurs in numerous beam designs and can be expressed in general terms using L for the span and w for the applied load/metre or W_{total} $(= wL)$ for the total applied load, as shown in Figure 4.36.

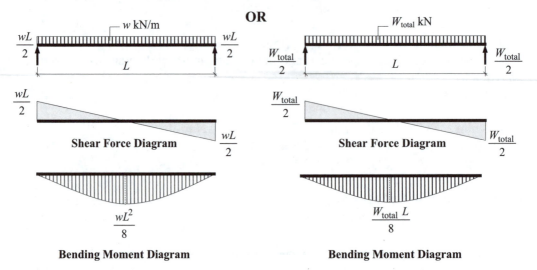

Figure 4.36

Clearly both give the same magnitude of support reactions, shear forces and bending moments.

In cantilever beams, all support restraints are provided at one location, i.e. an 'encastre' or 'fixed' support.

4.7.5 Example 4.5: Beam with Combined Point Loads and UDLs

A simply supported beam ABCD carries a uniformly distributed load of 3,0 kN/m between A and B, point loads of 4 kN and 6 kN at B and C respectively, and a uniformly distributed load of 5,0 kN/m between B and D, as shown in Figure 4.37. Determine the support reactions, sketch the shear force diagram, and determine the position and magnitude of the maximum bending moment.

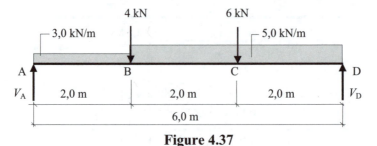

Figure 4.37

Consider the vertical equilibrium of the beam:

$$+ve \uparrow \Sigma F_y = 0$$

$$V_A - (3,0 \times 2,0) - 4,0 - 6,0 - (5,0 \times 4,0) + V_D = 0 \qquad \therefore \; V_A + V_D = 36,0 \text{ kN} \qquad \text{(i)}$$

Consider the rotational equilibrium of the beam:

$+ve \,\rangle\, \Sigma M_A = 0$

$(3,0 \times 2,0 \times 1,0) + (4,0 \times 2,0) + (6,0 \times 4,0) + (5,0 \times 4,0 \times 4,0) - (V_D \times 6,0) = 0$ (ii)

$\therefore V_D = 19,67$ kN

Substituting into equation (i) gives $\therefore V_A = 16,33$ kN

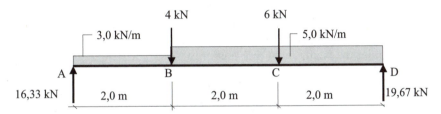

Shear force = algebraic sum of the vertical forces
Consider the shear force at a section 'x' from the left-hand end of the beam:

$x = 0$ $V_x = 16,33$ kN

At position x to the left of B before the 4,0 kN load
$\qquad V_x = + 16,33 - (3,0 \times 2,0) = + 10,33$ kN

At position x to the right of B after the 4,0 kN load
$\qquad V_x = + 10,33 - 4,0 \qquad = + 6,33$ kN

At position x to the left of C before the 6,0 kN load
$\qquad V_x = + 6,33 - (5,0 \times 2,0) \quad = - 3,67$ kN

At position x to the right of C after the 6,0 kN load
$\qquad V_x = - 3,67 - 6,0 \qquad = - 9,67$ kN

$x = 6,0$ m $V_x = - 9,67 - (5,0 \times 2,0) \quad = - 19,67$ kN

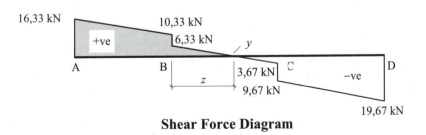

Shear Force Diagram

The maximum bending moment occurs at the position of zero shear, i.e. point y on the shear force diagram. The value of z can be determined from the shear force and applied loads:

$$z = \frac{6{,}33}{5{,}0} = 1{,}266 \text{ m} \qquad \text{(i.e. shear force/the value of the load per m length)}$$

Note: The slope of the shear force diagram between B and C is equal to the UDL of 5,0 kN/m.

Maximum bending moment M_y = shaded area of the shear force diagram
$$= [0{,}5 \times (16{,}33 + 10{,}33) \times 2{,}0] + [0{,}5 \times 1{,}266 \times 6{,}33]$$
$$= 30{,}67 \text{ kNm}$$

Alternatively, consider the beam cut at this section:

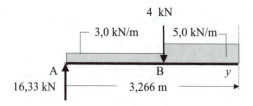

$$M_y = + (16{,}33 \times 3{,}266) - (3{,}0 \times 2{,}0 \times 2{,}266) - (4{,}0 \times 1{,}266) - [(5{,}0 \times 1{,}266) \times 0{,}633]$$
$$= + 30{,}67 \text{ kNm}$$

4.7.6 Secondary Bending

The top and bottom chords of trusses are normally continuous members. The points of application of the applied loads often do not coincide with the joint positions as shown in Figure 4.38.

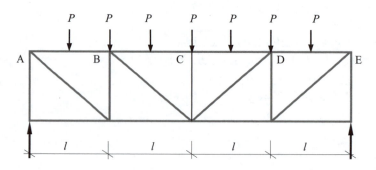

actual applied loading system

Figure 4.38

 In such circumstances the chords are subjected to combined axial and secondary bending effects. The magnitudes of the axial loads are determined using standard pin-jointed frame analysis assuming a simple static distribution of all the loads to adjacent nodes, as shown in Figure 4.39(a). The secondary bending moments can be determined assuming the chord to be a multi-span beam, as shown in Figure 4.39 (b).

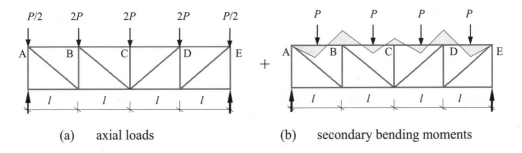

(a)   axial loads   (b)   secondary bending moments

Figure 4.39

In situations where the applied loads are due to numerous, closely spaced members or continuous decking, it is convenient to consider the chord to be supporting a uniformly distributed load of w kN/m as shown in Figure 4.40.

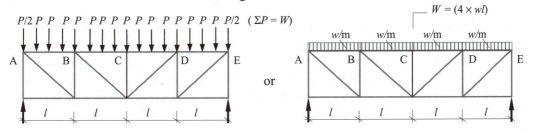

actual applied loading system

Figure 4.40

A similar analysis can be carried out assuming a multi-span beam with a distributed load to determine the secondary bending moments; the axial loads are evaluated as before, as indicated in Figure 4.41 (a) and (b).

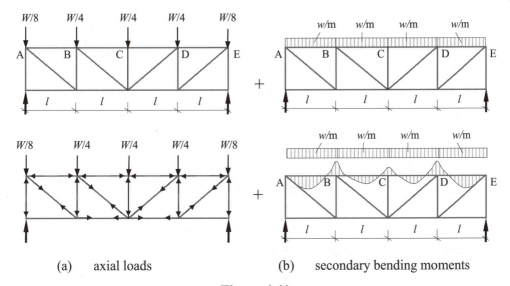

(a)   axial loads   (b)   secondary bending moments

Figure 4.41

4.8 *Frames with Sway*

One of the most important considerations in the design of any structure is that of overall lateral stability. There are numerous structural arrangements which can provide adequate lateral stability such as:

♦ rigid joints   (e.g. two-pinned and three-pinned portal frames),
♦ fixed bases   (frames with cantilever columns),
♦ braced frames   (frames with triangular bracing systems),
♦ membrane action within frames,

or combinations of the above.

The method adopted by a designer must be selected to accommodate the constraints imposed by materials and aesthetics and the client and site constraints for any particular structure.

Detailed consideration of lateral stability is not given in this text. A summary is given of an analysis technique which is sometimes used when designing frames subject to sway.

Consider the frame shown in Figure 4.42 in which a lattice girder is supported by simple connections to columns with pinned bases. The horizontal sway at the top of each column will be equal, irrespective of the column stiffnesses and relative horizontal loading. This deflection is imposed on the tops of the columns by the lattice girder, in which axial shortening of the members is negligible.

In column AB the horizontal deflection at B relative to the base at A is equal to the combined effects of a point load (H_1) and a UDL (W_1) acting on a cantilever.

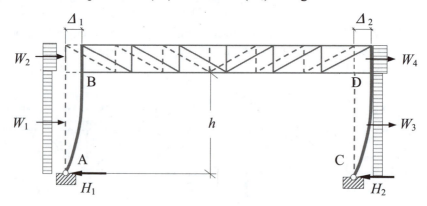

Figure 4.42

Relative deflection between A and B $\Delta_1 = \delta_1 - \delta_2$

$$\Delta_1 = \frac{H_1 h^3}{3EI} - \frac{W_1 h^3}{8EI}$$

Similarly for the relative deflection between C and D: $\Delta_2 = \dfrac{H_2 h^3}{3EI} - \dfrac{W_3 h^3}{8EI}$

The horizontal sway at the top of each column is equal, irrespective of the column stiffnesses and relative loadings.

$$\Delta_1 = \Delta_2$$

$$\frac{H_1 h^3}{3EI} - \frac{W_1 h^3}{8EI} = \frac{H_2 h^3}{3EI} - \frac{W_3 h^3}{8EI}$$

$$H_1 = \frac{3EI}{h^3}\left\{\frac{H_2 h^3}{3EI} - \frac{W_3 h^3}{8EI} + \frac{W_1 h^3}{8EI}\right\}$$

$$H_1 = (H_2 - 0{,}375 W_3 + 0{,}375 W_1) \qquad\qquad \text{Equation (1)}$$

In addition, consider the horizontal equilibrium of the entire frame:

+ve $\longrightarrow \Sigma F_x = 0$ $\qquad\qquad W_1 + W_2 - H_1 + W_3 + W_4 - H_2 = 0 \qquad$ Equation (2)

$$H_2 = (W_1 + W_2 + W_3 + W_4) - H_1$$

Substitute in equation (1)

$$H_1 = (W_1 + W_2 + W_3 + W_4 - H_1 - 0{,}375 W_3 + 0{,}375 W_1)$$
$$2H_1 = (1{,}375 W_1 + W_2 + 0{,}625 W_3 + W_4)$$
$$\mathbf{H_1 = (0{,}688 W_1 + 0{,}5 W_2 + 0{,}313 W_3 + 0{,}5 W_4)}$$
$$\mathbf{H_2 = (0{,}312 W_1 + 0{,}5 W_2 + 0{,}687 W_3 + 0{,}5 W_4)}$$

A similar analysis can be carried out for frames in which the bases are fixed. In this case a *point of contraflexure* (i.e. a point of zero moment) is assumed in each of the columns as shown in Figure 4.43.

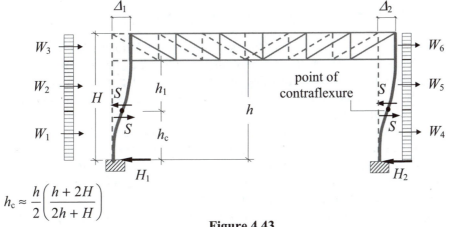

$$h_c \approx \frac{h}{2}\left(\frac{h + 2H}{2h + H}\right)$$

Figure 4.43

h_c is the assumed height of the point of contraflexure, above the fixed base. This assumption is sufficiently accurate for design purposes, even when the loading on each column is not equal.

The magnitude of the horizontal shear force (S) is equal to the sum of the horizontal loadings above the points of contraflexure, distributed between the columns in proportion to their stiffnesses. In most instances both columns are the same, resulting in the following relationships:

$$S = \frac{W_2 + W_3 + W_5 + W_6}{2}$$

$$\Delta_1 = \frac{1}{EI}\left\{\frac{Sh_1^3}{3} - \frac{W_2 h_1^3}{8} + \frac{Sh_c^3}{3} + \frac{W_1 h_c^3}{8}\right\}$$

$$\Delta_2 = \frac{1}{EI}\left\{\frac{Sh_1^3}{3} - \frac{W_5 h_1^3}{8} + \frac{Sh_c^3}{3} + \frac{W_4 h_c^3}{8}\right\}$$

As before:

$$\Delta_1 = \Delta_2 \quad \text{and} \quad \boldsymbol{H_1 = S + W_1} \quad \boldsymbol{H_2 = S + W_4}$$

The use of this analysis is illustrated in Example 7.2 in Chapter 7.

4.9 *Deflection of Beams*

Structural design encompasses a wide range of considerations in addition to strength criteria as evaluated using design stresses. One of the most important of these is the stiffness of a structure or structural element. The stiffness is reflected in the deformations and deflections induced by the applied load system. There are large variations in what are considered by practising engineers to be acceptable deflections for different circumstances, e.g. limitations on the deflections of beams are necessary to avoid consequences such as:

♦ damage to finishes, e.g. to brittle plaster or ceiling tiles,
♦ unnecessary alarm to occupants of a building,
♦ misalignment of door frames causing difficulty in opening.

If situations arise in which a designer considers the recommendations given in the design codes are too lenient or too severe (e.g. conflicting with the specification of suppliers or manufacturers), then individual engineering judgement must be used. There are well established analytical methods for calculating theoretical deflections, in most cases computer analysis is adopted.

In a simply supported beam, the maximum deflection induced by the applied loading always approximates the mid-span value if it is not equal to it. A number of standard, frequently used load cases for which the elastic deformation is required are given in Table 4.7 of this text. In the case indicated with '*' the actual maximum deflection will be approximately equal to the value given (i.e. within 2.5%).

In the design of timber beams the 'shear deformation' may be significant, particularly in the case of thin-webbed beams and box-beams, and should be considered. This is dealt with in Section 5.3.6 of Chapter 5 and in Appendix B.

Load Case	Deflection	Load Case	Deflection
W_{total} UDL over span L, simply supported	$\dfrac{5WL^3}{384EI}$	W_{total} UDL over span L, fixed ends	$\dfrac{WL^3}{384EI}$
W_{total} partial UDL (a, b, a), span L	$\dfrac{W}{384EI}\alpha_1$	P at midspan $L/2$, fixed ends	$\dfrac{PL^3}{192EI}$
P at midspan $L/2$, simply supported	$\dfrac{PL^3}{48EI}$	P at a, b ($b > a$), fixed ends	$\dfrac{2\,Pa^2b^3}{3\,EI\beta}$
* P at a, b ($b > a$), simply supported	$\approx \dfrac{PL^3}{48EI}\alpha_2$	** W_{total} UDL over a, with b cantilever	$-\dfrac{Wa^2b}{24EI}$
W_{total} cantilever UDL over a, b	$\dfrac{Wa^3}{8EI}\alpha_3$	W_{total} partial UDL on b region	$\dfrac{Wb^3}{8EI}+\dfrac{Wab^2}{6EI}$
P cantilever at a, b	$\dfrac{Pa^3}{3EI}\alpha_4$	P at end of overhang b	$\dfrac{Wb^3}{3EI}+\dfrac{Wab^2}{3EI}$

$$\alpha_1 = (8L^3 - 4Lb^2 - b^3) \qquad \alpha_2 = \left[\dfrac{3a}{L} - 4\left(\dfrac{a}{L}\right)^3\right] \qquad \alpha_3 = \left(1+\dfrac{4b}{3a}\right) \qquad \alpha_4 = \left(1+\dfrac{3b}{2a}\right)$$

$\beta = (3L - 2a)^2$ * value is within 2,5% of the maximum value. ** upwards at the end of the cantilever.

Table 4.7

4.10 Equivalent UDL Technique for the Deflection of Beams

In many cases beams support complex load arrangements which do not lend themselves to either an individual load case or a combination of the load cases given in Table 4.7. Provided that deflection is not the governing design criterion, a calculation which gives an approximate answer is usually adequate. The equivalent UDL method is a useful tool for estimating the deflection in a simply supported beam with a complex loading.

In a simply supported beam, the maximum deflection induced by the applied loading always approximates the mid-span value if it is not equal to it.

Consider a single-span, simply supported beam carrying a *non-uniform* loading which induces a maximum bending moment of *M* as shown in Figure 4.44.

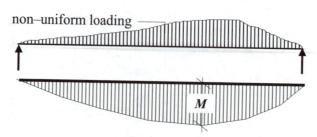

non–uniform loading

M

Bending Moment Diagram

Figure 4.44

The equivalent UDL (w_e) which would induce the same *magnitude* of maximum bending moment (**Note:** the position may be different) on a simply supported span carrying a *uniform* loading can be determined from:

Maximum bending moment
$$M = \frac{w_e L^2}{8}$$

$$\therefore \; w_e = \frac{8M}{L^2}$$

where w_e is the equivalent uniform distributed load.

The maximum deflection of the beam carrying the uniform loading will occur at the mid-span and will be equal to
$$\delta = \frac{5 w_e L^4}{384 EI}$$

Using this expression, the maximum deflection of the beam carrying the non-uniform loading can be estimated by substituting for the w_e term, i.e.

$$\delta \approx \frac{5 w_e L^4}{384 E I} = \frac{5 \times \left(\frac{8M}{L^2}\right) L^4}{384 \, EI} = \frac{0,104 \, M \, L^2}{EI}$$

The maximum bending moment in Example 4.5 is 30,67 kNm.

Using the equivalent UDL method to estimate the maximum deflection gives:

Example 4.5 $\delta_{maximum} \approx \dfrac{0,104 \, M \, L^2}{EI} = -\dfrac{114,9}{EI}$ m   (actual value $= \dfrac{114,4}{EI}$ m)

Note: The estimated deflection is more accurate for beams which are predominantly loaded with distributed loads. In addition this only includes bending effects and ignores the deformation due to shear effects. In the case of timber beams the shear deformation may be significant and should also be considered. This is dealt with in Section 5.3.6 of Chapter 5 and in Appendix B.

4.11 *Elastic Shear Stress Distribution*

The shear forces induced in a beam by an applied load system generate shear stresses in both the horizontal and vertical directions. At any point in an elastic body, the shear stresses in two mutually perpendicular directions are equal to each other in magnitude.

Consider an element of material subject to shear stresses along the edges, as shown in Figure 4.45.

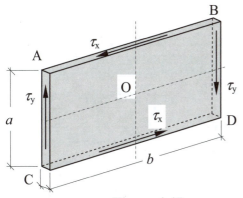

Force on surface AB $= F_{AB} = \tau_x \times bt$
Force on surface CD $= F_{CD} = \tau_x \times bt$
Force on surface AC $= F_{AC} = \tau_y \times at$
Force on surface BD $= F_{BD} = \tau_y \times at$

+ve $\uparrow \Sigma F_y = 0$

$$F_{AC} = F_{BD}$$

+ve $\longrightarrow \Sigma F_x = 0$

Figure 4.45

$$F_{AB} = F_{CD}$$

+ve $\curvearrowright \Sigma M_O = 0$

$$- (F_{AB} \times \frac{a}{2}) - (F_{CD} \times \frac{a}{2}) + (F_{AC} \times \frac{b}{2}) + (F_{BD} \times \frac{b}{2}) = 0$$

Substitute for F_{CD} and F_{BD}:
$$- (F_{AB} \times a) + (F_{AC} \times b) = 0 \qquad \therefore \ F_{AB}\,a = F_{AC}\,b$$
$$(\tau_x bt)\,a = (\tau_y at)\,b$$
$$\tau_x = \tau_y$$

The two shear stresses are equal and complementary. The magnitude of the shear stress at any vertical cross-section on a beam can be determined using:

$$\tau = \frac{V A \overline{z}}{I b}$$

where:
V the vertical shear force at the section being considered,
A the area of the cross-section above (or below) the 'horizontal' plane being considered (note that the shear stress varies throughout the depth of a cross-section for any given value of shear force),

$\bar{z}$ the distance from the elastic neutral axis to the centroid of the area A,
b the breadth of the beam at the level of the horizontal plane being considered,
I the second moment of area of the full cross-section about the elastic neutral axis.

The intensity of shear stress throughout the depth of a section is not uniform and is a maximum at the level of the neutral axis.

4.11.1 Example 4.6: Shear Stress Distribution in a Rectangular Beam

The rectangular beam shown in Figure 4.46 is subject to a vertical shear force of 3,0 kN. Determine the shear stress distribution throughout the depth of the section.

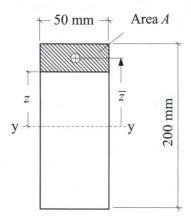

The shear stress at any horizontal level a distance z from the neutral axis is given by:

$$\tau = \frac{VA\bar{z}}{Ib}$$

V = shear force = 3,0 kN

$$I = \frac{bd^3}{12} = \frac{50 \times 200^3}{12} = 33,33 \times 10^6 \text{ mm}^4$$

$b = 50$ mm   (for all values of z)

Figure 4.46

Consider the shear stress at a number of values of y,

$z = 100$ mm $A\bar{z} = 0$ (since $A = 0$) $\tau_{100} = 0$

$z = 75$ mm

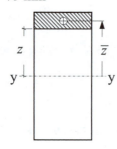

$A = 50 \times 25 = 1250 \text{ mm}^2$
$\bar{z} = 75 + 12,5 = 87,5$ mm
$A\bar{z} = 109,375 \times 10^3 \text{ mm}^3$

$$\tau_{75} = \frac{3 \times 10^3 \times 109,375 \times 10^3}{33,33 \times 10^6 \times 50} = 0,20 \text{ N/mm}^2$$

$z = 50$ mm

$A = 50 \times 50 = 2500 \text{ mm}^2$
$\bar{z} = 50 + 25 = 75$ mm
$A\bar{z} = 187,5 \times 10^3 \text{ mm}^3$

$$\tau_{50} = \frac{3 \times 10^3 \times 187,5 \times 10^3}{33,33 \times 10^6 \times 50} = 0,34 \text{ N/mm}^2$$

$z = 25$ mm

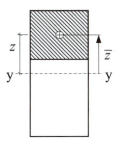

$A = 50 \times 75 = 3750$ mm^2
$\bar{z} = 25 + 37{,}5 = 62{,}5$ mm
$A\bar{z} = 234{,}375 \times 10^3$ mm^3

$$\tau_{25} = \frac{3 \times 10^3 \times 234{,}375 \times 10^3}{33{,}33 \times 10^6 \times 50} = 0{,}42 \text{ N/mm}^2$$

$z = 0$ mm

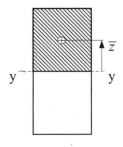

$A = 50 \times 100 = 5000$ mm^2
$\bar{z} = 50$ mm
$A\bar{z} = 250 \times 10^3$ mm^3

$$\tau_0 = \frac{3 \times 10^3 \times 250 \times 10^3}{33{,}33 \times 10^6 \times 50} = 0{,}45 \text{ N/mm}^2$$

$z = -25$ mm

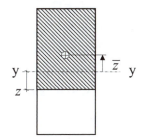

$A = 50 \times 125 = 6250$ mm^2
$\bar{z} = 37{,}5$ mm
$A\bar{z} = 234{,}375 \times 10^3$ mm^3

$$\tau_{75} = \frac{3 \times 10^3 \times 23375 \times 10^3}{33{,}33 \times 10^6 \times 50} = 0{,}42 \text{ N/mm}^2$$

This is the same value as for $z = +25$ mm

The cross-section (and hence the stress diagram) is symmetrical about the elastic neutral axis, as shown in Figure 4.47.

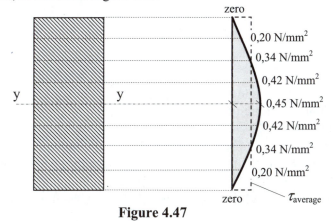

Shear Stress Distribution

Figure 4.47

The maximum value occurs at the same level as the elastic neutral axis. The 'average' shear stress for a cross-section is equal to the applied force distributed uniformly over the entire cross-section, i.e.

$$\tau_{average} = \frac{\text{Force}}{\text{Area}} = \frac{V}{A} = \frac{3,0\times10^3}{50\times200} = 0,3 \text{ N/mm}^2$$

For a *rectangular* section:

$$\tau_{maximum} = 1,5 \times \tau_{average} = \frac{1,5V}{A} = 1,5 \times 0,3 = 0,45 \text{ N/mm}^2$$

4.12 *Elastic Bending Stress Distribution*

The bending moments induced in a beam by an applied load system generate bending stresses in the material fibres which vary from a maximum in the extreme fibres to a minimum at the level of the neutral axis as shown in Figure 4.48.

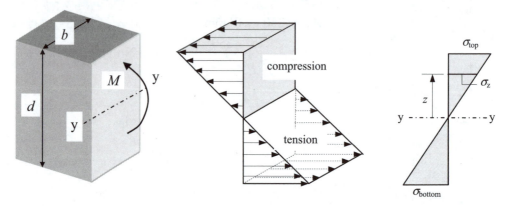

Bending Stress Distribution   **Stress Diagram**

Figure 4.48

The magnitude of the bending stresses at any vertical cross-section can be determined using the simple theory of bending from which the following equation is derived:

$$\frac{M}{I} = \frac{E}{R} = \frac{\sigma}{z} \qquad \therefore \ \sigma = \frac{Mz}{I}$$

where:
M the applied bending moment at the section being considered,
E the value of Young's modulus of elasticity,
R the radius of curvature of the beam,
σ the bending stress,
z the distance measured from the elastic neutral axis to the level on the cross-section at which the stress is being evaluated,
I the second moment of area of the full cross-section about the elastic neutral axis.

It is evident from the Equation that for any specified cross-section in a beam subject to a known value of bending moment (i.e. M and I constant), the bending stress is directly proportional to the distance from the neutral axis, i.e.

$$\sigma = \text{constant} \times z \qquad \therefore \qquad \sigma \, \alpha \, z$$

This is shown in Figure 4.48, in which the maximum bending stress occurs at the extreme fibres, i.e. $z_{max} = d/2$.

In design it is usually the extreme fibre stresses relating to the z_{max} values at the top and bottom which are critical. These can be determined using:

$$\sigma_{top} = \frac{M}{W_{top}} \qquad \text{and} \qquad \sigma_{bottom} = \frac{M}{W_{bottom}}$$

where:

σ and M are as before,

W_{top} is the elastic section modulus relating to the top fibres and defined as $\dfrac{I_{yy}}{z_{top}}$

W_{bottom} is the elastic section modulus relating to the top fibres and defined as $\dfrac{I_{yy}}{z_{bottom}}$

If a cross-section is symmetrical about the y–y axis then $W_{top} = W_{bottom}$. In asymmetric sections the maximum stress occurs in the fibres corresponding to the smallest W value. For a rectangular cross-section of breadth b and depth d subject to a bending moment M about the major y–y axis, the appropriate values of I, z and W are:

$$I_{yy} = \frac{bd^3}{12} \qquad z_{max} = \frac{d}{2} \qquad W_{y,min} = \frac{bd^2}{6}$$

In the case of bending about the minor z–z axis:

$$I_{zz} = \frac{db^3}{12} \qquad y_{max} = \frac{b}{2} \qquad W_{z,\,min} = \frac{db^2}{6}$$

The maximum stress induced in a cross-section subject to bi-axial bending is given by:

$$\sigma_{max} = \frac{M_y}{W_{y,\,min}} + \frac{M_z}{W_{z,\,min}}$$

where M_y and M_z are the applied bending moments about the y and z axes respectively.

4.12.1   Example 4.7: Bending Stress Distribution in a Rectangular Beam

The rectangular beam shown in Figure 4.49 is subject to a bending moment of 2,0 kNm. Determine the bending stress distribution throughout the depth of the section.

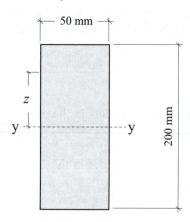

The bending stress at any horizontal level a distance z from the neutral axis is given by:

$$\sigma = \frac{Mz}{I_{yy}}$$

M = bending moment = 2,0 kNm

$$I_{yy} = \frac{bd^3}{12} = \frac{50 \times 200^3}{12} = 33,33 \times 10^6 \text{ mm}^4$$

Figure 4.49

Consider the bending stress at a number of values of z,

$z = 100$ mm $\qquad \sigma_{100} = \dfrac{2,0 \times 10^6 \times 100}{33,33 \times 10^6} = 6,0 \text{ N/mm}^2$

$z = 75$ mm $\qquad \sigma_{75} = \dfrac{2,0 \times 10^6 \times 75}{33,33 \times 10^6} = 4,5 \text{ N/mm}^2$

$z = 50$ mm $\qquad \sigma_{50} = \dfrac{2,0 \times 10^6 \times 50}{33,33 \times 10^6} = 3,0 \text{ N/mm}^2$

$z = 25$ mm $\qquad \sigma_{25} = \dfrac{2,0 \times 10^6 \times 25}{33,33 \times 10^6} = 1,5 \text{ N/mm}^2$

$z = 0$ $\qquad \sigma_0 = 0$

$z = -25$ mm $\qquad \sigma_{-25} = -\dfrac{2,0 \times 10^6 \times 25}{33,33 \times 10^6} = -1,5 \text{ N/mm}^2$

The cross-section (and hence the stress diagram) is symmetrical about the elastic neutral axis, as shown in Figure 4.50.

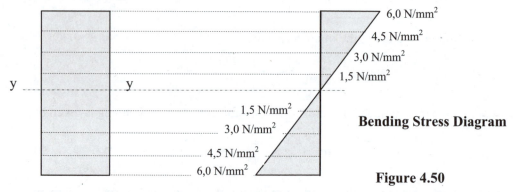

Bending Stress Diagram

Figure 4.50

4.13 Transformed Sections

Beams such as thin-webbed **I** and box-beams in timber are generally fabricated from different materials, e.g. plywood webs and softwood flanges fastened together. During bending, the stresses induced in such sections are shared among all the component parts.

The extent to which sharing occurs is dependent on the method of connection at the interfaces. This connection is normally designed such that *no slip* occurs between the different materials during bending. The resulting structural element is a composite section which is non-homogeneous. (**Note:** *This invalidates the simple theory of bending in which homogeneity is assumed.*)

A useful technique often used when analysing such composite sections is to obtain '*effective section properties*' using the *transformed section* method. When using this method, an equivalent homogeneous section is considered in which all components are assumed to be the same material. The simple theory of bending is then used to determine the stresses in the transformed sections, which are subsequently modified to determine the stresses in the *actual* materials.

Consider the composite section shown in Figure 4.51(a) in which a steel plate has been securely fastened to the underside face. There are two possible transformed sections which can be considered:

(i)   an equivalent section in terms of timber, Figure 4.51(b) or
(ii) an equivalent section in terms of steel, Figure 4.51(c).

To obtain an equivalent section made from timber, the same material as the existing timber must replace the steel plate. The dimension of the replacement timber must be modified to reflect the different material properties. The equivalent *transformed* section properties are shown in Figure 4.51(b).

The overall depth of both sections is the same $(d + t)$. The *strain* in element δA_s in the original section is equal to the strain in element δA_t^* of the transformed section:

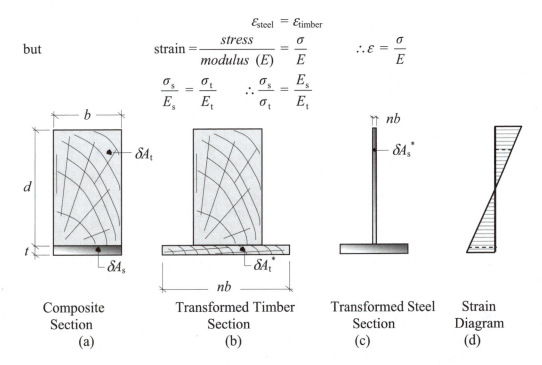

$$\varepsilon_{steel} = \varepsilon_{timber}$$

but $\qquad$ $strain = \dfrac{stress}{modulus \ (E)} = \dfrac{\sigma}{E}$ $\qquad \therefore \varepsilon = \dfrac{\sigma}{E}$

$$\dfrac{\sigma_s}{E_s} = \dfrac{\sigma_t}{E_t} \qquad \therefore \dfrac{\sigma_s}{\sigma_t} = \dfrac{E_s}{E_t}$$

| Composite
Section
(a) | Transformed Timber
Section
(b) | Transformed Steel
Section
(c) | Strain
Diagram
(d) |

Figure 4.51

The force in each element must also be equal:

$$\text{Force} = (\text{stress} \times \text{area}) \qquad P_s = P_t$$
$$\sigma_s \delta A_s = \sigma_t \delta A_t^*$$
$$\delta A_t^* = \frac{\sigma_s}{\sigma_t} \delta A_s = \frac{E_s}{E_t} \delta A_s$$

This indicates that in the transformed section:

Equivalent area of transformed timber = $(n \times \text{original area of steel})$

where: n is the '*modular ratio*' of the materials and is equal to $\dfrac{E_s}{E_t}$

The equivalent area of timber must be subject to the same value of strain as the original material it is replacing, i.e. it is positioned at the same distance from the elastic neutral axis. The simple elastic equation of bending (see Section 4.11) can be used with the transformed section properties to determine the bending stresses. The *actual* stresses in the steel will be equal to ($n \times$ equivalent timber stresses.) The use of this method is illustrated in Example 4.8. A similar, alternative analysis can be carried out using a transformed steel section, as shown in Figure 4.53.

4.13.1 Example 4.8: Composite Timber/Steel Section

A timber beam is enhanced by the addition of two steel plates as shown in Figure 4.52. Determine the maximum timber and steel stresses induced in the cross-section when the beam is subjected to a bending moment of 70 kNm.

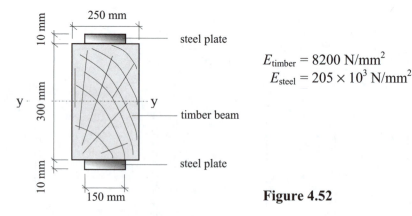

$$E_{timber} = 8200 \text{ N/mm}^2$$
$$E_{steel} = 205 \times 10^3 \text{ N/mm}^2$$

Figure 4.52

(a)   Transformed section based on *timber*
Equivalent width of timber to replace the steel plate = $(n \times 150)$ mm
where:

$$n = \frac{E_{steel}}{E_{timber}} = \frac{205 \times 10^3}{8200} = 25 \qquad\qquad nb = (25 \times 150) = 3750 \text{ mm}$$

The maximum stresses occur in the timber when $z = 150$ mm, and in the steel (or equivalent replacement timber) when $z = 160$ mm.

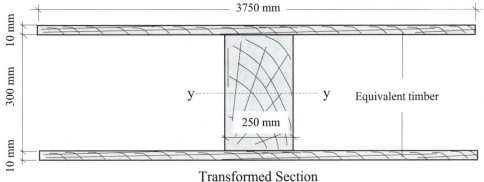

Transformed Section

Figure 4.53

$$I_{yy \text{ transformed}} = \left\{ \frac{3750 \times 320^3}{12} - \frac{3500 \times 300^3}{12} \right\} = 2,365 \times 10^9 \text{ mm}^4$$

Maximum bending stress in the *timber* is given by:

$$\sigma_{\text{timber}} = \frac{Mz_{150}}{I} = \frac{70 \times 10^6 \times 150}{2,365 \times 10^9} = 4,44 \text{ N/mm}^2$$

Maximum bending stress in the *equivalent timber* is given by:

$$\sigma = \frac{Mz_{160}}{I} = \frac{70 \times 10^6 \times 160}{2,365 \times 10^9} = 4,74 \text{ N/mm}^2$$

This value of stress represents a maximum value of stress in the steel plates given by:

$$\sigma_{\text{steel}} = n \times \sigma = (25 \times 4,74) = 118,5 \text{ N/mm}^2$$

(b)   Transformed section based on *steel*

Equivalent width of steel to replace the timber beam $= (n \times 150)$ mm

where:

$$n = \frac{E_{\text{timber}}}{E_{\text{steel}}} = \frac{1}{25} \qquad\qquad nb = \frac{1 \times 250}{25} = 10 \text{ mm}$$

The maximum stresses occur in the timber (or equivalent replacement steel) when $z = 150$ mm, and in the steel when $z = 160$ mm.

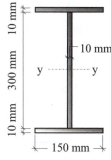

Figure 4.54

$$I_{yy,\text{transformed}} = \left\{ \frac{150 \times 320^3}{12} - \frac{140 \times 300^3}{12} \right\}$$
$$= 94,6 \times 10^6 \text{ mm}^4$$

Maximum bending stress in the *steel* is given by:

$$\sigma_{\text{steel}} = \frac{Mz_{160}}{I} = \frac{70 \times 10^6 \times 160}{94,6 \times 10^6} = 118,39 \text{ N/mm}^2$$

Maximum bending stress in the *equivalent steel* is given by:

$$\sigma = \frac{Mz_{150}}{I} = \frac{70 \times 10^6 \times 150}{94,6 \times 10^6} = 110,99 \text{ N/mm}^2$$

This value of stress represents a maximum value of stress in the timber given by:

$$\sigma_{\text{timber}} = n \times \sigma = \left(\frac{1}{25} \times 110,99 \right) = 4,44 \text{ N/mm}^2$$

The simplified method outlined in Section 4.12 assumes full composite action between the flanges and webs, e.g. as in glued construction. No allowance has been made for creep effects which can be significant in timber. The effective section properties should be modified when designing structural timber members.

    The equivalent section determined in terms of the flange material with the web stresses being modified accordingly resulting in the following expression for the effective cross-sectional area A_{ef} and second moment of area I_{ef}:

$$A_{\text{ef}} = A_{\text{flange}} + \left(\frac{E_{\text{web}}}{E_{\text{flange}}} \right) \left(\frac{1 + k_{\text{def,flange}}}{1 + k_{\text{def,web}}} \right) A_{\text{web}} \qquad I_{\text{ef}} = I_{\text{flange}} + \left(\frac{E_{\text{web}}}{E_{\text{flange}}} \right) \left(\frac{1 + k_{\text{def,flange}}}{1 + k_{\text{def,web}}} \right) I_{\text{web}}$$

where:

A_{flange} is the cross-sectional area of the flange(s),
I_{flange} is the second moment of area of the flange(s),
E_{flange} is the mean modulus of elasticity of the flange(s),
$k_{\text{def,flange}}$ is the deformation factor relating to the flange material,
A_{web} is the cross-sectional area of the web(s),
I_{web} is the second moment of area of the web(s),
E_{web} is the mean modulus of elasticity of the web(s),
$k_{\text{def,web}}$ is the deformation factor relating to the web material.

In the case of members comprising mechanically jointed elements allowance must also be made for the slip which occurs between the flanges and webs due to the horizontal shear forces induced by bending. This explained and included in Section 5.2 and Example 5.5 of Chapter 5 in relation to the requirements of Annex B (*Mechanically jointed beams*) of the code.

5. Flexural Members

> **Objective:** *to illustrate the design of flexural members considering solid, composite I and composite box - beams, thin-flanged beams and glued laminated beams.*

5.1 Introduction

Beams are the most commonly used structural elements, for example as floor joists, trimmer joists around openings, rafters, etc. The cross-section of a timber beam may be one of a number of frequently used sections such as those indicated in Figure 5.1.

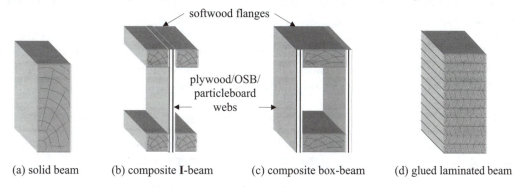

(a) solid beam   (b) composite **I**-beam   (c) composite box-beam   (d) glued laminated beam

Figure 5.1

The principal considerations in the design of all beams and floor systems comprise both ultimate and serviceability limit states as follows:

Ultimate Limit States:
- ◆ bending,   (Clause 6.1.6)
- ◆ shear,   (Clause 6.1.7)
- ◆ bearing,   (Clause 6.1.5)
- ◆ lateral torsional stability.   (Clause 6.3.3)

Serviceability Limit States:
- ◆ deflection,   (Clause 7.2)
- ◆ vibration.   (Clause 7.3)

Provision is also made for checking the ultimate limit state relating to torsion in Clause 6.1.8 if required.

In the case of glued thin-webbed beams such as the composite I and box beams as shown in Figures 5.1(b) and (c), the additional phenomenon of rolling shear, (which is specific to plywood and other wood-based materials), must also be considered.

The size of timber beams may be governed by the requirements of:

  ♦ the section modulus (W) necessary to limit the bending stresses and ensure that
    neither lateral torsional buckling of the compression flange nor fracture of the
    tension flange induce failure,
  ♦ the cross-sectional area necessary to ensure that the shear stresses parallel or
    perpendicular to the grain do not induce failure,
  ♦ the second moment of area necessary to limit the deflection induced by bending
    and/or shear action to acceptable levels,
  ♦ in the case of floor systems, the equivalent bending stiffness necessary to limit
    vibration to acceptable levels.

Generally, the bearing area actually provided at the ends of a beam is much larger than is
necessary to satisfy the bearing stress requirement. Whilst lateral stability should be
checked it is frequently provided to the compression flange of a beam by nailing of floor
boards, roof decking, etc. Similarly the proportions of solid timber beams are usually such
that lateral instability is unlikely.

   The detailed design of solid beams, glued thin-webbed beams, thin-flanged beams and
glued laminated beams is explained and illustrated in the examples given in Sections 5.4,
5.5 and 5.6 respectively. In each case the relevant modification factors (k values), their
application and value/location are summarized in each section.

5.2 Span

Most timber beams are designed as simply supported and the span is normally assumed to
be the distance measured between the centres of the bearings as shown in Figure 5.2.

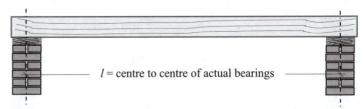

Figure 5.2

l = centre to centre of actual bearings

5.3 Solid Rectangular Beams

The modification factors, which are pertinent when designing solid timber beams, are
summarized in Table 5.1.

5.3.1 Bending (Clause 6.1.6 – Equations (6.11) and (6.12))

As indicated in Section 4.12 of Chapter 4, the applied bending stress is determined using
simple elastic bending theory:

$$\sigma_{m,d} = \frac{M_d}{W}$$

where:
$\sigma_{m,d}$ is the design bending stress parallel to the grain,
M_d is the design bending moment,
W is the elastic section modulus about the axis of bending, (usually the y-y axis).

Factors	Application	Clause Number	Value/Location
k_{mod}	Relates to all strength properties for service and load-duration classes.	3.1.3	Table 3.1
k_{def}	Relates to creep deformation.	2.3.2.2(4)	Table 3.2
k_h	Relates to the bending strength/tension strength parallel to the grain and to member depth/width.	3.2(3) 3.3(3) 3.4(3)	Equation (3.1) Equation (3.2) Equation (3.3)
$k_{c,90}$	Relates to compression strength perpendicular to the grain.	6.1.5	Equations (6.4) to (6.6) and (6.10)
k_m	Relates to bending strengths in a cross-section subject to biaxial bending.	6.1.6(2)	6.1.6(2)
k_{crit}	Relates to the reduction in the bending strength due to lateral torsional buckling.	6.3.3(3)	Equation (6.34)
$k_{m,\alpha}$	Relates to the bending strength of single-tapered beams.	6.4.2(2)	Equations (6.39) and (6.40)
k_l	Relates to the apex bending stress in double-tapered, curved and pitch cambered beams.	6.4.3(4)	Equation (6.43)
k_r	Relates to the bending strength of double-tapered, curved and pitch cambered beams.	6.4.3(5)	Equation (6.49)
k_{vol}	Relates to the tensile strength perpendicular to the grain in tapered, curved and cambered beams.	6.4.3(6)	Equation (6.51)
k_{dis}	Relates to the tensile strength perpendicular to the grain in tapered, curved and cambered beams.	6.4.3(6)	Equation (6.52)
k_p	Relates to the design apex tensile stress perpendicular to the grain in double-tapered, curved and pitch cambered beams.	6.4.3(8)	Equation (6.56)
k_v	Relates to the shear strength in notched beams.	6.5.2(2)	Equations (6.61) and (6.62)
k_n	Relates to the k_v factor in notched beams.	6.5.2(2)	Equation (6.63)
k_{sys}	Relates to lateral load distribution in a continuous deck or floor system.	6.6(2)	Generally 1,1 Figure 6.12 for laminated decks

Table 5.1   Modification Factors - solid beams

The governing Equations (6.11) and (6.12) given in EC5, relate to bi-axial bending about the major y-y axis and the minor z-z axis and are as follows:

$$\frac{\sigma_{m,y,d}}{f_{m,y,d}} + k_m \frac{\sigma_{m,z,d}}{f_{m,z,d}} \leq 1 \qquad \text{(Equation (6.11) in EC5)}$$

and

$$k_m \frac{\sigma_{m,y,d}}{f_{m,y,d}} + \frac{\sigma_{m,z,d}}{f_{m,z,d}} \leq 1 \qquad \text{(Equation (6.12) in EC5)}$$

where:

$\sigma_{m,y,d}$ is the design bending stress about the y-y axis,

$\sigma_{m,z,d}$ is the design bending stress about the z-z axis,

$f_{m,y,d}$ is the design bending strength about the y-y axis,
$f_{m,z,d}$ is the design bending strength about the z-z axis,
k_m is a factor which allows for the redistribution of stresses and inhomogeneities of the material in the cross-section.

In the case of uniaxial bending about the major y-y axis, Equation (6.12) is not required and Equation (6.11) reduces to:

$$\sigma_{m,y,d} \leq f_{m,y,d}$$

The design bending strength ($f_{m,y,d}$) is dependent on the strength class of the timber i.e. the characteristic bending strength ($f_{m,y,k}$) and the relevant modification factors as follows:

$$f_{m,y,d} = (f_{m,y,k} \times k_{mod} \times k_h \times k_{m,\alpha} \times k_{sys})/\gamma_M$$

where the k values are as indicated in Table 5.1, γ_M the partial material factor is given in Table NA.3 of the UK National Annex.

5.3.2 Shear (Clause 6.1.7 – Equation (6.13))

The characteristic shear strength ($f_{v,k}$) given in EN 338:2003 relates to the maximum shear stress parallel to the grain for each given strength class. As indicated in Section 4.11 of Chapter 4, in solid beams of rectangular cross-section the maximum horizontal shear stress occurs at the level of the neutral axis and is equal to $1.5 \times$ the average value:

$$\tau_d \leq f_{v,d} \qquad \text{(Equation (6.13) in EC5)}$$

where:

τ_d is the design shear stress $= \dfrac{1{,}5\,V_d}{bh}$

V_d is the design vertical shear force,
b is the width of the beam,
h is the depth of the beam,
$f_{v,d}$ is the design shear strength.

The value of $f_{v,d}$ is given by $f_{v,d} = (f_{v,k} \times k_{mod} \times k_v \times k_{sys})/\gamma_M$

where the k values are as indicated in Table 5.1 and γ_M is as before.

The case of beams with a notch at the support is given in Clause 6.5 of EC5, (see Section 5.3.5).

5.3.3 Bearing (Clause 6.1.5 – Equation (6.3))

The behaviour of timber under the action of concentrated loads, e.g. at positions of support, is complex and influenced by both the length and location of the bearing.
The design compressive strength perpendicular to the grain ($f_{c,90,d}$) is used to determine the suitability of the bearing strength as follows:

$$\sigma_{c,90,d} \leq (k_{c,90} \times f_{c,90,d}) \qquad \text{(Equation (6.3) in EC5)}$$

where:
$\sigma_{c,90,d}$ is the design compressive stress on the contact area perpendicular to the grain,
$f_{c,90,d}$ is the design compressive strength perpendicular to the grain,
$k_{c,90}$ is a factor which takes into account the load configuration, the possibility of splitting and the degree of compression deformation.

The design bearing stress is determined from: $\sigma_{c,90,d} = \dfrac{F_{c,d,90}}{(b \times l)}$

where:
$F_{c,90,d}$ is the design bearing force,
b is the breadth of the contact area,
l is the length of the contact area.

The value of the bearing strength is given by $f_{c,90,d} = (f_{c,90,d} \times k_{mod} \times k_{sys})/\gamma_M$

As indicated in Clause 6.1.5(2) the value of $k_{c,90}$ is equal to 1,0 unless the member arrangements satisfy the conditions illustrated in Figures 6.2, 6.3 or 6.4 in EC5, e.g. in the case of a beam resting on a discrete support as shown in Figure 5.3, the value of $k_{c,90}$ should be calculated from Equation (6.4) in EC5 as follows:

$$k_{c,90} = \left(2,38 - \frac{l}{250}\right)\left(1 + \frac{h}{12l}\right) \leq 4,0$$

provided that the end distance 'a' $\leq h/3$.

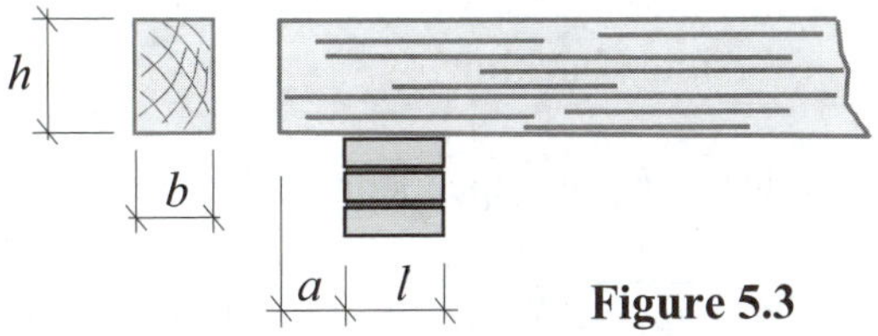

Figure 5.3

Note: the value of $k_{c,90}$ should not exceed 4,0 as indicated in Clause 6.1.5(2) of EC5.

The details given in Figures 6.2 to 6.4 of EC5 relate to both discrete and continuous support conditions and both end and internal supports.

The actual bearing area is the net area of the contact surface and allowance must be made for any reduction in the width of bearing due to wane, as shown in Figure 5.4. In timber engineering, the presence of wane is frequently excluded and consequently this can often be ignored.

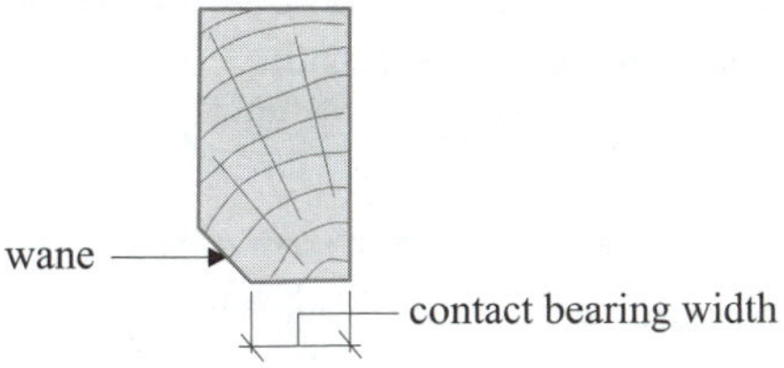

wane

Figure 5.4   contact bearing width

5.3.4 Lateral Stability (Clause 6.3.3 – Equation (6.33))

A beam in which the depth and length are large in comparison to the width (i.e. a slender cross-section) may fail at a lower bending stress value due to lateral torsional buckling, as shown in Figure 5.5.

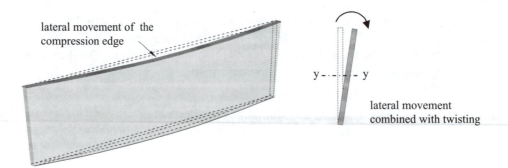

lateral movement of the
compression edge

y - - - | - - y

lateral movement
combined with twisting

Figure 5.5

In the case where only uniaxial bending about the y-y axis exists the following expression should be satisfied:

$\sigma_{m,d} \leq (k_{crit} \times f_{m,d})$ (Equation (6.33) in EC5)

where:

$\sigma_{m,d}$ is the design bending stress,
$f_{m,d}$ is the design bending strength,
k_{crit} is a factor which takes into account the reduction in bending strength due to lateral torsional buckling.

The value of k_{crit} is dependent on the relative slenderness for bending $\lambda_{rel,m}$ defined as:

$$\lambda_{rel,m} = \sqrt{\frac{f_{m,k}}{\sigma_{m,crit}}} \quad \text{where} \quad \sigma_{m,crit} = \frac{\pi\sqrt{E_{0,05}I_z G_{0,05}I_{tor}}}{l_{ef}W_y}$$ (Equations (6.31) and (6.32) in EC5)

and

$E_{0,05}$ is the fifth percentile value of the modulus of elasticity parallel to the grain,
$G_{0,05}$ is the fifth percentile value of the shear modulus parallel to the grain,
I_z is the second moment of area about the z-z axis (i.e. the weaker axis),
I_{tor} is the torsional moment of inertia,
W_y is the section modulus about the y-y axis (i.e. the stronger axis),
l_{ef} is the effective length of the beam as indicated in Table 6.1 of EC5.

Note that if the load is applied at the compression edge of a beam, l_{ef} should be increased by $2h$ and if applied at the tension edge may be decreased by $0,5h$.

As an alternative in the case of softwood, rectangular cross-sections the expression given for $\sigma_{m,crit}$ above may be replaced by:

$$\sigma_{m,crit} = \frac{0,78b^2}{hl_{ef}}E_{0,05}$$ (Equation (6.32) in EC5)

In cases where the compression edge of a beam is prevented throughout its length and where torsional rotation is prevented at its supports, k_{crit} may be taken as 1,0 otherwise:

$$k_{\mathrm{crit}} = \begin{cases} 1,0 & \text{for} \quad \lambda_{\mathrm{rel,m}} \leq 0,75 \\ 1,56 - 0.75\lambda_{\mathrm{rel,m}} & \text{for} \quad 0,75 < \lambda_{\mathrm{rel,m}} \leq 1,4 \\ \dfrac{1}{\lambda_{\mathrm{rel,m}}^{2}} & \text{for} \quad 1,4 < \lambda_{\mathrm{rel,m}} \end{cases} \qquad \text{(Equation (6.34) in EC5)}$$

5.3.5 Notched Beams (Clause 6.5 – Equation (6.60))

It is often necessary to create notches or holes in beams to accommodate fixing details such as gutters, reduced fascias and connections with other members. In such circumstances high stress concentrations occur at the locations of the notches/holes. Whilst notches and holes should be kept to a minimum, when they are necessary, cuts with square re-entrant corners should be avoided. This can be achieved by providing a fillet or taper or cutting the notch to a pre-drilled hole, typically of 8 mm diameter.

The effect of stress concentrations on the design shear strength of beams with a rectangular cross-section and where the grain is essentially parallel to the length of the member should be allowed for by ensuring that:

$$\tau_{\mathrm{d}} = \frac{1,5\,V_{\mathrm{d}}}{bh_{\mathrm{ef}}} \leq (k_{\mathrm{v}} \times f_{\mathrm{v,d}}) \qquad \text{(Equation (6.60) in EC5)}$$

where:
τ_{d} and $f_{\mathrm{v,d}}$ are as defined previously,
h_{ef} is the effective depth of the beam as shown in Figure 5.6.

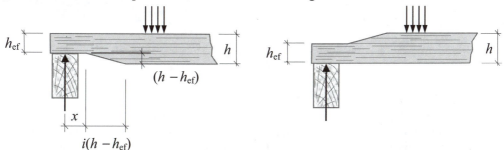

Figure 5.6

where:
i is the notch inclination as indicated in Figure 5.6(a),
h is the beam depth measured in mm,
x is the distance from the line of action of the support reaction to the corner of the notch,
$$\alpha = \frac{h_{\mathrm{ef}}}{h},$$
$$k_{\mathrm{n}} = \begin{cases} 4,5 & \text{for LVL (laminated veneer lumber)} \\ 5 & \text{for solid timber} \\ 6,5 & \text{for glued laminated timber} \end{cases}$$
k_{v} is a reduction factor defined as follows:

In **case (a)** indicated in Figure 5.6

$k_v \le 1,0$

and

$$k_v \le \frac{k_n\left(1+\dfrac{1,1\,i^{1,5}}{\sqrt{h}}\right)}{\sqrt{h}\left(\sqrt{\alpha(1-\alpha)}+0,8\dfrac{x}{h}\sqrt{\dfrac{1}{\alpha}-\alpha^2}\right)}$$

In **case (b)** indicated in Figure 5.6

$k_v = 1,0$

*Note: k_v is limited to a maximum value of 1,0 to avoid shear failure in the net cross-section.

The effects of stress concentrations due to notches as indicated above need not be considered in the following cases:

(i)   where there is tension parallel to the grain,
(ii) where there is compression parallel to the grain,
(iii) where there are tensile stresses at the notch due to bending and the taper is not steeper than 1: $i = 1:10$, i.e. when $i \ge 10$ as shown in Figure 5.7(a) and
(iv) where there are compressive stresses at the notch due to bending as shown in Figure 5.7(b).

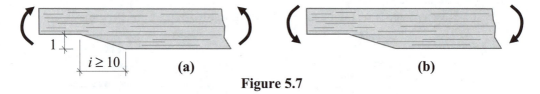

(a)   (b)

Figure 5.7

5.3.6 Deflection (Clause 7.2 – Equation (7.2))

In the absence of any special requirements for deflection in buildings, it is customary to adopt limiting values to minimize the risk of damage to non-structural elements and brittle finishes such as plastered ceilings in addition to consideration of aesthetic/visual effects, for example, unsightly sagging.

The controlling limits for deflection should be agreed with a client for each individual project. Guidance relating to limiting values for beams are given in Table NA.4 of the UK National Annex as indicated in Table 5.2 below.

The calculated deflection in any given case relating to the 'net final deflection' as defined in Clause 7.2(1) of EC5 should be determined from Equation (7.2) in the code, i.e.

$$w_{net,fin} = w_{inst} + w_{creep} - w_c = w_{fin} - w_c \qquad \text{(Equation (7.2) in EC5)}$$

where:
$w_{net,fin}$ is the net final deflection,
w_{inst} is the instantaneous deflection,
w_{creep} is the creep deflection,
w_c is the precamber,
w_{fin} is the final deflection as shown in Figure 5.8.

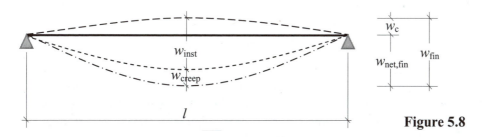

Figure 5.8

| Type of member | Limiting value for the net final deflection of beams, w_{fin} | |
	A member of span, l between two supports	A member with a cantilever of span, l
Roof or floor members with a plastered or plasterboard ceiling	$l/250$	$l/125$
Roof or floor members without a plastered or plasterboard ceiling	$l/150$	$l/75$
Note: w_{fin} should be calculated as u_{fin} in accordance with EN1995-1-1:2004 2.2.3(5)		

Table 5.2 – Limiting w_{net} values given in the UK National Annex Table NA.4

The final deflection comprises a reversible component which is present during limited periods and an irreversible component which continually increases with time, i.e. creep deformation. The reversible component is due to both permanent and variable actions.

In EC5 the final deflections should be determined using the simplified method given in Clause 2.2.3(5) and as indicated in Table NA.4 (see Table 5.2 above). This will be satisfactory in most cases. Equation (2.2) given in Clause 2.2.3(5) is in general terms relating to deformations (u values) as follows:

$$u_{fin} = u_{fin,G} + u_{fin,Q1} + u_{fin,Qi} \qquad \text{(Equation (2.2) in EC5)}$$
where:
u_{fin} is the final deformation,
$u_{fin,G}$ is the final deformation for the permanent actions G,
$u_{fin,Q1}$ is the final deformation for the leading variable action Q_1,
$u_{fin,Qi}$ is the final deformation for the accompanying actions Q_i ($i > 1$).

In Equation (2.2) each of the u_{fin} terms comprises two elements, one relating to the instantaneous deformation and the other to the long-term creep deformation, i.e.

$$u_{fin} = u_{inst} + u_{creep}$$

The creep deformation is dependent on the type of material, the service class and the load duration and is represented by the k_{def} factor from Table 3.2 in EC5 and the ψ_2 factor from Table NA.A1.1 in the UK National Annex to EN1990:2002. Each term in Equation (2.2) is defined as follows:

$$u_{fin,G} = u_{inst,G} + u_{creep,G} = u_{inst,G}(1 + k_{def})$$
$$u_{fin,Q1} = u_{inst,Q1} + u_{creep,Q1} = u_{inst,Q1}(1 + \psi_{2,1}k_{def})$$
$$u_{fin,Qi} = u_{inst,Qi} + u_{creep,Qi} = u_{inst,Qi}(\psi_{0,i} + \psi_{2,i}k_{def})$$

where:

$u_{inst,G}, u_{inst,Q1}$ and $u_{inst,Qi}$ are the instantaneous deformations for actions G, Q_1 and Q_i,

$\psi_{2,1}, \psi_{0,i}$ and $\psi_{2,i}$ are the factors for the quasi-permanent value of variable actions as given in Table NA.A1.1 of the National Annex to EC1990:2002,

k_{def} is the creep factor from Table 3.2 in EC5.

The values of u_{inst} should be calculated on the basis of the characteristic combination of actions using Equation (6.14(b)) in EN1990:2002 and using mean values of the appropriate modulii of elasticity, shear modulii and slip modulii, as indicated in Clause 2.2.3(2) of EC5.

Traditionally when calculating the deflection of beams, only the component due to the bending action is considered. This is due to the fact that in other materials, for example steel, the shear modulus is considerably higher as a percentage of the true elastic modulus than is the case in timber and similarly with the depth to span ratio of the cross-section. A consequence of this is that when considering the deflection of timber beams the effect of shearing deformation may be significant.

The deflection due to bending can be determined using the standard equations as indicated for various load cases in Table 4.7 of Chapter 4. The deflection due to shear is influenced by a number of factors such as the cross-sectional dimensions, the shear modulus of the web (G), and the position and intensity of the loads. A number of complex analytical expressions have been developed to determine the magnitude of the shear deflection, (see Appendix B, Young and Budynas (77), COFI (60)).

In the case of a simply-supported rectangular beam of span L, subjected to a uniformly distributed load (total load = W kN), the total deflection can be determined from:

$$u_{inst} = \frac{5WL^3}{384E_dI} + \frac{3WL}{20bhG_d}$$

In the case of a simply-supported rectangular beam of span L, subjected to a mid-span point load equal to P kN, the total deflection can be determined from:

$$u_{inst} = \frac{PL^3}{48E_dI} + \frac{3PL}{10bhG_d}$$

In general, an estimate of the deflection of a simply supported beam can be made based on the maximum bending moment using the following equations or alternatively, using the method given in Appendix B. Thin-webbed beams are considered in Section 5.4.

$$u_{inst} \approx \frac{0,104ML^2}{E_dI} + \frac{M}{bhG_d}$$

where:

E_d is the modulus of elasticity,
G_d is modulus of rigidity,
b is breadth of the beam,
h is the depth of the beam
I is the second moment of area about the axis of bending,

M is the maximum bending moment in the span,
L is the span of the beam.

5.3.7 Vibration (Clause 7.3.2 – Equations (7.3) and (7.4))

In general, the most significant sources of vibration in timber structures are installed machinery and those induced by human activity. The former is dealt with in Clause 7.3.2 of EC5 by reference to Appendix A of ISO 2631-2 assuming a multiplying factor of 1,0 to determine acceptable levels of vibration. The latter is considered in Clause 7.3.3 of EC5 in relation to residential floors.

The qualitative and quantitative assessment of human perception of, susceptibility to and acceptance of structural vibrations is very complex. These characteristics are dependent on many factors such as: the awareness of and proximity to a source of vibration, the duration of the source and whether or not physical activity is being undertaken at the time. In particular, human sensitivity to acceptance of vibration is influenced by the vibration frequency, acceleration, velocity and induced displacement.

Experimental research (STEP 1 (58)) has indicated that the dynamic footfall force induced by ordinary walking has two types of component:

(i)   low-frequency components, i.e. 0 – 8 Hz, originating from the step frequency and its harmonics,
(ii)   higher-frequency components, i.e. 8 – 40 Hz, originating from impacts when the heel comes into contact with the floor surface.

The design of timber floors in accordance with EC5 relates to those in which the fundamental frequency of vibration is greater than 8 Hz and in which it is not anticipated that vigorous physical activity such as dancing or undertaking of rhythmic exercises will occur. The acceptance criteria given in Clause 7.3.3 are based on a limiting maximum instantaneous vertical deflection induced by:

(i)   a unit vertical concentrated static force and
(ii)   a limiting value of unit impulse velocity response induced by an ideal unit impulse applied at the point of the floor giving maximum response.

The use of the static force is a direct consequence of the low-frequency vibrations in the case of floors with a fundamental frequency more than 8 Hz, whose amplitudes are governed by the structural stiffness, the effect of the mass being minimal. The unit impulse velocity response represents the high-frequency components in such cases. The requirements of Clause 7.3.3 are:

(i)   the fundamental frequency of the rectangular floor system $f_1 > 8$ Hz.

This can be checked using Equation (7.5) of the EC5 $f_1 = \dfrac{\pi}{2l^2}\sqrt{\dfrac{(EI)_l}{m}}$ where:

m is the mass per unit area in kg/m^2
l is the floor span in m,
$(EI)_l$ is the equivalent plate bending stiffness of the floor about an axis perpendicular to the beam direction (i.e. stiffer direction) in Nm^2/m,

(ii)   the maximum instantaneous vertical deflection 'w' caused by a vertical static force $F = 1,0$ applied at any point on the floor and taking account of load distribution does not exceed the limiting value 'a' i.e.

(iii) $\dfrac{w}{F} \le a$ mm/kN   (Equation (7.3) in EC5)

The values of a and w/F (equal to w for unit load F) can be determined as indicated in Table NA.5 and Clause NA.2.6 of the UK National Annex, i.e.

$l \le 4000$ mm $a = 1,8$ mm
$l > 4000$ mm $a = 16500//l^{1,1}$ mm where $l =$ joist span in mm   and

$$w = \frac{1000 k_{dist} l_{eq}^3 k_{amp}}{48 (EI)_{joist}} \text{ mm } \quad [k_{dist}, l_{eq}, k_{amp} \text{ and } (EI)_{joist} \text{ are defined in the Annex.}]$$

(iv)   the unit impulse velocity response, i.e. the maximum initial value of the vertical floor vibration velocity (in m/s) caused by an ideal unit impulse (1 Ns) applied at the point of the floor giving maximum response, where components above 40 Hz may be neglected, should satisfy the following equation:

$v \le b^{(f_1 \zeta - 1)}$ (Equation (7.4) in EC5)

where:
v is the unit impulse velocity
b is the constant for the control on unit impulse velocity response and can be determined from Table NA.5 of the UK National Annex, i.e.
$b = 180 - 60a$ m/Ns^2 for $a \le 1$ mm
$b = 160 - 40a$ m/Ns^2 for $a > 1$ mm
These values correspond to the values between the ranges $0 \le a \le 1$ and $1 \le a \le 2$ in Figure 7.2 of EC5 indicating the relationship between a and b,
f_1 is the fundamental frequency in Hz as calculated above,
ζ is the modal damping ratio (equal to 0,02 in the UK National Annex).

The value of v can be calculated from Equation (7.6) in EC5 as:
$$v = \frac{4(0,4 + 0,6 n_{40})}{mbl + 200}$$
where:
v, m and l are as above,
n_{40} is the number of first-order modes with natural frequencies up to 40 Hz,
b is the floor width in m.

This equation is based on the distributed mass of the 'bare' floor plus an additional 50 kg for each modal mass to represent a notional vibration portion of the body disturbed by the vibration.

5.3.8 Example 5.1: Suspended Timber Floor System

Consider the design of a suspended timber floor system in a domestic building in which the joists are simply supported on timber wall plates on load-bearing brickwork, as shown in Figure 5.9(a).

(a) Determine a suitable section size for the tongue and groove floor boards.
(b) Determine a suitable section size for the main joists.
(c) Assuming one end of the joists to be notched and supported by a wall plate as shown in Figure 5.9(b), check the shear capacity of the joists.

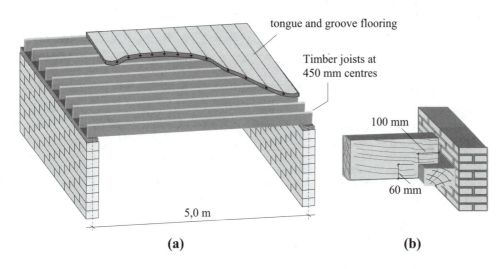

(a)   (b)

Figure 5.9

Design data:

Centres of timber joists (s)	450 mm
Distance between the centre-lines of the brickwork wall (span)	5,0 m
Breadth of floor (b)	4,05 m
Contact length of bearing	100 mm
Strength class of timber for joists and tongue and groove boarding and beams	C16
Characteristic variable action	4,0 kN/m^2
Timber Service Class	2

Assume:

(i) that the floor members do not have a plastered or plastered board ceiling, and
(ii) that torsional restraint (omitted from Figure 5.9(a)) is provided to the joists at the supports.

5.3.8.1 Solution to Example 5.1

References	Calculations	Output
Contract : Solid beams Job Ref. No. : Example 5.1 **Part of Structure : Suspended floor system** **Calculation Sheet No. : 1 of 12**		**Calculated by : W.McK.** **Checked by : B.Z.** **Date :**

References	Calculations	Output
BS EN 1990 BS EN 1991 BS EN 338:2003 BS EN 1995-1-1 2004	Eurocode: Basis of Structural Design Eurocode 1: Actions on Structures Structural Timber – Strength Classes Eurocode 5: Design of Timber Structures Part 1.1: General – Common Rules and Rules for Buildings.	
National Annex	UK - National Annex to BS EN 1995-1-1	
BS EN 338:2003 Table 1	Characteristic values for C16 Timber **Strength Properties:** Bending $\qquad\qquad\qquad\qquad\qquad f_{m,k} = 16{,}0 \text{ N/mm}^2$ Compression parallel to grain $\qquad f_{c,0,k} = 17{,}0 \text{ N/mm}^2$ Compression perpendicular to grain $\quad f_{c,90,k} = 2{,}2 \text{ N/mm}^2$ Shear $\qquad\qquad\qquad\qquad\qquad\quad f_{v,k} = 1{,}8 \text{ N/mm}^2$ **Stiffness Properties:** Mean modulus of elasticity parallel to the grain $\qquad\qquad\qquad\qquad\qquad\qquad E_{0,mean} = 8{,}0 \text{ kN/mm}^2$ 5% modulus of elasticity parallel to the grain $\qquad\qquad\qquad\qquad\qquad\qquad E_{0,05} = 5{,}4 \text{ kN/mm}^2$ Mean modulus of elasticity perpendicular to the grain $\qquad\qquad\qquad\qquad\qquad\qquad E_{90,mean} = 0{,}27 \text{ kN/mm}^2$ Mean shear modulus $\qquad\qquad G_{mean} = 0{,}5 \text{ kN/mm}^2$ Density $\qquad\qquad\qquad\qquad\quad \rho_k = 310{,}0 \text{ kg/m}^3$ Mean density $\qquad\qquad\qquad \rho_{mean} = 370{,}0 \text{ kg/m}^3$ **Note:** a value of characteristic density is also given for use when designing joints.	
BS EN 1995 Clause 2.3.1.2 Table NA.1 (EC5: Table 2.3)	Load-duration Classes: use Table NA.1 from National Annex Self-weight $\qquad$ - $\quad$ permanent Imposed floor loading - $\quad$ medium-term	
Clause 2.3.2.1	Load-duration and moisture influence on strength: k_{mod} The flooring and joists have the same time-dependent behaviour.	
Table 3.1	Solid timber: $\qquad\qquad$ Service Class 2 Permanent actions $\qquad k_{mod} = 0{,}6$ Medium-term actions $\quad k_{mod} = 0{,}8$	
Clause 3.1.3(2)	Use k_{mod} corresponding to action with the shortest duration.	$k_{mod} = 0{,}8$

Contract : Solid beams Job Ref. No. : Example 5.1 Part of Structure : Suspended floor system Calculation Sheet No. : 2 of 12	Calculated by : W.McK. Checked by : B.Z. Date :

References	Calculations	Output
National Annex Table NA.3 EC5 (Table 2.3)	Partial factor for material properties and resistance: γ_M For solid timber – untreated and preservative treated	$\gamma_M = 1,3$
Clause 3.2(3) Equation (3.1)	Member size: For timber in bending with $\rho_k \leq 700$ kg/m^3 where the depth is less than 150 mm $f_{m,k}$ can be multiplied by k_h where: $$k_h = \min\left\{\begin{array}{l}\left(\dfrac{150}{h}\right)^{0,2}\\[2mm] 1,3\end{array}\right.$$ Assume tongue and groove boarding 16 mm thick. $$k_h = \min\left\{\begin{array}{l}\left(\dfrac{150}{16}\right)^{0,2}\\[2mm] 1,3\end{array}\right. = \min\left\{\begin{array}{l}1,56\\[1mm]1,3\end{array}\right.$$	$k_h = 1,3$
Clause 6.6	System strength: k_{sys} Since the tongue and groove boarding has adequate provision for lateral distribution of loading the strength properties can be multiplied by k_{sys}. **(a) Tongue and groove floor boarding** Consider 1,0 m width of flooring and 16 mm thick boarding: ρ_{mean} = 370,0 kg/m^3 Self-weight $= \dfrac{(0,016 \times 370) \times 9,81}{10^3} = 0,06$ kN/m^2 Permanent action: $g_k = 0,06$ kN/m^2 Variable action: $q_k = 4,0$ kN/m^2	$k_{sys} = 1,1$
BS EN 1990 Equation (6.10)	Eurocode: Basis of Structural Design $$E_d = \sum_{j\geq 1}\gamma_{G,j}G_{k,j} + \gamma_{Q,1}Q_{k,1} + \sum_{i>1}\gamma_{Q,i}\psi_{0,i}Q_{k,i}$$	

Contract : Solid beams Job Ref. No. : Example 5.1 **Part of Structure : Suspended floor system** **Calculation Sheet No. : 3 of 12**		**Calculated by : W.McK.** **Checked by : B.Z.** **Date :**

References	Calculations	Output
    UK NA to BS EN 1990 Table NA.A1.2(B) BS EN 1995-1-1 Clause 6.1.6 Clause 7.2 Clause 2.2.3 Equation (2.2) Equation (2.3) Equation (2.4)	$j = 1$ and $i = 1$ $E_d = \left(\gamma_G G_k + \gamma_Q Q_k \right)$ where E_d is the design value of the effect of the actions $\gamma_G = 1{,}35$ and $\gamma_Q = 1{,}50$ The design load $= \left(\gamma_G G_k + \gamma_Q Q_k \right) = (1{,}35 \times 0{,}06) + (1{,}5 \times 4{,}0)$ $F_d = 6{,}1 \ \text{kN/m}^2$ **Bending:** Design bending moment $M_{y,d} \approx (wL^2/10)$ (Allowing for the continuity of the boards over the joists) $M_{y,d} = \dfrac{6{,}1 \times 0{,}45^2}{10} = 0{,}124 \ \text{kNm}$ Design bending strength $f_{m,y,d} = (f_{m,y,k} \times k_{mod} \times k_h \times k_{sys})/\gamma_M$ $\qquad\qquad\qquad\qquad\quad = (16{,}0 \times 0{,}8 \times 1{,}3 \times 1{,}1)/1{,}3$ $\qquad\qquad\qquad\qquad\quad = 14{,}1 \ \text{N/mm}^2$ Since uniaxial bending about the y-y axis only, is present: $\sigma_{m,y,d} \leq f_{m,y,d}$ $\sigma_{m,y,d} = \dfrac{M_{y,d}}{W_y} = \dfrac{0{,}124 \times 10^6}{W_y} \leq 14{,}1 \ \text{N/mm}^2$ $\therefore \ \ W_y \geq \dfrac{0{,}124 \times 10^6}{14{,}1} \geq 8{,}794 \times 10^3 \ \text{mm}^3 \ /\text{m width}$ $W_y = \dfrac{bh^2}{6} \ \ \therefore h \geq \sqrt{\dfrac{8{,}794 \times 10^3 \times 6}{1000}} = 7{,}3 \ \text{mm}$ Assume an additional 3 mm for wear $h = (7{,}3 + 3{,}0) = 10{,}3 \ \text{mm}$ $\qquad\qquad\qquad\qquad\qquad\qquad\qquad\qquad\qquad\qquad < 16{,}0 \ \text{mm}$ For calculations, $h = 16{,}0 - 3{,}0 = 13{,}0 \ \text{mm}$ is used. Limiting values for deflection of beams: $w_{net,fin} = w_{inst} + w_{creep} - w_c = w_{fin} - w_c$ Since there is no camber $w_c = 0$ and $w_{net,fin} = w_{inst} + w_{creep} = w_{fin}$ **Serviceability limit states** $u_{fin} = u_{fin,G} + u_{fin,Q1} + u_{fin,Qi}$ $\qquad$ since $Q_i = 0$ $\qquad\quad u_{fin} = u_{fin,G} + u_{fin,Q1}$ $u_{fin,G} \ = u_{inst,G} + u_{creep,G} \ = u_{inst,G} \, (1 + k_{def})$ $u_{fin,Q1} = u_{inst,Q1} + u_{creep,Q1} = u_{inst,Q1} \, (1 + \psi_{2,1} \, k_{def})$ where k_{def} in Table 3.2 is a factor to allow for creep deformation (u in Equations (2.3) and (2.4) represents w in Equation (7.2).)	 $F_d = 6{,}1 \ \text{kN/m}^2$ $M_{y,d} = 0{,}124 \ \text{kNm}$ $f_{m,y,d} = 14{,}1 \ \text{N/mm}^2$ **The tongue and groove** **boarding is adequate** **with respect to bending.**

References	Calculations	Output
	Contract : Solid beams Job Ref. No. : Example 5.1 **Part of Structure : Suspended floor system** **Calculation Sheet No. : 4 of 12** **Calculated by : W.McK.** **Checked by : B.Z.** **Date :**	

References	Calculations	Output
Table 3.2	For solid timber Service Class 2 $k_{def} = 0,8$	$k_{def} = 0,8$
	Permanent load deformation: $w_{inst,G}(1 + k_{def}) = (w_{inst,m} + w_{inst,v})_G (1 + k_{def})$ where:	
	$w_{inst,m}$ is the deformation due to bending $\approx \dfrac{5WL^3}{384E_d I}$ for a UDL	
	$w_{inst,v}$ is the deformation due to shear $\approx \dfrac{3WL}{20bhG_d}$ for a UDL	
	using 'mean' values of the appropriate modulus of elasticity.	
	(Note: $\gamma_G = \gamma_Q = 1,0$ for characteristic deformations)	
	$W = (0,06 \times 1,0 \times 0,45) = 0,027$ kN $L = 450$ mm $E_d = E_{0,mean} = 8,0$ kN/mm^2 $G_d = G_{mean} = 0,5$ kN/mm^2	
	$I_y = \dfrac{bh^3}{12} = \dfrac{1000 \times 13^3}{12} = 183,08 \times 10^3$ mm^4	
	$\dfrac{5WL^3}{384E_d I} = \dfrac{5 \times 0,027 \times 450^3}{384 \times 8,0 \times 183,08 \times 10^3} = 0,02$ mm	
	$\dfrac{3WL}{20bhG_d} = \dfrac{3 \times 0,027 \times 450}{20 \times 1000 \times 13 \times 0,5} = 0,0003$ mm	
	$w_{inst,G} = (w_{inst,m} + w_{inst,v})_G = (0,02 + 0,0003) = 0,02$ mm	
	$w_{fin,G} = w_{inst,G}(1 + k_{def}) = 0,02 \times (1 + 0,8) = 0,04$ mm	
	Variable load deformation: $W = (4,0 \times 1,0 \times 0,45) = 1,8$ kN $L = 450$ mm $E_d = E_{0,mean} = 8,0$ kN/mm^2 $G_d = G_{mean} = 0,5$ kN/mm^2	
UK NA to BS EN 1990 Table NA.A1.1	$\psi_{2,1} = 0,3$	$\psi_{2,1} = 0,3$
	$\dfrac{5WL^3}{384E_d I} = \dfrac{5 \times 1,8 \times 450^3}{384 \times 8,0 \times 183,08 \times 10^3} = 1,46$ mm	
	$\dfrac{3WL}{20bhG_d} = \dfrac{3 \times 1,8 \times 450}{20 \times 1000 \times 13 \times 0,5} = 0,02$	

References	Calculations	Output
	$w_{inst,Q1} = (w_{inst,m} + w_{inst,v})_{Q1} = (1,46 + 0,02) = 1,48$ mm	
	$w_{fin,Q1} = w_{inst,Q1}(1 + \psi_{2,1}k_{def}) = 1,48 \times [1 + (0,3 \times 0,8)]$ $= 1,84$ mm	
	$w_{fin} \doteq w_{fin,G} + w_{fin,Q1} = (0,04 + 1,84) = 1,88$ mm	
	Assume 50% due to continuity, i.e. $\approx 0,5$ mm	$w_{fin} \approx 1,88$ mm
UK NA to EC5 Table NA.4	Limiting ratio $= l/150 = 450/150 = 3,0$ mm $> 1,88$ mm	The t & g boarding is adequate with respect to deflection.
	(b) Joists at 450 mm centres Self-weight – assume a value $= 0,12$ kN/m Permanent action due to t &g boarding $= (0,06 \times 0,45)$ $= 0,03$ kN/m Total permanent action $g_k = (0,12 + 0,03) = 0,15$ kN/m Variable action: $q_k = (4,0 \times 0,45) = 1,8$ kN/m	
BS EN 1990 Equation (6.10) UK NA to EC Table NA.A1.2(B)	Eurocode: Basis of Structural Design $E_d = (\gamma_G G_k + \gamma_Q Q_k)$ $\gamma_G = 1,35$ and $\gamma_Q = 1,50$ The design load $= (\gamma_G G_k + \gamma_Q Q_k) = (1,35 \times 0,15) + (1,5 \times 1,8)$ $F_d = 2,90$ kN/m	$F_d = 2,90$ kN/m
	Bending: Design bending moment $M_{y,d} = (wL^2/8)$ $M_{y,d} = \dfrac{2,9 \times 5,0^2}{8} = 9,06$ kNm	$M_{y,d} = 9,06$ kNm
BS EN 1995 Clause 3.2(3) Clause 6.6	Design bending strength $f_{m,y,d} = (f_{m,y,k} \times k_{mod} \times k_h \times k_{sys})/\gamma_M$ Assume $h > 150$ mm therefore k_h does not apply. System strength: $k_{sys} = 1,1$ $f_{m,y,d} = (16,0 \times 0,8 \times 1,1)/1,3 = 10,83$ N/mm^2	$f_{m,y,d} = 10,83$ N/mm^2
BS EN 1995-1-1 Clause 6.1.6	Since there is only bending about the y-y axis: $\sigma_{m,y,d} \leq f_{m,y,d}$ $\sigma_{m,y,d} = \dfrac{M_{y,d}}{W_y} = \dfrac{9,06 \times 10^6}{W_y} \leq 10,83$ N/mm^2	

References	Calculations	Output
	$W_y \geq \dfrac{9,06 \times 10^6}{10,83} = 837,0 \times 10^3 \text{ mm}^3$ Try a 75 mm wide x 275 mm deep beam. Section properties: $W_y = \dfrac{75 \times 275^2}{6} = 945,31 \times 10^3 \text{ mm}^3$ $I_{yy} = \dfrac{75 \times 275^3}{12} = 129,98 \times 10^6 \text{ mm}^4$ $A = (75 \times 275) = 20,63 \times 10^3 \text{ mm}^2$ **Shear:** Design shear force $F_{v,d} = (F_d \times 5,0)/2 = (2,95 \times 5,0)/2$ $\qquad\qquad\qquad\qquad\qquad = 7,25 \text{ kN}$	$F_{v,d} = 7,25 \text{ kN}$
	Design shear strength $f_{v,d} = (f_{v,k} \times k_{mod} \times k_v \times k_{sys})/\gamma_M$	
Clause 6.5.2	k_v applies to notched beams only $f_{v,d} = (1,8 \times 0,8 \times 1,1)/1,3 = 1,22 \text{ N/mm}^2$	$f_{v,d} = 1,22 \text{ N/mm}^2$
Clause 6.1.7	Design shear stress $\tau_d = \dfrac{1,5 \times F_{v,d}}{A} = \dfrac{1,5 \times 7,25 \times 10^3}{20,63 \times 10^3}$ $\qquad\qquad\qquad = 0,53 \text{ N/mm}^2 < 1,22 \text{ N/mm}^2$	**The joist is adequate with respect to shear.**
Clause 6.1.5	**Bearing:** Design bearing force $F_{c,90,d} = (2,90 \times 5,0)/2 = 7,25 \text{ kN}$	$F_{c,90,d} = 7,25 \text{ kN}$
	Design bearing strength $f_{c,90,d} = (f_{c,90,k} \times k_{mod} \times k_{sys})/\gamma_M$ $\qquad\qquad\qquad\qquad\quad = (2,2 \times 0,8 \times 1,1)/1,3$ $\qquad\qquad\qquad\qquad\quad = 1,49 \text{ N/mm}^2$	$f_{c,90,d} = 1,49 \text{ N/mm}^2$
	Contact length of bearing $l = 100$ mm Design bearing stress $\sigma_{c,90,d} = \dfrac{F_{c,90,d}}{b \times l} = \dfrac{7,25 \times 10^3}{75 \times 100}$ $\qquad\qquad\qquad\qquad\qquad = 0,97 \text{ N/mm}^2$	
Equation (6.3) Clause 6.1.5(2) Clause 6.1.5(3)	$\sigma_{c,90,d} \leq (k_{c,90} \times f_{c,90,d})$ Since 'a' in Figure 6.2 of EC5 = 0 $k_{c,90} = \left(2,38 - \dfrac{l}{250}\right)\left(1 + \dfrac{h}{12l}\right)$ and $\leq 4,0$	

Contract : Solid beams Job Ref. No. : Example 5.1 Part of Structure : Suspended floor system Calculation Sheet No. : 7 of 12	Calculated by : W.McK. Checked by : B.Z. Date :

References	Calculations	Output
	$k_{c,90} = \left(2,38 - \dfrac{100}{250}\right)\left(1 + \dfrac{275}{12 \times 100}\right) = 2,43 \le 4,0$ $(k_{c,90} \times f_{c,90,d}) = (2,43 \times 1,49) = 3,62 \ \text{N/mm}^2 > 0,97 \ \text{N/mm}^2$	$k_{c,90} = \textbf{2,43}$ **The section is adequate with respect to bearing.**
Clause 6.3.3 Equation (6.33) Clause 6.3.3(5)	**Lateral Torsional Stability:** $\sigma_{m,y,d} \le (k_{crit} \times f_{m,d})$ Since the compression flange is fully restrained by the decking and torsional restraint is provided at the supports the value of $k_{crit} = 1,0$. Design strength $f_{m,d} = 10,83 \ \text{N/mm}^2$ $\qquad (k_{crit} \times f_{m,d}) = (1,0 \times 10,83) = 10,83 \ \text{N/mm}^2$ Design stress $\sigma_{m,y,d} = \dfrac{M_{y,d}}{W_y} = \dfrac{9,06 \times 10^6}{945,31 \times 10^3} = 9,58 \ \text{N/mm}^2$ $\qquad\qquad\qquad\qquad\qquad\qquad < 10,83 \ \text{N/mm}^2$	**The section is adequate with respect to lateral torsional stability.**
Clause 7.2 Table 3.2	**Serviceability Limit States** **Deflection:** Limiting values for deflection of beams: $w_{net,fin} = w_{inst} + w_{creep} - w_c = w_{fin} - w_c$ Since there is no camber $w_c = 0$ and $w_{net,fin} = w_{inst} + w_{creep} = w_{fin}$ Permanent load deformation: For solid timber Service Class 2 $k_{def} = 0,8$ $w_{inst,G}\,(1 + k_{def}) = (w_{inst,m} + w_{inst,v})_G\,(1 + k_{def})$ $W = (0,15 \times 5,0) = 0,75 \ \text{kN}$ $L = 5000 \ \text{mm}$ $E_d = E_{0,mean} = 8,0 \ \text{kN/mm}^2$ $G_d = G_{mean} = 0,5 \ \text{kN/mm}^2$ $I_y = 129,98 \times 10^3 \ \text{mm}^4$ $w_{inst,G,m} = \dfrac{5WL^3}{384 E_d I} = \dfrac{5 \times 0,75 \times 5000^3}{384 \times 8,0 \times 129,98 \times 10^6} = 1,17 \ \text{mm}$ $w_{inst,G,v} = \dfrac{3WL}{20 bh G_d} = \dfrac{3 \times 0,75 \times 5000}{20 \times 75 \times 275 \times 0,5} = 0,06 \ \text{mm}$ $w_{inst,G} = (w_{inst,m} + w_{inst,v})_G = (1,17 + 0,06) = 1,23 \ \text{mm}$ $w_{fin,G} = w_{inst,G}\,(1 + k_{def}) = 1,23 \times (1 + 0,8) = 2,21 \ \text{mm}$	$k_{def} = \textbf{0,8}$

Contract : Solid beams Job Ref. No. : Example 5.1 Part of Structure : Suspended floor system Calculation Sheet No. : 8 of 12	Calculated by : W.McK. Checked by : B.Z. Date :

References	Calculations	Output
	Variable load deformation: $W = (1,8 \times 5,0) = 9,0$ kN $L = 5000$ mm $E_d = E_{0,mean} = 8,0$ kN/mm^2 $G_d = G_{mean} = 0,5$ kN/mm^2 $w_{inst,Q1,m} = \dfrac{5WL^3}{384E_d I} = \dfrac{5 \times 9,0 \times 5000^3}{384 \times 8,0 \times 129,98 \times 10^6} = 14,09$ mm $w_{inst,Q1,v} = \dfrac{3WL}{20bhG_d} = \dfrac{3 \times 9,0 \times 5000}{20 \times 75 \times 275 \times 0,5} = 0,65$ mm	
UK NA to BS EN 1990 Table NA.A1.1	$\psi_{2,1} = 0,3$ $(w_{inst,m} + w_{inst,v})_{Q1} = (14,09 + 0,65) = 14,74$ mm $w_{fin,Q1} = w_{inst,Q1}(1 + \psi_{2,1} k_{def}) = 14,74 \times [1 + (0,3 \times 0,8)]$ $= 18,28$ mm $w_{fin} = w_{fin,G} + w_{fin,Q1} = (2,21 + 18,28) = 20,49$ mm	$\psi_{2,1} = 0,3$ $w_{fin} \approx 20,49$ mm
BS EN 1995 UK NA Table NA.4	Limiting ratio $= l/150 = 5000/150 = 33,33$ mm $> 21,32$ mm	**The joists are adequate with respect to deflection.**
Clause 7.3.3(1) Clause 7.3.3(4)	**Vibration:** Check the fundamental frequency of the floor system > 8 Hz. Fundamental frequency $f_1 \approx \dfrac{\pi}{2l^2}\sqrt{\dfrac{(EI)_l}{m}}$ Self-weight of flooring $= 0,06$ kN/m^2 (6,12 kg/m^2) Self-weight of joists $= (0,075 \times 0,275 \times 370,0) \times \dfrac{1000}{450} = 16,96$ Mass of floor $m = (6,12 + 16,96) = 23,08$ kg/m^2 Span $l = 5,0$ m $(EI)_l = (E_{0,mean} I_y)/s$ (where s is the joist spacing) $= \left(\dfrac{8,0 \times 10^9 \times 0,075 \times 0,275^3}{12 \times 0,45}\right) = 2,31 \times 10^6$ Nm^2/m $f_1 \approx \dfrac{\pi}{2l^2}\sqrt{\dfrac{(EI)_l}{m}} = \dfrac{\pi}{2 \times 5,0^2}\sqrt{\dfrac{2,31 \times 10^6}{23,08}} = 19,88$ Hz $> 8,0$ Hz	 **mass = 23,08 kg/m^2** $f_1 = 19,88$ Hz

Contract : Solid beams Job Ref. No. : Example 5.1 **Part of Structure : Suspended floor system** **Calculation Sheet No. : 9 of 12**	**Calculated by : W.McK.** **Checked by : B.Z.** **Date :**	

References	Calculations	Output
UK NA to EC5 NA.2.6.2	**Note:** no allowance is made for composite action when evaluating $(EI)_l$ unless the floor is designed in accordance with Clause 9.1.2 and with adhesives meeting the requirements of Clause 3.6. Since $f_1 > 8$ Hz, Clause 7.3.3 for residential floors can be used, i.e.	
Equation (7.3)	Unit load deflection: $\dfrac{w}{F} \leq a$ mm/kN and	
Equation (7.4)	Unit impulse velocity response: $v \leq b^{(f_1\zeta -1)}$ m/(Ns²)	
UK NA to EC5	$w = \dfrac{1000 k_{dist} l_{eq}^3 k_{amp}}{48(EI)_{joist}}$	
Clause NA 2.6	$k_{dist} = k_{strut}\left[0,38 - 0,08 ln\left(\dfrac{14(EI)_b}{s^4}\right)\right]$ $\geq 0,3$ Assume $k_{strut} = 1,0$ $(EI)_b = E_{0,mean}bt^3/12 = (8,0 \times 10^3 \times 1000 \times 13^3)/12$ $\qquad\qquad = 1,465 \times 10^9$ Nmm²/m $k_{dist} = 1,0\left[0,38 - 0,08 ln\left(\dfrac{14 \times 1,465 \times 10^9}{450^4}\right)\right] = 0,44$ $\geq 0,3$ $k_{amp} = 1,05$ for simply-supported solid timber joists. $l_{eq} = 5,000$ mm $(EI)_{joist} = \dfrac{8,000 \times 75 \times 275^3}{12} = 1,04 \times 10^{12}$ Nmm² $\therefore w = \dfrac{1000 k_{dist} l_{eq}^3 k_{amp}}{48(EI)_{joist}} = \dfrac{1,000 \times 0,44 \times 5,000^3 \times 1,05}{48 \times 1,04 \times 10^{12}}$ $\qquad\qquad = 1,16$ mm/kN	$k_{dist} = 0,44$
Table NA.5	For $l > 4,000$ mm: $\qquad a = 16,500/l^{1,1} = 16,500/5,000^{1,1} = 1,408$ mm $> 1,16$ mm	**The floor system is adequate with respect to unit load deflection.**

Contract : Solid beams Job Ref. No. : Example 5.1 **Part of Structure : Suspended floor system** **Calculation Sheet No. : 10 of 12**		**Calculated by : W.McK.** **Checked by : B.Z.** **Date :**

References	Calculations	Output
	Unit impulse velocity response: $v \le b^{(f_1\zeta - 1)}$ m/(Ns^2)	
Clause 7.3.7(5)	$v = \dfrac{4(0,4 + 0,6n_{40})}{mbl + 200}$	
	$n_{40} = \left\{\left[\left(\dfrac{40}{f_1}\right)^2 - 1\right]\left(\dfrac{b}{l}\right)^4 \dfrac{(EI)_l}{(EI)_b}\right\}^{0,25}$	
	$f_1 = 19,88$ Hz	
	b (floor width) = 4,05 m l (span of floor) = 5,0 m $(EI)_l = 2,31 \times 10^6$ Nm^2/m $(EI)_b = 1464,67$ Nm^2/m	
	$n_{40} = \left\{\left[\left(\dfrac{40}{19,88}\right)^2 - 1\right]\left(\dfrac{4,05}{5,0}\right)^4 \times \dfrac{2,31\times10^6}{1464,67}\right\}^{0,25} = 6,74$	
	$v = \dfrac{4\times\left[0,4 + (0,6\times6,74)\right]}{\left[(23,08\times4,05\times5,0) + 200\right]} = 0,0266$ m/Ns^2	
Table NA.5	For $a > 1$ mm $b = (160 - 40a) = 160 - (40 \times 1,408) = 103,68$ m/Ns^2	**The floor system is adequate with respect to unit impulse velocity response.**
	$\zeta = 0,02 \qquad \therefore (f_1\zeta - 1) = [(19,88 \times 0,02) - 1] = -0,602$	
	$v \le b^{(f_1\zeta - 1)} = 103,68^{-0,602} = 0,0612$ m/Ns^2 $> 0,0266$ m/Ns^2	
Clause 6.5	**(c) A notched beam at the end (assumed)**	
	100 x 75 wall plate	

References	Calculations	Output
	Contract : Solid beams Job Ref. No. : Example 5.1 **Part of Structure : Suspended floor system** **Calculation Sheet No. : 11 of 12** **Calculated by : W.McK.** **Checked by : B.Z.** **Date :**	

References	Calculations	Output
Figure 6.11(a)	$h = 275$ mm, $x = 50$ mm, $i = 0$, $h_{ef} = (275 - 60) = 215$ mm $$\tau_d = \frac{1,5V_d}{bh_{ef}} \le (k_v \times f_{v,d})$$ $$\tau_d = \frac{1,5 \times 7,25 \times 10^3}{75 \times 215} = 0,67 \text{ N/mm}^2$$	
Equation (6.62)	$k_v \le 1,0$ and $$k_v \le \frac{k_n\left(1 + \dfrac{1,1\, i^{1,5}}{\sqrt{h}}\right)}{\sqrt{h}\left(\sqrt{\alpha(1-\alpha)} + 0,8\dfrac{x}{h}\sqrt{\dfrac{1}{\alpha} - \alpha^2}\right)}$$ $$\alpha = \frac{h_{ef}}{h} = \frac{215}{275} = 0,782$$	
Equation (6.63)	$$k_n = \begin{cases} 4,5 \text{ for LVL} \\ 5 \text{ for solid timber} \\ 6,5 \text{ for glued laminated timber} \end{cases}$$ $$k_n\left(1 + \frac{1,1\, i^{1,5}}{\sqrt{h}}\right) = 5 \times (1 + 0) = 5,0$$ $$\sqrt{h}\left(\sqrt{\alpha(1-\alpha)} + 0,8\frac{x}{h}\sqrt{\frac{1}{\alpha} - \alpha^2}\right)$$ $$= \sqrt{275}\left(\sqrt{0,782(1-0,782)} + 0,8 \times \frac{50}{275} \times \sqrt{\frac{1}{0,782} - 0,782^2}\right)$$ $$= 16,583 \times [0,413 + (0,145 \times 0,817)] = 8,81$$	$k_n = 5$

References	Calculations	Output
	Contract : Solid beams Job Ref. No. : Example 5.1 **Part of Structure : Suspended floor system** **Calculation Sheet No. : 12 of 12**	**Calculated by : W.McK.** **Checked by : B.Z.** **Date :**

References	Calculations	Output
	$k_v \leq \dfrac{k_n\left(1+\dfrac{1,1\, i^{1,5}}{\sqrt{h}}\right)}{\sqrt{h}\left(\sqrt{\alpha(1-\alpha)}+0,8\dfrac{x}{h}\sqrt{\dfrac{1}{\alpha}-\alpha^2}\right)} = \dfrac{5,0}{8,81}=0,57$ $\leq 1,0$	$k_v = 0,57$
	$\begin{aligned}f_{v,d} &= (f_{v,k}\times k_{mod}\times k_{sys})/\gamma_m = (1,8\times 0,8\times 1,1)/1,3\\ &= 1,22\ \text{N/mm}^2\end{aligned}$ $k_v f_{v,d} = (0,57\times 1,22)=0,70\ \text{N/mm}^2 \qquad \therefore\ \tau_d < k_v f_{v,d}$	**Joist is adequate with respect to shear.**
Clause 6.5.1	**Note:** In certain circumstances the effects of stress concentrations at notches can be disregarded, see Section 5.3.5 of this text.	

5.3.9 Example 5.2: Trimmer Beam

Consider the design of a trimmer beam inserted to create an opening in an intermediate floor as shown in Figure 5.10. Using the design data given, check the suitability of the trimmer beam indicated. (**Note:** the joists supporting the trimmer beam are normally made 25 mm thicker than the other standard joists or doubled up.)

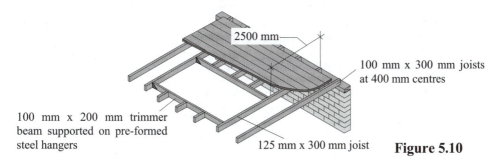

100 mm x 200 mm trimmer beam supported on pre-formed steel hangers

2500 mm

100 mm x 300 mm joists at 400 mm centres

125 mm x 300 mm joist   **Figure 5.10**

Design data:

Centres of timber joists (s)	400 mm
Timber species	UK Douglas fir
Visual grading	General structural
Characteristic variable action	3,0 kN/m^2
Assume the self-weight of the tongue and groove boarding	0,06 kN/m^2

5.3.9.1 Solution to Example 5.2

References	Calculations	Output
	Contract : Solid beams Job Ref. No. : Example 5.2 **Calculated by : W.McK.** **Part of Structure : Trimmer Beam** **Checked by : B.Z.** **Calculation Sheet No. : 1 of 6** **Date :**	

References	Calculations	Output
	Trimmer beam design The trimmed joists provide lateral restraint to the compression flange of the beam	
BS EN 1990 BS EN 1991 BS EN 1912:2004 BS EN 338:2003 BS EN 1995-1-1: 2004	Eurocode: Basis of Structural Design Eurocode 1: Actions on Structures Structural timber – Grading – Requirements for visual strength grading standards Structural Timber – Strength Classes Eurocode 5: Design of Timber Structures Part 1.1: General – Common Rules and Rules for Buildings.	
National Annex	UK - National Annex to BS EN 1995-1-1	
BS EN 1912:2004 Table 1	For visually graded- general structural Douglas fir: Strength class is C14.	
BS EN 338:2003 Clause 5.0 Table 1	Characteristic values for C14 Timber **Strength Properties:** Bending $\qquad\qquad\qquad\qquad\qquad f_{m,k} = 14{,}0$ N/mm^2 Compression parallel to grain $\qquad f_{c,0,k} = 16{,}0$ N/mm^2 Compression perpendicular to grain $\quad f_{c,90,k} = 2{,}0$ N/mm^2 Shear $\qquad\qquad\qquad\qquad\qquad\quad f_{v,k} = 1{,}7$ N/mm^2 **Stiffness Properties:** Mean modulus of elasticity parallel to the grain $\qquad\qquad\qquad\qquad\qquad E_{0,mean} = 7{,}0$ kN/mm^2 5% modulus of elasticity parallel to the grain $\qquad\qquad\qquad\qquad\qquad E_{0,05} = 4{,}7$ kN/mm^2 Mean modulus of elasticity perpendicular to the grain $\qquad\qquad\qquad\qquad\qquad E_{90,mean} = 0{,}23$ kN/mm^2 Mean shear modulus $\qquad\quad G_{mean} = 0{,}44$ kN/mm^2 Density $\qquad\qquad\qquad\qquad\quad \rho_k = 290{,}0$ kg/m^3 Mean density $\qquad\qquad\quad \rho_{mean} = 350{,}0$ kg/m^3	
Clause 2.3.1.2 Table NA.1 (EC5: Table 2.3)	Load-duration Classes: use Table NA.1 from National Annex Self-weight - permanent Imposed floor loading - medium-term	
National Annex (EC5: Cl. 2.3.1.3) Table NA.2	Assignment of timber construction to service classes For intermediate floors Service Class is 1.	

Contract : Solid beams Job Ref. No. : Example 5.2 Part of Structure : Trimmer Beam Calculation Sheet No. : 2 of 6	Calculated by : W.McK. Checked by : B.Z. Date :

References	Calculations	Output
Clause 2.3.2.1	Load-duration and moisture influence on strength: k_{mod} The flooring and joists have the same time-dependent behaviour.	
Table 3.1	Solid timber: Service Class 1 Permanent actions $k_{mod} = 0,6$ Medium-term actions $k_{mod} = 0,8$	
Clause 3.1.3(2)	Use k_{mod} corresponding to action with the shortest duration.	$k_{mod} = 0,8$
UK NA to EC5 Table NA.3 (Table 2.3)	Partial factor for material properties and resistance: γ_M For solid timber – untreated and preservative treated	$\gamma_M = 1,3$
	The trimmer is a 100 mm x 200 mm Douglas fir solid section Self-weight of trimmer $= \dfrac{(0,1 \times 0,2 \times 370 \times 9,81)}{10^3}$ $\qquad\qquad\qquad\qquad = 0,07$ kN/m P is the load imposed by each trimmed joist. Area of floor supported by *trimmed* joist $= 2,5 \times 0,4$ $\qquad\qquad\qquad\qquad\qquad = 1,0$ m^2	

Contract : Solid beams Job Ref. No. : Example 5.2	Calculated by : W.McK.
Part of Structure : Trimmer Beam	**Checked by : B.Z.**
Calculation Sheet No. : 3 of 6	**Date :**

References	Calculations	Output
	Permanent action due to t&g boarding $= (0,06 \times 1,0)$ $\qquad\qquad\qquad\qquad\qquad\qquad\qquad\qquad = 0,06$ kN	
	Variable action $\qquad\qquad\qquad\qquad = (3,0 \times 1,0)$ $\qquad\qquad\qquad\qquad\qquad\qquad\qquad\qquad = 3,0$ kN	
BS EN 1990 Equation (6.10)	Eurocode: Basis of Structural Design $E_d = \left(\gamma_G G_k + \gamma_1 Q_k\right)$ where E_d is the design value of the effect of the actions.	
UK NA to EC Table NA.A1.2(B)	$\gamma_G = 1,35$ and $\gamma_Q = 1,50$ The design load $= \left(\gamma_G G_k + \gamma_Q Q_k\right)$	
	UDL due to self-weight of trimmer $= (1,35 \times 0,07) = 0,09$ kN/m	
	Bending moment due to self-weight $= (wL^2/8) = 0,02$ kNm	
	Total load due to trimmed joist $\qquad = (1,35 \times 0,06) + (1,5 \times 3,0)$ $\qquad\qquad\qquad\qquad\qquad\qquad\qquad\qquad = 4,58$ kN	
	End reaction from trimmed joist on trimmer beam $= \dfrac{4,58}{2}$ $\qquad\qquad\qquad\qquad\qquad\qquad\qquad\qquad\qquad\qquad = 2,29$ kN	
	2,29 kN$\qquad$2,29 kN$\qquad$2,29 kN $\qquad$— 338 —$\quad$— 400 —$\quad$— 400 —$\quad$— 338 — $\qquad\qquad\qquad$— 1476 mm — 3,44 kN$\qquad\qquad\qquad\qquad\qquad\qquad\qquad$3,44 kN	
	Bending moment due to trimmed joists: $= (3,44 \times 0,738) - (2,29 \times 0,4) = 1,62$ kNm	
	Total design bending moment $M_{y,d} = (0,02 + 1,62) = 1,64$ kNm Total design shear force $F_{v,d} = (0,09 \times 0,738) + 3,44 = 3,51$ kN	$M_{y,d} = \mathbf{1,64}$ **kNm** $F_{v,d} = \mathbf{3,51}$ **kN**
Clause 2.4.1(1)	Design value for material property: X_d $X_d = k_{mod} \dfrac{X_k}{\gamma_m}$ This value can be modified to allow for additional factors.	
Clause 6.6	System strength: k_{sys} Load sharing does not apply in this case – ignore k_{sys}.	

References	Calculations	Output
	Contract : Solid beams Job Ref. No. : Example 5.2 Part of Structure : Trimmer Beam Calculation Sheet No. : 4 of 6	Calculated by : W.McK. Checked by : B.Z. Date :

References	Calculations	Output
Clause 3.2.(3) Equation (3.1)	Member size: Since $h > 150$ mm ignore k_h. **Bending:** Design bending moment $M_{y,d} = 1{,}64$ kNm Design bending strength $f_{m,y,d} = (f_{m,y,k} \times k_{mod} \times k_h \times k_{sys})/\gamma_M$	
Clause 3.2(3) Clause 6.6	Assume $h > 150$ mm therefore k_h does not apply. No load distribution therefore k_{sys} strength does not apply. $f_{m,y,d} = (14{,}0 \times 0{,}8)/1{,}3 = 8{,}62$ N/mm^2	$f_{m,y,d} = \mathbf{8{,}62}$ N/mm^2
BS EN 1995-1-1 Clause 6.1.6	Since uniaxial bending about the y-y axis only, is present: $\sigma_{m,y,d} \leq f_{m,y,d}$ $\sigma_{m,y,d} = \dfrac{M_d}{W_y}$ $W_y = \dfrac{b \times h^2}{6} = \dfrac{100 \times 200^2}{6} = 0{,}667 \times 10^6$ mm^3 $\sigma_{m,y,d} = \dfrac{1{,}64 \times 10^6}{0{,}667 \times 10^6} = 2{,}46$ N/mm$^2 < 8{,}62$ N/mm^2	**The trimmer beam is adequate with respect to bending.**
	Shear: Design shear force $F_{v,d} = 3{,}51$ kN Design shear strength $f_{v,d} = (f_{v,k} \times k_{mod} \times k_{sys})/\gamma_M$	
Clause 6.5.2	k_v applies to notched beams only $f_{v,d} = (1{,}7 \times 0{,}8)/1{,}3 = 1{,}05$ N/mm^2	$f_{v,d} = \mathbf{1{,}05}$ N/mm^2
Clause 6.1.7	Design shear stress $\tau_d = \dfrac{1{,}5 \times F_{v,d}}{A} = \dfrac{1{,}5 \times 3{,}51 \times 10^3}{100 \times 200}$ $ = 0{,}26$ N/mm$^2 < 1{,}05$ N/mm^2	**The joist is adequate with respect to shear.**
Clause 6.1.5	**Bearing:** Design bearing force $F_{c,90,d} = 3{,}49$ kN Design bearing strength $f_{c,90,d} = (f_{c,90,k} \times k_{mod} \times k_{sys})/\gamma_M$ $\phantom{Design bearing strength f_{c,90,d}} = (2{,}0 \times 0{,}8)/1{,}3$ $\phantom{Design bearing strength f_{c,90,d}} = 1{,}23$ N/mm^2 Design bearing stress $\sigma_{c,90,d} = \dfrac{F_{c,90,d}}{\text{Bearing Area}}$	$f_{c,90,d} = \mathbf{1{,}23}$ N/mm^2

References	Calculations	Output
Equation (6.3) Clause 6.1.5(2) Clause 6.1.5(3)	$\sigma_{c,90,d} \le (k_{c,90} \times f_{c,90,d})$ Assume $k_{c,90} = 1,0$ $(k_{c,90} \times f_{c,90,d}) = 1,23$ N/mm^2 Minimum bearing area required $= \dfrac{F_{c,90,d}}{\sigma_{c,90,d}} = \dfrac{3,51 \times 10^3}{1,23}$ $= 2,85 \times 10^3$ mm^2 Minimum bearing length $= (2,85 \times 10^3)/100 = 28,5$ mm	Provide a bearing length of 50 mm.
Clause 6.3.3 Equation (6.33) Clause 6.3.3(5)	**Lateral Torsional Stability:** $\sigma_{m,d} \le (k_{crit} \times f_{m,d})$ Since the compression flange is fully restrained by the decking and torsional restraint is provided at the supports the value of $k_{crit} = 1,0$. Design strength $f_{m,d} = 8,62$ N/mm^2 $(k_{crit} \times f_{m,d}) = (1,0 \times 8,62) = 8,62$ N/mm$^2 > \sigma_{m,d}$	The section is adequate with respect to lateral torsional buckling.
Clause 7.2 Table 3.2	**Serviceability Limit States** **Deflection:** Limiting values for deflection of beams: $w_{net,fin} = w_{inst} + w_{creep} - w_c = w_{fin} - w_c$ Since there is no camber $w_c = 0$ and $w_{net,fin} = w_{inst} + w_{creep} = w_{fin}$ For solid timber Service Class 1 $k_{def} = 0,6$ Permanent load deformation: $w_{fin,G} = w_{inst,G}(1 + k_{def}) = (w_{inst,m} + w_{inst,v})_G (1 + k_{def})$ where: $W \approx (0,07 \times 1,476) + (3 \times 0,03) = 0,19$ kN (due to self-weight) $L = 1476$ mm $E_d = E_{0,mean} = 7,0$ kN/mm^2 $G_d = G_{mean} = 0,44$ kN/mm^2 $I_y = \dfrac{bh^3}{12} = \dfrac{100 \times 200^3}{12} = 66,67 \times 10^6$ mm^4 $\dfrac{5WL^3}{384E_dI} = \dfrac{5 \times 0,19 \times 1476^3}{384 \times 7,0 \times 66,67 \times 10^6} = 0,017$ mm $\dfrac{3WL}{20bhG_d} = \dfrac{3 \times 0,19 \times 1476}{20 \times 100 \times 200 \times 0,44} = 0,005$ mm	$k_{def} = 0,6$

References	Calculations	Output
	Contract : Solid beams Job Ref. No. : Example 5.2 **Part of Structure : Trimmer Beam** **Calculation Sheet No. : 6 of 6** **Calculated by : W.McK.** **Checked by : B.Z.** **Date :**	

References	Calculations	Output
	$w_{inst,G} = (w_{inst,m} + w_{inst,v})_G = (0,017 + 0,005) = 0,022$ mm $w_{fin,G} = w_{inst,G}(1 + k_{def}) = 0,022 \times (1 + 0,6) = 0,04$ mm Variable load deformation: $P = 1,5$ kN $L = 1476$ mm $E_d = E_{0,mean} = 7,0$ kN/mm^2 $G_d = G_{mean} = 0,44$ kN/mm^2 The deflection for three point loads can be estimated using the equivalent UDL technique, i.e. Maximum bending moment $= (2,25 \times 0,738) - (1,5 \times 0,4)$ $\qquad = 1,06$ kNm $= (W_e L/8)$ where W_e is the total equivalent UDL required to produce the same maximum bending moment. $\therefore W_e = (1,06 \times 8,0)/1,476 = 5,75$ kN $w_{inst,Q1,m} = \dfrac{5W_e L^3}{384E_d I} = \dfrac{5 \times 5,75 \times 1476^3}{384 \times 7,0 \times 66,67 \times 10^6} = 0,52$ mm Using the method indicated in Example B.1(b) of Appendix B $w_{inst,Q1,v} = \dfrac{0,15W_e L}{bhG_d} = \dfrac{0,15 \times 5,75 \times 1476}{100 \times 200 \times 0,44} = 0,14$ mm	
UK NA to BS EN 1990 Table NA.A1.1	$\psi_{2,1} = 0,3$	$\psi_{2,1} = 0,3$
	$w_{inst,Q1} = (w_{inst,m} + w_{inst,v})_{Q1} = (0,52 + 0,14) = 0,66$ mm $w_{fin,Q1} = w_{inst,Q1}(1 + \psi_{2,1} k_{def}) = 0,66 \times [1 + (0,3 \times 0,6)]$ $\qquad = 0,78$ mm $w_{fin} = w_{fin,G} + w_{fin,Q1} = (0,04 + 0,78) = 0,82$ mm	$w_{fin} \approx 0,82$ mm
UK NA to EC5 Table NA.4	Limiting ratio $= l/150 = 1476/150 = 9,84$ mm $\gg 0,82$ mm More accurate shear deformation can be calculated based on Appendix B. **Note:** If the floor had a plastered ceiling, the permanent load effects would have been higher and the limiting ratio for deflection would have been equal to $l/250$, i.e. 5,9 mm. The reader should check the suitability considering bending, shear, bearing and deflection assuming a plastered ceiling.	**The trimmer beam is adequate with respect to deflection.**

5.4 Glued Thin-webbed Beams

In situations where heavy loads and/or long spans require beams of strength and stiffness which are not available as solid sections, glued thin-webbed beams constructed of I- or box-section are frequently used (see Figure 5.1). The increased size of thin-webbed beams (e.g. 500 mm deep) and consequent strength/weight characteristics permit larger spacing (typically 1,2 m to 4,0 m for ply-web beams) than solid beams, but can still be sufficiently close to enable the use of standard cladding and ceiling systems. In addition, they are frequently able to accommodate services, and insulation materials. The expansion of timber framed housing in the UK has resulted in the use of smaller beams with e.g. OSB or particleboard webs, typically 200 mm to 400 mm deep for floor joists and roof framing.

A considerable saving in weight can be achieved over solid timber joists, and problems often associated with warping, cupping, bowing, twisting and splitting of sawn timber joists can be significantly reduced.

In most cases, since glued thin-webbed beams are hidden, the surface finishes including features such as nail heads, holes and glue marks need not be disguised. If desired, surface treatment can be carried out to enhance the appearance, however this will incur additional cost. The construction of glued thin-webbed beams comprises four principal components:

- ◆   web,
- ◆   stiffeners,
- ◆   flanges,
- ◆   joints between flanges and the web.

The manufacture should comply with the requirements of *BS 6446:1997 'Manufacture of glued structural components of timber and wood based panel products.'*

The modification factors, which are pertinent when designing glued thin-webbed beams, are summarized in Table 5.3.

Factors	Application	Clause Number	Value/Location
k_{mod}	Relates to all strength properties for service and load-duration classes.	3.1.3	Table 3.1
k_{def}	Relates to creep deformation.	2.3.2.2(4)	Table 3.2
k_c	Relates to compression strength perpendicular to the grain.	6.3.2	Equations (6.22), (6.26), (6.28), and (9.5)
k_m	Relates to bending strengths in a cross-section subject to biaxial bending.	6.1.6(2)	1,0
k_{sys}	Relates to lateral load distribution in a continuous deck or floor system.	6.6(2)	Generally 1,1 Figure 6.12 for laminated decks

Table 5.3   Modification Factors – glued thin-webbed beams

5.4.1 Effective Cross-Section Properties

It is assumed that the connection between the flanges and the web are designed such that the component parts of the cross-section behave compositely in resisting both the design bending moment and design shear force. Since the flange and web materials have different elastic modulii and creep characteristics it is convenient to determine 'effective' cross-section properties. This is usually carried out in terms of the flange material with the web stresses being modified accordingly (see Section 4.13 of Chapter 4) leading to the expressions below for the effective cross-sectional area A_{ef} and second moment of area I_{ef}:

$$A_{ef} = A_{flange} + \left(\frac{E_{web}}{E_{flange}}\right)\left(\frac{1+k_{def,flange}}{1+k_{def,web}}\right)A_{web} \qquad I_{ef} = I_{flange} + \left(\frac{E_{web}}{E_{flange}}\right)\left(\frac{1+k_{def,flange}}{1+k_{def,web}}\right)I_{web}$$

Generally, the creep in the flange material is less than that in the web material and consequently the stresses in the web decrease with time whilst those in the flanges increase with time. It is necessary therefore to check the flange stresses corresponding with the final deformations and the web stresses corresponding with the instantaneous deformations.

The final deformations may comprise several components due to a number of actions i.e. permanent, medium-term, short-term etc. which are reflected in the k_{def} values. Consider a glued thin-webbed I-beam for domestic floors with solid timber as flanges and OSB/3 as web. Service Class 2 is assumed here. The values of k_{def} for flanges and web are given from Table 3.2 of EC5 and the value of ψ_2 is given from Table NA.A1.1 of the UK NA to EC, respectively, as

30,0% permanent loading   and   70,0 % variable loading

$k_{def,flange} = k_{def,timber} = 0,80,$ $k_{def,web} = k_{def,OSB/3} = 2,25,$ $\psi_2 = 0,3$

$k_{def,flange,u} = [(k_{def,flange} \times \%\text{ permanent load}) + (\psi_2 \times k_{def,flange} \times \%\text{ variable load})]$
$k_{def,flange,u} = [(0,8 \times 0,3)_{\text{permanent load}} + (0,3 \times 0,8 \times 0,7)_{\text{variable load}}] = 0,41$

$k_{def,web,u} = [(k_{def,web material} \times \%_{\text{permanent load}}) + (\psi_2 \times k_{def,web material} \times \%_{\text{variable load}})]$
$k_{def,web,u} = [(2,25 \times 0,3)_{\text{permanent load}} + (0,3 \times 2,25 \times 0,7)_{\text{variable load}}] = 1,15$

$$\left(\frac{1+k_{def,flange}}{1+k_{def,web}}\right)_u = \left(\frac{1+0,41}{1+1,15}\right) = 0,66$$

The instantaneous deformations do not include the effects of creep, i.e. $k_{def} = $ zero, so:

$$\left(\frac{1+k_{def,flange}}{1+k_{def,web}}\right)_{inst} = 1,0$$

5.4.2 Flanges (Clause 9.1.1(1))

The primary purpose of the **flanges** is to resist tensile and compressive stresses induced by bending effects and/or axial loads. Their construction is normally carried out using continuous or finger-jointed structural timber such as European whitewood, Douglas fir-larch or redwoods; the first of these being the most commonly used. Alternatively plywood or glued-laminated components can be used.

The stresses in the flanges should be checked in accordance with Clause 9.1.1 of EC5, i.e.

$$\sigma_{f,c,max,d} \leq f_{m,d} \qquad \text{Equation (9.1)}$$
$$\sigma_{f,t,max,d} \leq f_{m,d} \qquad \text{Equation (9.2)}$$
$$\sigma_{f,c,d} \leq k_c\, f_{c,0,d} \qquad \text{Equation (9.3)}$$
$$\sigma_{f,t,d} \leq f_{t,0,d} \qquad \text{Equation (9.4)}$$

in EC5

where:

$$f_{m,d} = (f_{m,k} \times k_{mod} \times k_{sys})/\gamma_M$$
$$f_{c,0,d} = (f_{c,0,k} \times k_{mod} \times k_{sys})/\gamma_M$$
$$f_{t,0,d} = (f_{t,0,k} \times k_{mod} \times k_{sys})/\gamma_M$$

k_c is a factor which takes into account lateral instability. In situations where the *compression flange* is not fully restrained against lateral torsional buckling then k_c is evaluated according to Clause 6.3.2 as follows. For example, the factor about z-axis, $k_{c,z}$, can be determined as

$$k_{c,z} = \frac{1}{k_z + \sqrt{k_z^2 - \lambda_{rel,z}^2}} \qquad \text{Equation (6.26) in EC5}$$

where

$$k_z = 0{,}5\left[1 + \beta_c\left(\lambda_{rel,z} - 0{,}3\right) + \lambda_{rel,z}^2\right] \qquad \text{Equation (6.28) in EC5}$$

$$\lambda_{rel,z} = \frac{\lambda_z}{\pi}\sqrt{\frac{f_{c,0,k}}{E_{0,05}}} \qquad \text{Equation (6.22) in EC5}$$

$$\lambda_z = \sqrt{12}\left(\frac{l_c}{b}\right) \qquad \text{Equation (9.5) in EC5}$$

and l_c is equal to the distance between lateral restaints of the compression flange.

If the flange is fully restrained $k_c = 1{,}0$. The stresses for ULS are calculated as

$$\sigma_{f,c,max,d} = \sigma_{f,t,max,d} = \frac{M_{y,d} \times z_{max}}{I_{ef,u}}$$

$$\sigma_{f,c,d} = \sigma_{f,t,d} = \frac{M_{y,d} \times z_{mean}}{I_{ef,u}}$$

For symmetrical beams subject to pure bending.

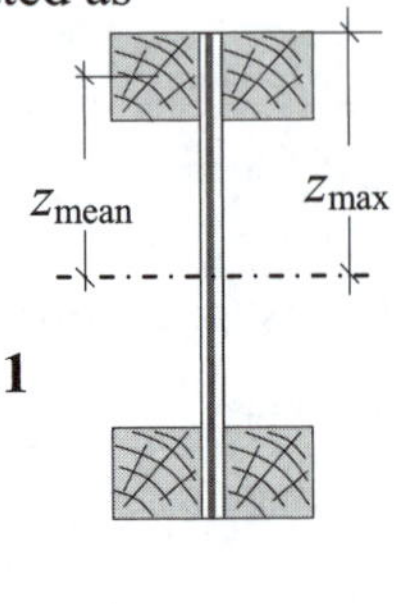

z_{mean} z_{max}

Figure 5.11

In cases where an axial load F_d is also present:

$$\sigma_{f,c,max,d} = \sigma_{f,t,max,d} = \left\{\pm\frac{M_{y,d} \times z_{max}}{I_{ef,u}} \pm \frac{F_d}{A_{ef,u}}\right\}$$

$$\sigma_{f,c,d} = \sigma_{f,t,d} = \left\{\pm\frac{M_{y,d} \times z_{mean}}{I_{ef,u}} \pm \frac{F_d}{A_{ef,u}}\right\}$$

5.4.3 Web

The primary purpose of the **web** is to resist stresses induced by shear forces but it also contributes to the bending strength of the cross-section. In many cases the material adopted is plywood. Developments in other wood-based products such as OSB, particleboard and fibreboards have resulted in an increasing use of these materials for the web. The strength and stiffness properties required when designing the web are:

♦ compression parallel to the grain $(f_{c,0,k})$,
♦ tension parallel to the grain $(f_{t,0,k})$,
♦ panel shear $(f_{v,0,k})$,
♦ planar (rolling) shear $(f_{v,90,k})$,
♦ mean modulus of elasticity parallel to the grain $(E_{0,mean})$ and
♦ modulus of rigidity – panel shear (G_v).

With the exception of plywood, the characteristic values for these variables can be found in BS EN 12369-1:2001 in relation to the classification of the material as given in the following codes: for OSB – BS EN 300:2006, for particleboard – BS EN 312:2003 and for fibreboard and MDF – BS EN 622:1997.

In the case of plywood, characteristic values for bending, panel and planar shear strengths can be found in BS EN 12369-2:2004(E). The values given are in relation to the classification of the material in accordance with either the bending strength and the mean bending modulus as given in BS EN 636:2003 – Plywood Specifications, e.g. a given plywood with the following characteristics:

$$f_{m,0,05} = 25,5 \text{ N/mm}^2 \quad f_{m,90,05} = 42,3 \text{ N/mm}^2 \quad E_{m,0,05} = 3750 \text{ N/mm}^2 \quad E_{m,90,05} = 4130 \text{ N/mm}^2$$

is classified in BS EN 636, Tables 1 and 2 as: ***F15/25 E30/40***
where ***F*** relates to the 5th percentile bending strength and ***E*** relates to the 5th percentile bending modulus.

These values do not represent the characteristic values to be used for design purposes; they are used to classify the plywood and in quality control procedures during manufacture. The characteristic values for design purposes may be determined from Tables 2 and 3 in BS EN 12369-2 in accordance with the classification obtained from BS EN 636, i.e. for ***F15/25 E30/40 plywood***

From BS EN 12369-2   Table 2 $f_{m,0,k} = 15,0 \text{ N/mm}^2$ $f_{m,90,k} = 25,0 \text{ N/mm}^2$
                     Table 3 $E_{m,0,mean} = 3000 \text{ N/mm}^2$ $E_{m,90,mean} = 4000 \text{ N/mm}^2$
                               $E_{m,0,05} = 2400 \text{ N/mm}^2$ $E_{m,90,05} = 3200 \text{ N/mm}^2$
                     Table 4 $f_{v,k,panel} = 3,0 \text{ N/mm}^2$ $G_{v,k,panel} = 300 \text{ N/mm}^2$
                               $f_{v,k,planar} = 0,5 \text{ N/mm}^2$ $G_{v,k,planar} = 20,0 \text{ N/mm}^2$

The characteristic values of tensile and compressive strength are not given in the design codes and should be determined by mechanical testing in accordance with BS EN 789:2004 – Timber Structures – Test methods – Mechanical properties of wood-based panels. The values, as presented in the code, are only of limited use to designers and use of an alternative to the code values may be more convenient.

Characteristic values can be obtained from the manufacturer of the material being used or, for specific types of plywood, can be estimated by modifying the values given in BS 5268-2:2002 as follows:

Characteristic strength properties (X_k) = (2,7×grade strength property from BS 5268-2)
i.e. 2,7 × the values given in Tables 40 to 56

Characteristic stiffness properties (E_k) = (1,8×grade strength property from BS 5268-2)
i.e. 1,8 × the values given in Tables 40 to 56

The construction of the webs is normally carried out such that butt end joints do not occur at the mid-span location and full 2440 mm panels are used where possible. In Finnish birch-faced plywood the face grain is normally perpendicular to the span whilst in cases where Douglas fir is used the face grain is normally parallel to the span.

Since the web also contributes to the bending strength in addition to the shear strength, there are four stress conditions which must be checked according to EC5:

♦ normal maximum compressive stress,
♦ normal maximum tensile stress,
♦ panel shear stress, and
♦ planar shear stress (i.e. rolling shear).

The critical normal and panel web stresses correspond with the instantaneous deformations (i.e. no creep with k_{def} = zero) and are checked as indicated in Sections 5.4.3.1 and 5.4.3.2.

5.4.3.1 Normal Maximum Compressive and Tensile Stresses (Clause 9.1.1(4))

$\sigma_{w,c,d} \leq f_{c,w,d}$ Equation (9.6) in EC 5
$\sigma_{w,t,d} \leq f_{t,w,d}$ Equation (9.7) in EC5

where:
$f_{c,w,d} = (f_{c,0,k} \times k_{mod} \times k_{sys})/\gamma_M$
$f_{t,w,d} = (f_{t,0,k} \times k_{mod} \times k_{sys})/\gamma_M$

$$\sigma_{w,c,d} = \sigma_{w,t,d} = \frac{M_{y,d} \times z_{max}}{I_{ef,inst}}$$

In cases where an axial load F_d is also present:

$$\sigma_{w,c,d} = \left\{ \pm \frac{M_{y,d} \times z_{max}}{I_{ef,inst}} \pm \frac{F_d}{A_{ef,inst}} \right\}$$

$$\sigma_{w,t,d} = \left\{ \pm \frac{M_{y,d} \times z_{max}}{I_{ef,inst}} \pm \frac{F_d}{A_{ef,inst}} \right\}$$

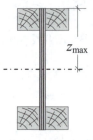

z_{max}

Figure 5.12

5.4.3.2 Panel Shear (Clause 9.1.1(7))

The aspect ratio of the web should be limited to 70 unless a detailed analysis is carried out to assess the resistance to web shear buckling. This is reflected in the code, i.e.

$h_w \leq 70b_w$ Equation (9.8) in EC5

The panel shear strength (of each web) should be checked such that:

$$F_{v,w,Ed} \leq \begin{cases} b_w h_w \left(1 + \dfrac{0,5\left(h_{f,t} + h_{fc}\right)}{h_w} \right) f_{v,0,d} & \text{for} \quad h_w \leq 35b_w \\[4mm] 35b_w^2 \left(1 + \dfrac{0,5\left(h_{f,t} + h_{fc}\right)}{h_w} \right) f_{v,0,d} & \text{for} \quad 35b_w \leq h_w \leq 70b_w \end{cases}$$

Equation (9.9) in EC5

where:
$F_{v,w,Ed}$ is the design shear force on each web,
h_w is the clear distance between the flanges,
b_w is the thickness of each web,
h_{fc} is the compression flange depth,
h_{ft} is the tension flange depth,
$f_{v,0,d} = (f_{v,0,k} \times k_{mod} \times k_{sys})/\gamma_M$.

Figure 5.13

The maximum horizontal shear stress induced in a beam subjected to bending and vertical shear forces occurs at the level of the neutral axis.

5.4.3.3 Planar (Rolling Shear) (Clause 9.1.1(8))

The physical construction of some materials e.g. plywood, where alternate veneers have grain directions which are mutually perpendicular, enables a mode of failure called '*rolling shear*' to occur (see Figure 5.14); there is a tendency for the material fibres to roll across each other, creating a horizontal shear failure plane. This phenomenon can occur at locations where plywood is joined to other members/materials, either at the interface with the plywood or between adjacent veneers of the plywood.

In glued thin-webbed beams the rolling shear must be checked at the connection of the web to the flanges and sufficient thickness (h_f) of flange must be available to transfer the horizontal shear force at this location.

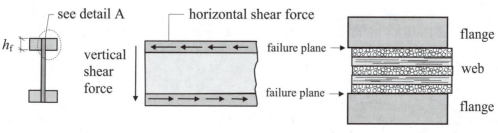

Figure 5.14   Detail A – Plan View

The rolling shear must satisfy the following condition:

$$\tau_{mean,d} \leq \begin{cases} f_{v,90,d} & \text{for } h_f \leq 4b_{ef} \\ f_{v,90,d}\left(\dfrac{4b_{ef}}{h_f}\right)^{0,8} & \text{for } h_f > 4b_{ef} \end{cases}$$ Equation (9.10) in EC5

where:

$f_{v,90,d} = (f_{v,90,k} \times k_{mod} \times k_{sys}) / \gamma_M$

$$b_{ef} = \begin{cases} b_w & \text{for boxed beams} \\ b_w/2 & \text{for I - beams} \end{cases}$$ Equation (9.11) in EC5

The magnitude of the rolling shear stress for ULS can be determined for the critical section x-x at the interface between the web and the flange, as shown in Figure 5.15.

$$\tau_{mean,d} = \frac{V_d Az}{I_{ef,u} \times h_f}$$

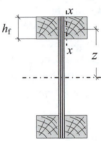

where:

$\tau_{mean,d}$ is the mean design horizontal shear stress,
V_d is the design vertical shear force,   **Figure 5.15**
Az is the first moment of area of the material above the neutral axis,
$I_{ef,u}$ is the effective second moment of area of the cross-section for ULS,
b is the thickness of the web at the position of the section being considered.

5.4.3.4 Web Stiffeners

Where webs are slender and at locations such as supports and points of application of concentrated load, there is the possibility of failure caused by buckling of the web. EC5 does not give any guidance on the design of either non-load bearing (intermediate) or load bearing web stiffeners.

Most proprietary suppliers of glued thin-webbed beams advise web stiffener details based on the results of full scale tests of their product.

Design methods are illustrated in various publications, notably by the Council of Forest Industries of British Columbia (COFI) publication *Fir Plywood Web Beam Design* (60) *and Timber Designers' Manual* (73). Reference should be made to these publications for further information regarding stiffeners.

5.4.3.5 Deflection (Clause 7.2)

As indicated in Section 5.3.6 both bending and shear deflection should be checked; this is particularly important in glued thin-webbed beams where the shear deflection can be a significantly higher proportion of the overall value than in solid beams. The deflection is calculated with respect to permanent and variable loads and compared with the recommended limiting values given in Table NA.4 of the UK National Annex.

Note: the values are not mandatory and should be agreed with the client at the beginning of the project.

The final value, $w_{fin} = w_{fin,G} + w_{fin,Q1}$, is calculated on the basis of the instantaneous values which are subsequently modified using k_{def} and ψ_2 to allow for creep and load duration effects, i.e.

For permanent load deformations:

$$w_{fin,G} = w_{inst,G}\,(1 + k_{def})$$
$$= [w_{inst,G}\,(1 + k_{def})]_m + [w_{inst,G}\,(1 + k_{def})]_v$$

Equation (2.3) in EC5

For variable deformations:

$$w_{fin,Q1} = w_{inst,Q1}\,(1 + \psi_{2,1}\,k_{def})$$
$$= [w_{inst,Q1}\,(1 + \psi_{2,1}\,k_{def})]_m + [w_{inst,Q1}\,(1 + \psi_{2,1}\,k_{def})]_v$$

Equation (2.4) in EC5

where $\psi_{2,1}$ is the combination factor given in the Table NA.A1.1 of the UK National Annex to BS EN 1990.

When calculating the value of w_{inst}, k_{def} = zero and consequently the appropriate section property for the bending deflection is:

$$I_{ef,inst} = I_{flange} + \left(\frac{E_{web}}{E_{flange}}\right) I_{web}$$

as indicated above and for the shear deflection is:
$$A_{web,inst} = (h \times t)_{web}$$

When calculating the value of w_{fin}, the I_{ef} for the bending deflection comprises values from two wood-based materials having different time-dependent behaviour and the k_{def} value should be determined from Equation (2.13) in EC5, i.e.

$$k_{def} = 2 \times \sqrt{k_{def,flange} \times k_{def,web}}$$

Since the shear deflection does not include the flange material properties the k_{def} value adopted is that appropriate to the web material.

The resulting expressions for $w_{fin,G}$ and $w_{fin,Q1}$ are:

$$w_{fin,G} = [w_{inst,G}\,(1 + 2 \times \sqrt{k_{def,flange} \times k_{def,web}}\,)]_m + [w_{inst,G}\,(1 + k_{def})]_v$$

$$w_{fin,Q1} = [w_{inst,Q1}(1 + \psi_{2,1} \times 2 \times \sqrt{k_{def,flange} \times k_{def,web}}\,)]_m + [w_{inst,Q1}(1 + \psi_{2,1}\,k_{def})]_v$$

5.4.3.6 Lateral Stability

The suitability of glued thin-walled beams with respect to lateral stability is checked by ensuring that Equation (9.3) of EC5 is satisfied, i.e.

$$\sigma_{f,c,d} \le k_c\, f_{f,c,d}$$

where k_c is a factor to take into account lateral instability as defined previously in Section 5.4.2.

5.4.4 Example 5.3: Glued Thin-webbed Beam - Roof Beam Design

A local sports club is to be extended to accommodate two squash courts and changing rooms, as shown in Figures 5.16(a) and (b). The roof construction is to be of traditional flat roof design comprising felt, insulation and sarking, supported by a series of timber glued thin-webbed I-beam sections sitting on block walls.

Check the suitability of the proposed beam section for a typical internal beam (snow and wind loading is not considered).

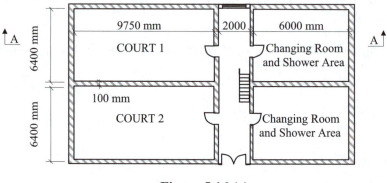

Figure 5.16 (a)

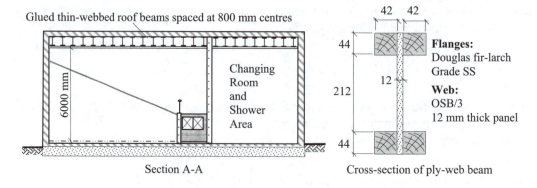

Figure 5.16 (b)

5.4.4.1 Solution to Example 5.3

Contract : I-beams Job Ref. No. : Example 5.3	Calculated by : W.McK.
Part of Structure : Glued thin-webbed roof beam	Checked by : B.Z.
Calculation Sheet No. : 1 of 9	Date :

References	Calculations	Output
BS EN 1990	Eurocode: Basis of Structural Design	
BS EN 1991	Eurocode 1: Actions on Structures	
BS EN 338:2002	Structural Timber - Strength Classes	
BS EN 1912:2004	Structural Timber - Strength Classes – Assignment of visual grades and species	
BS EN 1995-1-1: 2004	Eurocode 5: Design of Timber Structures Part 1.1: General - Common Rules and Rules for Buildings.	
BS EN 12369-1: 2001	Wood-based panels – Characteristic values for structural design Part 1: OSB, particleboards and fibreboards	
National Annex	UK National Annex to BS EN 1995-1-1	

Dimensions: 42 | 42 at top flange width, 44 top flange thickness, 212 web height, 12 web thickness, 44 bottom flange thickness.

Flanges: Douglas fir-larch Grade SS

Web: OSB/3 12 mm thick panel

BS EN 1912:2004 Table 1	Characteristic values for Douglas fir-Larch timber Grade SS Strength class = C24	
BS EN 338:2002 Table 1	Characteristic values for C24 Timber **Strength Properties:**	
	Bending $\qquad f_{m,k} = 24,0 \text{ N/mm}^2$	
	Compression parallel to grain $\qquad f_{c,0,k} = 21,0 \text{ N/mm}^2$	
	Tension parallel to grain $\qquad f_{t,0,k} = 14,0 \text{ N/mm}^2$	
	Stiffness Properties:	
	Mean modulus of elasticity parallel to the grain $\qquad E_{0,mean} = 11,0 \text{ kN/mm}^2$	
	Modulus of rigidity $\qquad G_{mean} = 0,69 \text{ kN/mm}^2$	
BS EN 12369-1 Table 2	Characteristic values for OSB/3 12mm thick panel **Strength Properties:**	
	Compression parallel to grain $\qquad f_{c,0,k} = 15,4 \text{ N/mm}^2$	
	Tension parallel to grain $\qquad f_{t,0,k} = 9,4 \text{ N/mm}^2$	
	Panel shear $\qquad f_{v,k} = 6,8 \text{ N/mm}^2$	
	Planar shear $\qquad f_{v,90,k} = 1,0 \text{ N/mm}^2$	
	Stiffness Properties:	
	Mean modulus of elasticity in bending parallel to the grain $\qquad E_{0,mean} = 4,93 \text{ kN/mm}^2$	
	Modulus of rigidity – panel shear $\qquad G_v = 1,08 \text{ kN/mm}^2$	

Contract : I-beams Job Ref. No. : Example 5.3 Part of Structure : Glued thin-webbed roof beam Calculation Sheet No. : 2 of 9	Calculated by : W.McK. Checked by : B.Z. Date :

References	Calculations	Output
Clause 2.3.1.2 Table NA.1 (EC5: Table 2.3)	Load-duration Classes: use Table NA.1 from National Annex Self-weight - permanent Imposed floor loading - medium-term	
Clause 2.3.2.1	Load-duration and moisture influence on strength: k_{mod} The flanges and web have different time-dependent behaviour.	
National Annex Table NA. 2 Table 3.1 Clause 3.1.3(2)	Service classes: Assume Service Class 2 Solid timber: Service Class 2 Permanent actions $k_{mod} = 0,6$ Medium-term actions $k_{mod} = 0,8$ Use k_{mod} corresponding to action with the shortest duration.	$k_{mod} = 0,8$
Table 3.1 Clause 3.1.3(2)	**OSB/3:** Service Class 2 Permanent actions $k_{mod} = 0,3$ Instantaneous actions $k_{mod} = 0,9$ Use k_{mod} corresponding to action with the shortest duration.	$k_{mod} = 0,9$
National Annex Table NA.3	Partial factors for material properties: γ_M Solid timber: $\gamma_M = 1,3$ OSB/3 $\gamma_M = 1,2$	$\gamma_{M,\text{solid timber}} = 1,3$ $\gamma_{M,\text{OSB}} = 1,2$
Clause 6.6	System strength: k_{sys} Assume the sarking has adequate provision for lateral distribution of loading the strength properties can be multiplied by k_{sys}.	$k_{sys} = 1,1$
	Loading: Permanent actions 3 layers bituminous felt $= 0,11$ kN/m^2 Fibreboard insulation $= 0,20$ kN/m^2 Timber sarking $\underline{= 0,07}$ kN/m^2 Total $= 0,38$ kN/m^2 Allowing for self-weight assume $= 0,40$ kN/m^2	
EN 1991-1-1 National Annex Tables 6.9/6.10	Variable actions Category of loaded area is H Imposed roof load $q_k = 0,6$ kN/m^2	
BS EN 1990 Equation (6.10)	Eurocode: Basis of Structural Design $E_d = \sum_{j\geq1} \gamma_{G,j} G_{k,j} + \gamma_{Q,1} Q_{k,1} + \sum_{i>1} \gamma_{Q,i} \psi_{0,i} Q_{k,i}$ $E_d = \left(\gamma_G G_k + \gamma_{Q1} Q_{k,1} \right)$ where E_d is the design value of the effect of the actions.	

References	Calculations	Output
Contract : I-beams Job Ref. No. : Example 5.3 **Part of Structure : Glued thin-webbed roof beam** **Calculation Sheet No. : 3 of 9**		**Calculated by : W.McK.** **Checked by : B.Z.** **Date :**

References	Calculations	Output
UK NA to BS EN 1990 Table NA.A1.2(B)	$\gamma_G = 1{,}35$ and $\gamma_{Q,1} = 1{,}50$ The design load $F_d = \gamma_G g_k + \gamma_{Q,1} q_{k,1}$ Permanent design loading $= (1{,}35 \times 0{,}4) = 0{.}54 \text{ kN/m}^2$ Variable design loading $= (1{,}5 \times 0{,}6) = \underline{0{,}90 \text{ kN/m}^2}$ Total design loading $F_d = (0{,}54 + 0{,}90) = 1{,}44 \text{ kN/m}^2$ Roof area supported by one internal beam = shaded area 800 mm spacing Shaded area $= 0{,}8 \times (6{,}4 + 0{,}1) = 5{,}2 \text{ m}^2$ Permanent load $= (0{,}54 \times 5{,}2) = 2{,}81 \text{ kN}$ Variable loading $= (0{,}9 \times 5{,}2) = \underline{4{,}68 \text{ kN}}$ Total design loading $\qquad W_d = 7{,}49 \text{ kN}$ Permanent load $= (2{,}81/7{,}49) \times 100 = 37{,}5 \%$ of the total load Variable load $= (4{,}68/7{,}49) \times 100 = 62{,}5 \%$ of the total load Design bending moment $M_{y,d} = \dfrac{W_d L}{8} = \dfrac{7{,}49 \times 6{,}5}{8} = 6{,}09 \text{ kNm}$ Design shear force $V_d = (W_d/2) = (0{,}5 \times 7{,}49) = 3{,}75 \text{ kN}$	$F_d = 1{,}44 \text{ kN/m}^2$ $M_{y,d} = 6{,}09 \text{ kNm}$ $V_d = 3{,}75 \text{ kN}$
See Section 4.13 Chapter 4 	Effective cross-sectional properties in terms of flange material for UDL: $$A_{ef,u} = A_{flange} + \left(\dfrac{E_{web}}{E_{flange}}\right)\left(\dfrac{1 + k_{def,flange}}{1 + k_{def,web}}\right) A_{web}$$ $$I_{ef,u} = I_{flange} + \left(\dfrac{E_{web}}{E_{flange}}\right)\left(\dfrac{1 + k_{def,flange}}{1 + k_{def,web}}\right) I_{web}$$ When considering instantaneous deformations the effect of creep can be ignored and $k_{def} = 0$. When considering final deformations the effects of creep must be included and values of k_{def} can be determined from Table 3.2 of EC5.	

References	Calculations	Output
Contract : I-beams Job Ref. No. : Example 5.3 Part of Structure : Glued thin-webbed roof beam Calculation Sheet No. : 4 of 9		Calculated by : W.McK. Checked by : B.Z. Date :

References	Calculations	Output
BS EN 1995-1-1 Clause 2.3.2.2(2)	For ultimate limit states the final mean value of modulus of elasticity, $E_{mean,fin}$ should be calculated from: $$E_{mean,fin} = \frac{E_{mean}}{(1+\psi_2 k_{def})}$$ In the case of permanent actions ψ_2 should be replaced by 1,0	
National Annex BS EN 1990 Table NA.A1.1	For imposed loads on roofs as defined in the National Annex to BS EN 1991-1-:2002 (i.e. category H in Table 6.9 of EC1-1-1) $\psi_2 = 0,0$ k_{def} comprises two components, one due to the permanent load and the other due to the variable load, i.e. $k_{def,flange} = (k_{def,flange})_{permanent} + (\psi_2\, k_{def,flange})_{variable}$ $\qquad k_{def,web} = (k_{def,web})_{permanent} + (\psi_2\, k_{def,web})_{variable}$ The components are evaluated in proportion to the % load in each case and in accordance with the appropriate ψ_2 values as indicated above.	
BS EN 1995-1-1 Table 3.2	Values of k_{def} for Service Class 2 Flange: Solid timber, $k_{def} = 0,8$ $(k_{def,flange})_{permanent} \quad = (k_{def,flange} \times \text{proportion of load} \times \psi_2)$ $\qquad\qquad\qquad\qquad = (0,8 \times 0,375 \times 1,0) = 0,30$ $(k_{def,flange})_{variable} \quad = (k_{def,flange} \times \text{proportion of load} \times \psi_2)$ $\qquad\qquad\qquad\qquad = (0,8 \times 0,625 \times 0,0) = 0$ $k_{def,flange} = (0,30 + 0) = 0,30$	$k_{def,flange} = 0,30$
Table 3.2	Web: OSB/3, $k_{def} = 2,25$ $(k_{def,web})_{permanent} \quad = (k_{def,web} \times \text{proportion of load} \times \psi_2)$ $\qquad\qquad\qquad\qquad = (2,25 \times 0,375 \times 1,0) = 0,84$ $(k_{def,web})_{variable} \quad = (k_{def,web} \times \text{proportion of load} \times \psi_2)$ $\qquad\qquad\qquad\qquad = (2,25 \times 0,625 \times 0,0) = 0$ $k_{def,web} = (0,84 + 0) = 0,84$ $$\left(\frac{1+k_{def,flange}}{1+k_{def,web}}\right)_u = \left(\frac{1+0,3}{1+0,84}\right) = 0,71$$ $$\left(\frac{E_{web}}{E_{flange}}\right) = \left(\frac{4,93}{11,0}\right) = 0,448$$ Creep in the solid timber flanges is less than that in the OSB web, i.e. the normal flange stresses increase with time whilst those in the web decrease. The flange stresses corresponding with the final deformations and the web stresses corresponding with the instantaneous deformations should therefore be checked.	$k_{def,web} = 0,84$

Contract : I-beams Job Ref. No. : Example 5.3	Calculated by : W.McK.
Part of Structure : Glued thin-webbed roof beam	Checked by : B.Z.
Calculation Sheet No. : 5 of 9	Date :

References	Calculations	Output
	Effective properties for ultimate limit states corresponding with the instantaneous deformations: $k_{def} = 0$ $$A_{flange} = [4 \times (42 \times 44)] = 7392,0 \text{ mm}^2$$ $$I_{flange} = 4 \times (I_{centroid} + Az^2) = 4 \times \left(\frac{bh^3}{12} + bhz^2 \right)$$ $$= 4 \times \left\{ \frac{42 \times 44^3}{12} + \left[42 \times 44 \times (150 - 22)^2 \right] \right\} = 122,303 \times 10^6 \text{ mm}^4$$ $$A_{web} = (300 \times 12) = 3600,0 \text{ mm}^2$$ $$I_{web} = \left(\frac{bh^3}{12} \right) = \left(\frac{12 \times 300^3}{12} \right) = 27,0 \times 10^6 \text{ mm}^4$$ Since $k_{def} = 0 \quad \left(\dfrac{1 + k_{def,flange}}{1 + k_{def,web}} \right) = 1,0$ $$A_{ef,inst} = A_{flange} + \left(\frac{E_{web}}{E_{flange}} \right) \left(\frac{1 + k_{def,flange}}{1 + k_{def,web}} \right) A_{web}$$ $$= [7392,0 + (0,448 \times 1,0 \times 3600,0)] = 9,005 \times 10^3 \text{ mm}^2$$ $$I_{ef,inst} = I_{flange} + \left(\frac{E_{web}}{E_{flange}} \right) \left(\frac{1 + k_{def,flange}}{1 + k_{def,web}} \right) I_{web}$$ $$= [122,303 \times 10^6 + (0,448 \times 1,0 \times 27,0 \times 10^6)]$$ $$= 134,40 \times 10^6 \text{ mm}^4$$ **Effective properties for ultimate limit states corresponding with the final deformations:** $k_{def,flange} = 0,3$ and $k_{def,web} = 0,84$ $$A_{ef,u} = [7392,0 + (0,448 \times 0,71 \times 3600,0)] = 8537,1 \times 10^3 \text{ mm}^2$$ $$I_{ef,u} = [122,303 \times 10^6 + (0,448 \times 0,71 \times 27,0 \times 10^6)]$$ $$= 130,89 \times 10^6 \text{ mm}^4$$	$A_{ef,inst} = 9,005 \times 10^6 \text{ mm}^4$ $I_{ef,inst} = 134,40 \times 10^6 \text{ mm}^4$ $I_{ef,u} = 130,89 \times 10^6 \text{ mm}^4$
Clause 9.1.1(1) Equation (9.1) Equation (9.2)	**Check flange normal stresses:** Design bending strength of the flange $$f_{m,d} = (f_{m,k} \times k_{mod} \times k_{sys})/\gamma_M = (24,0 \times 0,8 \times 1,1)/1,3$$ $$= 16,25 \text{ N/mm}^2$$ $\sigma_{f,c,max,d} < f_{m,d}$ and $z_{max} = (300/2) = 150 \text{ mm}$ $$\sigma_{f,c,max,d} = \frac{M_{y,d} \times z_{max}}{I_{ef,u}} = \left(\frac{6,09 \times 10^6 \times 150}{130,89 \times 10^6} \right) = 6,98 \text{ N/mm}^2$$ $$< f_{m,d}$$ Similarly for $\sigma_{f,t,max,d} < f_{m,d}$	$f_{m,d} = 16,25 \text{ N/mm}^2$

Contract : I-beams Job Ref. No. : Example 5.3	Calculated by : W.McK.
Part of Structure : Glued thin-webbed roof beam	Checked by : B.Z.
Calculation Sheet No. : 6 of 9	Date :

References	Calculations	Output
	Design compressive strength of the flange parallel to the grain $f_{c,0,d} = (f_{c,0,k} \times k_{mod} \times k_{sys})/\gamma_M = (21,0 \times 0,8 \times 1,1)/1,3$ $\qquad = 14,22$ N/mm² Assume that the compression flange is fully restrained by the sarking and $k_c = 1.0$ $\therefore k_c f_{c,0,d} = 14,22$ N/mm² (k_c is a factor taking into account lateral instability.)	$k_c f_{c,0,d} = 14,22$ N/mm²
Equation (9.3)	$\sigma_{f,c,d} \le k_c f_{c,0,d}$ where $\sigma_{f,c,d}$ is the mean flange design compressive stress and $z_{mean} = (150 - 22) = 128$ mm $\sigma_{f,c,max,d} = \dfrac{M_{y,d} \times z_{mean}}{I_{ef,u}} = \left(\dfrac{6,09\times10^6 \times 128}{130,89\times10^6}\right) = 5,96$ N/mm² $\qquad\qquad < k_c f_{c,0,d}$	
Equation (9.4)	Design tensile strength of the flange parallel to the grain $f_{t,0,d} = (f_{t,0,k} \times k_{mod} \times k_{sys})/\gamma_M = (14,0 \times 0,8 \times 1,1)/1,3$ $\qquad = 9,48$ N/mm² $\sigma_{f,t,d} \le f_{t,0,d}$ where $\sigma_{f,t,d}$ is the mean flange design tensile stress $\sigma_{f,t,max,d} = \dfrac{M_{y,d} \times z_{mean}}{I_{ef,u}} = \left(\dfrac{6,09\times10^6 \times 128}{130,89\times10^6}\right) = 5,96$ N/mm² $\qquad\qquad < f_{t,0,d}$	$f_{t,0,d} = 9,48$ N/mm² Flanges are adequate
Clause 9.1.1(4)	**Check web normal stresses:** (converted to web material) Design compressive strength of the web parallel to the grain $f_{c,w,d} = (f_{c,0,k} \times k_{mod} \times k_{sys})/\gamma_M = (15,4 \times 0,9 \times 1,1)/1,2$ $\qquad = 12,71$ N/mm²	$f_{c,w,d} = 12,71$ N/mm²
Equation (9.6)	$\sigma_{w,c,d} \le f_{c,w,d}$ where $\sigma_{w,c,d}$ is the design compressive stress in the web $\sigma_{w,c,d} = \dfrac{M_{y,d} \times z_{max}}{I_{ef,inst}} \times \dfrac{E_{web}}{E_{flange}} = \left(\dfrac{6,09\times10^6 \times 150 \times 0,448}{134,40\times10^6}\right)$ $\qquad = 3,05$ N/mm² $< f_{c,w,d}$	Web is adequate with respect to compression.
Equation (9.7)	Design tensile strength of the web parallel to the grain $f_{t,w,d} = (f_{t,0,k} \times k_{mod} \times k_{sys})/\gamma_M = (9,4 \times 0,9 \times 1,1)/1,2$ $\qquad = 7,76$ N/mm² $\sigma_{w,t,d} \le f_{t,w,d}$ where $\sigma_{w,t,d}$ is the design tensile stress in the web $\sigma_{w,t,d} = \dfrac{M_{y,d} \times z_{max}}{I_{ef,inst}} \times \dfrac{E_{web}}{E_{flange}} = \left(\dfrac{6,09\times10^6 \times 150 \times 0,448}{134,40\times10^6}\right)$ $\qquad = 3,05$ N/mm² $< f_{t,w,d}$	$f_{t,w,d} = 7,76$ N/mm² Web is adequate with respect to tension.

References	Calculations	Output
	Contract : I-beams Job Ref. No. : Example 5.3 **Part of Structure : Glued thin-webbed roof beam** **Calculation Sheet No. : 7 of 9**	**Calculated by : W.McK.** **Checked by : B.Z.** **Date :**

References	Calculations	Output
Clause 9.1.1(7) Equation (9.8)	**Check web aspect ratio and panel shear strength** Since no detailed buckling analysis of the web is being carried out, the aspect ratio should be checked such that $h_w \leq 70b_w$ $h_w = 212$ mm, and $70b_w = (70 \times 12) = 840$ mm $> h_w$	**Aspect ratio of web is satisfactory.**
Equation (9.9)	$F_{v,w,Ed} \leq \begin{cases} b_w h_w \left(1 + \dfrac{0,5\left(h_{f,t} + h_{fc}\right)}{h_w}\right) f_{v,0,d} & \text{for } h_w \leq 35b_w \\[3mm] 35b_w^2 \left(1 + \dfrac{0,5\left(h_{f,t} + h_{fc}\right)}{h_w}\right) f_{v,0,d} & \text{for } 35b_w \leq h_w \leq 70b_w \end{cases}$	
	Design panel shear strength of the web $f_{v,0,d} = (f_{v,0,k}\, k_{mod}\, k_{sys})/\gamma_M = (6,8 \times 0,9 \times 1,1)/1,2 = 5,61$ N/mm^2 $35b_w = (35 \times 12) = 420$ mm $> h_w$ $\left[12 \times 212 \times \left(1 + \dfrac{0,5(44+44)}{212}\right) \times 5,61\right] / 10^3 = 17,23$ kN $F_{v,w,Ed} = 3,75$ kN $< 17,23$ kN	$f_{v,0,d} = \mathbf{5,61}$ **N/mm^2** **Web is adequate with respect to panel shear.**
Clause 9.1.1(8)	**Check the planar (rolling) stresses at the glue lines:**	
Equation (9.10)	$\tau_{mean,d} \leq \begin{cases} f_{v,90,d} & \text{for } h_f \leq 4b_{ef} \\[3mm] f_{v,90,d}\left(\dfrac{4b_{ef}}{h_f}\right)^{0,8} & \text{for } h_f > 4b_{ef} \end{cases}$	
Equation (9.11)	$b_{ef} = b_w/2$ for I-beams $\therefore b_{ef} = 12/2 = 6,0$ mm, $4b_{ef} = 24,0$ mm $h_f = 44,0$ mm $> 4b_{ef}$ $f_{v,90,d} = (f_{v,90,k} \times k_{mod} \times k_{sys})/\gamma_M = (1,0 \times 0,9 \times 1,1)/1,2$ $= 0,83$ N/mm^2 $f_{v,90,d}\left(\dfrac{4b_{ef}}{h_f}\right)^{0,8} = 0,83 \times \left(\dfrac{4 \times 6,0}{44,0}\right)^{0,8} = 0,51$ N/mm^2	$f_{v,90,d} = \mathbf{0,83}$ **N/mm^2**
Figure 9.1 See Chapter 4 Section 4.11	The design shear stress at the joint between the flanges and the web is given by: $\tau_{mean,d} = \dfrac{V_d A z}{I_{ef,u}\, b}$ The critical shear stresses on the gluelines correspond with the critical flange stresses, i.e. when final deformations occur.	

References	Calculations	Output
	$Az = 2 \times [42 \times 44 \times (150 - 22)] = 473,1 \times 10^3 \text{ mm}^3$	
	$I_{ef,u} = 130,89 \times 10^6 \text{ mm}^4$	
	$b = (2 \times 44) = 88,0 \text{ mm}$	**Web is adequate with**
	$V_d = 3,75 \text{ kN}$	**respect to planar shear.**
	$\tau_{mean,d} = \dfrac{3,75 \times 10^3 \times 473,1 \times 10^3}{130,89 \times 10^6 \times 88,0} = 0,15 \text{ N/mm}^2 < 0,51 \text{ N/mm}^2$	
	Serviceability Limit States	
	Deformation:	
Clause 7.2	Limiting values for deflection of beams:	
	$w_{net,fin} = w_{inst} + w_{creep} - w_c = w_{fin} - w_c$	
	Since there is no camber $w_c = 0$ and $w_{net,fin} = w_{inst} + w_{creep} = w_{fin}$	
	(Note: $\gamma_G = \gamma_Q = 1,0$ for deformations)	
	Permanent design loading $= (1,0 \times 0,4) = 0,4 \text{ kN/m}^2$	
	Variable design loading $= (1,0 \times 0,6) = 0,6 \text{ kN/m}^2$	
	Shaded area $= 0,8 \times (6,4 + 0,1) = 5,2 \text{ m}^2$	
	Permanent load $= (0,4 \times 5,2) = 2,08 \text{ kN}$	
	Variable loading $= (0,6 \times 5,2) = 3,12 \text{ kN}$	
BS EN 1995-1-1	Values of k_{def} for Service Class 2	
Table 3.2	Flange: Solid timber, $k_{def,flange} = 0,8$	$k_{def,flang} = 0,8$
Table 3.2	Web: OSB/3, $k_{def,web} = 2,25$	$k_{def,web} = 2,25$
	Permanent load deformation:	
Equation (2.3)	$w_{fin,G} = w_{inst,G}(1 + k_{def}) = (w_{inst,m} + w_{inst,v})_G (1 + k_{def})$	
	Total deformation due to (bending + shear effects) is given by:	
	$w_{inst,G} = \dfrac{5WL^3}{384 E_f I_{ef}} + \dfrac{0,68WL}{AG}$ (see Example B.2 of Appendix B)	
Appendix B	Transformed area $A = 20,09 \times 10^3 \text{ mm}^2$	
	$W = 2,08 \text{ kN}$	
	$L = 6500 \text{ mm}$	
	$E_f = E_{0,mean} = 11,0 \text{ kN/mm}^2$	
	$G_w = G_v = 1,08 \text{ kN/mm}^2$	
	For instantaneous deformation (w_{inst}): $k_{def} = 0$ i.e. no creep	
	$I_{ef,inst} = I_{flange} + \left(\dfrac{E_{web}}{E_{flange}} \right)\left(\dfrac{1 + k_{def,flange}}{1 + k_{def,web}} \right) I_{web} = 134,40 \times 10^6 \text{ mm}^4$	
	$A_{web} = (300 \times 12) = 3600,0 \text{ mm}^2$	

Contract : I-beams Job Ref. No. : Example 5.3	Calculated by : W.McK.
Part of Structure : Glued thin-webbed roof beam	Checked by : B.Z.
Calculation Sheet No. : 8 of 9	Date :

References	Calculations	Output
	Contract : I-beams Job Ref. No. : Example 5.3 **Part of Structure : Glued thin-webbed roof beam** **Calculation Sheet No. : 9 of 9**	**Calculated by : W.McK.** **Checked by : B.Z.** **Date :**

References	Calculations	Output
	Bending deformation: $$w_{inst,G,m} = \frac{5WL^3}{384E_f I_{ef,inst}} = \frac{5 \times 2,08 \times 6500^3}{384 \times 11,0 \times 134,4 \times 10^6} = 5,03 \text{ mm}$$ Shear deformation: $$w_{inst,G,v} = \frac{0,68WL}{A_{web}G_w} = \frac{0,68 \times 2,08 \times 6500}{20,09 \times 10^3 \times 1,08} = 0,42 \text{ mm}$$ $$w_{fin,G} = [w_{inst,G,m}(1 + k_{def})] + [w_{inst,G,v}(1 + k_{def})]$$ Since the deflection is based on two wood-based elements having different time-dependent behaviour, the k_{def} value is calculated using Equation (2.13):	
Equation (2.13)	$$k_{def} = 2 \times \sqrt{k_{def,flange}\ k_{def,web}} = 2 \times \sqrt{(0,8 \times 2,25)} = 2,68$$ $$w_{fin,G} = [5,03 \times (1 + 2,68)] + [0,42 \times (1 + 2,68)] = 20,06 \text{ mm}$$ **Variable load deformation:** $W = 3,12 \text{ kN}$ $L = 6500 \text{ mm}$ $E_f = E_{0,mean} = 11,0 \text{ kN/mm}^2$ $G_w = G_v = 1,08 \text{ kN/mm}^2$ Bending deformation: $$w_{inst,Q1,m} = \frac{5WL^3}{384E_f I_{ef,inst}} = \frac{5 \times 3,12 \times 6500^3}{384 \times 11,0 \times 134,4 \times 10^6} = 7,55 \text{ mm}$$ Shear deformation: $$w_{inst,Q1,v} = \frac{0,68WL}{A_{web}G_w} = \frac{0,68 \times 3,12 \times 6500}{20,09 \times 10^3 \times 1,08} = 0,64 \text{ mm}$$	
National Annex to BS EN 1990 Table NA.A1.1	$\psi_{2,1} = 0,0$ for roof $$w_{fin,Q1} = [w_{inst,Q1,m}(1 + \psi_{2,1} k_{def})] + [w_{inst,Q1,v}(1 + \psi_{2,1} k_{def})]$$ $$= 7,55 \times [1 + (0,0 \times 2,68)] + 0,64 \times [1 + (0,0 \times 2,68)]$$ $$= 8,19 \text{ mm}$$ **Final Deformation:** $$w_{fin} = w_{fin,G} + w_{fin,Q1} = (20,06 + 8,19) = 28,25 \text{ mm}$$	w_{fin} = **28,25 mm**
National Annex Table NA.4	Limiting ratio = $l/150$ = 6500/150 = 43,33 mm > 28,25 mm **Note:** acceptable limits should be agreed by the client at the beginning of the project; the code values are recommended, not mandatory.	**The joists are adequate with respect to deflection.**

5.4.5 *Example 5.4: Access Walkway in Exhibition Centre*

A temporary timber walkway is required within an exhibition hall, as indicated in Figure 5.17. Details of the proposed cross-section of the structure are indicated in Figures 5.18(a) and (b). Using this information, check the suitability of the following timber members:

i)   the boarding, allowing 5 mm for wear,
ii) a typical central stringer,
iii) a typical cross-beam,
iv) a main ply-web box beam.

Boards are to be strength class C45, and stringers and cross-beams are to be strength class C24. Assume that the structure satisfies Service Class 2 requirements.

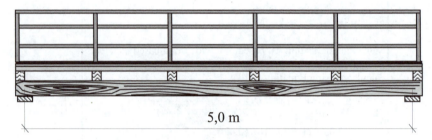

Figure 5.17

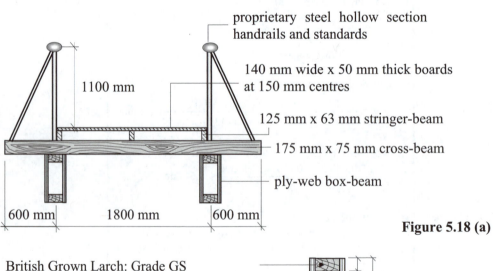

proprietary steel hollow section handrails and standards

140 mm wide x 50 mm thick boards at 150 mm centres

125 mm x 63 mm stringer-beam

175 mm x 75 mm cross-beam

ply-web box-beam

Figure 5.18 (a)

British Grown Larch: Grade GS

1,4 mm veneer sanded Finnish combi plywood (9 plies), face ply perpendicular to span

British Grown Larch: Grade GS

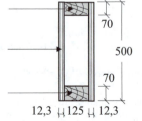

Figure 5.18 (b)

5.4.5.1 Solution to Example 5.4

Contract : **Walkway** Job Ref. No. : **Example 5.4**	Calculated by : **W.McK.**
Part of Structure : Timber Properties	**Checked by : B.Z.**
Calculation Sheet No. : 1 of 30	**Date :**

References	Calculations	Output
BS EN 1990 BS EN 1991 BS EN 338:2003 BS EN 1995-1-1: 2004 BS EN 1995-2: 2004 National Annex	Eurocode: Basis of Structural Design Eurocode 1: Actions on Structures Structural Timber - Strength Classes Eurocode 5: Design of Timber Structures Part 1.1: General - Common Rules and Rules for Buildings. Eurocode 5: Design of Timber Structures Part 2: Bridges. UK - National Annex to BS EN 1995-1-1 Plywood manufacturer's catalogue	
BS EN 338:2003 Clause 5.0 Table 1	Characteristic values for C45 Timber **Strength Properties:** Bending $\quad\quad f_{m,k} = 45{,}0 \text{ N/mm}^2$ Compression parallel to grain $\quad f_{c,0,k} = 27{,}0 \text{ N/mm}^2$ Compression perpendicular to grain $\quad f_{c,90,k} = 3{,}1 \text{ N/mm}^2$ Shear $\quad\quad f_{v,k} = 3{,}8 \text{ N/mm}^2$ **Stiffness Properties:** Mean modulus of elasticity parallel to the grain $\quad\quad E_{0,mean} = 15{,}0 \text{ kN/mm}^2$ 5% modulus of elasticity parallel to the grain $\quad\quad E_{0,05} = 10{,}0 \text{ kN/mm}^2$ Mean modulus of elasticity perpendicular to the grain $\quad\quad E_{90,mean} = 0{,}50 \text{ kN/mm}^2$ Mean shear modulus $\quad G_{mean} = 0{,}94 \text{ kN/mm}^2$ Characteristic density $\quad \rho_k = 440{,}0 \text{ kg/m}^3$ Mean density $\quad \rho_{mean} = 520{,}0 \text{ kg/m}^3$	
BS EN 338:2003 Clause 5.0 Table 1	Characteristic values for C24 Timber **Strength Properties:** Bending $\quad\quad f_{m,k} = 24{,}0 \text{ N/mm}^2$ Compression parallel to grain $\quad f_{c,0,k} = 21{,}0 \text{ N/mm}^2$ Compression perpendicular to grain $\quad f_{c,90,k} = 2{,}5 \text{ N/mm}^2$ Shear $\quad\quad f_{v,k} = 2{,}5 \text{ N/mm}^2$ **Stiffness Properties:** Mean modulus of elasticity parallel to the grain $\quad\quad E_{0,mean} = 11{,}0 \text{ kN/mm}^2$ 5% modulus of elasticity parallel to the grain $\quad\quad E_{0,05} = 7{,}4 \text{ kN/mm}^2$ Mean modulus of elasticity perpendicular to the grain $\quad\quad E_{90,mean} = 0{,}37 \text{ kN/mm}^2$ Mean shear modulus $\quad G_{mean} = 0{,}69 \text{ kN/mm}^2$ Characteristic density $\quad \rho_k = 350{,}0 \text{ kg/m}^3$ Mean density $\quad \rho_{mean} = 420{,}0 \text{ kg/m}^3$	

Contract : Walkway Job Ref. No. : Example 5.4 Part of Structure : Timber Properties Calculation Sheet No. : 2 of 30	Calculated by : W.McK. Checked by : B.Z. Date :

References	Calculations	Output
BS EN 1912 Table 1 BS EN 338:2003	British grown Larch, Grade GS: Strength Class C16 Characteristic values for C16 Timber	

Clause 5.0
Table 1
	Strength Properties:
	Bending $f_{m,k} = 16{,}0$ N/mm^2
	Compression parallel to grain $f_{c,0,k} = 17{,}0$ N/mm^2
	Tension parallel to grain $f_{t,0,k} = 10{,}0$ N/mm^2
	
	Stiffness Properties:
	Mean modulus of elasticity parallel to the grain
	                                        $E_{0,mean} = 8{,}0$ kN/mm^2
	5% modulus of elasticity parallel to the grain
	                                        $E_{0,05} = 5{,}4$ kN/mm^2
	Mean shear modulus $G_{mean} = 0{,}5$ kN/mm^2
	Mean density $\rho_{mean} = 370{,}0$ kg/m^3	

Manufacturer's Catalogue	Characteristic values: Finnish combi-plywood wood (12,3 mm thick, 9 plies) **Strength Properties:** Compression parallel to grain $f_{c,0,k} = 22{,}1$ N/mm^2 Tension parallel to grain $f_{t,0,k} = 27{,}7$ N/mm^2 Panel shear $f_{v,0,k} = 9{,}5$ N/mm^2 Planar shear $f_{v,90,k} = 1{,}7$ N/mm^2 **Stiffness Properties:** Mean modulus of elasticity in bending parallel to the grain $E_{0,mean} = 9{,}68$ kN/mm^2 Modulus of rigidity – panel shear $G_v = 0{,}67$ kN/mm^2 Mean density $\rho_{mean} = 650{,}0$ kg/m^3	
Clause 2.3.1.2 Table NA.1 BS EN 1995-2 Clause 2.3.1.2(1)	Load-duration Classes: use Table NA.1 from National Annex Self-weight - permanent Pedestrian traffic - short-term	
Clause 2.3.2.1	Load-duration and moisture influence on strength: k_{mod} The deck members and the main beams have different time-dependent behaviour.	
Table 3.1	Consider solid timber members: Solid timber Service Class 2 Permanent actions $k_{mod} = 0{,}6$ Short-term actions $k_{mod} = 0{,}9$	
Clause 3.1.3(2)	Use k_{mod} corresponding to action with the shortest duration.	$k_{mod} = 0{,}9$

References	Calculations	Output
	Contract : Walkway Job Ref. No. : Example 5.4 **Part of Structure : Timber Properties** **Calculation Sheet No. : 3 of 30**	**Calculated by : W.McK.** **Checked by : B.Z.** **Date :**

References	Calculations	Output
Table 3.1	Consider the box-beams: Plywood web: Service Class 2 Permanent actions $k_{mod} = 0,6$ Instantaneous actions $k_{mod} = 1,1$	
Clause 3.1.3(2)	Use k_{mod} corresponding to action with the shortest duration.	$k_{mod} = 1,1$
National Annex Table NA.3 EC5 (Table 2.3)	Partial factor for material properties and resistance: γ_M For solid timber – untreated and preservative treated For plywood	$\gamma_{M,solid} = 1,3$ $\gamma_{M,plywood} = 1,2$
Clause 6.6	System strength: k_{sys} There is no load sharing in this case.	$k_{sys} = 1,0$

	Loading
	Permanent actions:
	Boards: $\rho_{mean,C45,solid} = 520,0$ kg/m³
	                                = 5,10 kN/m³
	Stringers and cross-beams: $\rho_{mean,C24,solid} = 420,0$ kg/m³
	                                = 4,12 kN/m³	

	Self-weight due to boards: 140 mm x 50 mm Volume/m² of plan area = $1,0 \times 0,05 \times \dfrac{140}{150} = 0,047$ m³/m Dead load = $0,047 \times 5,10 = 0,24$ kN/m²	**Boarding: 0,24 kN/m²**
	Self-weight due to stringer section: 125 mm x 63 mm Volume/m length = $0,125 \times 0,063 = 7,9 \times 10^{-3}$ m³/m Dead load = $7,9 \times 10^{-3} \times 4,12 = 0,03$ kN/m	**Stringers: 0,03 kN/m**
	Self-weight due to cross-beams: 175 mm x 75 mm Volume/m length = $0,175 \times 0,075 = 13,0 \times 10^{-3}$ m³/m Dead load = $13,0 \times 10^{-3} \times 4,12 = 0,05$ kN/m	**Cross-beams:0,05 kN/m**
	Assume an allowance for handrails, fitments, etc. = 1,0 kN/m²	**Handrails, etc.:** **1,0 kN/m²**
	Self-weight of ply-web beams: $\rho_{mean,C16,solid} = 370,0$ kg/m³ $\rho_{mean,plywood} = 650,0$ kg/m³ C16 timber = 3,63 kN/m³; Plywood 12,3 mm thick = 0,08 kN/m²	

Contract : Walkway Job Ref. No. : Example 5.4 Part of Structure : Actions Calculation Sheet No. : 4 of 30	Calculated by : W.McK. Checked by : B.Z. Date :

References	Calculations	Output
	Volume/m length/flange $= 0,125 \times 0,07 \times 1,0 = 0,009 \text{ m}^3$ Self-weight due to flanges $= 2 \times 0,009 \times 3,63 = 0,07 \text{ kN/m}$ Self-weight due to webs $= 2 \times 0,5 \times 1,0 \times 0,08 = 0,08 \text{ kN/m}$ Dead load $= (0,07 + 0,08)$ $= 0,15 \text{ kN/m}$	**Main beams: 0,15 kN/m**
NA EN 1991-1-1 Table NA.2	Variable actions: In this case assume the structure is Category C38: Walkway	
Table NA.3	Category C38: Uniformly distributed load q_{fwk} $= 7,5 \text{ kN/m}^2$ Assume a concentrated load Q_{fwk} $= 4,5 \text{ kN}$ The concentrated load is considered for local effects only.	$q_{\text{fwk}} = 7,5 \text{ kN/m}^2$ $Q_{\text{fwk}} = 4,5 \text{ kN}$
Table NA.8	A line force of 1,5 kN/m acting as a variable load, at the top of the parapet.	$Q_{\text{k,parapet}} = 1,5 \text{ kN/m}$
BS EN 1990 Equation (6.10)	Eurocode: Basis of Structural Design $E_d = \sum_{j\geq1} \gamma_{G,j} G_{k,j} + \gamma_{Q1} Q_{k,1} + \sum_{i>1} \gamma_{Q,i} \psi_{0,i} Q_{k,i}$ $E_d = \left(\gamma_G G_k + \gamma_{Q,1} Q_{k,1} \right)$	
UK NA to BS EN 1990 Table NA.A1.2(B)	where E_d is the design value of the effect of the actions $\gamma_G = 1,35$ and $\gamma_{Q,1} = 1,50$	
BS 6399 : Part 1 Table 1	**(a) Boarding:** Consider one board with 0,15 m width Permanent action $= 0,24 \times 0,15 = 0,04 \text{ kN/m}$ Variable action $= 7,5 \times 0,2 = 1,5 \text{ kN/m}$ or a 4,5 kN concentrated load based on the board. Design loads: Minimum permanent action $= (1,0 \times 0,04) = 0,04 \text{ kN/m}$ Maximum permanent action $= (1,35 \times 0,04) = 0,05 \text{ kN/m}$ Maximum variable action $= (1,5 \times 1,5) = 2,25 \text{ kN/m}$ Maximum concentrated action $= (1,5 \times 4,5) = 6,75 \text{ kN}$	

References	Calculations	Output
	Contract : **Walkway** **Job Ref. No. : Example 5.4** **Calculated by : W.McK.** **Part of Structure :** **Boarding** **Checked by : B.Z.** **Calculation Sheet No. : 5 of 30** **Date :**	

References	Calculations	Output
	Bending: Three load cases should be considered: (i) maximum bending moment with both spans fully loaded, (ii) maximum bending moment with one span fully loaded and the other span with only permanent load, (iii) maximum moment with concentrated load at mid-span with permanent load. Load case (i): Total design load = $(0,05 + 2,25) = 2,30$ kN/m	
	0,05 kN/m 2,25 kN/m 900 mm 900 mm	

Appendix A of this text	Maximum bending moment $= M_{y,d} = -0,125\ WL$
	    (occurs over support) $= -0,125 \times (2,30 \times 0,9) \times 0,9$
	                  $= -0,23$ kNm
	Load case (ii): maximum span moment
		Load case (i)
		$M_{y,d} = 0,23$ **kNm**

	2,30 kN/m 0,04 kN/m 900 mm 900 mm 2,30 kN/m 2,26 kN/m (a) = (b) + (c)	
Appendix A of this text	Continuity moments M $0,125WL$ $0.063WL$ (a) = (b) + (c) $M_{y,d} = -[(0,125 \times 2,30 \times 0,9^2) - (0,063 \times 2,26 \times 0,9^2)]$ $= -0,12$ kNm	

Contract : **Walkway**　　Job Ref. No. : **Example 5.4**	Calculated by : **W.McK.**
Part of Structure :　**Boarding**	Checked by : **B.Z.**
Calculation Sheet No. : **6 of 30**	Date :

References	Calculations	Output

	2,30 kN/m　　0,12 kNm
	
	2,30 kN/m　　0,12 kNm
	
	Consider the left-hand span
	Sum of the moments to the left-hand side:
	
	$+\,0{,}12 + (0{,}9 \times V) - (2{,}30 \times 0{,}9 \times \dfrac{0{,}9}{2}) = 0 \quad \therefore\ V = 0{,}90$ kN
	
	  0,90 kN　　part shear force diagram
	
	position of zero shear $= x = \dfrac{0{,}90}{2{,}30} = 0{,}39$ m
	
	Maximum bending moment = area under shear force diagram
	　　　　　　$M_{y,d} = (0{,}5 \times 0{,}39 \times 0{,}90) = 0{,}18$ kNm
	
	Load case (iii): due to concentrated load and permanent load
	
	  6,75 kN　0,05 kN/m　0,04 kN/m　450 mm　450 mm　900 mm
		Load case (ii)
		$M_{y,d} = \mathbf{0{,}18}$ **kNm**

Appendix A of this text	$M_{y,G,d} = 0{,}070 \times 0{,}04 \times 0{,}9^2 + 0{,}096 \times 0{,}01 \times 0{,}9^2 = 0{,}00$ kNm $M_{y,Q,d} = 0{,}203\ WL = 0{,}203 \times 6{,}75 \times 0{,}9 = 1{,}23$ kNm $M_{y,d} = M_{y,G,d} + M_{y,Q,d} = 0{,}00 + 1{,}23 = 1{,}23$ kNm **Load case (iii) is the critical case** Bending: Design bending moment $M_{y,d} = 1{,}23$ kNm/board Design bending strength $f_{m,y,d} = (k_{mod} \times f_{m,y,k} \times k_h \times k_{sys})/\gamma_M$ Section properties: $h = 45$ mm thick (allow 5 mm for wear) $W_y = \dfrac{140 \times 45^2}{6} = 47{,}25 \times 10^3$ mm^3 $I_{yy} = \dfrac{140 \times 45^3}{12} = 1063{,}13 \times 10^3$ mm^4	**Load case (iii)** $M_{y,d} = \mathbf{1{,}23}$ **kNm**

References	Calculations	Output

Contract : Walkway   Job Ref. No. : Example 5.4
Part of Structure :   Boarding
Calculation Sheet No. : 7 of 30

Calculated by : W.McK.
Checked by : B.Z.
Date :

Clause 3.1.3(2)

Use k_{mod} corresponding to action with the shortest duration.

$k_{mod} = 0,9$

Clause 3.2(3)
Equation (3.1)

Member size: $h < 150$ mm therefore k_h applies.
For timber in bending with $\rho_k \leq 700$ kg/m^3 where the depth is less than 150 mm $f_{m,k}$ can be multiplied by k_h where:

$$k_h = \min\left\{\left(\frac{150}{h}\right)^{0,2}\right\} = \min\left\{\left(\frac{150}{50}\right)^{0,2}\right\} = \min\left\{\begin{array}{l}1,25\\1,3\end{array}\right.$$
$$1,3 \qquad\qquad 1,3$$

$k_h = 1,25$

Clause 6.6

System strength: $k_{sys} = 1,0$

$k_{sys} = 1,0$

Table NA.3

For solid timber – untreated and preservative treated
$f_{m,y,d} = (0,9 \times 45,0 \times 1,25 \times 1,0)/1,3 = 38,94$ N/mm^2

$\gamma_{M,solid} = 1,3$
$f_{m,y,d} = 38,94$ N/mm^2

BS EN 1995-1-1
Clause 6.1.6

Since uniaxial bending about the y-y axis only, is present:
$\sigma_{m,y,d} \leq f_{m,y,d}$

$$\sigma_{m,y,d} = \frac{M_{y,d}}{W_y} = \frac{1,23\times10^6}{47,25\times10^3} = 26,03 \text{ N/mm}^2 < 38,94 \text{ N/mm}^2$$

The joist is adequate with respect to bending

Shear:

6,75 kN

0,05 kN/m   0,04 kN/m

V_A

450 mm | 450 mm | 900 mm

Case (iii) gives the largest shear force at the middle support B.

Appendix A of this text

$V_A = (0,375 \times 0,04 \times 0,9) + (0,438 \times 0,01 \times 0,90 + (0,406 \times 6,75)$
$= 2,76$ kN
$F_{v,d(B)} = 2,76 - 0,05 \times 0,9 - 6,75 = -4,04$ kN
Design shear force $F_{v,d} = 4,04$ kN

$F_{v,d} = 4,04$ kNm

Design shear strength $f_{v,d} = (f_{v,k} \times k_{mod} \times k_{sys})/\gamma_M$

Clause 6.5.2

k_v applies to notched beams only
$f_{v,d} = (3,8 \times 0,9 \times 1,0)/1,3 = 2,63$ N/mm^2

$f_{v,d} = 2,63$ N/mm^2

Clause 6.1.7

Design shear stress $\tau_d = \frac{1,5\times F_{v,d}}{A} = \frac{1,5\times4,04\times10^3}{140\times45}$
$= 0,96$ N/mm$^2 < 2,63$ N/mm^2

The joist is adequate with respect to shear.

References	Calculations	Output
	Contract : Walkway Job Ref. No. : Example 5.4 **Part of Structure : Boarding** **Calculation Sheet No. : 8 of 30**	**Calculated by : W.McK.** **Checked by : B.Z.** **Date :**

References	Calculations	Output
Clause 6.1.5	**Bearing:** Case (iii) also causes the largest reaction at middle support as $V_B = 1,25 \times 0,04 \times 0,9 + 0,625 \times 0,01 \times 0,9 + 0,688 \times 6,75$ $\quad = 4,69$ kN Design bearing force $F_{c,90,d} = 4,69$ kN	$F_{c,90,d} = \textbf{4,69 kN}$
	Design bearing strength $f_{c,90,d} = (f_{c,90,k} \times k_{mod} \times k_{sys})/\gamma_M$ $\qquad\qquad\qquad\qquad = (3,1 \times 0,9 \times 1,0)/1,3$ $\qquad\qquad\qquad\qquad = 2,15$ N/mm^2	$f_{c,90,d} = \textbf{2,15 N/mm}^2$
	Contact length of bearing $l = 63$ mm Design bearing stress $\sigma_{c,90,d} = \dfrac{F_{c,90,d}}{b \times l} = \dfrac{4,69 \times 10^3}{140 \times 63}$ $\qquad\qquad\qquad\qquad\qquad = 0,53$ N/mm^2	
Equation (6.3) Clause 6.1.5(2) Clause 6.1.5(3)	$\sigma_{c,90,d} \leq (k_{c,90} \times f_{c,90,d})$ At internal supports,	
Equation (6.5)	$k_{c,90} = \left(2,38 - \dfrac{l}{250}\right)\left(1 + \dfrac{h}{6l}\right)$ and $\leq 4,0$ $k_{c,90} = \left(2,38 - \dfrac{63}{250}\right)\left(1 + \dfrac{45}{6 \times 63}\right) = 2,38$ $(k_{c,90} \times f_{c,90,d}) = (2,38 \times 2,15) = 5,12$ N/mm$^2 > 0,53$ N/mm^2	**The section is adequate with respect to bearing**
	Serviceability Limit States: **Deflection:** Case (iii) gives the largest deflection at the middle of left span.	
Clause 7.2	Limiting values for deflection of beams: $w_{net,fin} = w_{inst} + w_{creep} - w_c = w_{fin} - w_c$ Since there is no camber $w_c = 0$ and $w_{net,fin} = w_{inst} + w_{creep} = w_{fin}$.	
Table 3.2	For solid timber Service Class 2 $k_{def} = 0,8$ $w_{fin} = w_{inst}(1 + k_{def}) = (w_{inst,m} + w_{inst,v})(1 + k_{def})$	$k_{def} = \textbf{0,8}$

Contract : Walkway Job Ref. No. : Example 5.4	Calculated by : W.McK.
Part of Structure : Boarding	Checked by : B.Z.
Calculation Sheet No. : 9 of 30	Date :

References	Calculations	Output
	Permanent load deformation:	

Permanent load deformation:

$W = (0,24 \times 0,9 \times (150/1000)) = 0,03$ kN/board/span

$L = 900$ mm

$E_d = E_{0,mean} = 15,0$ kN/mm^2

$G_d = G_{mean} = 0,94$ kN/mm^2

$I_y = 1063,13 \times 10^3$ mm^4

Deformation due to bending is calculated by Unit Load Method.
Deformation due to shear is calculated by Appendix B.

Unit Load Method

$$w_{inst,G,m} = \frac{0,0054WL^3}{E_dI} = \frac{0,0054 \times 0,03 \times 900^3}{15,0 \times 1063,13 \times 10^3} = 0,007 \text{ mm}$$

Appendix B

$$w_{inst,G,v} = \frac{0,178WL}{bhG_d} = \frac{0,178 \times 0,03 \times 900}{140 \times 45 \times 0,94} = 0,001 \text{ mm}$$

$w_{inst,G} = (w_{inst,m} + w_{inst,v})_G = (0,007 + 0,001) = 0,01$ mm

$w_{fin,G} = w_{inst,G} (1 + k_{def}) = 0,01 \times (1 + 0,8) = 0,02$ mm

The deformation due to permanent action can be ignored.

Variable load deformation:

$P = 4,5$ kN/board

Unit Load Method

$$w_{inst,Q1,m} = \frac{0,015PL^3}{E_dI} = \frac{0,015 \times 4,5 \times 900^3}{15,0 \times 1063,13 \times 10^3} = 3,09 \text{ mm}$$

Appendix B

$$w_{inst,Q1,v} = \frac{0,321PL}{bhG_d} = \frac{0,321 \times 4,5 \times 900}{140 \times 45 \times 0,94} = 0,22 \text{ mm}$$

UK NA to
BS EN 1990
Table NA.A1.1

$\psi_{2,1} = 0,6$ for congregation areas

$w_{inst,Q1} = (w_{inst,m} + w_{inst,v})_{Q1} = (3,09 + 0,22) = 3,31$ mm

$w_{fin,Q1} = w_{inst,Q1} (1 + \psi_{2,1} k_{def}) = 3,31 \times [1 + (0,6 \times 0,8)]$
$\qquad = 4,90$ mm

$w_{fin} = w_{in,G} + w_{fin,Q1} = (0,02 + 4,90) = 4,92$ mm

National Annex
Table NA.4

Limiting ratio = $l/150 = 900/150 = 6,0$ mm $> 4,92$ mm

The coefficients for calculating the deformations due to bending
and shear are left to readers to prove.

Output column:

$\psi_{2,1} = 0,6$

$w_{fin} = 4,92$ mm

**Adopt 140 mm x 50 mm
boarding at 150 mm
centres: Class C45.**

Contract : Walkway Job Ref. No. : Example 5.4	Calculated by : W.McK.
Part of Structure : Stringers	Checked by : B.Z.
Calculation Sheet No. : 10 of 30	Date :

References	Calculations	Output

	(b) Central Stringers:
	The load imposed on a central stringer is equal to the support reaction from the two span boarding + self-weight, i.e.
	
	Consider 1 metre width
	Permanent action = 0,24 kN/m
	Variable action = 7,5 kN/m or a 4,5 kN concentrated load at the mid-span of the middle stringer.
	
	Design loads:
	Maximum permanent action = $(1,35 \times 0,24) = 0,32$ kN/m
	
	Maximum variable action = $(1,5 \times 7,5) = 11,25$ kN/m
	Maximum concentrated action = $(1,5 \times 4,5) = 6,75$ kN
	
	Case (i): UDL permanent and variable loads
	
	0,32 kN/m 11,25 kN/m
	
	900 mm 900 mm
	
	The maximum value of the reaction 'V' occurs when the full load is applied to both spans of the boarding.
	
	Self-weight of stringer = $(0,03 \times 1,35) = 0,04$ kN/m
	Maximum total load on central stringer
	   = $1,25 \times (11,57 \times 0,9) + 0,04 = 13,06$ kN/m
	
	Maximum total load on outer stringers
	   = $0,375 \times (11,57 \times 0,9) + 0,04 = 3,94$ kN/m
	
	Assuming stringers to be single-span and simply supported.
	
	Design bending moment $M_{y,d} = \dfrac{wL^2}{8} = \dfrac{13,06 \times 1,0^2}{8} = 1,63$ kNm
	
	Case (ii): UDL permanent load and concentrated load at mid-span of the central stringer
	
	6,75 kN
	0,32 kN/m
	
	900 mm 900 mm
		Load case (i)
		
		$M_{y,d} = 1,63$ kNm

Appendix A (in References column)

References	Calculations	Output
	Contract : Walkway Job Ref. No. : Example 5.4 **Part of Structure : Stringers** **Calculation Sheet No. : 11 of 30**	**Calculated by : W.McK.** **Checked by : B.Z.** **Date :**

References	Calculations	Output
Appendix A	Maximum total UDL load on central stringer $w = 1{,}25 \times (0{,}32 \times 0{,}9) + 0{,}04 = 0{,}40$ kN/m	
	Maximum concentrated load on mid-span of central stringers $P = 6{,}75$ kN	
	Assuming stringers to be single-span and simply supported. Design bending moment	**Load case (ii)**
	$M_{y,d} = \dfrac{wL^2}{8} + \dfrac{PL}{4} = \dfrac{0{,}40 \times 1{,}0^2}{8} + \dfrac{6{,}75 \times 1{,}0}{4} = 1{,}74$ kNm	$M_{y,d} = \mathbf{1{,}74}$ **kNm**
	Load case (ii) is the critical case.	
	Section properties:	
	$W_y = \dfrac{63 \times 125^2}{6} = 164{,}1 \times 10^3$ mm^3	
	$I_{yy} = \dfrac{63 \times 125^3}{12} = 10{,}25 \times 10^6$ mm^4	
Clause 3.1.3(2)	Use k_{mod} corresponding to action with the shortest duration.	$k_{mod} = \mathbf{0{,}9}$
Clause 3.2.(3) Equation (3.1)	Member size: $h < 150$ mm therefore k_h applies. For timber in bending with $\rho_k \leq 700$ kg/m^3 where the depth is less than 150 mm $f_{m,k}$ can be multiplied by k_h where:	
	$k_h = \min\left\{ \begin{array}{l} \left(\dfrac{150}{h}\right)^{0{,}2} \\ 1{,}3 \end{array} \right. = \min\left\{ \begin{array}{l} \left(\dfrac{150}{125}\right)^{0{,}2} \\ 1{,}3 \end{array} \right. = \min\left\{ \begin{array}{l} 1{,}04 \\ 1{,}3 \end{array} \right.$	$k_h = \mathbf{1{,}04}$
Clause 6.6	System strength: $k_{sys} = 1{,}0$	$k_{sys} = \mathbf{1{,}0}$
Table NA.3	For solid timber – untreated and preservative treated	$\gamma_{M,solid} = \mathbf{1{,}3}$
	Design bending strength $f_{m,y,d} = (f_{m,y,k} \times k_{mod} \times k_h \times k_{sys})/\gamma_M$ $f_{m,y,d} = (24{,}0 \times 0{,}9 \times 1{,}04 \times 1{,}0)/1{,}3 = 17{,}28$ N/mm^2	$f_{m,y,d} = \mathbf{17{,}28}$ **N/mm^2**
BS EN 1995-1-1	Since uniaxial bending about the y-y axis only, is present:	
Clause 6.1.6	$\sigma_{m,y,d} \leq f_{m,y,d}$ $\sigma_{m,y,d} = \dfrac{M_{y,d}}{W_y} = \dfrac{1{,}74 \times 10^6}{164{,}1 \times 10^3} = 10{,}60$ N/mm$^2 \leq 17{,}28$ N/mm^2	
		The stringers are adequate in bending.

References	Calculations	Output
	Contract : Walkway Job Ref. No. : Example 5.4 **Part of Structure : Stringers** **Calculation Sheet No. : 12 of 30**	**Calculated by : W.McK.** **Checked by : B.Z.** **Date :**

References	Calculations	Output
	Shear: **Case (i) is the critical case.** Design shear force $V_d = (F_d \times 1,0)/2,0 = (13,06 \times 1,0)/2,0$ $\qquad = 6,53$ kN	$V_d = 6{,}53$ kN
	Design shear strength $f_{v,d} = (f_{v,k} \times k_{mod} \times k_v \times k_{sys})/\gamma_M$	
Clause 6.5.2	k_v applies to notched beams only $f_{y,d} = (2,5 \times 0,9 \times 1,0)/1,3 = 1,73$ N/mm²	$f_{v,d} = 1{,}73$ N/mm²
Clause 6.1.7	Design shear stress $\tau_d = \dfrac{1,5 \times F_{v,d}}{A} = \dfrac{1,5 \times 6,53 \times 10^3}{125 \times 63}$ $\qquad = 1,24$ N/mm² $< 1,73$ N/mm²	The stringers are adequate with respect to shear.
Clause 6.3.3 Equation (6.33)	**Lateral Torsional Stability:** $\sigma_{m,d} \le (k_{crit} \times f_{m,d})$	
Clause 6.3.3(5)	Since the compression flange is fully restrained by the decking and torsional restraint is provided at the supports the value of $k_{crit} = 1,0$.	The section is adequate with respect to lateral torsional buckling.
Clause 6.1.5	**Bearing:** Design bearing force $F_{c,90,d} = 6,53$ kN	$F_{c,90,d} = 6{,}53$ kN
	Design bearing strength $f_{c,90,d} = (f_{c,90,k} \times k_{mod} \times k_{sys})/\gamma_M$ $\qquad = (2,5 \times 0,9 \times 1,0)/1,3$ $\qquad = 1,73$ N/mm²	$f_{c,90,d} = 1{,}73$ N/mm²
	Assume a contact length of bearing $l = 35$ mm Design bearing stress $\sigma_{c,90,d} = \dfrac{F_{c,90,d}}{b \times l} = \dfrac{6,53 \times 10^3}{35 \times 63}$ $\qquad = 2,96$ N/mm²	
Equation (6.3) Clause 6.1.5(2) Clause 6.1.5(3)	$\sigma_{c,90,d} \le (k_{c,90} \times f_{c,90,d})$ Since 'a' in Figure 6.4 of EC5 is equal to zero: $k_{c,90} = \left(2,38 - \dfrac{l}{250}\right)\left(1 + \dfrac{h}{12l}\right)$ and $\le 4,0$ $k_{c,90} = \left(2,38 - \dfrac{35}{250}\right)\left(1 + \dfrac{125}{12 \times 35}\right) = 2,91$ $(k_{c,90} \times f_{c,90,d}) = (2,91 \times 1,73) = 5,03$ N/mm² $> 2,96$ N/mm²	The section is adequate with respect to bearing.

Contract : Walkway Job Ref. No. : Example 5.4	Calculated by : W.McK.
Part of Structure : Stringers	Checked by : B.Z.
Calculation Sheet No. : 13 of 30	Date :

References	Calculations	Output
Table 3.2	**Serviceability Limit States:**	$k_{def} = 0,8$

Serviceability Limit States:

Deflection:

Case (i) is the critical case. This leaves readers to prove.

For solid timber Service Class 2 $k_{def} = 0,8$

0,24 kN/m   7,5 kN/m

900 mm   900 mm

The maximum value of the reaction 'V' occurs when the full load is applied to both spans of the boarding.

Appendix A

Maximum load on central stringer = $(1,25 \times w)$

Due to permanent load = $1,25 \times (0,24 \times 0,9) + 0,03 = 0,30$ kN/m

Due to variable load = $1,25 \times (7,5 \times 0,9) = 8,44$ kN/m

Maximum load on outer stringer = $(0,375 \times w)$

Due to permanent load = $0,375 \times (0,24 \times 0,9) + 0,03 = 0,11$ kN/m

Due to variable load = $0,375 \times (7,5 \times 0,9) = 2,53$ kN/m

$w_{fin,G} = w_{inst,G} (1 + k_{def}) = (w_{inst,m} + w_{inst,v})_G (1 + k_{def})$

where:

Permanent load deformation:

$W = (0,3 \times 1,0) = 0,3$ kN

$L = 1000$ mm

$E_d = E_{0,mean} = 11,0$ kN/mm^2

$G_d = G_{mean} = 0,69$ kN/mm^2

$I_{yy} = 10,25 \times 10^6$ mm^4

The bending and shear deflections can be estimated using:

$$w_{inst,G,m} = \frac{5WL^3}{384 E_d I} = \frac{5 \times 0,3 \times 1000^3}{384 \times 11,0 \times 10,25 \times 10^6} = 0,03 \text{ mm}$$

$$w_{inst,G,v} = \frac{3WL}{20 bh G_d} = \frac{3 \times 0,3 \times 1000}{20 \times 63 \times 125 \times 0,69} = 0,01 \text{ mm}$$

$w_{inst,G} = (w_{inst,m} + w_{inst,v})_G = (0,03 + 0,01) = 0,04$ mm

$w_{fin,G} = w_{inst,G}(1 + k_{def}) = 0,04 \times (1 + 0,8) = 0,07$ mm

Variable load deformation:

$W = (8,44 \times 1,0) = 8,44$ kN

Contract : Walkway Job Ref. No. : Example 5.4	**Calculated by : W.McK.**	
Part of Structure : Cross-beams	**Checked by : B.Z.**	
Calculation Sheet No. : 14 of 30	**Date :**	

References	Calculations	Output
	$w_{inst,Q1,m} = \dfrac{5WL^3}{384E_dI} = \dfrac{5 \times 8,44 \times 1000^3}{384 \times 11,0 \times 10,25 \times 10^6} = 0,97$ mm	
	$w_{inst,Q1,v} = \dfrac{3WL}{20bhG_d} = \dfrac{3 \times 8,44 \times 1000}{20 \times 63 \times 125 \times 0,69} = 0,23$ mm	
UK NA to BS EN 1990 Table NA.A1.1	$\psi_{2,1} = 0,6$ $w_{inst,Q1} = (w_{inst,m} + w_{inst,v})_{Q1} = (0,97 + 0,23) = 1,20$ mm $w_{fin,Q1} = w_{inst,Q1}(1 + \psi_{2,1}k_{def}) = 1,20 \times [1 + (0,6 \times 0,8)]$ $= 1,78$ mm	$\psi_{2,1} = 0,6$
	$w_{fin} = w_{in,G} + w_{fin,Q1} = (0,07 + 1,78) = 1,85$ mm	$w_{fin} = 1,85$ mm
National Annex Table NA.4	Limiting ratio $= l/150 = 1000/150 = 6,67$ mm $>$ 1,85 mm	**Deflection is acceptable. Adopt 125 mm x 63 mm stringer beams.**
	(c) Cross-beams: Beams at 1,0 m centres Two load cases should be considered (i) A horizontal force (*F*) which is considered to act at a height of 1,1 m above the datum level. $F = 1,5$ kN/m. (ii) The vertical loading due to permanent and variable load. In this case the two load cases compensate for each other and should be considered separately. Load case (i):	
	Since the vertical standards are spaced at 1,0 m centres Horizontal load/post = 1,5 kN	
UK NA to BS EN 1990 Table NA.A1.2(B)	$\gamma_G = 1,35$ and $\gamma_Q = 1,50$ Design load: Variable action = $(1,5 \times 1,5) = 2,25$ kN/post Bending moment due to load = $2,25 \times (1,1 + 0,263) = 3,07$ kNm	

Contract : Walkway Job Ref. No. : Example 5.4 Part of Structure : Cross-beams Calculation Sheet No. : 15 of 30	Calculated by : W.McK. Checked by : B.Z. Date :

References	Calculations	Output

Load case (ii):

Load due to stringer = 13,06 kN

Design load due to self-weight of cross-beam $= (0,05 \times 1,35)$
$$= 0,07 \text{ kN/m}$$

Maximum bending moment $= [(6,64 \times 0,9) - (0,07 \times 1,5^2)/2]$
$$= 5,90 \text{ kNm}$$

Maximum shear force $= (-0,07 \times 0,6) + 10,58 - 3,94$
$$= 6,60 \text{ kN}$$

Critical load case is due to the vertical loading

Section properties:

$$W_y = \frac{75 \times 175^2}{6} = 382,81 \times 10^3 \text{ mm}^3$$

$$I_{yy} = \frac{75 \times 175^3}{12} = 33,50 \times 10^6 \text{ mm}^4$$

$$A = (75 \times 175) = 13,13 \times 10^3 \text{ mm}^2$$

Bending:
Design bending moment $M_{y,d} = 5,90$ kNm
Design bending strength $f_{m,y,d} = (f_{m,y,k} \times k_{mod} \times k_h \times k_{sys})/\gamma_M$

Clause 3.2(3) $h > 150$ mm therefore k_h does not apply.

Clause 6.6   System strength: $k_{sys} = 1,0$

$f_{m,y,d} = (24,0 \times 0,9 \times 1,0 \times 1,0)/1,3 = 16,62 \text{ N/mm}^2$

Output:

$M_{y,d} = \mathbf{5,90 \text{ kNm}}$

$V_d = \mathbf{6,60 \text{ kN}}$

$f_{m,y,d} = \mathbf{16,62 \text{ N/mm}^2}$

References	Calculations	Output

Contract : Walkway   Job Ref. No. : Example 5.4
Part of Structure :   Cross-beams
Calculation Sheet No. : 16 of 30

Calculated by : W.McK.
Checked by : B.Z.
Date :

BS EN 1995-1-1
Clause 6.1.6

Since uniaxial bending about the y-y axis only, is present:

$\sigma_{m,y,d} \leq f_{m,y,d}$

$\sigma_{m,y,d} = \dfrac{M_{y,d}}{W_y} = \dfrac{5,90 \times 10^6}{382,81 \times 10^3} = 15,41 \text{ N/mm}^2 \leq 16,62 \text{ N/mm}^2$

The cross-beams are adequate in bending.

Shear:
Design shear force $V_d = 6,60$ kN

Clause 6.5.2

Design shear strength $f_{v,d} = (f_{v,k} \times k_{mod} \times k_v \times k_{sys})/\gamma_M$
k_v applies to notched beams only

$f_{v,d} = (2,5 \times 0,9 \times 1,0)/1,3 = 1,73 \text{ N/mm}^2$

$f_{v,d} = \mathbf{1,73 \text{ N/mm}^2}$

Clause 6.1.7

Design shear stress $\tau_d = \dfrac{1,5 \times F_{v,d}}{A} = \dfrac{1,5 \times 6,60 \times 10^3}{75 \times 175}$

$= 0,75 \text{ N/mm}^2 < 1,73 \text{ N/mm}^2$

The joist is adequate with respect to shear.

Clause 6.3.3
Equation (6.33)

Lateral Torsional Stability:
$\sigma_{m,d} \leq (k_{crit} \times f_{m,d})$

Equation (6.34)

$$k_{crit} = \begin{cases} 1 & \text{for} & \lambda_{rel,m} \leq 0,75 \\ 1,56 - 0,75\lambda_{rel,m} & \text{for} & 0,75 < \lambda_{rel,m} \leq 1,4 \\ \dfrac{1}{\lambda_{rel,m}^2} & \text{for} & 1,4 < \lambda_{rel,m} \end{cases}$$

Equations (6,31)
and (6,32)

$\lambda_{rel,m} = \sqrt{\dfrac{f_{m,k}}{\sigma_{m,crit}}}$ where $\sigma_{m,crit} = \dfrac{0,78b^2}{hl_{ef}}E_{0,05}$

Table 6.1

Assume $l_{ef}/l \approx 1,0 \therefore l_{ef} = (1,0 \times 1800) = 1800$ mm

Since the load is applied to the compression edge of the beam, increase l_{ef} by $2h \therefore l_{ef} = [1800 + (2 \times 175)] = 2150$ mm

$\sigma_{m,crit} = \dfrac{0,78b^2}{hl_{ef}}E_{0,05} = \dfrac{0,78 \times 75^2}{175 \times 2150} \times \left(7,4 \times 10^3\right)$

$= 86,29 \text{ N/mm}^2$

$\lambda_{rel,m} = \sqrt{\dfrac{f_{m,k}}{\sigma_{m,crit}}} = \sqrt{\dfrac{24,0}{86,29}} = 0,53$

Contract : Walkway Job Ref. No. : Example 5.4	Calculated by : W.McK.
Part of Structure : Cross-beams	Checked by : B.Z.
Calculation Sheet No. : 17 of 30	Date :

References	Calculations	Output
Equation (6.34)	$k_{crit} = 1,0$	
	Design strength $f_{m,y,d} = 16,62$ N/mm^2 $(k_{crit} \times f_{m,d}) = (1,0 \times 16,62) = 16,62$ N/mm^2	
	Design stress $\sigma_{m,d} = 15,41$ N/mm^2 < 16,62 N/mm^2	**The section is adequate with respect to lateral torsional buckling.**
Clause 6.1.5	**Bearing:** Design bearing force $F_{c,90,d} = 10,58$ kN	$F_{c,90,d} = \mathbf{10,58}$ **kN**
	Design bearing strength $f_{c,90,d} = (f_{c,90,k} \times k_{mod} \times k_{sys})/\gamma_M$ $= (2,5 \times 0,9 \times 1,0)/1,3$ $= 1,73$ N/mm^2	$f_{c,90,d} = \mathbf{1,73}$ **N/mm^2**
	Contact length of bearing $l = [(2 \times 12,3) + 125] = 149,6$ mm	
	Design bearing stress $\sigma_{c,90,d} = \dfrac{F_{c,90,d}}{b \times l} = \dfrac{10,58 \times 10^3}{75 \times 149,6}$ $= 0,94$ N/mm^2	
Equation (6.3) Clause 6.1.5(2) Clause 6.1.5(3)	$\sigma_{c,90,d} \leq (k_{c,90} \times f_{c,90,d})$ In Figure 6.2 of EC5 'a' = 600 mm > $h/3$ $\quad \therefore k_{c,90} = 1,0$	
	$(k_{c,90} \times f_{c,90,d}) = (1,0 \times 1,73) = 1,73$ N/mm^2 > 0,94 N/mm^2	**The section is adequate with respect to bearing.**
	Deflection: Permanent load due to central stinger = 0,30 kN Permanent load due to cross-beam = 0,05 kN/m Variable load due to central stinger = 8,44 kN	
Table 3.2	For solid timber Service Class 2 $k_{def} = 0,8$ Neglect the outer cantilever loads and assume span of 1,8 m $w_{fin,G} = w_{inst,G} (1 + k_{def}) = (w_{inst,m} + w_{inst,v})_G (1 + k_{def})$	$k_{def} = \mathbf{0,8}$

References	Calculations	Output
	Contract : Walkway Job Ref. No. : Example 5.4 **Part of Structure : Cross-beams** **Calculation Sheet No. : 18 of 30**	**Calculated by : W.McK.** **Checked by : B.Z.** **Date :**

References	Calculations	Output
	Permanent load deformation: $W = (0,05 \times 1,8) = 0,09$ kN $P = 0,30$ kN $L = 1800$ mm $E_d = E_{0,mean} = 11,0$ kN/mm^2 $G_d = G_{mean} = 0,69$ kN/mm^2 $I_{yy} = 33,50 \times 10^6$ mm^4 The bending and shear deflections can be estimated using: $$w_{inst} \approx \left(\frac{5WL^3}{384E_dI} + \frac{3WL}{20bhG_d} \right)_{UDL} + \left(\frac{PL^3}{48E_dI} + \frac{3PL}{10bhG_d} \right)_{point\ load}$$ $$= \left(\frac{5,0 \times 0,09 \times 1800^3}{384 \times 11,0 \times 33,5 \times 10^6} + \frac{3,0 \times 0,09 \times 1800}{20 \times 75 \times 175 \times 0,69} \right)_{UDL}$$ $$+ \left(\frac{0,30 \times 1800^3}{48 \times 11,0 \times 33,5 \times 10^6} + \frac{3 \times 0,30 \times 1800}{10 \times 75 \times 175 \times 0,69} \right)_{point\ load}$$ $$= (0,019 + 0,003) + (0,099 + 0,018)$$ $w_{inst,G} = (w_{inst,m} + w_{inst,v})_G = (0,022 + 0,117) = 0,14$ mm $w_{fin,G} = w_{inst,G}(1 + k_{def}) = 0,14 \times (1 + 0,8) = 0,25$ mm Variable load deformation: $P = 8,44$ kN $$w_{inst} \approx \left(\frac{PL^3}{48E_dI} + \frac{3PL}{10bhG_d} \right)_{point\ load}$$ $$+ \left(\frac{8,44 \times 1800^3}{48 \times 11,0 \times 33,5 \times 10^6} + \frac{3 \times 8,44 \times 1800}{10 \times 75 \times 175 \times 0,69} \right)_{point\ load}$$ $$= (2,78 + 0,50)$$	
UK NA to BS EN 1990 Table NA.A1.1	$\psi_{2,1} = 0,6$ $w_{inst,Q1} = (w_{inst,m} + w_{inst,v})_{Q1} = (2,78 + 0,50) = 3,28$ mm $w_{fin,Q1} = w_{inst,Q1}(1 + \psi_{2,1}k_{def}) = 3,28 \times [1 + (0,6 \times 0,8)]$ $\qquad = 4,85$ mm $w_{fin} = w_{in,G} + w_{fin,Q1} = (0,25 + 4,85) = 5,10$ mm	$\psi_{2,1} = 0,6$ $w_{fin} \approx 5,10$ mm
National Annex Table NA.4	Limiting ratio $= l/150 = 1800/150 = 12,0$ mm $> 5,10$ mm	**The cross-beams are** **acceptable with respect** **to deflection.**

References	Calculations	Output
	Contract : Walkway Job Ref. No. : Example 5.4	

Contract : Walkway   Job Ref. No. : Example 5.4
Part of Structure :   Box beams
Calculation Sheet No. : 19 of 30

Calculated by : W.McK.
Checked by : B.Z.
Date :

References	Calculations	Output
	(d) Box-Beams: The box-beam supports cross-beams at 1,0 m centres. Characteristic cross-beam loading: $G_k = 0,11$ kN $G_k = 0,30$ kN $G_k = 0,11$ kN $Q_k = 2,53$ kN $Q_k = 8,44$ kN $Q_k = 2,53$ kN $g_k = 0,05$ kN/m 0,34 kN 0,34 kN 6,75 kN 6,75 kN 600 mm 1800 mm 600 mm Permanent loads: Minimum reaction from a cross-beam $= (0,34 \times 1,0) = 0,34$ kN Maximum reaction from a cross-beam $= (0,34 \times 1,35) = 0,46$ kN Variable loads: Minimum reaction from a cross-beam $= (6,75 \times 1,0) = 6,75$ kN Maximum reaction from a cross-beam $= (6,75 \times 1,5) = 10,13$ kN Design load due to self- weight of ply-web beam $= (0,15 \times 1,35)$ $= 0,2$ kN/m Ultimate design loads: Reaction due to cross section $= (0,43 + 10,13) = 10,56$ kN 5,30 kN 10,59 kN 10,59 kN 10,59 kN 10,59 kN 5,30 kN 0,2 kN/m 5 @ 1,0 m centres = 5,0 m 26,98 kN 26,98 kN Design bending moment due to permanent action $M_{y,d} = [(26,98 - 5,30) \times 2,5] - [10,59 \times (1,5 + 0,5)]$ $- [0,2 \times 2,5 \times 1,25)] = 32,35$ kNm Design shear force due to permanent action $V_d = (26,98 - 5,30) = 21,68$ kN	 $M_{y,d} = $ **32,35 kNm** $V_{G,d} = $ **21,68 kN**

References	Calculations	Output
	Contract : Walkway Job Ref. No. : Example 5.4 **Part of Structure : Box beams** **Calculation Sheet No. : 20 of 30**	**Calculated by : W.McK.** **Checked by : B.Z.** **Date :**

References	Calculations	Output
	Characteristic permanent load/cross-beam = 2,45 kN Characteristic variable load/cross-beam = 33,75 kN Total load = 36,1 kN/cross beam Permanent load = $(2,45/36,2) \times 100 = 6,8\%$ of the total load Variable load = $(33,75/36,2) \times 100 = 93,2\%$ of the total load Effective cross-sectional properties in terms of flange material: $$A_{ef} = A_{flange} + \left(\frac{E_{web}}{E_{flange}}\right)\left(\frac{1+k_{def,flange}}{1+k_{def,web}}\right)A_{web}$$ $$I_{ef} = I_{flange} + \left(\frac{E_{web}}{E_{flange}}\right)\left(\frac{1+k_{def,flange}}{1+k_{def,web}}\right)I_{web}$$ When considering permanent action induced stress, $\psi_2 = 1,0$. When considering instantaneous deformations the effect of creep can be ignored and $k_{def} = 0$. When considering final deformations the effects of creep must be included and values of k_{def} can be determined from Table 3.2 of EC5.	
BS EN 1995-1-1 Clause 2.3.2.2(2)	For ultimate limit states the final mean value of modulus of elasticity, $E_{mean,fin}$ should be calculated from: $$E_{mean,fin} = \frac{E_{mean}}{(1+\psi_2 k_{def})}$$ In the case of permanent actions ψ_2 should be replaced by 1,0. k_{def} comprises two components, one due to the permanent load and the other due to the variable load, i.e. $k_{def,flange} = (k_{def,flange})_{permanent} + (\psi_2 k_{def,flange})_{variable}$ $\quad k_{def,web} = (k_{def,web})_{permanent} + (\psi_2 k_{def,web})_{variable}$ The components are evaluated in proportion to the % load in each case and in accordance with the appropriate ψ_2 values as indicated above.	
UK NA to EC Table NA.A1.1	Category C3: $\psi_2 = 0,6$	$\psi_2 = 0,6$
BS EN 1995-1-1 Table 3.2	Values of k_{def} for Service Class 2 Flange: Solid timber, $k_{def} = 0,8$ $(k_{def,flange})_{permanent} = (k_{def,flange} \times$ proportion of load $\times \psi_2)$ $\qquad\qquad\qquad = (0,8 \times 0,068 \times 1,0) = 0,05$ $(k_{def,flange})_{variable} = (k_{def,flange} \times$ proportion of load $\times \psi_2)$ $\qquad\qquad\qquad = (0,8 \times 0,932 \times 0,6) = 0,45$ $k_{def,flange} = (0,05 + 0,45) = 0,50$	$k_{def,flange} = 0,50$

References	Calculations	Output
	Contract : Walkway Job Ref. No. : Example 5.4	**Calculated by : W.McK.**
	Part of Structure : Box beams	**Checked by : B.Z.**
	Calculation Sheet No. : 21 of 30	**Date :**

References	Calculations	Output
Table 3.2	Web: Plywood, $k_{def} = 1,0$ $(k_{def,web})_{permanent} = (k_{def,web} \times \text{proportion of load} \times \psi_2)$ $\qquad\qquad\qquad = (1,0 \times 0,068 \times 1,0) = 0,07$ $(k_{def,web})_{variable} = (k_{def,web} \times \text{proportion of load} \times \psi_2)$ $\qquad\qquad\qquad = (1,0 \times 0,932 \times 0,6) = 0,56$ $k_{def,web} = (0,07 + 0,56) = 0,63$ $\left(\dfrac{1 + k_{def,flange}}{1 + k_{def,web}}\right)_u = \left(\dfrac{1 + 0,50}{1 + 0,63}\right) = 0,92$ $\left(\dfrac{E_{web}}{E_{flange}}\right) = \left(\dfrac{9,68}{8,0}\right) = 1,21$ Creep in the solid timber flanges is less than that in the OSB web, i.e. the normal flange stresses increase with time whilst those in the web decrease. The flange stresses corresponding with the ultimate limit state limits and the web stresses corresponding with the instantaneous states should therefore be checked. **Effective properties for ultimate limit states corresponding with the instantaneous deformations: $k_{def} = 0$** $A_{flange} = (125 \times 70) = 8750,0 \text{ mm}^2$ $I_{flange} = (I_{centroid} + Az^2) = \left(\dfrac{bh^3}{12} + bhz^2\right)$ $= \left\{\dfrac{125 \times 70^3}{12} + \left[125 \times 70 \times (250 - 35)^2\right]\right\} = 408,04 \times 10^6 \text{ mm}^4$ $A_{web} = (500 \times 2 \times 12,3) = 12,3 \times 10^3 \text{ mm}^2$ $I_{web} = \left(\dfrac{bh^3}{12}\right) = \left(\dfrac{24,6 \times 500^3}{12}\right) = 256,25 \times 10^6 \text{ mm}^4$ Since $k_{def} = 0$ $\quad \left(\dfrac{1 + k_{def,flange}}{1 + k_{def,web}}\right) = 1,0$ $A_{ef,inst} = A_{flange} + \left(\dfrac{E_{web}}{E_{flange}}\right)\left(\dfrac{1 + k_{def,flange}}{1 + k_{def,web}}\right)A_{web}$ $= [2 \times 8750,0 + (1,21 \times 1,0 \times 12,3 \times 10^3)] = 32,38 \times 10^3 \text{ mm}^2$	$k_{def,web} = 0,63$ $A_{ef,inst} = 32,38 \times 10^3 \text{ mm}^2$

Contract : Walkway Job Ref. No. : Example 5.4 Part of Structure : Box beams Calculation Sheet No. : 22 of 30	Calculated by : W.McK. Checked by : B.Z. Date :

References	Calculations	Output
	$I_{\text{ef,inst}} = I_{\text{flange}} + \left(\dfrac{E_{\text{web}}}{E_{\text{flange}}} \right) \left(\dfrac{1 + k_{\text{def,flange}}}{1 + k_{\text{def,web}}} \right) I_{\text{web}}$ $= [2 \times 408{,}04 \times 10^6 + (1{,}21 \times 1{,}0 \times 256{,}25 \times 10^6)]$ $= 1126{,}14 \times 10^6 \text{ mm}^4$	$I_{\text{ef,inst}} = 1126{,}14 \times 10^6 \text{ mm}^4$
	Effective properties for ultimate limit states corresponding with the final deformations: $k_{\text{def,flange}} = 0{,}03$ and $k_{\text{def,web}} = 0{,}04$ $A_{\text{ef,u}} = [2 \times 8750{,}0 + (1{,}21 \times 0{,}92 \times 12{,}3 \times 10^3)] = 31{,}19 \times 10^3 \text{ mm}^2$ $I_{\text{ef,u}} = [(2 \times 408{,}04 \times 10^6) + (1{,}21 \times 0{,}92 \times 256{,}25 \times 10^6)]$ $= 1101{,}34 \times 10^6 \text{ mm}^4$	$A_{\text{ef,u}} = 31{,}19 \times 10^3 \text{ mm}^2$ $I_{\text{ef,u}} = 1101{,}34 \times 10^6 \text{ mm}^4$
Clause 9.1.1(1)	**Check flange normal stresses:** Design bending strength of the flange $f_{\text{m,d}} = (f_{\text{m,k}} \times k_{\text{mod}} \times k_{\text{sys}}) / \gamma_M = (16{,}0 \times 0{,}9 \times 1{,}0)/1{,}3$ $= 11{,}08 \text{ N/mm}^2$	$f_{\text{md}} = 11{,}08 \text{ N/mm}^2$
Equation (9.1)	$\sigma_{\text{f,c,max,d}} < f_{\text{m,d}}$ and $(z_{\max} = 500/2) = 250 \text{ mm}$ $\sigma_{\text{f,c,max,d}} = \dfrac{M_{\text{y,d}} \times z_{\max}}{I_{\text{ef,u}}} = \left(\dfrac{32{,}35 \times 10^6 \times 250}{1101{,}34 \times 10^6} \right) = 7{,}34 \text{ N/mm}^2$ $\qquad\qquad\qquad\qquad\qquad\qquad\qquad\qquad < f_{\text{m,d}}$	
Equation (9.2)	Similarly for $\sigma_{\text{f,t,max,d}} < f_{\text{m,d}}$ Design compressive strength of the flange parallel to the grain $f_{\text{c,0,d}} = (f_{\text{c,0,k}} \times k_{\text{mod}} \times k_{\text{sys}}) / \gamma_M = (17{,}0 \times 0{,}9 \times 1{,}0)/1{,}3$ $= 11{,}77 \text{ N/mm}^2$	
Equation (9.3)	$\sigma_{\text{f,c,d}} \le k_c f_{\text{c,0,d}}$ where $\sigma_{\text{f,c,d}}$ is the mean flange design compressive stress and k_c is a factor taking into account lateral instability. $z_{\text{mean}} = (250 - 35) = 215 \text{ mm}$ $\sigma_{\text{f,c,d}} = \dfrac{M_{\text{y,d}} \times z_{\text{mean}}}{I_{\text{ef,u}}} = \left(\dfrac{32{,}35 \times 10^6 \times 215}{1101{,}34 \times 10^6} \right) = 6{,}32 \text{ N/mm}^2$ The compression flange is restrained by the cross-beams at 1,0 m centres.	
Clause 9.1.1(3) Equation (9.5) Figure 9.1	Determine approximate value of k_c using Clause 6.3.2 with the slenderness $\lambda_z = \sqrt{12} \left(\dfrac{l_c}{b} \right)$, where $l_c = 1000 \text{ mm}$ and $b = [(2 \times 12{,}3) + 125] = 149{,}6 \text{ mm}$	

Contract : Walkway Job Ref. No. : Example 5.4	Calculated by : W.McK.
Part of Structure : Box beams	Checked by : B.Z.
Calculation Sheet No. : 23 of 30	Date :

References	Calculations	Output
	$\lambda_z = \sqrt{12}\left(\dfrac{1000}{149,6}\right) = 23,16$	
Equation (6.22)	$\lambda_{rel,z} = \dfrac{\lambda_z}{\pi}\sqrt{\dfrac{f_{c,0,k}}{E_{0,05}}} = \dfrac{23,16}{\pi}\sqrt{\dfrac{17,0}{5400,0}} = 0,41$	
Equation (6.28)	$k_z = 0,5(1 + \beta_c(\lambda_{rel,z} - 0,3) + \lambda_{rel,z}^2)$	
Equation (6.29)	Since k_c is related to the flange material (solid timber) use $\beta_c = 0,2$	
	$k_z = 0,5 \times [1 + 0,2 \times (0,41 - 0,3) + 0,41^2] = 0,60$	
	$k_{cz} = \dfrac{1}{k_z + \sqrt{k_z^2 - \lambda_{rel,z}^2}} = \dfrac{1}{0,60 + \sqrt{0,60^2 - 0,41^2}} = 0,96$	
	$k_c f_{c,0,d} = (0,96 \times 11,77) = 11,30 \text{ N/mm}^2 > \sigma_{f,c,d}$	$k_c f_{c,0,d} = 11,30 \text{ N/mm}^2$
	Design tensile strength of the flange parallel to the grain $f_{t,0,d} = (f_{t,0,k} \times k_{mod} \times k_{sys})/\gamma_M = (10,0 \times 0,9 \times 1,0)/1,3$ $= 6,92 \text{ N/mm}^2$	$f_{t,0,d} = 6,92 \text{ N/mm}^2$
Equation (9.4)	$\sigma_{f,t,d} \le f_{t,0,d}$ where $\sigma_{f,t,d}$ is the mean flange design tensile stress $\sigma_{f,t,d} = \sigma_{c,t,d} = 6,30 \text{ N/mm}^2 < f_{t,0,d}$	Flanges are adquate with respect to stresses.
Clause 9.1.1(4)	**Check web normal stresses:** (converted to web material) Design compressive strength of the web parallel to the grain $f_{c,w,d} = (f_{c,0,k} \times k_{mod} \times k_{sys})/\gamma_M = (22,0 \times 1,1 \times 1,0)/1,2$ $= 20,17 \text{ N/mm}^2$	$f_{c,w,d} = 20,17 \text{ N/mm}^2$
Equation (9.6)	$\sigma_{w,c,d} \le f_{c,w,d}$ where $\sigma_{w,c,d}$ is the design compressive stress in the web $\sigma_{w,c,d} = \dfrac{M_{y,d} \times z_{max}}{I_{ef,inst}} \times \dfrac{E_{web}}{E_{flange}}$ $= \left(\dfrac{32,35 \times 10^6 \times 250 \times 1,21}{1126,14 \times 10^6}\right) = 8,69 \text{ N/mm}^2 < f_{c,w,d}$	Web is adequate with respect to compression.
	Design tensile strength of the web parallel to the grain $f_{t,w,d} = (f_{t,0,k} \times k_{mod} \times k_{sys})/\gamma_M = (27,7 \times 1,1 \times 1,0)/1,2$ $= 25,39 \text{ N/mm}^2$	$f_{t,w,d} = 25,39 \text{ N/mm}^2$

Contract : Walkway Job Ref. No. : Example 5.4	Calculated by : W.McK.
Part of Structure : Box beams	Checked by : B.Z.
Calculation Sheet No. : 24 of 30	Date :

References	Calculations	Output
Equation (9.7)	$\sigma_{w,t,d} \le f_{t,w,d}$ where $\sigma_{w,t,d}$ is the design tensile stress in the web $\sigma_{w,t,d} = \dfrac{M_{y,d} \times z_{max}}{I_{ef,inst}} \times \dfrac{E_{web}}{E_{flange}} = \left(\dfrac{32,35\times10^6 \times 250 \times 1,21}{1126,14\times10^6} \right)$ $= 8,69 \text{ N/mm}^2 < f_{t,w,d}$	**Web is adequate with respect to tension.**
Clause 9.1.1(7) Equation (9.8)	**Check web aspect ratio and panel shear strength:** Since no detailed buckling analysis of the web is being carried out, the aspect ratio should be checked such that $h_w \le 70b_w$ $h_w = 360$ mm, and $70b_w = (70 \times 12,3) = 861$ mm $> h_w$	**Aspect ratio of web is satisfactory.**
Equation (9.9)	$F_{v,w,Ed} \le \begin{cases} b_w h_w \left(1 + \dfrac{0,5\left(h_{f,t} + h_{fc}\right)}{h_w} \right) f_{v,0,d} \text{ for } h_w \le 35b_w \\\\ 35b_w^2 \left(1 + \dfrac{0,5\left(h_{f,t} + h_{fc}\right)}{h_w} \right) f_{v,0,d} \text{ for } 35b_w \le h_w \le 70b_w \end{cases}$ Design panel shear strength of the web. $f_{v,0,d} = (f_{v,0,k} \times k_{mod} \times k_{sys})/\gamma_M = (9,5 \times 1,1 \times 1,0)/1,2 = 8,71 \text{ N/mm}^2$ $35b_w = (35 \times 12,3) = 430,5$ mm $> h_w$ $\left[12,3 \times 360 \times \left(1 + \dfrac{0,5\times(70+70)}{360} \right) \times 8,71 \right] \Big/ 10^3 = 46,07 \text{ kN}$ $F_{v,w,Ed} = [0,5 \times 21,68] = 10,84 \text{ kN/web} < 46,07 \text{ kN}$	$f_{v,0,d} = \mathbf{8,71 \text{ N/mm}^2}$ **Webs are adequate with respect to panel shear.**
Clause 9.1.1(8)	**Check the planar (rolling) stresses at the glue lines:**	
Equation (9.10)	$\tau_{mean,d} \le \begin{cases} f_{v,90,d} & \text{for } h_f \le 4b_{ef} \\\\ f_{v,90,d} \left(\dfrac{4b_{ef}}{h_f} \right)^{0,8} & \text{for } h_f > 4b_{ef} \end{cases}$	
Equation (9.11)	$b_{ef} = b_w$ for box-beams $\therefore b_{ef} = 12,3$ mm, $4b_{ef} = 49,2$ mm $h_f = 70,0$ mm $> 4b_{ef}$ $f_{v,90,d} = (f_{v,90,k} \times k_{mod} \times k_{sys})/\gamma_M = (1,7 \times 1,1 \times 1,0)/1,2$ $= 1,56 \text{ N/mm}^2$ $f_{v,90,d}\left(\dfrac{4b_{ef}}{h_f} \right)^{0,8} = 1,56 \times \left(\dfrac{49,2}{70,0} \right)^{0,8} = \mathbf{1,18 \text{ N/mm}^2}$	$f_{v,90,d} = \mathbf{1,56 \text{ N/mm}^2}$

References	Calculations	Output

Contract : Walkway   Job Ref. No. : Example 5.3
Part of Structure :   Box beams
Calculation Sheet No. : 25 of 30

Calculated by : W.McK.
Checked by : B.Z.
Date :

Figure 9.1
See Chapter 4

Section 4.13

The design shear stress at the joint between the flanges and the web is given by:

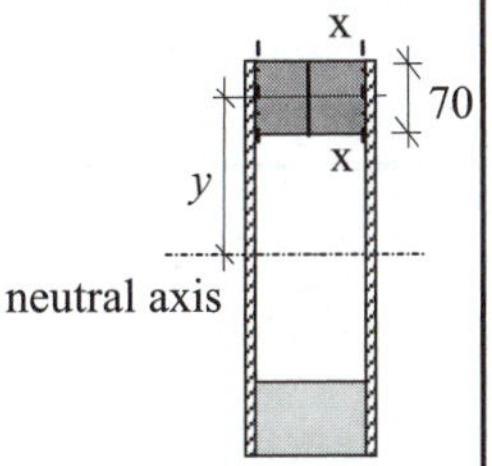

$$\tau_{mean,d} = \frac{V_d Az}{I_{ef} b}$$

The critical shear stresses on the gluelines correspond with the critical flange stresses, i.e. when final deformations occur.

$Az = [125 \times 70 \times (250 - 35)] = 1881{,}25 \times 10^3 \text{ mm}^3$
$I_{ef,u} = 1101{,}34 \times 10^9 \text{ mm}^4$
$b = (2 \times 70) = 140 \text{ mm}$
$V_d = 21{,}68 \text{ kN}$

$$\tau_{mean,d} = \left(\frac{21{,}68 \times 10^3 \times 1881{,}25 \times 10^3}{1101{,}34 \times 10^6 \times 140{,}0} \right) = 0{,}26 \text{ N/mm}^2$$

$< 1{,}18 \text{ N/mm}^2$

Web is adequate with respect to planar shear.

Serviceability Limit States:
Deformation:

Clause 7.2

Limiting values for deflection of beams:

$w_{net,fin} = w_{inst} + w_{creep} - w_c = w_{fin} - w_c$

Since there is no camber $w_c = 0$ and $w_{net,fin} = w_{inst} + w_{creep} = w_{fin}$

(**Note:** $\gamma_G = \gamma_Q = 1{,}0$ for characteristic deformations)

BS EN 1995-1-1
Table 3.2
Table 3.2

Values of k_{def} for Service class 2
Flange: Solid timber, $k_{def} = 0{,}8$
Web: Plywood, $k_{def} = 1{,}0$

$k_{def,flange} = 0{,}8$
$k_{def,web} = 1{,}0$

The permanent load deflection comprises both bending and shear components for the distributed self-weight and the point loads due to the cross-beams.
The variable load deflection comprises both bending and shear components for the point loads due to the cross-beams.

Permanent load due to the self-weight:

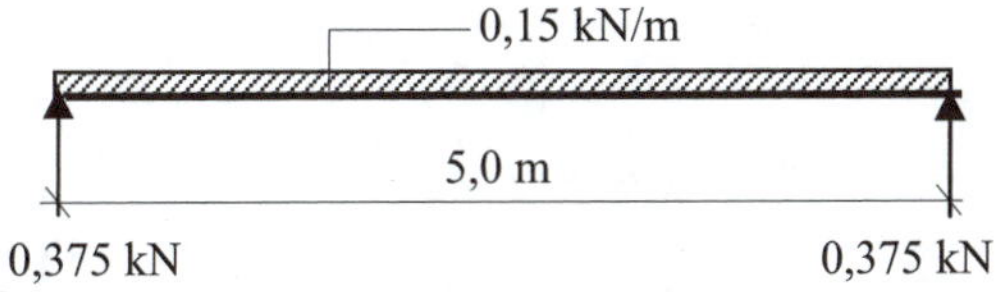

0,15 kN/m

5,0 m

0,375 kN   0,375 kN

Contract : **Walkway** **Job Ref. No. : Example 5.4** **Part of Structure :** **Box beams** **Calculation Sheet No. : 26 of 30**	**Calculated by : W.McK.** **Checked by : B.Z.** **Date :**

References	Calculations	Output
	Permanent loads due to cross-beams: 0,17 kN 0,34 kN 0,34 kN 0,34 kN 0,34 kN 0,17 kN 5 @ 1,0 m centres = 5,0 m 0,85 kN 0,85 kN Maximum bending moment due to cross-beams $M_{y,G,d} = (0,15 \times 5^2/8) + [(0,85 - 0,17) \times 2,5)]$ $\qquad - [0,34 \times (0,5 + 1,5)] = 1,49$ kNm Variable loads due to cross-beams: 3,38 kN 6,75 kN 6,75 kN 6,75 kN 6,75 kN 3,38 kN 5 @ 1,0 m centres = 5,0 m 16,88 kN 16,88 kN Maximum bending moment $M_{y,Q,d} = (16,88 \times 2,5) - (3,38 \times 2,5) - [6,75 \times (0,5 + 1,5)]$ $\qquad = 20,25$ kNm Bending deflection due to UDL $= \dfrac{5WL^3}{384 E_f I_{ef}}$ Shear deflection due to UDL $= k \times \dfrac{0,125WL}{AG}$ where k is a constant for a given shape, i.e. for a rectangle = 1,2. Bending deflection due to cross-beams $\approx \dfrac{0,104 ML^2}{E_f I_{ef}}$ Shear deflection due to the cross-beams can be estimated using the method given in Appendix B as follows: Shear deflection $\approx \dfrac{k}{AG_w} \displaystyle\sum_{i=1}^{n} T_i v_i$ In the case of this box-beam the values of k and $\displaystyle\sum_{i=1}^{n} T_i v_i$ are calculated as indicated below.	
Appendix B		

References	Calculations	Output

Contract : Walkway   Job Ref. No. : Example 5.4
Part of Structure :   Box beams
Calculation Sheet No. : 27 of 30

Calculated by : W.McK.
Checked by : B.Z.
Date :

Chapter 4
Section 4.13

The cross-section must be transformed into one material (see Chapter 4, Section 4.13)
Consider a transformed section in terms of the *web* material as follows:

Modular ratio $n = \dfrac{E_{\text{flange}}}{E_{\text{web}}} = \dfrac{8,0}{9,68} = 0,83$

$\beta = \dfrac{G_{\text{flange}}}{G_{\text{web}}} = \dfrac{0,5}{0,67} = 0,75$

The transformed width of the flanges $= (125 \times 0,83)$
$= 103,8$ mm

$b_1 = (2 \times 12,3) = 24,6$ mm
$b_2 = (103,8 + 24,6) = 128,4$ mm

$d_1 = (250,0 - 70,0) = 180,0$ mm
$d_2 = 250,0$ mm

$p = b_1/b_2 = (24,6/103,8) = 0,19$
$t = d_1/d_2 = (180,0/250,0) = 0,72$

Transformed area:
$A = (500 \times 128,4) - (2 \times 180,0 \times 103,8)$
$= 26,83 \times 10^3$ mm^2

$A = 26,83 \times 10^3$ mm^2

$$k = \dfrac{4,5\left[\dfrac{1}{p}(1-t)+t\right] \times \left\{\dfrac{1}{p^2}\left(\dfrac{t^5}{2}-t^3+\dfrac{t}{2}\right)+\dfrac{1}{p}\left[-t^5\left(\dfrac{3}{30\beta}+\dfrac{2}{3}\right)+t^3\left(\dfrac{1}{3\beta}+\dfrac{2}{3}\right)-\dfrac{t}{2\beta}+\dfrac{8}{30\beta}\right]+\dfrac{8t^5}{30}\right\}}{\left[\dfrac{1}{p}(1-t^3)+t^3\right]^2}$$

$\left[\dfrac{1}{p}(1-t)+t\right] = \left[\dfrac{1}{0,19}\times(1-0,72)+0,72\right] = 2,194$

$\dfrac{1}{p^2}\left(\dfrac{t^5}{2}-t^3+\dfrac{t}{2}\right) = \dfrac{1}{0,19^2}\times\left(\dfrac{0,72^5}{2,0}-0,72^3+\dfrac{0,72}{2,0}\right) = 2,313$

$-t^5\left(\dfrac{3}{30\beta}+\dfrac{2}{3}\right) = -0,72^5\times\left(\dfrac{3,0}{30\times0,75}+\dfrac{2,0}{3,0}\right) = -0,155$

References	Calculations	Output

Contract : Walkway   Job Ref. No. : Example 5.4   Calculated by : W.McK.
Part of Structure :   Box beams   Checked by : B.Z.
Calculation Sheet No. : 28 of 30   Date :

$$t^3\left(\frac{1}{3\beta}+\frac{2}{3}\right)= 0,72^3\times\left(\frac{1}{3,0\times0,75}+\frac{2,0}{3,0}\right)=0,415$$

$$\frac{t}{2\beta}=\frac{0,72}{2,0\times0,75}=0,480,\qquad \frac{8}{30\beta}=\frac{8,0}{30,0\times0,75}=0,356$$

$$\frac{8t^5}{30}=\frac{8,0\times0,72^5}{30,0}=0,0516$$

$$\left[\frac{1}{p}\left(1-t^3\right)+t^3\right]^2=\left[\frac{1}{0,19}\left(1-0,72^3\right)+0,72^3\right]^2=13,483$$

$$k=\frac{(4,5\times2,194)\times\left[2,313+\frac{1}{0,19}(-0,155+0,415-0,48+0,356)+0,0516\right]}{13,483}=2,26$$

References	Calculations	Output

Contract : **Walkway**   Job Ref. No. : **Example 5.4**
Part of Structure :   **Box beams**
Calculation Sheet No. : **29 of 30**

Calculated by : **W.McK.**
Checked by : **B.Z.**
Date :

$$\sum_{i=1}^{n} T_i v_i = 2 \times [(2P \times 0,2L)(0,5) + (P \times 0,2L)(0,5)] = 0,6PL$$

Shear deflection $= k \times \dfrac{0,6PL}{AG_w} = 2,26 \times \dfrac{0,6PL}{AG_w} = \dfrac{1,36PL}{AG_w}$

Permanent load deformation:

Equation (2.3)

$w_{fin,G} = w_{inst,G}(1 + k_{def}) = (w_{inst,m} + w_{inst,v})_G (1 + k_{def})$
$W = (0,15 \times 5,0) = 0,75$ kN
$P = 0,34$ kN
$M = 1,49$ kNm
$L = 5000$ mm
$E_f = E_{0,mean} = 8,0$ kN/mm^2
$G_w = G_{mean} = 0,67$ kN/mm^2
$A = 26,83 \times 10^3$ mm^2

For instantaneous deformation (w_{inst}): $k_{def} = 0$ i.e. no creep

$$I_{ef inst} = I_{flange} + \left(\frac{E_{web}}{E_{flange}} \right) \left(\frac{1 + k_{def,flange}}{1 + k_{def,web}} \right) I_{web}$$

$= 1126,14 \times 10^6$ mm^4

$w_{fin,G} = [w_{inst,G,m}(1 + k_{def})] + [w_{inst,G,v}(1 + k_{def})]$

Since the bending deflection is based on two wood-based elements having different time-dependent behaviour, the k_{def} value is calculated using Equation (2.13):

Equation (2.13)

$k_{def} = 2 \times \sqrt{k_{def,flange}\ k_{def,web}} = 2 \times \sqrt{(0,8 \times 1,0)} = 1,79$

Bending deformation:

$$w_{inst,G,m} = \frac{5WL^3}{384 E_f I_{ef,inst}} + \frac{0,104 ML^2}{E_f I_{ef,inst}}$$

$$= \frac{5 \times 0,75 \times 5000^3}{384 \times 8,0 \times 1,126 \times 10^9} + \frac{0,104 \times 1,49 \times 10^3 \times 5000^2}{8,0 \times 1,127 \times 10^9}$$

$= (0,14 + 0,43) = 0,57$ mm

Shear deformation:

$$w_{inst,G,v} = 2,26 \times \frac{0,125 WL}{AG} + \frac{1,36 PL}{AG_w}$$

$$= \frac{0,28 \times 0,75 \times 5000}{26,83 \times 10^3 \times 0,67} + \frac{1,36 \times 0,34 \times 5000}{26,83 \times 10^3 \times 0,67}$$

$= (0,06 + 0,13) = 0,19$ mm

Contract : Walkway Job Ref. No. : Example 5.4	Calculated by : W.McK.
Part of Structure : Box beams	Checked by : B.Z.
Calculation Sheet No. : 30 of 30	Date :

References	Calculations	Output
	$w_{fin,G} = [0{,}57 \times (1 + 1{,}79) + 0{,}19 \times (1 + 1{,}79)] = 2{,}12$ mm	
Equation (2.3)	**Variable load deformation:** $w_{fin,Q1} = w_{inst,Q1} (1 + k_{def}) = (w_{inst,m} + w_{inst,v})_{Q1} (1 + k_{def})$ $P = 6{,}75$ kN $M = 20{,}25$ kNm $L = 5000$ mm $E_f = E_{0,mean} = 8{,}0$ kN/mm^2 $G_w = G_{mean} = 0{,}67$ kN/mm^2 $A = 26{,}83 \times 10^3$ mm^2	
	Bending deformation: $w_{inst,Q1,m} = \dfrac{0{,}104 ML^2}{E_f I_{ef,inst}} = \dfrac{0{,}104 \times 20{,}25 \times 10^3 \times 5000^2}{8{,}0 \times 1{,}126 \times 10^9} = 5{,}84$ mm	
	Shear deformation: $w_{inst,Q1,v} = \dfrac{1{,}36 PL}{A G_w} = \dfrac{1{,}36 \times 6{,}75 \times 5000}{26{,}83 \times 10^3 \times 0{,}67} = 2{,}55$ mm	
National Annex to BS EN 1990 Table NA.A1.1	$\psi_{2,1} = 0{,}6$ for congregation areas $w_{fin,Q1} = [w_{inst,Q1,m} (1 + \psi_{2,1} k_{def})] + [w_{inst,Q1,v} (1 + \psi_{2,1} k_{def})]$ $= 5{,}84 \times [1 + (0{,}6 \times 1{,}79)] + 2{,}55 \times [1 + (0{,}6 \times 1{,}79)]$ $= (12{,}11 + 5{,}29) = 17{,}40$ mm	$\psi_{2,1} = 0{,}6$
	Final Deformation: $w_{fin} = w_{fin,G} + w_{fin,Q1} = (2{,}12 + 17{,}40) = 19{,}52$ mm	$w_{fin} = 19{,}52$ **mm**
National Annex Table NA.4	Limiting ratio $= l/150 = 5000/150 = 33{,}33$ mm $> 19{,}52$ mm	**The ply-web are adequate with respect to deflection.**
	Note: acceptable limits should be agreed by the client at the beginning of the project; the code values are recommended, not mandatory.	

5.5 Thin-flanged Beams (Stressed-skin Panels)

A typical cross-section of a stressed-skin panel is shown in Figure 5.19. The construction consists of solid longitudinal members (webs/ribs/stringers) and normally plywood skins, connected to form a composite structural section. The skins may consist of other material such as OSB, particle board or fibre board and the connecting mechanism may be glue, nails, screws or proprietary mechanical fasteners.

The composite section acts as a series of **T**-beams if only one side is covered with a skin or a series of **I**-beams if there are skins on both sides.

T-beam construction   **I**-beam construction

Figure 5.19

The primary objective of stressed-skin panels is to make structural use of the skin in addition to the webs. In traditional construction using standard timber floors, the joists carry the full load without assistance from the flooring. The efficient use of materials using stressed-skin panels results in smaller quantities of timber being required, a reduction in weight to be handled and an improvement in efficiency of construction.

The most common uses for stressed panels are for prefabricated roof or floor panels in timber frame construction. In addition, they can be used to resist compression, bending and racking forces when used as wall units.

Stressed-skin panels are prefabricated and transported to site for erection. Generally roof and floor panels are 1,25 m to 2,5 m wide and between 5 m and 6 m long. Panels used as walls are normally about 2,5 m high with a maximum length of 10 m.

The webs of panels are normally solid sawn timber but can be made from other alternatives such as wood-based panels, glulam or prefabricated I-section. In roofs and floors sawn timber webs vary in width between 38 mm and 63 mm and between 150 mm and 300 mm in depth. Similarly for wall panels, the corresponding dimensions are between 38 mm and 80 mm for the width and 80 mm and 200 mm for the depth. The web spacing varies between 300 mm and 625 mm depending on the type and size of sheeting adopted for the flanges.

The flanges of panels are usually between 10 mm and 19 mm thick. Generally the face grain of Douglas fir plywood flanges runs parallel to the webs and the permissible stress parallel to the face grain is used when assessing the composite strength. In some Finnish plywood sheets, the face grain runs in the short direction and more economic and efficient design may result from this face grain running perpendicular to the webs. The arrangement improves the ability of the plywood to span between the webs, allowing a greater web spacing, and in addition reduces the number of joints required in the length of the panel. In this case the stress perpendicular to the face grain is used when evaluating the composite strength of the panel.

The effect of shear deformation is complex and reduces the contribution of the flange to the bending strength and bending stiffness of the composite unit. In design this is usually dealt with by considering an effective width of flange, similar to the practice adopted in the design of reinforced concrete beam/slab floor systems. A detailed discussion of the effects of shear deformation and its influence on the behaviour of the flange is beyond the scope of this text and further reference should be made to the *Timber Designers' Manual* (73), and *STEP 1* and *STEP 2: Structural Timber Education Programme* (58, 59).

5.5.1 Glued Thin-flanged Beams (Clause 9.1.2)

The design of glued thin-webbed beams is carried out on the assumption of a linear variation of strain over the depth of the beam. The top flange can be regarded as a continuous span beam with supports at the web locations as indicated in Figure 5.20.

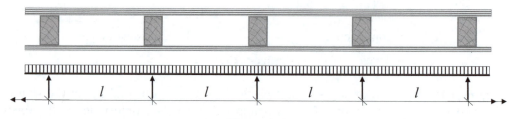

Figure 5.20

The suitability of the section is checked with respect to bending, shear and deflection using Clauses 6.1.6, 6.1.7 and 7.2 respectively.

The composite section is checked in accordance with the requirements of Clause 9.1.2 by assuming a modified cross-section with effective flange widths allowing for shear lag and plate-buckling (determined using Equations (9.12) and (9.13) and Table 9.1) as follows.

For internal beams:

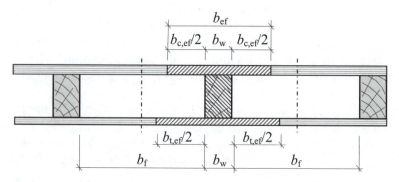

Figure 5.21

In the case of the compression flange: $b_{ef} = b_{c,ef} + b_w$
where:
$b_{c,ef} \leq$ Table 9.1 value for shear lag,
    $\leq$ Table 9.1 value for plate buckling.

In the case of the tension flange: $b_{ef} = b_{t,ef} + b_w$
where:
$b_{t,ef} \leq$ Table 9.1 value for shear lag.

For edge beams:

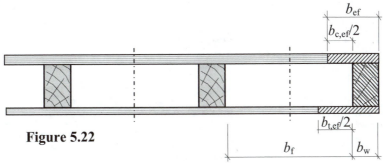

Figure 5.22

In the case of the compression flange: $b_{ef} = 0,5b_{c,ef} + b_w$
where:
$b_{c,ef} \leq$ Table 9.1 value for shear lag,
$\quad\quad\leq$ Table 9.1 value for plate buckling.

In the case of the tension flange: $b_{ef} = 0,5b_{t,ef} + b_w$
where:
$b_{t,ef} \leq$ Table 9.1 value for shear lag.

Note: The top and bottom flanges are not necessary to have the same thickness of material or have the same b_{ef} value and consequently the resulting modified cross-section can be asymmetric.

As indicated in Clause 9.1.2(5), the unrestrained flange width (b_f) should not be greater than twice the value of $b_{c,ef}$ due to plate buckling, i.e.

$b_f \leq (2 \times b_{c,ef})$ from Table 9.1 for plate buckling.

Since the flanges and webs are of different materials with different elastic properties, the stress calculations should be determined on the basis of a transformed section. The different creep characteristics of the materials result in web stresses corresponding with the final deformations and flange stresses corresponding with the instantaneous deformations being critical, (i.e. assuming $k_{def,web} < k_{def,flanges}$).

The mean flange design compressive and tensile stresses should satisfy Equations (9.15) and (9.16), i.e.

$\sigma_{f,c,d} \leq f_{f,c,d}$ Equation (9.15) and
$\sigma_{f,t,d} \leq f_{f,t,d}$ Equation (9.16)

where:

$\sigma_{f,c,d} = \dfrac{M_{y,d} \times z_{f,c,mean}}{I_{ef}}$ and $f_{f,c,d} = (f_{f,c,k} \times k_{mod})/\gamma_M$

$\sigma_{f,t,d} = \dfrac{M_{y,d} \times z_{f,t,mean}}{I_{ef}}$ and $f_{f,t,d} = (f_{f,t,k} \times k_{mod})/\gamma_M$

I_{ef} is the effective second moment of area for the transformed section corresponding with the instantaneous deformations.

The normal stresses in the web should satisfy Equations (9.6) and (9.7), i.e.

$$\sigma_{w,c,d} \leq f_{c,w,d} \quad \text{Equation (9.6)} \quad \text{and}$$
$$\sigma_{w,t,d} \leq f_{t,w,d} \quad \text{Equation (9.7)}$$

where:

$$\sigma_{w,c,d} = \frac{M_{y,d} \times z_{web,top}}{I_{ef}} \times \frac{E_{web}}{E_{flange}} \quad \text{and} \quad f_{c,w,d} = (f_{m,0,k} \times k_{mod})/\gamma_M$$

$$\sigma_{w,t,d} = \frac{M_{y,d} \times z_{web,bottom}}{I_{ef}} \times \frac{E_{web}}{E_{flange}} \quad \text{and} \quad f_{t,w,d} = (f_{m,0,k} \times k_{mod})/\gamma_M$$

I_{ef} is the effective second moment of area for the transformed section corresponding with the final deformations.

The mean design shear stress $\tau_{mean,d}$ (assuming a uniform stress distribution) in the web should be checked in accordance with Equation (9.14) such that:

$$\tau_{mean,d} \leq \begin{cases} f_{v,90,d} & \text{for } b_w \leq 8h_f \\[2ex] f_{v,90,d}\left(\dfrac{8h_f}{b_w}\right)^{0,8} & \text{for } b_w > 8h_f \end{cases} \qquad \text{Equation (9.14)}$$

where:
$f_{v,90,d}$ is the design planar (rolling) shear strength of the flange,
h_f is the flange thickness,
b_w is the web thickness.

In the case of asymmetric sections this should be checked at both connections between the web and the flanges as shown in Figure 5.23.

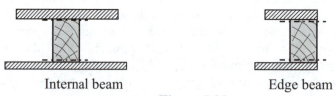

Internal beam   Edge beam

Figure 5.23

In the case of an edge beam the following expression should be satisfied:

$$\tau_{mean,d} \leq \begin{cases} f_{v,90,d} & \text{for } b_w \leq 4h_f \\[2ex] f_{v,90,d}\left(\dfrac{4h_f}{b_w}\right)^{0,8} & \text{for } b_w > 4h_f \end{cases} \qquad \text{Equation (9.14)}$$

where $\quad \tau_{mean,d} = \dfrac{V_d}{A_{web}}$

The serviceability criteria regarding deflection should be checked in a similar manner to that indicated for glued thin-webbed beams in Section 5.43.5. In this case however, it is likely that an asymmetric cross-section will occur because of the different effective compression and tension flange widths.

In Example 5.5 an approximate value for the shear deflection has been determined using Chart 4 from Appendix B based on $p = b_1/b_2$ where b_2 is based on the larger effective flange width and consequently the larger 'k' value in relation to flange widths.

5.5.2 Mechanically Jointed Beams (Clause 9.1.3 and Annex B)

In glued composite sections the joints between the flanges and the web(s) are assumed to be rigid. In the case of mechanically jointed beams an allowance must be made for the slip which occurs between the flanges and web due to the horizontal shear forces induced during bending. A simplified analysis neglecting shear deformations is given in Annex B of EC5 in which a number of equations are given to determine the 'effective stiffness $(EI)_{ef}$', the 'normal stresses (σ and σ_m)', and the fastener load (F).

Normal stresses:

$$\sigma_i = \frac{\gamma_i E_i a_i M}{(EI)_{ef}} \qquad \text{(Equation (B.7) of EC 5)}$$

$$\sigma_{m,i} = \frac{0,5 E_i h_i M}{(EI)_{ef}} \qquad \text{(Equation (B.8) of EC 5)}$$

Maximum shear stress in the web:

$$\tau_{2,max} = \frac{\gamma_3 E_3 A_3 a_3 + 0,5 E_2 b_2 h_2^2}{b_2 (EI)_{ef}} V \qquad \text{(Equation (B.9) of EC 5)}$$

Fastener load:

$$F_i = \frac{\gamma_i E_i A_i a_i s_i}{(EI)_{ef}} V \qquad \text{(Equation (B.10) of EC5)}$$

where the variables γ_i, E_i, A_i, a_i, h_i and s_i with $i = 1$, 2 and 3 are defined in the Annex.

Typical bending stress distribution diagrams for an I-beam (or box-beam) in which the web does not extend through the flanges, an I-beam in which the web does extend through the flanges and for a T-section are given in Figure B.1. of Annex B.

In Figure B.1 σ_1 and σ_3 relate to the mean flange design stresses which should satisfy Equations (9.15) and (9.16) of the code.

The maximum web stress is given by ($\sigma_2 + \sigma_{m,2}$) and should not exceed the design bending stress of the web material.

The design fastener load in the flange to web connections should not exceed the design load-carrying capacity per shear plane per fastener in accordance with Section 8 of EC5, i.e.

$$F_i \leq F_{v,Rd}$$

where $F_{v,Rd} = (k_{mod} \times F_{v,Rk})/\gamma_M$.

5.5.3 Example 5.5: Stressed-skin Floor Panel

A stressed-skin flat floor panel as shown in Figure 5.24 is required to span 4,0 m and support the loads indicated. Using the data given check the suitability of the proposed section assuming:

(i)   the flanges and webs are glued together, and
(ii) the flanges and webs are nailed together using 60 mm long x 4,0 mm diameter smooth round nails at 40 mm spacing.

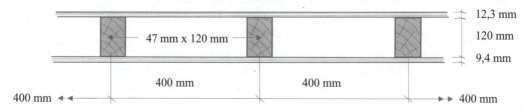

Figure 5.24

Top flange: Finnish combi-plywood wood/face grain parallel to the webs
Bottom flange: Finnish combi-plywood wood/face grain parallel to the webs
Webs: Whitewood   Strength Class C24
Characteristic permanent load: (excluding self-weight):   0,6 kN/m^2
Characteristic variable load: 1,5 kN/m^2

5.5.3.1 Solution to Example 5.5

Contract : Floor Panel Job Ref. No. : Example 5.5 Part of Structure : Stressed-skin Panel Calculation Sheet No. : 1 of 28	Calculated by : W.McK. Checked by : B.Z. Date :

References	Calculations	Output
BS EN 1990 BS EN 1991 BS EN 338:2003 BS EN 1912:2004 BS EN 1995-1-1: 2004 National Annex	Eurocode: Basis of Structural Design Eurocode 1: Actions on Structures Structural Timber - Strength Classes Structural Timber - Strength Classes – Assignment of visual grades and species Eurocode 5: Design of Timber Structures Part 1.1: General – Common Rules and Rules for Buildings. UK - National Annex to BS EN 1995-1-1 Plywood manufacturer's catalogue The plywood in the top flange should be checked as a continuous beam spanning between the webs using the strength values perpendicular to the grain. The composite panel is checked as an I-section using the transformed section properties.	

Contract : Floor Panel Job Ref. No. : Example 5.5	Calculated by : W.McK.
Part of Structure : Stressed-skin Panel	Checked by : B.Z.
Calculation Sheet No. : 2 of 28	Date :

References	Calculations	Output
BS EN 338:2002	Characteristic values for C24 Timber	
Clause 5.0 Table 1	**Strength Properties:** Bending $\qquad f_{m,k} = 24{,}0$ N/mm^2 Compression parallel to grain $\qquad f_{c,0,k} = 21{,}0$ N/mm^2 Compression perpendicular to grain $\qquad f_{c,90,k} = 2{,}5$ N/mm^2 Tension parallel to grain $\qquad f_{t,0,k} = 14{,}0$ N/mm^2 Shear $\qquad f_{v,k} = 2{,}5$ N/mm^2	

Stiffness Properties:
Mean modulus of elasticity parallel to the grain
$$E_{0,mean} = 11{,}0 \text{ kN/mm}^2$$
5% modulus of elasticity parallel to the grain
$$E_{0,05} = 7{,}4 \text{ kN/mm}^2$$
Mean shear modulus $\qquad G_{mean} = 0{,}69$ kN/mm^2
Density $\qquad \rho_k = 350{,}0$ kg/m^3
Mean density $\qquad \rho_{mean} = 420{,}0$ kg/m^3

| Manufacturer's
Catalogue | Characteristic values:
Finnish combi-plywood wood (9 plies)
Thickness $t_{mean} = 12{,}3$ mm | |

Strength Properties:
Bending parallel to grain $\qquad f_{m,0,k} = 42{,}2$ N/mm^2
Bending perpendicular to grain $\qquad f_{m,90,k} = 31{,}6$ N/mm^2
Compression parallel to grain $\qquad f_{c,0,k} = 22{,}1$ N/mm^2
Tension parallel to grain $\qquad f_{t,0,k} = 27{,}7$ N/mm^2
Panel shear $\qquad f_{v,0,k} = 9{,}5$ N/mm^2
Planar shear $\qquad f_{v,90,k} = 1{,}7$ N/mm^2

Stiffness Properties:
Mean modulus of elasticity in bending parallel to the grain
$$E_{0,mean} = 9{,}68 \text{ kN/mm}^2$$

Mean modulus of elasticity in bending perpendicular to the grain
$$E_{90,mean} = 6{,}44 \text{ kN/mm}^2$$

Mean modulus of elasticity in tension and compression
parallel to the grain $\qquad E_{0,t} = E_{0,c} = 7{,}66$ kN/mm^2

Mean modulus of elasticity in tension and compression
perpendicular to the grain $\qquad E_{90,t} = E_{90,c} = 7{,}77$ kN/mm^2

Modulus of rigidity – panel shear $\qquad G_{0,mean} = 0{,}67$ kN/mm^2
Modulus of rigidity – planar shear $\qquad G_{90,mean} = 0{,}08$ kN/mm^2
Mean density $\qquad \rho_{mean} = 650{,}0$ kg/m^3

Contract : Floor Panel Job Ref. No. : Example 5.5	Calculated by : W.McK.
Part of Structure : Stressed-skin Panel	Checked by : B.Z.
Calculation Sheet No. : 3 of 28	Date :

References	Calculations	Output
	Characteristic values: Finnish combi-plywood wood (7 plies) Thickness $t_{mean} = 9{,}4$ mm **Strength Properties:** Bending parallel to grain $f_{m,0,k} = 46{,}0$ N/mm^2 Compression parallel to grain $f_{c,0,k} = 23{,}0$ N/mm^2 Tension parallel to grain $f_{t,0,k} = 30{,}8$ N/mm^2 Panel shear $f_{v,0,k} = 9{,}5$ N/mm^2 Planar shear $f_{v,90,k} = 1{,}7$ N/mm^2 **Stiffness Properties:** Mean modulus of elasticity in bending parallel to the grain $\qquad\qquad\qquad\qquad\qquad E_{0,mean} = 10{,}56$ kN/mm^2 Mean modulus of elasticity in bending perpendicular to the grain $\qquad\qquad\qquad\qquad\qquad E_{90,mean} = 5{,}79$ kN/mm^2 Mean modulus of elasticity in tension and compression parallel to the grain $E_{0,t} = E_{0,c} = 7{,}99$ kN/mm^2 Mean modulus of elasticity in tension and compression perpendicular to the grain $E_{0,t} = E_{0,c} = 7{,}60$ kN/mm^2 Modulus of rigidity – panel shear $G_{0,mean} = 0{,}67$ kN/mm^2 Modulus of rigidity – planar shear $G_{90,mean} = 0{,}08$ kN/mm^2 Mean density $\rho_{mean} = 650{,}0$ kg/m^3	
Clause 2.3.1.2 Table NA.1 (EC5: Table 2.3)	Load-duration Classes: use Table NA.1 from National Annex Self-weight - permanent Imposed floor loading - variable	
Table NA.2	Intermediate floor: Service Class 1	
Table 3.1	Solid timber: Service Class 1 Permanent actions $k_{mod} = 0{,}6$ Variable actions $k_{mod} = 0{,}8$	
Clause 3.1.3(2)	Use k_{mod} corresponding to action with the shortest duration.	**$k_{mod,solid} = 0{,}8$**
	Plywood: Service Class 1 Permanent actions $k_{mod} = 0{,}6$ Variable actions $k_{mod} = 0{,}8$	
Clause 3.1.3(2)	Use k_{mod} corresponding to action with the shortest duration.	**$k_{mod,plywood} = 0{,}8$**

Contract : Floor Panel Job Ref. No. : Example 5.5	Calculated by : W.McK.
Part of Structure : Stressed-skin Panel	Checked by : B.Z.
Calculation Sheet No. : 4 of 28	Date :

References	Calculations	Output
National Annex Table NA.3 (Table 2.3)	Partial factor for material properties and resistance: γ_M For solid timber – untreated and preservative treated For plywood	$\gamma_{M,solid} = 1,3$ $\gamma_{M,plywood} = 1,2$

National Annex
Table NA.3
(Table 2.3)

Partial factor for material properties and resistance: γ_M
For solid timber – untreated and preservative treated
For plywood

$\gamma_{M,solid} = 1,3$
$\gamma_{M,plywood} = 1,2$

BS EN 1990

Eurocode: Basis of Structural Design

Equation (6.10)

$$E_d = \sum_{j\geq1}\gamma_{G,j}G_{k,j} + \gamma_{Q,1}Q_{k,1} + \sum_{i>1}\gamma_{Q,i}\psi_{0,i}Q_{k,i}$$

$$E_d = \left(\gamma_G G_k + \gamma_{Q,1}Q_{k,1}\right)$$

where E_d is the design value of the effect of the actions

BS EN 1990
Table A1.2(B)

$\gamma_G = 1,35$ and $\gamma_{Q,1} = 1,50$

Top Flange:
Consider continuous beam of 1,0 m width spanning between the webs.

Mean thickness = 12,3 mm
Mean density = 650 kg/m^3 ($\equiv$ 6,38 kN/m^3)
Self-weight of top flange = (0,0123 × 6,38) = 0,08 kN/m^2

Cross-sectional area $A = 12,3 \times 10^3$ mm^2/metre width

Elastic section modulus = $W_y = \dfrac{1000 \times 12,3^2}{6}$
$$= 25,22 \times 10^3 \text{ mm}^3/\text{metre width}$$

Second moment of area = $I_y = \dfrac{1000 \times 12,3^3}{12}$
$$= 155,07 \times 10^3 \text{ mm}^4/\text{metre width}$$

Since the grain is parallel to the span of the webs, the stresses perpendicular to the grain should be used to determine the strength of the panel spanning between the webs.

Permanent action = (0,08 + 0,6) × 1,0 = 0,68 kN/m
Variable action = 1,5 × 1,0 = 1,5 kN/m

Design loads:
Minimum permanent action = (1,0 × 0,68)   = 0,68 kN/m
Maximum permanent action = (1,35 × 0,68)   = 0,92 kN/m
Maximum variable action   = (1,5 × 1,5)   = 2,25 kN/m
Total design load   = (0,92 + 2,25)   = 3,17 kN/m

Contract : Floor Panel Job Ref. No. : Example 5.5		**Calculated by : W.McK.**
Part of Structure : Stressed-skin Panel		**Checked by : B.Z.**
Calculation Sheet No. : 5 of 28		**Date :**

References	Calculations	Output
Appendix A this textbook	**Bending:** Design bending moment $M_{y,d} \approx 0,121 wL^2$ $M_{y,d} = (0,121 \times 3,17 \times 0,4^2) = 0,06$ kNm	$M_{y,d} = 0,06$ **kNm**
	Design bending strength $f_{m,y,d} = (f_{m,90,k} \times k_{mod} \times k_{sys})/\gamma_M$	
Clause 6.6	System strength: $k_{sys} = 1,0$ $f_{m,y,d} = (31,6 \times 0,8)/1,2 = 21,07$ N/mm^2	$f_{m,y,d} = 21,07$ **N/mm^2**
BS EN 1995-1-1 Clause 6.1.6	Since uniaxial bending about the y-y axis only, is present: $\sigma_{m,y,d} \leq f_{m,y,d}$ $\sigma_{m,y,d} = \dfrac{M_{y,d}}{W_y} = \dfrac{0,06 \times 10^6}{25,22 \times 10^3} = 2,38$ N/mm$^2 \ll 21,07$ N/mm^2	**The panel is adequate with respect to bending.**
Appendix A this textbook	**Shear:** Design shear force $F_{v,d} \approx (1 - 0,38) wL = (0,62 \times 3,17 \times 0,4)$ $\qquad\qquad\qquad = 0,79$ kN	$F_{v,d} = 0,79$ **kN**
	Design shear strength $f_{v,d} = (f_{m,y,k} \times k_{mod} \times k_{sys})/\gamma_M$	
	$f_{v,d} = f_{v,0,d} = (9,5 \times 0,8 \times 1,0)/1,2 = 6,33$ N/mm^2	$f_{v,d} = 6,33$ **N/mm^2**
Clause 6.1.7	Design shear stress $\tau_d = \dfrac{1,5 \times F_{v,d}}{A} = \dfrac{1,5 \times 0,79 \times 10^3}{1000 \times 12,3}$ $\qquad\qquad = 0,10$ N/mm$^2 \ll 6,33$ N/mm^2	**The panel is adequate with respect to shear**
Clause 7.2	**Deflection:** Limiting values for deflection of beams: $w_{net,fin} = w_{inst} + w_{creep} - w_c = w_{fin} - w_c$ Since there is no camber $w_c = 0$ and $w_{net,fin} = w_{inst} + w_{creep} = w_{fin}$	
Table 3.2	For Plywood Service Class 1 $k_{def} = 0,8$ $w_{fin,G} = w_{inst,G}(1 + k_{def}) = (w_{inst,m} + w_{inst,v})_G (1 + k_{def})$ where:	
	Permanent load deformation: $W = (0,68 \times 0,4) = 0,27$ kN $L = 400$ mm $E_d = E_{90,mean} = 6,44$ kN/mm^2 $G_d = G_{90,mean} = 0,08$ kN/mm^2 $I_y = 155,07 \times 10^3$ mm^4/metre width	

References	Calculations	Output
	$w_{inst,G,m} = \dfrac{5WL^3}{384E_d I} = \dfrac{5 \times 0,27 \times 400^3}{384 \times 6,44 \times 155,07 \times 10^3} = 0,23$ mm	

$$w_{inst,G,v} = \frac{3WL}{20bhG_d} = \frac{3 \times 0,27 \times 400}{20 \times 1000 \times 12,3 \times 0,08} = 0,02 \text{ mm}$$

$w_{inst,G} = (w_{inst,m} + w_{inst,v})_G = (0,23 + 0,02) = 0,25$ mm

$w_{fin,G} = w_{inst,G}(1 + k_{def}) = 0,25 \times (1 + 0,8) = 0,45$ mm

Variable load deformation:

$W = (1,5 \times 0,4) = 0,6$ kN

$L = 400$ mm

$E_d = E_{90,mean} = 6,44$ kN/mm^2

$G_d = G_{90,mean} = 0,08$ kN/mm^2

$I_y = 155,07 \times 10^3$ mm^4/metre width

$$w_{inst,Q1,m} = \frac{5WL^3}{384E_d I} = \frac{5 \times 0,6 \times 400^3}{384 \times 6,44 \times 155,07 \times 10^3} = 0,50 \text{ mm}$$

$$w_{inst,Q1,v} = \frac{3WL}{20bhG_d} = \frac{3 \times 0,6 \times 400}{20 \times 1000 \times 12,3 \times 0,08} = 0,04 \text{ mm}$$

UK NA to
BS EN 1990
Table NA.A1.1

$\psi_{2,1} = 0,3$ for domestic, residential and office areas

$w_{inst,Q1} = (w_{inst,m} + w_{inst,v})_{Q1} = (0,50 + 0,04) = 0,54$ mm

$w_{fin,Q1} = w_{inst,Q1}(1 + \psi_{2,1} k_{def}) = 0,54 \times [1 + (0,3 \times 0,8)]$
$\qquad\qquad\qquad\qquad\qquad\qquad\qquad = 0,67$ mm

$w_{fin} = w_{in,G} + w_{fin,Q1} = (0,45 + 0,67) = 1,12$ mm

National Annex
Table NA.4

Limiting ratio = $l/150 = 400/150 = 2,67$ mm $> 1,12$ mm

(i) Composite Panel (assuming flanges and webs are glued):
The composite panel spans 4,0 m. Consider a width equal to
400 mm comprising one web and a width of flange on each side
as a composite I-beam section.

Output column:

$\psi_{2,1} = 0,3$

$w_{fin} \approx 1,12$ mm

The panel is adequate with respect to deflection.

47 mm × 120 mm
12,3 mm
120 mm
9,4 mm
400 mm   400 mm

Contract : Floor Panel Job Ref. No. : Example 5.5	Calculated by : W.McK.
Part of Structure : Stressed-skin Panel	Checked by : B.Z.
Calculation Sheet No. : 7 of 28	Date :

References	Calculations	Output
UK NA to BS EN 1990 Table NA.A1.2(B)	**Loading:** Self-weight of top flange = $(0,0123 \times 6,38) = 0,08$ kN/m^2 **Bottom flange:** Mean flange thickness = 9,4 mm Mean density = 650 kg/m^3 ($\equiv 6,38$ kN/m^3) Self-weight of bottom flange = $(0,0094 \times 6,38) = 0,06$ kN/m^2 **Web:** Cross-sectional area = $(0,047 \times 0,12) = 0,006$ m^2 Mean density = 420 kg/m^3 ($\equiv 4,12$ kN/m^3) Self-weight of webs = $(0,006 \times 4,12) = 0,025$ kN/m length $\gamma_G = 1,35$ and $\gamma_{Q,1} = 1,50$ The design load = $\gamma_G g_k + \gamma_{Q,1} q_k$ Floor area supported by one **I**-beam section $= (0,4 \times 1,0) = 0,4$ m^2/m Load due to top flange = $(0,08 \times 0,4) = 0,032$ kN/m Load due to bottom flange = $(0,06 \times 0,4) = 0,024$ kN/m Load due to web = 0,025 kN/m Total load due to self-weight = $(0,032 + 0,024 + 0,025)$ $= 0,081$ kN/m Permanent applied load = $(0,6 \times 0,4) = 0,24$ kN/m Variable applied load = $(1,5 \times 0,4) = 0,60$ kN/m Minimum permanent action = $1,0 \times (0,081 + 0,24)$ $= 0,32$ kN/m Maximum permanent action = $1,35 \times (0,081 + 0,24)$ $= 0,43$ kN/m Maximum variable action = $(1,5 \times 0,6) = 0,9$ kN/m Maximum design load = $(0,43 + 0,9) = 1,33$ kN/m length Permanent load= $(0,43/1,33) \times 100 = 32,3\%$ of the total load Variable load = $(0,9/1,33) \times 100 = 67,7\%$ of the total load Span = 4,0 m Design bending moment $M_{y,d} = \dfrac{wL^2}{8} = \dfrac{1,33 \times 4,0^2}{8} = 2,66$ kNm Design shear force $V_d = (wL/2) = (1,33 \times 4,0)/2 = 2,66$ kN	$M_{y,d} = \mathbf{2,66}$ **kNm** $V_d = \mathbf{2,66}$ **kN**

References	Calculations	Output
Contract : Floor Panel Job Ref. No. : Example 5.5 Part of Structure : Stressed-skin Panel Calculation Sheet No. : 8 of 28		Calculated by : W.McK. Checked by : B.Z. Date :

References	Calculations	Output
Clause 9.1.2(3)	**Effective flange widths:** **Compression (top) flange** $b_{ef,top\ flange} = (b_{c,ef} + b_w)$	
Table 9.1	Plywood with grain direction parallel to the webs.	
	Allowing for shear lag effects $b_{c,ef} \le 0,1l$ where l = span = 4000 mm $b_{c,ef} \le (0,1 \times 4000) = 400,0$ mm	
	Allowing for plate buckling $b_{c,ef} \le 20h_f$ where h_f = compression flange thickness = 12,3 mm $b_{c,ef} \le (20,0 \times 12,3) = 246,0$ mm	
Clause 9.1.2(5)	Unrestrained flange width = 400 mm $\le (2 \times 246,0)$ $b_w = 47,0$ mm $b_{ef,top\ flange} = (b_{c,ef} + b_w) = (246,0 + 47,0) = 293,0$ mm	**Top flange is acceptable.** $b_{ef,top\ flange} = $ **293,0 mm**
Table 9.1	**Tension (bottom) flange** $b_{ef,bottom\ flange} = (b_{t,ef} + b_w)$ $\le$ actual flange width Allowing for shear lag effects $b_{t,ef} \le 0,1l$ $b_{t,ef} \le (0,1 \times 4000) = 400,0$ mm $b_w = 47,0$ mm $b_{ef,bottom\ flange} = (b_{t,ef} + b_w) = (400,0 + 47,0) = 447,0$ mm ≤ 400 mm	$b_{ef,bottom\ flange} = $ **400,0 mm**
	Transformed section in terms of top flange material - position of centroid Transformed width of web $= \left(\dfrac{E_{0,mean,web}}{E_{0,mean,top\ flange}} \times b_w \right) = \dfrac{11,0}{9,68} \times 47,0 = 53,4$ mm Transformed width of bottom flange $= \left(\dfrac{E_{0,mean,bottom\ flange}}{E_{0,mean,top\ flange}} \times b_w \right) = \dfrac{10,56}{9,68} \times 400,0 = 436,4$ mm	

293,0 mm | 12,3 mm | 53,4 mm | 120,0 mm | $\overline{y}$ | 9,4 mm | 436,4 mm

References	Calculations	Output

Contract : Floor Panel **Job Ref. No. : Example 5.5** **Calculated by : W.McK.**
Part of Structure : Stressed-skin Panel **Checked by : B.Z.**
Calculation Sheet No. : 9 of 28 **Date :**

$$\bar{y} = \frac{\left[\left(293,0\times12,3\times135,55\right)+\left(53,4\times120,0\times69,4\right)+\left(436,4\times9,4\times4,7\right)\right]}{\left(293,0\times12,3\right)+\left(53,4\times120,0\right)+\left(436,4\times9,4\right)} = 67,5 \text{ mm}$$

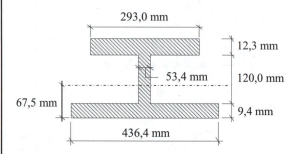

Effective cross-sectional properties in terms of flange material:

$$A_{\text{ef}} = A_{\text{top flange}} + \left(\frac{E_{\text{web}}}{E_{\text{top flange}}}\right)\left(\frac{1+k_{\text{def,top flange}}}{1+k_{\text{def,web}}}\right)A_{\text{web}}$$

$$+ \left(\frac{E_{\text{bottom flange}}}{E_{\text{top flange}}}\right)\left(\frac{1+k_{\text{def,top flange}}}{1+k_{\text{def,bottom flange}}}\right)A_{\text{bottom flange}}$$

$$I_{\text{ef}} = I_{\text{top flange}} + \left(\frac{E_{\text{web}}}{E_{\text{top flange}}}\right)\left(\frac{1+k_{\text{def,top flange}}}{1+k_{\text{def,web}}}\right)I_{\text{web}}$$

$$+ \left(\frac{E_{\text{bottom flange}}}{E_{\text{top flange}}}\right)\left(\frac{1+k_{\text{def,top flange}}}{1+k_{\text{def,bottom flange}}}\right)I_{\text{bottom flange}}$$

When considering instantaneous deformations the effect of creep can be ignored and $k_{\text{def}} = 0$.

When considering final deformations the effects of creep must be included and values of k_{def} can be determined from Table 3.2 of EC5.

BS EN 1995-1-1
Clause 2.3.2.2(2)

For ultimate limit states the final mean value of modulus of elasticity, $E_{\text{mean,fin}}$ should be calculated from:

$$E_{\text{mean,fin}} = \frac{E_{\text{mean}}}{\left(1+\psi_2 k_{\text{def}}\right)}$$

In the case of permanent actions ψ_2 should be replaced by 1,0.

References	Calculations	Output
National Annex BS EN 1990 Table NA.A1.1	For imposed loads on floors as defined in the National Annex to BS EN 1991-1-:2002 (i.e. category A) $\psi_2 = 0,3$	$\psi_2 = 0,3$
	k_{def} comprises two components, one due to the permanent load and the other due to the variable load, i.e.	
	$k_{def,top\ flange} = k_{def,bottom\ flange}$ $\qquad = (k_{def,top\ flange})_G + (\psi_2\ k_{def,top\ flange})_Q$ $k_{def,web} = (k_{def,web})_G + (\psi_2\ k_{def,web})_Q$	
	The components are evaluated in proportion to the % load in each case and in accordance with the appropriate ψ_2 values as indicated above.	
BS EN 1995-1-1 Table 3.2	Values of k_{def} for Service Class 1 Flange: Plywood, $k_{def} = 0,8$ $(k_{def,top\ flange})_G = (k_{def,top\ flange} \times \text{proportion of load} \times \psi_2)_G$ $\qquad = (0,8 \times 0,323 \times 1,0) = 0,258$ $(k_{def,top\ flange})_Q = (k_{def,top\ flange} \times \text{proportion of load} \times \psi_2)_Q$ $\qquad = (0,8 \times 0,677 \times 0,3) = 0,162$ $k_{def,top\ flange} = k_{def,bottom\ flange} = (0,258 + 0,162) = 0,42$	$k_{def,top\ flange} = 0,42$ $k_{def,bottom\ flange} = 0,42$
Table 3.2	Web: Solid timber, $k_{def} = 0,6$ $(k_{def,web})_G = (k_{def,web} \times \text{proportion of load} \times \psi_2)_G$ $\qquad = (0,6 \times 0,323 \times 1,0) = 0,194$ $(k_{def,web})_Q = (k_{def,web} \times \text{proportion of load} \times \psi_2)_Q$ $\qquad = (0,6 \times 0,677 \times 0,3) = 0,122$ $k_{def,web} = (0,194 + 0,122) = 0,32$	$k_{def,web} = 0,32$
	$\left(\dfrac{1 + k_{def,top\ flange}}{1 + k_{def,web}} \right) = \left(\dfrac{1 + 0,42}{1 + 0,32} \right) = 1,08$	
	$\left(\dfrac{1 + k_{def,top\ flange}}{1 + k_{def,bottom\ flange}} \right) = \left(\dfrac{1 + 0,42}{1 + 0,42} \right) = 1,0$	
	$\left(\dfrac{E_{web}}{E_{top\ flange}} \right) = \left(\dfrac{11,0}{9,68} \right) = 1,14$	
	$\left(\dfrac{E_{bottom\ flange}}{E_{top\ flange}} \right) = \left(\dfrac{10,56}{9,68} \right) = 1,09$	

Contract : Floor Panel   **Job Ref. No. :** Example 5.5
Part of Structure : Stressed-skin Panel
Calculation Sheet No. : 10 of 28

Calculated by : W.McK.
Checked by : B.Z.
Date :

Contract : Floor Panel Job Ref. No. : Example 5.5	Calculated by : W.McK.
Part of Structure : Stressed-skin Panel	Checked by : B.Z.
Calculation Sheet No. : 11 of 28	Date :

References	Calculations	Output
	Creep in the solid timber web is less than that in the top plywood flange and slightly less than that in the bottom plywood flange, i.e. the normal web stresses increase with time whilst those in the flanges decrease. The web stresses corresponding with the final deformations and the flange stresses corresponding with the instantaneous deformations should therefore be checked.	

Effective properties for ultimate limit states corresponding with the instantaneous deformations: $k_{def} = 0$

$A_{\text{top flange}} = (293,0 \times 12,3) = 3603,90 \text{ mm}^2$

$A_{\text{web}} = (53,4 \times 120,0) = 6408,00 \text{ mm}^2$

$A_{\text{bottom flange}} = (436,4 \times 9,4) = 4102,16 \text{ mm}^2$

$$I_{\text{top flange}} = (I_{\text{centroid}} + Az^2)_{\text{top flange}} = \left(\frac{bh^3}{12} + bhz^2 \right)_{\text{top flange}}$$

$$= \left\{ \frac{293,0 \times 12,3^3}{12} + \left[293,0 \times 12,3 \times (135,55 - 67,5)^2 \right] \right\}$$

$$= 16,734 \times 10^6 \text{ mm}^4$$

$$I_{\text{web}} = (I_{\text{centroid}} + Az^2)_{\text{web}} = \left(\frac{bh^3}{12} + bhz^2 \right)_{\text{web}}$$

$$= \left\{ \frac{53,4 \times 120,0^3}{12} + \left[53,4 \times 120,0 \times (60,0 + 9,4 - 67,5)^2 \right] \right\}$$

$$= 7,713 \times 10^6 \text{ mm}^4$$

$$I_{\text{bottom flange}} = (I_{\text{centroid}} + Az^2)_{\text{bottom flange}} = \left(\frac{bh^3}{12} + bhz^2 \right)_{\text{bottom flange}}$$

$$= \left\{ \frac{436,4 \times 9,4^3}{12} + \left[436,4 \times 9,4 \times (67,5 - 4,7)^2 \right] \right\}$$

$$= 16,208 \times 10^6 \text{ mm}^4$$

Since $k_{def} = 0$ $\left(\dfrac{1 + k_{def,\text{top flange}}}{1 + k_{def,\text{web}}} \right) = \left(\dfrac{1 + k_{def,\text{top flange}}}{1 + k_{def,\text{bottom flange}}} \right) = 1,0$

$$A_{\text{ef,inst}} = A_{\text{flange}} + \left(\frac{E_{\text{web}}}{E_{\text{top flange}}} \right) A_{\text{web}} + \left(\frac{E_{\text{bottom flange}}}{E_{\text{top flange}}} \right) A_{\text{bottom flange}}$$

$$= [3603,90 + (1,14 \times 6408,0) + (1,09 \times 4102,16)]$$

$$= 15,380 \times 10^3 \text{ mm}^2$$

Output: $A_{\text{ef,inst}} = 15,380 \times 10^3 \text{ mm}^2$

References	Calculations	Output
	$I_{\text{ef,inst}} = I_{\text{flange}} + \left(\dfrac{E_{\text{web}}}{E_{\text{top flange}}}\right) I_{\text{web}} + \left(\dfrac{E_{\text{bottom flange}}}{E_{\text{top flange}}}\right) I_{\text{bottom flange}}$	
	$= [16{,}734 \times 10^6 + (1{,}14 \times 7{,}713 \times 10^6) + (1{,}09 \times 16{,}208 \times 10^6)]$	
	$= 43{,}194 \times 10^6 \text{ mm}^2$	$I_{\text{ef,inst}} = 43{,}194{\times}10^6 \text{ mm}^4$
	Effective properties for ultimate limit states corresponding with the final deformations:	
	$k_{\text{def,top flange}} = k_{\text{def,bottom flange}} = 0{,}42$ and $k_{\text{def,web}} = 0{,}32$	
	$A_{\text{ef,fin}} = A_{\text{top flange}} + \left(\dfrac{E_{\text{web}}}{E_{\text{top flange}}}\right)\left(\dfrac{1 + k_{\text{def,top flange}}}{1 + k_{\text{def,web}}}\right) A_{\text{web}}$	
	$\qquad + \left(\dfrac{E_{\text{bottom flange}}}{E_{\text{top flange}}}\right)\left(\dfrac{1 + k_{\text{def,top flange}}}{1 + k_{\text{def,bottom flange}}}\right) A_{\text{bottom flange}}$	
	$= [3603{,}90 + (1{,}14 \times 1{,}08 \times 6408{,}0) + (1{,}09 \times 1{,}0 \times 4102{,}16)]$	
	$= 15{,}965 \times 10^3 \text{ mm}^2$	$A_{\text{ef,fin}} = 15{,}965{\times}10^3 \text{ mm}^2$
	$I_{\text{ef,fin}} = I_{\text{top flange}} + \left(\dfrac{E_{\text{web}}}{E_{\text{top flange}}}\right)\left(\dfrac{1 + k_{\text{def,top flange}}}{1 + k_{\text{def,web}}}\right) I_{\text{web}}$	
	$\qquad + \left(\dfrac{E_{\text{bottom flange}}}{E_{\text{top flange}}}\right)\left(\dfrac{1 + k_{\text{def,top flange}}}{1 + k_{\text{def,bottom flange}}}\right) I_{\text{bottom flange}}$	
	$= [16{,}734 \times 10^6 + (1{,}14 \times 1{,}08 \times 7{,}713 \times 10^6)$	
	$\qquad + (1{,}09 \times 1{,}0 \times 16{,}208 \times 10^6)]$	
	$= 43{,}897 \times 10^6 \text{ mm}^2$	$I_{\text{ef,fin}} = 43{,}897{\times}10^6 \text{ mm}^4$
Clause 9.1.2(7)	**Check flange normal stresses:**	
	Design compressive strength of the top flange	
	$f_{\text{f,c,d}} = (f_{\text{c,0,k}} \times k_{\text{mod}})/\gamma_M = (22{,}1 \times 0{,}8)/1{,}2$	
	$\qquad = 14{,}73 \text{ N/mm}^2$	$f_{\text{f,c,d}} = 14{,}73 \text{ N/mm}^2$
Equation (9.15)	$\sigma_{\text{f,c,d}} < f_{\text{f,c,d}}$	
	$z_{\text{mean}} = (9{,}4 + 120 + 6{,}15 - 67{,}5) = 68{,}05 \text{ mm}$	
	$\sigma_{\text{f,c,d}} = \dfrac{M_{\text{y,d}} \times z_{\text{mean}}}{I_{\text{ef,inst}}} = \left(\dfrac{2{,}66 \times 10^6 \times 68{,}05}{43{,}194 \times 10^6}\right) = 4{,}19 \text{ N/mm}^2$	
	$\qquad\qquad\qquad\qquad\qquad\qquad\qquad\qquad\quad \ll f_{\text{f,c,d}}$	**Top flange is adequate in compression.**
	Design tensile strength of the bottom flange (converted to bottom flange material):	
	$f_{\text{f,t,d}} = (f_{\text{t,0,k}} \times k_{\text{mod}})/\gamma_M = (30{,}8 \times 0{,}8)/1{,}2$	
	$\qquad = 20{,}53 \text{ N/mm}^2$	$f_{\text{f,t,d}} = 20{,}53 \text{ N/mm}^2$
Equation (9.16)	$\sigma_{\text{f,t,d}} < f_{\text{f,t,d}}$	
	$z_{\text{mean}} = (67{,}5 - 4{,}7) = 62{,}8 \text{ mm}$	

Contract : Floor Panel Job Ref. No. : Example 5.5	Calculated by : W.McK.
Part of Structure : Stressed-skin Panel	Checked by : B.Z.
Calculation Sheet No. : 13 of 28	Date :

References	Calculations	Output
	$\sigma_{f,c,d} = \left(\dfrac{M_{y,d} \times z_{mean}}{I_{ef,inst}} \right) \times \left(\dfrac{E_{bottom\ flange}}{E_{top\ flange}} \right)$	
	$= \left(\dfrac{2,66 \times 10^6 \times 62,8}{43,194 \times 10^6} \right) \times 1,09 = 4,22 \text{ N/mm}^2 \ll f_{f,t,d}$	**Bottom flange is adequate in tension.**
Clause 9.1.1(4)	**Check web normal stresses:** (converted to web material) Design compressive strength of the web parallel to the grain $f_{c,w,d} = (f_{m,0,k} \times k_{mod})/\gamma_M = (24,0 \times 0,8)/1,3$ $\qquad = 14,77 \text{ N/mm}^2$	$f_{c,w,d} = 14,77 \text{ N/mm}^2$
Equation (9.6)	$\sigma_{w,c,d} \leq f_{c,w,d}$ where $\sigma_{w,c,d}$ is the design compressive bending stress in the web. $z_{top\ of\ web} = (9,4 + 120,0 - 67,5) = 61,9 \text{ mm}$ $\sigma_{w,c,d} = \left(\dfrac{M_{y,d} \times z_{top\ of\ web}}{I_{ef,fin}} \right) \times \left(\dfrac{E_{web}}{E_{top\ flange}} \right)$	
	$= \left(\dfrac{2,66 \times 10^6 \times 61,9 \times 1,14}{43,897 \times 10^6} \right) = 4,28 \text{ N/mm}^2 \ll f_{c,w,d}$	**Web is adequate with respect to compression.**
	Design tensile strength of the web parallel to the grain $f_{t,w,d} = (f_{m,0,k} \times k_{mod})/\gamma_M = (24,0 \times 0,8)/1,3$ $\qquad = 14,77 \text{ N/mm}^2$	$f_{t,w,d} = 14,77 \text{ N/mm}^2$
Equation (9.7)	$\sigma_{w,t,d} \leq f_{t,w,d}$ where $\sigma_{w,t,d}$ is the design tensile bending stress in the web. $z_{bottom\ of\ web} = (67,5 - 9,4) = 58,1 \text{ mm}$ $\sigma_{w,t,d} = \left(\dfrac{M_{y,d} \times z_{bottom\ of\ web}}{I_{ef,fin}} \right) \times \left(\dfrac{E_{web}}{E_{top\ flange}} \right)$	
	$= \left(\dfrac{2,66 \times 10^6 \times 58,1 \times 1,09}{43,897 \times 10^6} \right) = 3,84 \text{ N/mm}^2 \ll f_{t,w,d}$	**Web is adequate with respect to tension.**
Clause 6.1.7(1)	**Check the direct shear stresses on the web:** Design shear strength $f_{v,d} = (f_{v,k} \times k_{mod} \times k_{sys})/\gamma_M$ $f_{v,d} = (2,5 \times 0,8)/1,3 = 1,54 \text{ N/mm}^2$	$f_{v,d} = 1,54 \text{ N/mm}^2$
	For shear stress, untransformed web thickness shuld be used.	

References	Calculations	Output

Contract : Floor Panel   Job Ref. No. : Example 5.5   **Calculated by : W.McK.**
Part of Structure : Stressed-skin Panel   **Checked by : B.Z.**
Calculation Sheet No. : 14 of 28   **Date :**

Clause 6.1.7

Design shear stress $\tau_d = \dfrac{1{,}5 \times F_{v,d}}{A_w} = \dfrac{1{,}5 \times 2{,}66 \times 10^3}{47 \times 120}$

$$= 0{,}71 \text{ N/mm}^2 < 1{,}54 \text{ N/mm}^2$$

The web is adequate with respect to shear.

Clause 9.1.2(6)

Check the planar (rolling) stresses at the flange glue lines:

Equation (9.14)

$$\tau_{mean,d} \leq \begin{cases} f_{v,90,d} & \text{for } b_w \leq 8h_f \\[2ex] f_{v,90,d}\left(\dfrac{8h_f}{b_w}\right)^{0,8} & \text{for } b_w > 8h_f \end{cases}$$

$b_w = 41{,}36 \text{ mm}$ $8h_f = (8{,}0 \times 12{,}3) = 98{,}4 \text{ mm} > b_w$

$f_{v,90,d} = (f_{v,90,k} \times k_{mod})/\gamma_M = (1{,}7 \times 0{,}8)/1{,}2$
$\qquad\quad = 1{,}13 \text{ N/mm}^2$

$f_{v,90,d} = \textbf{1,13 N/mm}^2$

$\tau_{mean,d} \leq f_{v,90,d} = 1{,}13 \text{ N/mm}^2$

Figure 9.2

The mean design shear stress at the joint between the flanges
and the web is given by: $\tau_{mean,d} = \dfrac{V_d}{A_w}$

$\tau_{mean,d} = \dfrac{2{,}66 \times 10^3}{120{,}0 \times 53{,}4} = 0{,}42 \text{ N/mm}^2 < 1{,}13 \text{ N/mm}^2$

Web is adequate with respect to planar shear.

Serviceability Limit States:
Deformation:

Clause 7.2

Limiting values for deflection of beams:
$w_{net,fin} = w_{inst} + w_{creep} - w_c = w_{fin} - w_c$
Since there is no camber $w_c = 0$ and $w_{net,fin} = w_{inst} + w_{creep} = w_{fin}$

(**Note:** $\gamma_G = \gamma_Q = 1{,}0$ for characteristic deformations)

Permanent design loading $= (0{,}32 \times 4{,}0) = 1{,}28 \text{ kN}$
Variable design loading $= (0{,}6 \times 4{,}0) = 2{,}4 \text{ kN}$

References	Calculations	Output
	Contract : Floor Panel Job Ref. No. : Example 5.5 **Part of Structure : Stressed-skin Panel** **Calculation Sheet No. : 15 of 28**	**Calculated by : W.McK.** **Checked by : B.Z.** **Date :**

References	Calculations	Output
Appendix B	**Bending deflection** $= \dfrac{5WL^3}{384E_f I_{ef}}$ **Shear deflection** $= k \times \dfrac{0{,}125WL}{AG_w}$ In Appendix B the method given to determine the shear deflection is intended for I-sections with equal flanges. It is being used here to estimate the values for an asymmetric section with different flange widths. Using Chart 4 in this case: $\beta = \dfrac{G_{\text{flange}}}{G_{\text{web}}} = \dfrac{0{,}67}{0{,}69} = 0{,}97, \quad p = b_1/b_2 \approx 53{,}4/436{,}4 = 0{,}122$ and $t = d_1/d_2 \approx 60{,}0/70{,}85 = 0{,}85$ Using the chart gives: $k \approx 2{,}2$. 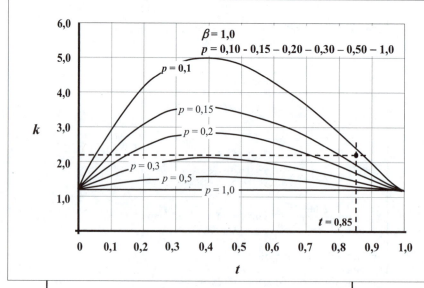 **Shear deflection** $= 2{,}2 \times \dfrac{0{,}125WL}{AG} = \dfrac{0{,}275WL}{AG}$	
BS EN 1995-1-1 Table 3.2 Table 3.2	Values of k_{def} for Service Class 1 Flange: $k_{\text{def}} = 0{,}8$ Web: $k_{\text{def}} = 0{,}6$	

References	Calculations	Output
	Contract : Floor Panel Job Ref. No. : Example 5.5 **Calculated by : W.McK.** **Part of Structure : Stressed-skin Panel** **Checked by : B.Z.** **Calculation Sheet No. : 16 of 28** **Date :**	

References	Calculations	Output
Equation (2.3)	**Permanent load deformation:** $w_{\text{fin,G}} = w_{\text{inst,G}}(1 + k_{\text{def}}) = (w_{\text{inst,m}} + w_{\text{inst,v}})_G (1 + k_{\text{def}})$ Total deformation due to (bending + shear effects) is given by: $$w_{\text{inst,G}} = \frac{5WL^3}{384 E_f I_{\text{ef,inst}}} + \frac{0,275WL}{A_{\text{ef,inst}} G_w}$$ $W = 1,28$ kN $L = 4000$ mm $E_f = E_{0,\text{mean}} = 9,68$ kN/mm^2 $G_w = G_{\text{mean}} = 0,69$ kN/mm^2 **For instantaneous deformation (w_{inst}):** $A_{\text{ef,inst}} = 15,965 \times 10^3$ mm^2 $I_{\text{ef,inst}} = 43,897 \times 10^6$ mm^4 Bending deformation: $$w_{\text{inst,G,m}} = \frac{5WL^3}{384 E_f I_{\text{ef,inst}}} = \frac{5 \times 1,28 \times 4000^3}{384 \times 9,68 \times 43,897 \times 10^6} = 2,51 \text{ mm}$$ Shear deformation: $$w_{\text{inst,G,v}} = \frac{0,275WL}{A_{\text{ef,inst}} G_w} = \frac{0,275 \times 1,28 \times 4000}{15,965 \times 10^3 \times 0,69} = 0,13 \text{ mm}$$ Since the deflection is based on two wood-based elements having different time-dependent behaviour, the k_{def} value is calculated using Equation (2.13):	
Equation (2.13)	$k_{\text{def}} = 2 \times \sqrt{k_{\text{def,flange}}\, k_{\text{def,web}}} = 2 \times \sqrt{(0,8 \times 0,6)} = 1,39$ $w_{\text{fin,G}} = [2,51 \times (1 + 1,39) + 0,13 \times (1 + 1,39)] = 6,31 \text{ mm}$	
Equation (2.4)	**Variable load deformation:** $w_{\text{fin,Q1}} = w_{\text{inst,Q1}}(1 + \psi_{2,1} k_{\text{def}}) = (w_{\text{inst,m}} + w_{\text{inst,v}})_{Q1}(1 + \psi_{2,1} k_{\text{def}})$ $W = 2,4$ kN $L = 4000$ mm $E_f = E_{0,\text{mean}} = 9,68$ kN/mm^2 $G_w = G_{\text{mean}} = 0,69$ kN/mm^2 Bending deformation: $$w_{\text{inst,Q1,m}} = \frac{5WL^3}{384 E_f I_{\text{ef,inst}}} = \frac{5 \times 2,4 \times 4000^3}{384 \times 9,68 \times 43,897 \times 10^6} = 4,71 \text{ mm}$$	

Contract : Floor Panel Job Ref. No. : Example 5.5 Part of Structure : Stressed-skin Panel Calculation Sheet No. : 17 of 28	Calculated by : W.McK. Checked by : B.Z. Date :

References	Calculations	Output
	Shear deformation: $w_{\text{inst,Q1,v}} = \dfrac{0,275WL}{A_{\text{ef,inst}}G_{\text{w}}} = \dfrac{0,275 \times 2,4 \times 4000}{15,965 \times 10^3 \times 0,69} = 0,24$ mm	
National Annex to BS EN 1990 Table NA.A1.1	$\psi_{2,1} = 0,3$ $w_{\text{fin,Q1}} = [w_{\text{inst,Q1}}(1 + \psi_{2,1}k_{\text{def}})]_{\text{m}} + [w_{\text{inst,Q1}}(1 + \psi_{2,1}k_{\text{def}})]_{\text{v}}$ $= 4,71 \times [1 + (0,3 \times 1,39)] + 0,24 \times [1 + (0,3 \times 1,39)] = 7,01$ mm **Final Deformation:** $w_{\text{fin}} = w_{\text{fin,G}} + w_{\text{fin,Q1}} = (6,31 + 7,01) = 13,32$ mm	$w_{\text{fin}} = 13,32$ mm
National Annex Table NA.4	Limiting ratio $= l/150 = 4000/150 = 26,67$ mm $> 13,32$ mm	The panel is adequate with respect to deflection.
BS EN 1995-1-1 Clause 9.1.3(1)P Annex B	**(ii) Assuming flanges and web are nailed together:** Mechanically jointed beams – See Annex B of EC5 which indicates a simplified analysis for the calculation of the load carrying capacity of mechanically jointed beams allowing for slip. 293,0 mm 12,3 mm 47,0 mm 120,0 mm $\bar{y}$ 9,4 mm 400,0 mm $$\bar{y} = \dfrac{\left[(293,0 \times 12,3 \times 135,55) + (47,0 \times 120,0 \times 69,4) + (400,0 \times 9,4 \times 4,7)\right]}{(293,0 \times 12,3) + (47,0 \times 120,0) + (400,0 \times 9,4)} = 69,03 \text{ mm}$$	
Annex B	Figure B.1 in EC 5 indicates the dimensions to be used when calculating the applied stresses in mechanically jointed beams. The dimension 'a_2' indicated in the diagram is assumed +ve when the centroid of the web is below the level of the neutral axis as shown; the values of all other variables are +ve.	

Contract : Floor Panel Job Ref. No. : Example 5.5 Part of Structure : Stressed-skin Panel Calculation Sheet No. : 18 of 28	Calculated by : W.McK. Checked by : B.Z. Date :

References	Calculations	Output

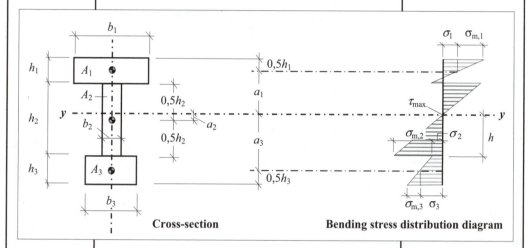

Cross-section   **Bending stress distribution diagram**

The following values apply in relation to the stressed-skin panel in Example 5.5.

$a_1 = (9,4 + 120 + 6,15 - 69,03) = 66,52$ mm
a_2 is calculated in accordance with Equation (B.6)
$a_3 = (69,03 - 4,7) = 64,33$ mm

$b_1 = 293,0$ mm
$b_2 = 47,0$ mm
$b_3 = 400,0$ mm

$h_1 = 12,3$ mm
$h_2 = 120,0$ mm
$h_3 = 9,4$ mm

$h = (69,03 - 9,4) = 59,63$ mm

Equation (B.2) $A_1 = (293,0 \times 12,3) = 3603,9$ mm
$A_2 = (47,0 \times 120,0) = 5640,0$ mm
$A_3 = (400,0 \times 9,4) \ = 3760,0$ mm

Equation (B.3) $I_1 = \dfrac{293,0 \times 12,3^3}{12} = 45,436 \times 10^3$ mm^4

$I_2 = \dfrac{47,0 \times 120,0^3}{12} = 6,768 \times 10^6$ mm^4

$I_3 = \dfrac{400,0 \times 9,4^3}{12} = 27,686 \times 10^3$ mm^4

Contract : Floor Panel Job Ref. No. : Example 5.5	Calculated by : W.McK.
Part of Structure : Stressed-skin Panel	Checked by : B.Z.
Calculation Sheet No. : 19 of 28	Date :

References	Calculations	Output
Table 7.1	**Slip modulus: use 4,0 x 60,0 mm nails with predrilling** $K_{ser,i} = \rho_m^{1,5} d / 23$ where $\rho_m = \sqrt{\rho_{m1}\rho_{m2}}$; $d = 4,0$ mm $\rho_m = \sqrt{650 \times 420} = 522,49$ kg/m^3 $K_{ser,1} = K_{ser,3} = (522,49^{1,5} \times 4,0)/23 = 2077,06$ N/mm	
Equation (2.1)	$K_{u,i} = 0,67 \times K_{ser,i}$ $K_{u,1} = K_{u,3} = (0,67 \times 2077,06) = 1391,63$ N/mm Spacing of fasteners $s_1 = s_3 = 40$ mm $E_1 = 9,68$ kN/mm^2 $E_2 = 11,0$ kN/mm^2 $E_3 = 10,56$ kN/mm^2	
Equation (B.4)	$\gamma_2 = 1,0$	
Equation (B.5)	For ultimate limit state calculations: $\gamma_1 = [1 + \pi^2 E_1 A_1 s_1/(K_{u,1}\, l^2)]^{-1}$ $= [1 + (\pi^2 \times 9,68 \times 10^3 \times 3603,9 \times 40)/(1391,63 \times 4000^2)]^{-1}$ $= 0,62$ $\gamma_3 = [1 + \pi^2 E_3 A_3 s_3/(K_{u,3}\, l^2)]^{-1}$ $= [1 + (\pi^2 \times 10,56 \times 10^3 \times 3760,0 \times 40)/(1391,63 \times 4000^2)]^{-1}$ $= 0,59$	
Equation (B.6)	$a_2 = \dfrac{\gamma_1 E_1 A_1 \left(h_1 + h_2\right) - \gamma_3 E_3 A_3 \left(h_2 + h_3\right)}{2\displaystyle\sum_{i=1}^{3} \gamma_i E_i A_i}$ $= \dfrac{0,62 \times 9,68 \times 3603,9 \times (12,3 + 120,0) - 0,59 \times 10,56 \times 3760,0 \times (120,0 + 9,4)}{2 \times \left[(0,62 \times 9,68 \times 3603,9) + (1,0 \times 11,0 \times 5640,0) + (0,59 \times 10,56 \times 3760,0) \right]}$ $a_2 = -0,79$ mm $a_3 = 69,03 - (0,5 \times 9,4) = 64,33$ mm	
Equation (B.1)	$(EI)_{ef,u} = \displaystyle\sum_{1}^{3} \left(E_i I_i + \gamma_i E_i A_i a_i^2 \right)$	

References	Calculations	Output

Contract : Floor Panel Job Ref. No. : Example 5.5
Part of Structure : Stressed-skin Panel
Calculation Sheet No. : 20 of 28

Calculated by : W.McK.
Checked by : B.Z.
Date :

$$(EI)_{ef,u} = (9{,}68 \times 10^3 \times 45{,}436 \times 10^3) + (0{,}62 \times 9{,}68 \times 10^3 \times 3603{,}9 \times 66{,}52^2)$$
$$+ (11{,}0 \times 10^3 \times 6{,}768 \times 10^6) + (1{,}0 \times 11{,}0 \times 10^3 \times 5640{,}0 \times 0{,}79^2)$$
$$+ (10{,}56 \times 10^3 \times 27{,}686 \times 10^3) + (0{,}59 \times 10{,}56 \times 10^3 \times 3760{,}0 \times 64{,}33^2)$$
$$= 267{,}87 \times 10^9 \text{ Nmm}^2$$

Clause 9.1.2(7)

Check flange normal stresses:
Design compressive strength of the top flange
$f_{f,c,d} = 14{,}73$ N/mm^2

Annex B

As indicated in Figure B.1 the mean stress in the top and bottom flanges is given by Equation (B.7) for normal stresses.

Equation (B.7)

$$\sigma_{1,d} = \frac{\gamma_1 E_1 a_1 M_d}{(EI)_{ef,u}} = \frac{0{,}62 \times 9{,}68 \times 10^3 \times 66{,}52 \times 2{,}66 \times 10^6}{267{,}87 \times 10^9}$$
$$= 3{,}96 \text{ N/mm}^2 \ll f_{f,c,d}$$

Design tensile strength of the bottom flange
$f_{f,t,d} = 20{,}53$ N/mm^2

Equation (B.7)

$$\sigma_{3,d} = \frac{\gamma_3 E_3 a_3 M_d}{(EI)_{ef,u}} = \frac{0{,}59 \times 10{,}56 \times 10^3 \times 64{,}33 \times 2{,}66 \times 10^6}{267{,}87 \times 10^9}$$
$$= 3{,}98 \text{ N/mm}^2 \ll f_{f,t,d}$$

Clause 9.1.1(1)

Check flange bending stresses:
Design bending strength of the top flange
$f_{m,d} = 28{,}13$ N/mm^2

Annex B

As indicated in Figure B.1 the extra bending stress in the top and bottom flanges is given by Equation (B.8).

Equation (B.8)

$$\sigma_{m,1} = \frac{0{,}5 E_1 h_1 M_d}{(EI)_{ef,u}} = \frac{0{,}5 \times 9{,}68 \times 10^3 \times 12{,}3 \times 2{,}66 \times 10^6}{267{,}87 \times 10^9} = 0{,}59$$
$$\sigma_{m,1,d} = \sigma_{1,d} + \sigma_{m,1} = 3{,}96 + 0{,}59 = 4{,}55 \text{ N/mm}^2 \ll f_{m,d}$$

Design bending strength of the bottom flange
$f_{m,d} = 30{,}67$ N/mm^2

Equation (B.8)

$$\sigma_{m,3} = \frac{0{,}5 E_3 h_3 M_d}{(EI)_{ef,u}} = \frac{0{,}5 \times 10{,}56 \times 10^3 \times 9{,}4 \times 2{,}66 \times 10^6}{267{,}87 \times 10^9} = 0{,}49$$
$$\sigma_{m,3,d} = \sigma_{3,d} + \sigma_{m,3} = 3{,}98 + 0{,}49 = 4{,}47 \text{ N/mm}^2 \ll f_{m,d}$$

Output:

Top flange is adequate with respect to axial compressive stress.

Bottom flange is adequate with respect to axial compressive stress.

Top flange is adequate with respect to bending.

Bottom flange is adequate with respect to bending.

References	Calculations	Output
Contract : Floor Panel Job Ref. No. : Example 5.5 **Part of Structure : Stressed-skin Panel** **Calculation Sheet No. : 21 of 28**		**Calculated by : W.McK.** **Checked by : B.Z.** **Date :**

References	Calculations	Output
Clause 9.1.1(4)	**Check web bending stresses:** Design bending strength of the web. $f_{m,d} = 14,77$ N/mm^2	
Figure B.1	In Figure B.1 the maximum web stresses are given as: $$\sigma_{m,2,d} = (\sigma_{2,d} + \sigma_{m,2}) = \frac{\gamma_2 E_2 a_2 M_d}{(EI)_{ef,u}} + \frac{0,5 E_2 h_2 M_d}{(EI)_{ef,u}}$$	
Equations (B.7) and (B.8)	$$= \frac{1,0 \times 11,0 \times 10^3 \times (-0,79) \times 2,66 \times 10^6}{267,87 \times 10^9} + \frac{0,5 \times 11,0 \times 10^3 \times 120,0 \times 2,66 \times 10^6}{267,87 \times 10^9}$$	
	$= -0,09 + 6,55 = 6,46$ N/mm$^2 < f_{m,d}$	**Web is adequate with respect to bending.**
	Maximum shear stress in the web: Design shear strength: Design shear force $V_d = 2,66$ kN Design shear strength $f_{v,d} = (f_{v,k} \times k_{mod})/\gamma_M$ $f_{v,d} = (2,5 \times 0,8)/1,3 = 1,54$ N/mm^2	$f_{v,d} = 1,54$ N/mm^2
Equation (B.9)	$$\tau_{2,max} = \frac{\gamma_3 E_3 A_3 a_3 + 0,5 E_2 b_2 h_2^2}{b_2 (EI)_{ef,u}} V_d$$	
	$$= \frac{\left(0,59 \times 10,56 \times 10^3 \times 3760,0 \times 64,33\right) + \left(0,5 \times 11,0 \times 10^3 \times 47,0 \times 120,0^2\right)}{47,0 \times 267,87 \times 10^9} \times 2660$$	
	$= 1,10$ N/mm$^2 < 1,54$ N/mm^2	
Equation (B.10)	**Fastener load for top flange:** $$F_{1,d} = \frac{\gamma_1 E_1 A_1 a_1 s_1}{(EI)_{ef,u}} V$$	
	$$= \frac{0,62 \times 9,68 \times 10^3 \times 3603,9 \times 66,52 \times 40,0}{267,87 \times 10^9} \times 2660,0 = 571,5 \text{ N}$$	$F_{1,d} = 571,5$ N/nail
	Design load-carrying capacity per shear plane per fastener $F_{i,d} \leq F_{v,Rd}$ where $F_{v,Rd} = (k_{mod} \times F_{v,Rk})/\gamma_M$.	
Clause 8.2.2 (1)	For fasteners in single shear $F_{v,Rk}$ is equal to the minimum value obtained from Equations (8.6(a)) to (8.6(f)) relating to the six failure modes indicated in Figure 8.2 of the code.	

References	Calculations	Output

Contract : Floor Panel   Job Ref. No. : Example 5.5
Part of Structure : Stressed-skin Panel
Calculation Sheet No. : 22 of 28

Calculated by : W.McK.
Checked by : B.Z.
Date :

$$f_{h,1,k}t_1 d \qquad\qquad \text{(a)}$$

$$f_{h,2,k}t_2 d \qquad\qquad \text{(b)}$$

$$\frac{f_{h,1,k}t_1 d}{1+\beta}\left[\sqrt{\beta+2\beta^2\left[1+\frac{t_2}{t_1}+\left(\frac{t_2}{t_1}\right)^2\right]+\beta^3\left(\frac{t_2}{t_1}\right)^2}-\beta\left(1+\frac{t_2}{t_1}\right)\right]+\frac{F_{ax,Rk}}{4} \qquad \text{(c)}$$

$$1,05\frac{f_{h,1,k}t_1 d}{2+\beta}\left[\sqrt{2\beta(1+\beta)+\frac{4\beta(2+\beta)M_{y,Rk}}{f_{h,1,k}d t_1^2}}-\beta\right]+\frac{F_{ax,Rk}}{4} \qquad \text{(d)}$$

$$1,05\frac{f_{h,1,k}t_2 d}{1+2\beta}\left[\sqrt{2\beta^2(1+\beta)+\frac{4\beta(1+2\beta)M_{y,Rk}}{f_{h,1,k}d t_2^2}}-\beta\right]+\frac{F_{ax,Rk}}{4} \qquad \text{(e)}$$

$$1,15\sqrt{\frac{2\beta}{1+\beta}}\sqrt{2M_{y,Rk}f_{h,1,k}d}+\frac{F_{ax,Rk}}{4} \qquad\qquad \text{(f)}$$

References	Calculations	Output
Clause 8.2.2 (2)	The 'rope effect' as determined from the F_{axRK} term in the above equations should be limited to 15% of the value from Johansen yield theory. Where the characteristic withdrawal capacity ($F_{ax,Rk}$) is unknown, the contribution from the effect should be taken as zero.	
	For axially loaded nails $F_{ax,Rk}$ can be determined in accordance with Clause 8.3.2 and Equations (8.23) to (8.26).	
	In this case neglect the contribution from the rope effect and assume a value of zero.	
	Characteristic densities of flanges and web $\rho_{k,1}=\rho_{k,3}=590$ kg/m^3 $\rho_{k,2}=350$ kg/m^3	
Clause 8.3.1.1(5) Equation (8.20)	Equation (8.6(a)): $t_1 = 12,3$ mm Assume with pre-drilled holes $f_{h,1,k}=0,11\,\rho_{k,1}\,d^{-0,3}=(0,11\times590,0\times4,0^{-0,3})=42,82$ N/mm^2 $f_{h,1,k}\,t_1\,d=(42,82\times12,3\times4,0)=2106,7$ N	

Contract : Floor Panel Job Ref. No. : Example 5.5	Calculated by : W.McK.
Part of Structure : Stressed-skin Panel	Checked by : B.Z.
Calculation Sheet No. : 23 of 28	Date :

References	Calculations	Output

References: Clause 8.3.1.1(5) Equation (8.15)

Equation (8.6(b)): $t_2 = (60 - 12,3) = 47,7$ mm

Assume pre-drilling

$f_{h,2,k} = 0,082\,(1 - 0,01\,d)\,\rho_{k,2} = [0,082 \times (1 - 0,01 \times 4,0) \times 350,0]$
$\qquad = 27,55$ N/mm^2

$f_{h,2,k}\,t_2\,d = (27,55 \times 47,7 \times 4,0) = 5256,5$ N

Equation (8.6(c)):

$t_2/t_1 = 47,7/12,3 = 3.88$

References: Equation (8.8)

$\beta = \dfrac{f_{h,2,k}}{f_{h,1,k}} = \dfrac{27,55}{42,82} = 0,64 \qquad$ Assume $F_{ax,Rk} = 0$

$$\frac{f_{h,1,k}t_1 d}{1+\beta}\left[\sqrt{\beta + 2\beta^2\left[1 + \frac{t_2}{t_1} + \left(\frac{t_2}{t_1}\right)^2\right] + \beta^3\left(\frac{t_2}{t_1}\right)^2} - \beta\left(1 + \frac{t_2}{t_1}\right)\right]$$

$$= \frac{2106,7}{1+0,64}\times\left[\sqrt{0,64 + 2\times0,64^2\times\left(1+3,88+3,88^2\right) + 0,64^3\times3,88^2} - 0,64\times\left(1+3,88\right)\right]$$

$= 1863,0$ N

References: Equation (8.14)

Equation (8.6(d)):

Characteristic yield moment for round nails $M_{y,Rk} = 0,3\,f_u\,d^{2,6}$

Assume $f_u = 600$ N/mm^2

$M_{y,Rk} = (0,3 \times 600 \times 4,0^{2,6}) = 6616,5$ Nmm

$$1,05\frac{f_{h,1,k}t_1 d}{2+\beta}\left[\sqrt{2\beta(1+\beta) + \frac{4\beta(2+\beta)M_{y,Rk}}{f_{h,1,k}dt_1^2}} - \beta\right] + \frac{F_{ax,Rk}}{4}$$

$$= 1,05\times\frac{2106,7}{2+0,64}\times\left[\sqrt{2\times0,64\times(1+0,64) + \frac{4\times0,64\times(2+0,64)\times6616,5}{42,82\times4,0\times12,3^2}} - 0,64\right]$$

$= 1102,4$ N

Equation (8.6(e)):

$$1,05\frac{f_{h,1,k}t_2 d}{1+2\beta}\left[\sqrt{2\beta^2(1+\beta) + \frac{4\beta(1+2\beta)M_{y,Rk}}{f_{h,1,k}dt_2^2}} - \beta\right] + \frac{F_{ax,Rk}}{4}$$

References	Calculations	Output
	Contract : Floor Panel Job Ref. No. : Example 5.5 **Part of Structure : Stressed-skin Panel** **Calculation Sheet No. : 24 of 28**	**Calculated by : W.McK.** **Checked by : B.Z.** **Date :**	

References	Calculations	Output
	$$= 1{,}05 \times \frac{42{,}82 \times 47{,}7 \times 4{,}0}{1+2\times 0{,}64} \times \left[\sqrt{2\times 0{,}64^2 \times (1+0{,}64) + \frac{4\times 0{,}64 \times (1+2\times 0{,}64)\times 6616{,}5}{42{,}82\times 4{,}0\times 47{,}7^2}} - 0{,}64 \right]$$	
	$= 2111{,}1$ N	
	Equation (8.6(f)):	
	$$1{,}15 \sqrt{\frac{2\beta}{1+\beta}} \sqrt{2 M_{y,Rk} f_{h,1,k} d} + \frac{F_{ax,Rk}}{4}$$	
	$$= 1{,}15 \times \sqrt{\frac{2\times 0{,}44}{1+0{,}44}} \times \sqrt{2\times 6616{,}5\times 42{,}82\times 4{,}0} = 1529{,}5 \text{ N}$$	
	Hence $F_{v,Rk} = 1102{,}4$ N	
	$F_{v,Rd} = (k_{mod} \times F_{v,Rk})/\gamma_M = (0{,}8\times 1102{,}4)/1{,}3 = 678{,}4$ N $\qquad\qquad\qquad\qquad\qquad\qquad\qquad > F_{1,d} = 571{,}5$ N	$F_{v,Rd} = \mathbf{678{,}4}$ N
Equation (B.10)	**Fastener load for bottom flange:** $$F_{3,d} = \frac{\gamma_3 E_3 A_3 a_3 s_3}{(EI)_{ef,u}} V_d$$	
	$$= \frac{0{,}59\times 10{,}56\times 10^3 \times 3760{,}0\times 64{,}33\times 40{,}0}{267{,}87\times 10^9} \times 2660{,}0 = 598{,}6 \text{ N}$$	$F_{3,d} = \mathbf{598{,}6}$ N
Clause 8.3.1.1(5) Equation (8.20)	Equation (8.6(a)): $t_3 = 9{,}4$ mm Assume with pre-drilled holes $f_{h,1,k} = 0{,}11\, \rho_{k,1}\, d^{-0,3} = (0{,}11\times 590{,}0\times 4{,}0^{-0,3}) = 42{,}82$ N/mm^2 $f_{h,1,k}\, t_1\, d = (42{,}82\times 9{,}4\times 4{,}0) = 1610{,}0$ N	
Clause 8.3.1.1(5) Equation (8.20)	Equation (8.6(b)): $t_2 = (60 - 9{,}4) = 50{,}6$ mm Assume pre-drilling $f_{h,2,k} = 0{,}082\,(1 - 0{,}01\, d)\, \rho_k = [0{,}082\times (1-0{,}01\times 4{,}0)\times 350{,}0]$ $\qquad\quad = 27{,}55$ N/mm^2 $f_{h,2,k}\, t_2\, d = (27{,}55\times 50{,}6\times 4{,}0) = 5576{,}1$ kN	
	Equation (8.6(c)): $t_2/t_1 = 50{,}6/9{,}4 = 5{,}38$	
Equation (8.8)	$\beta = \dfrac{f_{h,2,k}}{f_{h,1,k}} = \dfrac{27{,}55}{42{,}82} = 0{,}64 \qquad$ Assume $F_{ax,Rk} = 0$	

Contract : Floor Panel Job Ref. No. : Example 5.5 Part of Structure : Stressed-skin Panel Calculation Sheet No. : 25 of 28	Calculated by : W.McK. Checked by : B.Z. Date :

References	Calculations	Output
	$$\frac{f_{h,1,k}t_1 d}{1+\beta}\left[\sqrt{\beta+2\beta^2\left[1+\frac{t_2}{t_1}+\left(\frac{t_2}{t_1}\right)^2\right]+\beta^3\left(\frac{t_2}{t_1}\right)^2}-\beta\left(1+\frac{t_2}{t_1}\right)\right]$$ $$=\frac{1610,0}{1+0,64}\times\left[\sqrt{0,64+2\times0,64^2\times\left(1+5,38+5,38^2\right)+0,64^3\times\left(5,38\right)^2}-0,64\times\left(1+5,38\right)\right]$$ $$= 1976,3 \text{ N}$$	
Equation (8.14)	Equation (8.6(d)): Characteristic yield moment for round nails $M_{y,Rk}=0,3 f_u d^{2,6}$ Assume $f_u=600$ N/mm^2 $M_{y,Rk}=(0,3\times600\times4,0^{2,6})=6616,5$ Nmm $$1,05\frac{f_{h,1,k}t_1 d}{2+\beta}\left[\sqrt{2\beta(1+\beta)+\frac{4\beta(2+\beta)M_{y,Rk}}{f_{h,1,k}dt_1^2}}-\beta\right]+\frac{F_{ax,Rk}}{4}$$ $$=1,05\times\frac{1610,0}{2+0,64}\left[\sqrt{2\times0,64\times(1+0,64)+\frac{4\times0,64\times(2+0,64)\times6616,5}{42,82\times4,0\times9,4^2}}-0,64\right]$$ $$= 1029,7 \text{ N}$$ Equation (8.6(e)): $$1,05\frac{f_{h,1,k}t_2 d}{1+2\beta}\left[\sqrt{2\beta^2(1+\beta)+\frac{4\beta(1+2\beta)M_{y,Rk}}{f_{h,1,k}dt_2^2}}-\beta\right]+\frac{F_{ax,Rk}}{4}$$ $$=1,05\times\frac{42,82\times50,6\times4,0}{1+2\times0,64}\left[\sqrt{2\times0,64^2\times(1+0,64)+\frac{4\times0,64\times(1+2\times0,64)\times6616,5}{42,82\times4,0\times50,6^2}}-0,64\right]$$ $$= 2221,0 \text{ N}$$ Equation (8.6(f)): $$1,15\sqrt{\frac{2\beta}{1+\beta}}\sqrt{2M_{y,Rk}f_{h,1,k}d}+\frac{F_{ax,Rk}}{4}$$	

References	Calculations	Output
	$= 1{,}15 \times \sqrt{\dfrac{2 \times 0{,}64}{1+0{,}64}} \times \sqrt{2 \times 6616{,}5 \times 42{,}82 \times 4{,}0} = 1529{,}5$ N	
	Hence $F_{v,Rk} = 1029{,}7$ N	
	$F_{v,Rd} = (k_{mod} \times F_{v,Rk})/\gamma_M = (0{,}8 \times 1029{,}7)/1{,}3 = 633{,}7$ N $> F_{3,d} = 598{,}6$ N	$F_{v,Rd} = 633{,}7$ N
	Minimum spacings and edge distances for nails with predrilled holes in timber:	
Table 8.2	Minimum spacing parallel to the grain $a_{1,min}$ (with $a_{1,min} = (4 + \lvert \cos 0° \rvert)\, d = 5 \times 4{,}0 = 20{,}0$ mm < 40 mm	
Table 8.2	Minimum distance to the loaded ends $a_{4,t,min}$: $a_{4,t,min} = (3 + 2 \sin 0°)\, d = 3 \times 4{,}0 = 12{,}0$ mm $< 23{,}5$ mm	
Table 8.2	Minimum distance to the unloaded ends $a_{4,c,min}$: $a_{4,c,min} = 3\, d = 3 \times 4{,}0 = 12{,}0$ mm $< 23{,}5$ mm	**Nail connection is adequate for both top and bottom flanges.**
	Serviceability Limit States: Deformation:	
Clause 7.2	Limiting values for deflection of beams: $w_{net,fin} = w_{inst} + w_{creep} - w_c = w_{fin} - w_c$ Since there is no camber $w_c = 0$ and $w_{net,fin} = w_{inst} + w_{creep} = w_{fin}$	
	(**Note:** $\gamma_G = \gamma_Q = 1{,}0$ for characteristic deformations) Permanent design loading $= (0{,}32 \times 4{,}0) = 1{,}28$ kN Variable design loading $= (0{,}6 \times 4{,}0) = 2{,}4$ kN	
Appendix B		
Clause B.1.4	Bending deflection $= \dfrac{5WL^3}{384\,(EI)_{ef,ser}}$	
	The value of $(EI)_{ef,ser}$ should be calculated using the slip modulus for serviceability, i.e $K_{ser,1} = K_{ser,3} = 2077{,}06$ N/mm and nail spacing $s_1 = s_3 = 40$ mm.	
Equation (B.5)	For serviceability limit state calculations: $\gamma_1 = [1 + \pi^2 E_1 A_1 s_1/(K_{ser,1}\, l^2)]^{-1}$ $= [1 + (\pi^2 \times 9{,}68 \times 10^3 \times 3603{,}9 \times 40)/(2077{,}06 \times 4000^2)]^{-1}$ $= 0{,}71$ $\gamma_3 = [1 + \pi^2 E_3 A_3 s_3/(K_{ser,3}\, l^2)]^{-1}$ $= [1 + (\pi^2 \times 10{,}56 \times 10^3 \times 3760{,}0 \times 40)/(2077{,}06 \times 4000^2)]^{-1}$ $= 0{,}68$	

Contract : Floor Panel Job Ref. No. : Example 5.5
Part of Structure : Stressed-skin Panel
Calculation Sheet No. : 26 of 28

Calculated by : W.McK.
Checked by : B.Z.
Date :

Contract : Floor Panel Job Ref. No. : Example 5.5 **Part of Structure : Stressed-skin Panel** **Calculation Sheet No. : 27 of 28**		**Calculated by : W.McK.** **Checked by : B.Z.** **Date :**

References	Calculations	Output
Equation (B.6)	$a_1 = (9,4 + 120 + 6,15 - 69,03) = 66,52$ mm $a_2 = \dfrac{\gamma_1 E_1 A_1 \left(h_1 + h_2\right) - \gamma_3 E_3 A_3 \left(h_2 + h_3\right)}{2\displaystyle\sum_{i=1}^{3} \gamma_i E_i A_i}$ $= \dfrac{0,71\times9,68\times3603,9\times\left(12,3+120,0\right)-0,68\times10,56\times3760,0\times\left(120,0+9,4\right)}{2\times\left[\left(0,71\times9,68\times3603,9\right)+\left(1,0\times11,0\times5640,0\right)+\left(0,68\times10,56\times3760,0\right)\right]} = -0,95$ mm $a_2 = -\,0,95$ mm $a_3 = 69,03 - (0,5 \times 9,4) = 64,33$ mm	
Equation (B.1)	$(EI)_{\text{ef,ser}} = \displaystyle\sum_{1}^{3} \left(E_i I_i + \gamma_i E_i A_i a_i^2\right)$ $\begin{aligned}(EI)_{\text{ef,ser}} =\ & (9,68 \times 10^3 \times 45,436 \times 10^3) + (0,71 \times 9,68 \times 10^3 \times 3603,9 \times 66,52^2)\\ & + (11,0 \times 10^3 \times 6,768 \times 10^6) + (1,0 \times 11,0 \times 10^3 \times 5640,0 \times 0,95^2)\\ & + (10,56 \times 10^3 \times 27,686 \times 10^3) + (0,68 \times 10,56 \times 10^3 \times 3760,0 \times 64,33^2)\\ =\ & 296,57 \times 10^9 \text{ Nmm}^2\end{aligned}$	
	Estimate the shear deflection based on the web only. **Shear deflection** $\approx \dfrac{3WL}{20bhG_d}$ $b = 47$ mm; $h = 120$ mm; $G = 0,69$ kN/mm^2	
Equation (2.3)	**Permanent load deformation:** $w_{\text{fin,G}} = w_{\text{inst,G}} (1 + k_{\text{def}}) = (w_{\text{inst,m}} + w_{\text{inst,v}})_G (1 + k_{\text{def}})$ Total deformation due to (bending + shear effects) is approximately given by: $w_{\text{inst,G}} = \dfrac{5WL^3}{384(EI)_{\text{ef,ser}}} + \dfrac{3WL}{20bhG_d}$ **For instantaneous deformation (w_{inst}):** $k_{\text{def}} = 0$ i.e. no creep Bending deformation: $w_{\text{inst,G,m}} = \dfrac{5WL^3}{384(EI)_{\text{ef,ser}}} = \dfrac{5\times1,28\times10^3\times4000^3}{384\times296,57\times10^9} = 3,60$ mm	

Contract : Floor Panel Job Ref. No. : Example 5.5 Part of Structure : Stressed-skin Panel Calculation Sheet No. : 28 of 28		Calculated by : W.McK. Checked by : B.Z. Date :
References	**Calculations**	**Output**
	Shear deformation: $w_{\text{inst,v,m}} = \dfrac{3WL}{20bhG_{\text{d}}} = \dfrac{3,0\times1,28\times4000}{20\times47\times120\times0,69} = 0,20 \text{ mm}$ $w_{\text{fin,G}} = [w_{\text{inst,G}}\,(1 + k_{\text{def}})]_{\text{m}} + [w_{\text{inst,G}}\,(1 + k_{\text{def}})]_{\text{v}}$ Since the deflection is based on two wood-based elements having different time-dependent behaviour, the k_{def} value is calculated using Equation (2.13):	
Equation (2.13)	$k_{\text{def}} = 2\times\sqrt{k_{\text{def,1}}\,k_{\text{def,2}}} = 2\times\sqrt{(0,8\times0,6)} = 1,39$ $w_{\text{fin,G}} = [3,60\times(1 + 1,39) + 0,20\times(1 + 1,39)] = 9,08 \text{ mm}$ **Variable load deformation:** $W = 2,4 \text{ kN}$ $L = 4000 \text{ mm}$ $E_{\text{f}} = E_{0,\text{mean}} = 9,68 \text{ kN/mm}^2$ $G_{\text{w}} = G_{\text{mean}} = 0,69 \text{ kN/mm}^2$ $w_{\text{inst,Q1,m}} = \dfrac{5WL^3}{384(EI)_{\text{ef,ser}}} = \dfrac{5\times2,4\times10^3\times4000^3}{384\times296,57\times10^9} = 6,74 \text{ mm}$ Shear deformation: $w_{\text{inst,Q1,v}} = \dfrac{3WL}{20bhG_{\text{d}}} = \dfrac{3,0\times2,4\times4000}{20\times47\times120\times0,69} = 0,37 \text{ mm}$	
National Annex to BS EN 1990 Table NA.A1.1	$\psi_{2,1} = 0,3$ $w_{\text{fin,Q1}} = [w_{\text{inst,Q1}}\,(1 + \psi_{2,1}\,k_{\text{def}})]_{\text{m}} + [w_{\text{inst,Q1}}\,(1 + \psi_{2,1}\,k_{\text{def}})]_{\text{v}}$ $= 6,74\times[1 + (0,3\times1,39)] + 0,37\times[1 + (0,3\times1,39)]$ $= 10,07 \text{ mm}$ **Final Deformation:** $w_{\text{fin}} = w_{\text{fn,G}} + w_{\text{fin,Q1}} = (9,08 + 10,07) = 19,15 \text{ mm}$	$\boldsymbol{w_{\text{fin}} = 19,15}$ **mm**
National Annex Table NA.4	Limiting ratio = $l/150 = 4000/150 = 26,67 \text{ mm} > 19,15 \text{ mm}$	**The panel is adequate with respect to deflection.**

5.6 Glued Laminated Beams (Glulam)

The concept of using laminated timber as indicated in Figure 5.1(d) has been used for many years. In the early 19th century mechanically laminated timber structures were utilised throughout Europe. The development of synthetic resin adhesives in the 20th century presented the opportunity for extensive development of production techniques.

One of the fastest growing and most successful structural material industries in the UK is that related to the use of glulam. Traditionally until the 1970's and with the exception of specialist uses such as in aircraft and marine components, glulam was purpose made for a limited number of types of structure such as swimming pools, churches or footbridges. The availability of standard glulam components such as straight, tapered, curved or cambered members has been made possible by the introduction of improved, modern high-volume production plants. This has resulted in an ever expanding range of uses, e.g. timber lintel beams in domestic housing, large portal frames in conference and leisure centres, to structures such as the 162 m diameter dome of the Tacoma Sports and Convention Centre in Washington State, USA.

Glulam has many advantages such as:

♦ members can be straight or curved in profile and uniform or variable in cross-section,

♦ the strength:weight ratio is high, enabling the dead load due to the superstructure to be kept to a minimum with a consequent saving in foundation construction,

♦ factory production allows a high standard of material quality to be achieved,

♦ timbers of large cross-section have a superior performance in fire than alternatives such as concrete and steel,

♦ when treated with appropriate preservatives, softwood laminated timber is very durable in wet exposure situations; in addition it also has a high resistance to chemical attack and aggressive/polluted environments,

♦ there is no need for expansion joints because of the low coefficient of thermal expansion,

♦ defects such as knots are restricted to the thickness of one lamination and their effect on overall structural behaviour is significantly reduced,

♦ large spans are possible within the constraints of transportation to site.

An indication of the range of structures for which glulam is suitable is given in Figures 5.25(a) and (b).

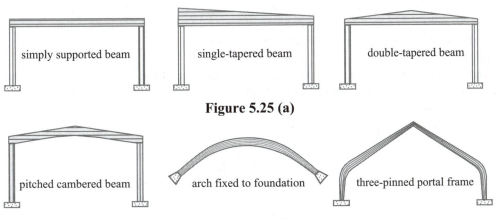

Figure 5.25 (a)

Figure 5.25 (b)

5.6.1 Vertically Laminated Beams

Vertically laminated beams in which the applied load is parallel to the laminate joints usually occur in elements such as laminated columns subjected to both axial and flexural loads. They are dealt with in Chapter 7 – *Members subject to combined axial and flexural loads.*

5.6.2 Horizontally Laminated Beams

Horizontally laminated beams are beams in which the laminations are parallel to the neutral plane. The loading is applied in a direction perpendicular to the plane of the laminations.

5.6.3 Manufacture of Glulam

Glulam members are fabricated by gluing together accurately prepared timber laminations in which the grain is in the longitudinal direction. BS EN 1995-1-1, Clause 3.3(1) states that '*Glued laminate timber shall comply with BS EN 14080.*' This code indicates general requirements with numerous references to other codes where more detailed information can be found regarding performance and production requirements, strength classes, adhesives, durability, fire classification, etc.

The thickness and cross-sectional area of individual laminates should not be greater than the values given in Table 3 of BS EN 386:2001(E) as indicated in Table 5.5 below.

	Service Class 1		Service Class 2		Service Class 3	
	Thickness (mm)	Area (mm^2)	Thickness (mm)	Area (mm^2)	Thickness (mm)	Area (mm^2)
Conifers	45	12000	45	12000	35	10000
Broad leaf	40	7500	40	7500	35	6000

Table 5.5

Members may be horizontally laminated from a single strength class of timber and are referred to as homogeneous glulam beams or alternatively from two strength classes of timber, which are referred to as combined glulam beams. As indicated in Annex A of EN 1194:1999 it is assumed that zones of different lamination grades amount to at least 1/6 of the beam depth or two laminations, whichever is the greater as shown in Figure 5.26.

When using mixed grade laminations it is evident that the higher strength laminations are used to resist the larger bending stresses in the outer fibres of the cross-section.

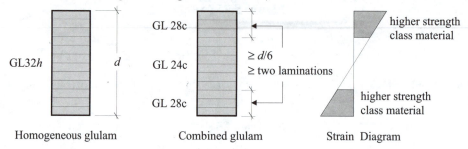

Figure 5.26

The characteristic strength, stiffness and density properties for four standard strength classes of glulam, i.e. GL 24h, GL 28h, GL 32h and GL 36h for homogeneous glulam and GL 24c, GL 28c, GL 32c and GL 36c for combined glulam are given in Tables 1 and 2 of EN 1194:1999.

5.6.3.1 Sequence of Production Operations

A typical sequence of operations for the production of glulam is as follows:

- selection of laminations,

- the laminations are dried in a kiln to achieve a moisture content of approximately 12% to ensure a maximum bond strength and glulam stability,

- individual laminations are finger-jointed to produce a continuous lamination of the appropriate stress grade (see Section 5.6.3.2),

- each finger-jointed laminate is accurately planed to the required thickness and cut to the required length,

- a bonding adhesive is carefully applied to the faces of each laminate,

- all laminations are placed in a mechanical or hydraulic jig as shown in Figures 5.27(a) and (b) and pressure applied,

- when the glue has cured, the glulam member is trimmed to size and surfaces finished,

- the bond strength is checked by mechanical testing (e.g. bending tests on the member), to ensure adequate bonding has been achieved.

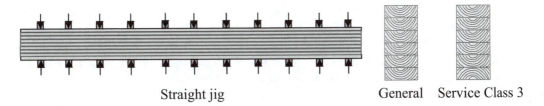

Straight jig   General   Service Class 3

Figure 5.27 (a)

The laminations generally have the pith facing the same side as indicated in Figure 5.27(a); except in the case of Service Class 3 glulam where the extreme laminations should have the pith facing outwards on either edge as specified in Clause 6.4.2.3 of BS EN 386:2001(E).

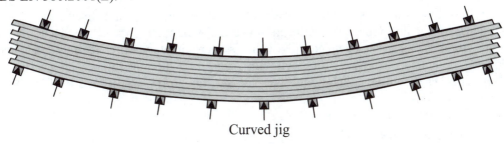

Curved jig

Figure 5.27 (b)

For curved members the maximum thickness of the laminations is also governed by their radius of curvature 'r' and the characteristic bending strength of the end joints ($f_{m,dc,k}$) as indicated in Clause 6.2.3 of BS EN 386:2001(E), i.e. $t \le \dfrac{r}{250}\left(1+\dfrac{f_{m,dc,k}}{80}\right)$.

5.6.3.2 Finger-joints

The length of glulam members frequently exceeds the length of commercially available solid timber, resulting in the need to finger-joint together individual planks to make laminations of the required length.

A typical finger-joint is shown in Figure 5.28:

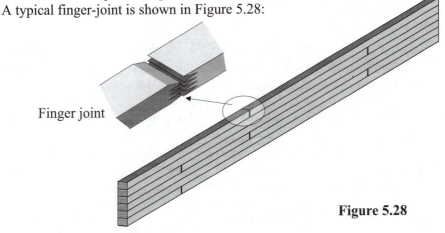

Finger joint

Figure 5.28

The finger-joint is cut into the end-grain of each plank and the planks are pressed together after applying adhesive.

Finger joints should be manufactured in accordance with BS EN 385:2001 and in the case of large-finger joints, with BS EN 387:2001. A large finger joint is defined in BS EN 387:2001 as one in which the length of the finger is at least 45 mm.

5.6.3.3 Adhesives

The most widely used adhesives are the phenol-resorcinol-formaldehyde (PRF) type. The adhesive, which is a liquid, is used with a 'hardener' containing formaldehyde and various inert fillers. Pure resorcinal, which is expensive, is usually replaced with alternative cheaper chemicals (phenols), which also react with formaldehyde. The chemical bond formed during this reaction is a carbon-to-carbon type which is very strong and durable.

The resulting adhesives are fully water, boil and weather resistant. In addition, PRF adhesives do not decompose or ignite in a fire and delamination will not occur under normal circumstances. The chemical 'pH' value is approximately 7 (i.e. neutral) and hence alkaline or acidic damage/corrosion will not occur in either the timber or any metal components. The requirements for adhesives are specified in BS EN 301 and BS EN 302.

5.6.3.4 Surface Finishes

Finished glulam beams are normally planed on their sides to remove any residual adhesive which has been squeezed out of their joints during manufacture. Modern planing machinery usually provides a finish of sufficient quality that subsequent sanding is unnecessary.

5.6.3.5 Metal Fasteners

The use of metal fasteners and connectors should satisfy the requirements for durability as given in of Section 4 of the code.

5.6.3.6 Preservation Treatment

The preservative treatment to glulam members tends to be applied to the finished product rather than individual laminations prior to gluing. Preservative treatments should comply with the requirements of BS EN 350-2 and BS EN 335.

5.6.3.7 Modification Factors

The modification factors which apply to glued laminated beams are as indicated in Table 5.1. The design for prismatic glulam beams is essentially the same as that for solid timber beams. The characteristic strength and stiffness properties can be found in Tables 1 and 2 of BS EN 1194-1999 for homogeneous glulam and combined glulam respectively. The values of the following factors differ in glulam and solid timber beams:

(i)   the partial factor for material properties and resistance γ_m: Table 2.3,

(ii)   the modification factor for service and load-duration classes k_{mod}: Table 3.1,

(iii)   the modification factor for member size k_h: Equation (3.2).

In the case of rectangular solid timber beams with $\rho_k < 700$ kg/m^3 the reference depth in bending and width in tension is 150 mm; the k_h factor being applied to $f_{m,k}$ and $f_{t,0,k}$ in cross-section with a depth or width *less than* this value.

In the case of glulam beams the corresponding reference depth is 600 mm, (there is no restriction on ρ_k).

In the case of laminated veneer lumber (LVL) the reference depth for bending is 300 mm; the k_h factor is given in Equation (3.3). The reference length in tension is 3000 mm; the k_h factor is given in Equation (3.4) In both cases the factor is applied to a section in which the depth for bending and the length for tension is *not equal to* the reference values,

(iv)   the modification factor for shear strength of notched beams k_n: Equation (6.63),

(v)   the modification factor for system strength k_{sys}: Clause 6.6.4 and Figure 6.12.

The following strength and stiffness criteria should be satisfied;

(i)   bending strength: Equations (6.11) and (6.12) (see Section 5.1.1),

(ii)   shear strength: Equations (6.13) and (6.60) for beams with a notch at the support, (see Sections 5.3.2 and 5.3.5),

(iii)   bearing strength: Equation (6.3)   (see Section 5.3.3),

(iv)   lateral stability: Equation (6.33)   (see Section 5.3.4),

(v)   deflection: Equation (7.2)   (see Section 5.3.6).

The design procedure for non-prismatic beams, e.g. single-tapered, double-tapered, curved and pitched cambered beams (see Figure 5.25), is more complex and dealt with in Section 6.4 of BS EN 1995-1-1:2004(E). The non-linearity in the distribution of the bending stresses and the co-existence of shear stresses, stresses perpendicular to the grain and radial stresses in curved members results in the requirement to check the following criteria:

(a)   the design bending stress at an angle to the grain $\sigma_{m,\alpha,d}$: Clause 6.4.2.
        This applies to single-tapered beams and those in which there are parts having a single taper. The following equation should be satisfied:

$$\sigma_{m,\alpha,d} \leq k_{m,\alpha}\, f_{m,y,d} \qquad \text{(Equation (6.38) in EC5)}$$

Design bending strength $f_{m,y,d} = (f_{m,y,k} \times k_{mod} \times k_h \times k_{sys})/\gamma_M$

For tensile stresses parallel to the tapered edge: (Equation (6.39) in EC5)

$$k_{m,\alpha} = \cfrac{1}{\sqrt{1+\left(\cfrac{f_{m,d}}{0,75 f_{v,d}}\tan\alpha\right)^2+\left(\cfrac{f_{m,d}}{f_{t,90,d}}\tan^2\alpha\right)^2}}$$

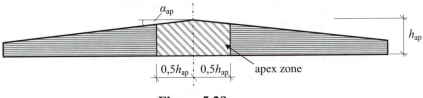

For compressive stresses parallel to the tapered edge: (Equation (6.40) in EC5)

$$k_{m,\alpha} = \cfrac{1}{\sqrt{1+\left(\cfrac{f_{m,d}}{1,5 f_{v,d}}\tan\alpha\right)^2+\left(\cfrac{f_{m,d}}{f_{c,90,d}}\tan^2\alpha\right)^2}}$$

(b)   The apex zone for double-tapered, curved and pitched cambered beams is critical with respect to combined stress conditions and is defined in Figure 6.9 of the code, e.g. in the case of double-tapered beams as shown below in Figure 5.29. A number of design checks should be carried out in this zone.

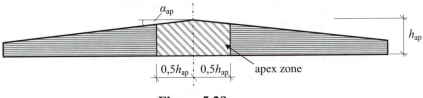

Figure 5.29

Apex zone checks:

(i)   the bending stress $\sigma_{m,\alpha,d}$ should satisfy the following equation:

$$\sigma_{m,d} \le k_r f_{m,d} \qquad \text{(Equation (6.41) in EC5)}$$

where:

$$\sigma_{m,d} = k_l \frac{6 M_{ap,d}}{b h_{ap}^2}$$

$M_{ap,d}$ is the design moment at the apex

$$k_l = k_1 + k_2\left(\frac{h_{ap}}{r}\right) + k_3\left(\frac{h_{ap}}{r}\right)^2 + k_4\left(\frac{h_{ap}}{r}\right)^3 \qquad \text{(Equation (6.43) in EC5)}$$

k_1, k_2, k_3 and k_4 are defined in Equations (6.44) to (6.47) in EC5.
r is defined in Equation (6.48) as equal to $r_{in} + 0,5 h_{ap}$
(For double-tapered beams $r = \infty$)
k_r is equal to 1,0 for double-tapered beams and is defined in Equation (6.49) of EC5 for curved and pitched cambered beams. This value is a strength reduction factor to allow for the bending of laminations during production.

(ii)   the greatest tensile stress perpendicular to the grain $\sigma_{t,90,d}$ should satisfy the following equation:

$$\sigma_{t,90,,d} \leq k_{dis}\, k_{vol}\, f_{t,90,d} \qquad \text{(Equation (6.50) in EC5)}$$

where k_{dis} and k_{vol} are given in Equations (6.51) and (6.52) of EC5. Their values are derived from consideration of effects of the non-linear stress distribution and using the '*weakest link*' theory for brittle materials.

    The value of $\sigma_{t,90,d}$ can be determined using either Equation (6.54) or Equation (6.55) in EC5. The UK National Annex indicates in Clause NA.2.4 that Equation (6.54) should be used, i.e.

$$\sigma_{t,90,,d} = k_p \frac{6M_{ap,d}}{bh_{ap}^2}$$

where:

$$k_p = k_5 + k_6\left(\frac{h_{ap}}{r}\right) + k_7\left(\frac{h_{ap}}{r}\right)^2 \qquad \text{(Equation (6.56) in EC5)}$$

k_5, k_6 and k_7, are defined in Equations (6.57) to (6.59) in EC5.

(iii)   the combined shear stress and tension stress perpendicular to the grain should satisfy the following interaction equation:

$$\frac{\tau_d}{f_{v,d}} + \frac{\sigma_{t,90,d}}{k_{dis}k_{vol}f_{t,90,d}} \leq 1,0 \qquad \text{(Equation (6.53) in EC5)}$$

(c)   The deflection.
    As with thin-webbed beams the shear deflection in tapered-beams is generally greater than in prismatic beams. The determination of the total deflection of tapered-beams is relatively complex and can be carried out using energy methods such as virtual work or finite element techniques.
    In most cases an approximate solution which includes both bending and shear effects, can be determined using the standard equations for prismatic members assuming an equivalent depth of beam. A value for the equivalent depth (h_{eq}) can be calculated for *uniformly distributed loads* using the equations given by Faherty and Williamson (63), i.e.

For a single-tapered, straight beam:

$$h_{eq} = \left[1 + 0,46 \times \left(\frac{h_{ap} - h_e}{h_e}\right)h_e\right] \quad \text{when } 0 < \left(\frac{h_{ap} - h_e}{h_e}\right) < 1,1$$

$$h_{eq} = \left[1 + 0,43 \times \left(\frac{h_{ap} - h_e}{h_e}\right)h_e\right] \quad \text{when } 1,1 < \left(\frac{h_{ap} - h_e}{h_e}\right) < 2,0$$

For a double-tapered, straight beam:

$$h_{eq} = \left[1 + 0,66 \times \left(\frac{h_{ap} - h_e}{h_e}\right)h_e\right] \quad \text{when } 0 < \left(\frac{h_{ap} - h_e}{h_e}\right) < 1,0$$

$$h_{eq} = \left[1 + 0,62 \times \left(\frac{h_{ap} - h_e}{h_e}\right)h_e\right] \quad \text{when } 1,0 < \left(\frac{h_{ap} - h_e}{h_e}\right) < 3,0$$

where:
h_{ap} is the depth of the beam at the apex,
h_e is the depth of the beam at the end.

The value of deflection can be estimated using: $u_{inst} = \dfrac{5WL^3}{384E_d I_{eq}} + \dfrac{3WL}{20bh_{eq}G_d}$

The more advanced analysis methods mentioned above should be used for other load cases.

5.6.4 Example 5.6: Glulam Roof Beam Design

An exhibition hall is to be designed with glued laminated timber beams at 3,0 m centres with a 10,0 m span, supporting the roof structure. The roofing is to be 63 mm tongue and groove boarding which is exposed on the underside and covered on the top side with insulation, felt and chippings.

   Check the suitability of the proposed section for the glulam beam indicated in Figure 5.30.

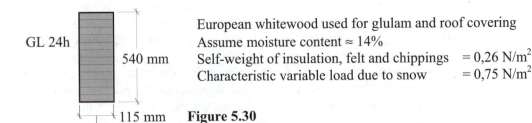

GL 24h

540 mm

115 mm   **Figure 5.30**

European whitewood used for glulam and roof covering
Assume moisture content ≈ 14%
Self-weight of insulation, felt and chippings $= 0,26$ N/m²
Characteristic variable load due to snow $= 0,75$ N/m²

5.6.4.1 Solution to Example 5.6

References	Calculations	Output
	Contract : Exhibition Hall Job Ref. No. : Example 5.6 **Part of Structure : Glulam Roof Beam** **Calculation Sheet No. : 1 of 5** · **Calculated by : W.McK.** **Checked by : B.Z.** **Date :**	

References	Calculations	Output
BS EN 1990 BS EN 1991 BS EN 338:2003 BS EN 1995-1-1: 2004 National Annex BS EN 1194: 1999	Eurocode: Basis of Structural Design Eurocode 1: Actions on Structures Structural Timber - Strength Classes Eurocode 5: Design of Timber Structures Part 1.1: General - Common Rules and Rules for Buildings. UK - National Annex to BS EN 1995-1-1 Glued Laminated Timber – Strength Classes and Characteristic Values	
BS EN 1194 Table 1	Characteristic values for GL 24h glulam **Strength Properties:** Bending $f_{m,g,k} = 24{,}0$ N/mm^2 Compression parallel to grain $f_{c,0,g,k} = 24{,}0$ N/mm^2 Compression perpendicular to grain $f_{c,90,g,k} = 2{,}7$ N/mm^2 Shear $f_{v,g,k} = 2{,}7$ N/mm^2 **Stiffness Properties:** Mean modulus of elasticity parallel to the grain $E_{0,g,mean} = 11{,}6$ kN/mm^2 5% modulus of elasticity parallel to the grain $E_{0,g,05} = 9{,}4$ kN/mm^2 Mean modulus of elasticity perpendicular to the grain $E_{90,g,mean} = 0{,}39$ kN/mm^2 Mean shear modulus $G_{g,mean} = 0{,}72$ kN/mm^2 Characteristic density $\rho_{g,k} = 380{,}0$ kg/m^3	
BS EN 1995 Clause 2.3.1.2 Table NA.1 (EC5: Table 2.3)	Load-duration Classes: use Table NA.1 from National Annex Self-weight - permanent Snow - short-term	
Table NA.2	Service Class 2 (cold roof)	
Clause 2.3.2.1	Load-duration and moisture influence on strength: k_{mod}	
Table 3.1	Glued-laminated timber: Service Class 2 Permanent actions $k_{mod} = 0{,}6$ Short-term actions $k_{mod} = 0{,}9$	
Clause 3.1.3(2)	Use k_{mod} corresponding to action with the shortest duration.	$k_{mod} = 0{,}9$
National Annex Table NA.3 EC5 (Table 2.3)	Partial factor for material properties and resistance: γ_M For glued-laminated timber	$\gamma_M = 1{,}25$

References	Calculations	Output
Contract : Exhibition Hall Job Ref. No. : Example 5.6 **Part of Structure : Glulam Roof Beam** **Calculation Sheet No. : 2 of 5**		**Calculated by : W.McK.** **Checked by : B.Z.** **Date :**

References	Calculations	Output
Clause 3.3 Clause 3.3(3)	Member size: For glued-laminated timber in bending where the depth is less than 600 mm $f_{m,k}$ can be multiplied by k_h where: $k_h = \min \begin{cases} \left(\dfrac{600}{h}\right)^{0,1} \\ \\ 1,1 \end{cases}$ $k_h = \min \begin{cases} \left(\dfrac{600}{540}\right)^{0,1} \\ \\ 1,1 \end{cases} = \min \begin{cases} 1,01 \\ \\ 1,1 \end{cases}$	$k_h = 1,01$
Clause 6.6(4) Figure 6.12	System strength: k_{sys} Assume laminations 45 mm thick: Number of laminations $= (540/45) = 12 > 8$	$k_{sys} = 1,2$
	Section properties: Area $= 115 \times 540 \quad = 62,1 \times 10^3$ mm^2 $I_{yy} = \dfrac{115 \times 540^3}{12} = 1509 \times 10^6$ mm^4 $Z_{yy} = \dfrac{115 \times 540^2}{6} = 5,59 \times 10^6$ mm^3	
	Loading: Characteristic density $= 380$ kg/m^3 Mean density $= 1,2 \times 380 = 456$ kg/m^3 (compared with C24) Assume 530 kg/m^3 to allow for adhesive, etc. $= 5,2$ kN/m^3 Self-weight of beam $\quad = (0,115 \times 0,54 \times 5,2) = 0,32$ kN/m Self-weight of decking $= (0,063 \times 4,5) \quad\quad = 0,28$ kN/m^2 Self-weight of insulation, felt and chippings $\quad = 0,26$ kN/m^2 Permanent load $g_k = [(0,26 + 0,28) \times 3,0] + 0,32 = 1,94$ kN/m Variable load $\quad q_k = (0,75 \times 3,0) \quad\quad\quad = 2,25$ kN/m	
BS EN 1990 Equation (6.10) UK NA to EC Table NA.A1.2(B)	Eurocode: Basis of Structural Design $E_d = \left(\gamma_G G_k + \gamma_Q Q_k\right)$ $\gamma_G = 1,35$ and $\quad \gamma_Q = 1,50$ The design load $= \left(\gamma_G G_k + \gamma_Q Q_k\right) = (1,35 \times 1,94) + (1,5 \times 2,25)$ $F_d = 5,99$ kN/m	$F_d = 5,99$ kN/m

References	Calculations	Output
	Contract : Exhibition Hall Job Ref. No. : Example 5.6 **Part of Structure : Glulam Roof Beam** **Calculation Sheet No. : 3 of 5**	**Calculated by : W.McK.** **Checked by : B.Z.** **Date :**

References	Calculations	Output
	Bending: Design bending moment $M_{y,d} = (wL^2/8)$ $M_{y,d} = \dfrac{5,99 \times 10,0^2}{8} = 74,88$ kNm Design bending strength $f_{m,y,d} = (f_{m,y,k} \times k_{mod} \times k_h \times k_{sys})/\gamma_M$ $f_{m,y,d} = (24,0 \times 0,9 \times 1,01 \times 1,2)/1,25 = 20,94$ N/mm^2	$M_{y,d} = \textbf{74,88 kNm}$ $f_{m,y,d} = \textbf{20,94 N/mm}^2$
BS EN 1995-1-1 Clause 6.1.6	Since uniaxial bending about the y-y axis only, is present: $\sigma_{m,y,d} \leq f_{m,y,d}$ $\sigma_{m,y,d} = \dfrac{M_{y,d}}{W_y} = \dfrac{74,88 \times 10^6}{5,59 \times 10^6} = 13,40$ N/mm$^2 \leq 20,94$ N/mm^2	**The joist is adequate with respect to bending.**
	Shear: Design shear force $F_{v,d} = (F_d \times 10,0)/2,0 = (5,99 \times 10,0)/2,0$ $= 29,95$ kN Design shear strength $f_{v,d} = (f_{v,k} \times k_{mod} \times k_v \times k_{sys})/\gamma_M$	$F_{v,d} = \textbf{29,95 kN}$
Clause 6.5.2	k_v applies to notched beams only $f_{v,d} = (2,7 \times 0,9 \times 1,2)/1,25 = 2,33$ N/mm^2 Design shear stress $\tau_d = \dfrac{1,5 \times F_{v,d}}{A} = \dfrac{1,5 \times 29,95 \times 10^3}{62,1 \times 10^3}$ $= 0,72$ N/mm$^2 < 2,33$ N/mm^2	$f_{v,d} = \textbf{2,33 N/mm}^2$ **The joist is adequate with respect to shear.**
Clause 6.1.5	**Bearing:** Design bearing force $F_{c,90,d} = 29,95$ kN Design bearing strength $f_{c,90,d} = (f_{c,90,k} \times k_{mod} \times k_{sys})/\gamma_M$ $= (2,7 \times 0,9 \times 1,2)/1,25$ $= 2,33$ N/mm^2 Assume a contact length of bearing $l = 50$ mm Design bearing stress $\sigma_{c,90,d} = \dfrac{F_{c,90,d}}{b \times l} = \dfrac{29,95 \times 10^3}{115 \times 50}$ $= 5,21$ N/mm^2	$F_{c,90,d} = \textbf{29,95 kN}$ $f_{c,90,d} = \textbf{2,33 N/mm}^2$

Contract : Exhibition Hall Job Ref. No. : Example 5.6 Part of Structure : Glulam Roof Beam Calculation Sheet No. : 4 of 5		Calculated by : W.McK. Checked by : B.Z. Date :
References	**Calculations**	**Output**

References	Calculations	Output
Equation (6.3) Clause 6.1.5(2) Clause 6.1.5(3)	$\sigma_{c,90,d} \leq (k_{c,90} \times f_{c,90,d})$ Assume 'a' in Figure 6.2 of EC5 = 0 $k_{c,90} = \left(2,38 - \dfrac{l}{250}\right)\left(1 + \dfrac{h}{12l}\right)$ and $\leq 4,0$ $k_{c,90} = \left(2,38 - \dfrac{50}{250}\right)\left(1 + \dfrac{540}{12 \times 50}\right) = 4,14 > 4,0$ $(k_{c,90} \times f_{c,90,d}) = (4,0 \times 2,33) = 9,32 \text{ N/mm}^2 \geq 5,23 \text{ N/mm}^2$	$k_{c,90} = 4,0$ The section is adequate with respect to bearing.
Clause 6.3.3 Equation (6.33) Clause 6.3.3(5)	**Lateral Torsional Stability:** $\sigma_{m,d} \leq (k_{crit} \times f_{m,d})$ Since the compression flange is fully restrained by the decking and torsional restraint is provided at the supports the value of $k_{crit} = 1,0$. Design strength $f_{m,d} = 20,94 \text{ N/mm}^2$ $(k_{crit} \times f_{m,d}) = (1,0 \times 20,94) = 20,94 \text{ N/mm}^2$ Design stress $\sigma_{m,d} = \dfrac{M_{y,d}}{W_y} = 13,40 \text{ N/mm}^2 < 20,94 \text{ N/mm}^2$	The section is adequate with respect to lateral torsional stability.
Clause 7.2 Table 3.2	**Serviceability Limit States:** **Deflection:** Limiting values for deflection of beams: $w_{net,fin} = w_{inst} + w_{creep} - w_c = w_{fin} - w_c$ Since there is no camber $w_c = 0$ and $w_{net,fin} = w_{inst} + w_{creep} = w_{fin}$ For glulam Service Class 2 $k_{def} = 0,8$ $w_{inst,G}(1 + k_{def}) = (w_{inst,m} + w_{inst,v})_G (1 + k_{def})$ Permanent load deformation: $W = (1,94 \times 10,0) = 19,4 \text{ kN}$ $L = 10000 \text{ mm}$ $E_d = E_{0,g,mean} = 11,6 \text{ kN/mm}^2$ $G_d = G_{g,mean} = 0,72 \text{ kN/mm}^2$ $I_y = 1509 \times 10^6 \text{ mm}^4$ $w_{inst,G,m} = \dfrac{5WL^3}{384 E_d I} = \dfrac{5 \times 19,4 \times 10000^3}{384 \times 11,6 \times 1509 \times 10^6} = 14,43 \text{ mm}$	$k_{def} = 0,8$

References	Calculations	Output
	$w_{\text{inst,G,v}} = \dfrac{3WL}{20bhG_d} = \dfrac{3 \times 19,4 \times 10000}{20 \times 115 \times 540 \times 0,72} = 0,65 \text{ mm}$	

Contract : Exhibition Hall Job Ref. No. : Example 5.6
Part of Structure : Glulam Roof Beam
Calculation Sheet No. : 5 of 5
Calculated by : W.McK.
Checked by : B.Z.
Date :

$w_{\text{inst,G}} = (w_{\text{inst,m}} + w_{\text{inst,v}}) = (14,43 + 0,65) = 15,08 \text{ mm}$
$w_{\text{fin,G}} = w_{\text{inst,G}}(1 + k_{\text{def}}) = 15,08 \times (1 + 0,8) = 27,14 \text{ mm}$

Variable load deformation:
$W = (2,25 \times 10,0) = 22,5 \text{ kN}$
$L = 5000 \text{ mm}$
$E_d = E_{0,g,\text{mean}} = 11,6 \text{ kN/mm}^2$
$G_d = G_{g,\text{mean}} = 0,72 \text{ kN/mm}^2$

Using proportion:

$w_{\text{inst,Q1,m}} = \dfrac{5WL^3}{384E_d I} = 14,51 \times \dfrac{22,5}{19,5} = 16,74 \text{ mm}$

$w_{\text{inst,Q1,v}} = \dfrac{3WL}{20bhG_d} = 0,65 \times \dfrac{22,5}{19,5} = 0,75 \text{ mm}$

UK NA to BS EN 1990 Table NA.A1.1: $\psi_{2,1} = 0,0$ for snow loads at for sites located at altitude H < 1000 m a.s.l. → $\psi_{2,1} = 0,0$

$w_{\text{inst,Q1}} = (w_{\text{inst,m}} + w_{\text{inst,v}})_{Q1} = (16,74 + 0,75) = 17,49 \text{ mm}$

$w_{\text{fin,Q1}} = w_{\text{inst,Q1}}(1 + \psi_{2,1}k_{\text{def}}) = 17,49 \times [1 + (0,0 \times 0,8)] = 17,49 \text{ mm}$

$w_{\text{fin}} = w_{\text{fin,G}} + w_{\text{fin,Q1}} = (27,14 + 17,49) = 44,63 \text{ mm}$ → $w_{\text{fin}} = 44,63 \text{ mm}$

National Annex Table NA.4: Limiting ratio = $l/150 = 10000/150 = 66,67 \text{ mm} > 44,78 \text{ mm}$

The joists are adequate with respect to deflection.

5.6.5 Example 5.7: Double-tapered Beam Design

A double-tapered, straight glulam roof beam is shown in Figure 5.31. Using the design data given, check the suitability of the section indicated. The roof decking provides continuous lateral restraint to the compression flange and the ends are restrained against rotation.

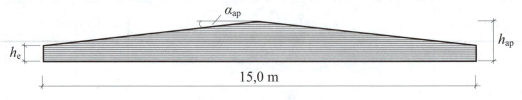

Figure 5.31

Design data:

End depth (h_e)	540 mm
Apex depth (h_{ap})	1350 mm
Beam breath (b)	115 mm
Timber strength class	GL 28h
Service Class	2
Characteristic permanent action including self-weight (g_k)	2,0 kN/m
Characteristic variable action due short-term wind load (q_k)	6,0 kN/m

5.6.5.1 Solution to Example 5.7

Contract : Glulam Beams Job Ref. No. : Example 5.7 Part of Structure : Double-tapered Roof Beam Calculation Sheet No. : 1 of 7	Calculated by : W.McK. Checked by : B.Z. Date :

References	Calculations	Output
BS EN 1990 BS EN 1991 BS EN 338:2003 BS EN 1995-1-1 2004 National Annex BS EN 1194: 1999 BS EN 1194 Table 1	Eurocode: Basis of Structural Design Eurocode 1: Actions on Structures Structural Timber - Strength Classes Eurocode 5: Design of Timber Structures Part 1.1: General - Common Rules and Rules for Buildings. UK National Annex to BS EN 1995-1-1 Glued Laminated Timber – Strength Classes and Characteristic Values Characteristic values for GL 28h glulam **Strength Properties:** Bending $f_{m,g,k} = 28,0$ N/mm^2 Tension perpendicular to grain $f_{t,90,g,k} = 0,45$ N/mm^2 Compression parallel to grain $f_{c,0,g,k} = 26,5$ N/mm^2 Compression perpendicular to grain $f_{c,90,g,k} = 3,0$ N/mm^2 Shear $f_{v,g,k} = 3,2$ N/mm^2	

References	Calculations	Output
	Contract : Glulam Beams Job Ref. No. : Example 5.7 **Part of Structure : Double-tapered Roof Beam** **Calculation Sheet No. : 2 of 7**	**Calculated by : W.McK.** **Checked by : B.Z.** **Date :**

References	Calculations	Output
	Stiffness Properties: Mean modulus of elasticity parallel to the grain $\qquad E_{0,g,mean} = 12,6 \text{ kN/mm}^2$ Mean shear modulus $\qquad G_{g,mean} = 0,78 \text{ kN/mm}^2$ Characteristic density $\qquad \rho_{g,k} = 410,0 \text{ kg/m}^3$	
BS EN 1995 Clause 2.3.2.1	Load-duration and moisture influence on strength: k_{mod}	
Table 3.1	Glued-laminated timber: Service Class 2 Permanent actions $\qquad k_{mod} = 0,6$ Short-term actions $\qquad k_{mod} = 0,9$	
Clause 3.1.3(2)	Use k_{mod} corresponding to action with the shortest duration.	$k_{mod} = 0,9$
National Annex Table NA.3 EC5 (Table 2.3)	Partial factor for material properties and resistance: γ_M For glued-laminated timber	$\gamma_M = 1,25$
Clause 3.3 Clause 3.3(3)	Member size: In this case enhancement does not apply since the depth is greater than 600 mm.	
Clause 6.6(4) Figure 6.12	System strength: k_{sys} Assume laminations 45 mm thick: Number of laminations > 8	$k_{sys} = 1,2$
	Cross-sectional area at the end: $A_e = (115 \times 540) = 62,1 \times 10^3 \text{ mm}^2$ **Cross-sectional area at the apex:** $A_{ap} = (115 \times 1350) = 155,3 \times 10^3 \text{ mm}^2$	
BS EN 1990 Equation (6.10) BS EN 1990 Table NA.A1.2(B)	**Loading:** Eurocode: Basis of Structural Design $E_d = \left(\gamma_G G_k + \gamma_Q Q_k \right)$ $\gamma_G = 1,35$ and $\gamma_Q = 1,50$ The design load $= \left(\gamma_G G_k + \gamma_Q Q_k \right) = (1,35 \times 2,0) + (1,5 \times 6,0)$ $F_d = 11,7 \text{ kN/m}$	$F_d = 11,7 \text{ kN/m}$
	Design bending moment $M_{y,d} = (wL^2/8)$ $M_{y,d} = \dfrac{11,7 \times 15,0^2}{8} = 329,1 \text{ kNm}$	$M_{y,d} = 329,1 \text{ kNm}$

References	Calculations	Output
Contract : Glulam Beams Job Ref. No. : Example 5.7 Part of Structure : Double-tapered Roof Beam Calculation Sheet No. : 3 of 7		Calculated by : W.McK. Checked by : B.Z. Date :

References	Calculations	Output
Clause 6.4.3(2) Equation (6.37) Clause 6.4.2(2) Equation (6.38) Equation (6.40) Clause 6.4.3(3) Equation (6.41) Clause 6.4.3(4) Equation (6.43) Equation (6.44)	**Design Bending stress:** The requirements of 6.4.2 (for single tapered beams) apply to the parts of the beam which have a single taper. $\sigma_{m,\alpha,d} = \sigma_{m,0,d} = \dfrac{6M_{y,d}}{bh^2} = \dfrac{6\times329{,}1\times10^6}{115\times1350^2} = 9{,}42 \text{ N/mm}^2$ $\sigma_{m,\alpha,d} \leq k_{m,\alpha}\, f_{m,y,d}$ Design bending strength $f_{m,y,d} = (f_{m,y,k}\times k_{mod}\times k_h \times k_{sys})/\gamma_M$ $k_h = 1{,}0$ $f_{m,y,d} = (28{,}0\times0{,}9\times1{,}0\times1{,}2)/1{,}25 = 24{,}19 \text{ N/mm}^2$ For compressive stresses parallel to the tapered edge: $k_{m,\alpha} = \dfrac{1}{\sqrt{1+\left(\dfrac{f_{m,d}}{1{,}5f_{v,d}}\tan\alpha\right)^2 + \left(\dfrac{f_{m,d}}{f_{c,90,d}}\tan^2\alpha\right)^2}}$ Design shear strength $f_{v,d} = (f_{v,k}\times k_{mod}\times k_{sys})/\gamma_M$ $f_{v,d} = (3{,}2\times0{,}9\times1{,}2)/1{,}25 = 2{,}76 \text{ N/mm}^2$ Design compressive strength $f_{c,90,d} = (f_{c,90k}\times k_{mod}\times k_{sys})/\gamma_M$ $f_{c,90,d} = (3{,}0\times0{,}9\times1{,}2)/1{,}25 = 2{,}59 \text{ N/mm}^2$ $\tan\alpha = (1350-540)/7500 = 0{,}108$ $k_{m,\alpha} = \dfrac{1}{\sqrt{1+\left(\dfrac{24{,}19}{1{,}5\times2{,}76}\times0{,}108\right)^2 + \left(\dfrac{24{,}19}{2{,}59}\times0{,}108^2\right)^2}} = 0{,}842$ $k_{m,\alpha}\, f_{m,y,d} = (0{,}842\times24{,}19) = 20{,}37 \text{ N/mm}^2 > 9{,}42 \text{ N/mm}^2$ **Bending stress in the apex zone:** $\sigma_{m,d} \leq k_r\, f_{m,y,d}$ $\sigma_{m,d} = k_l\, \dfrac{6M_{ap,d}}{bh_{ap}^2}$ $k_l = k_1 + k_2\left(\dfrac{h_{ap}}{r}\right) + k_3\left(\dfrac{h_{ap}}{r}\right)^2 + k_4\left(\dfrac{h_{ap}}{r}\right)^3$ For double-tapered beams: $r = \infty$ $\therefore k_l = k_1$ $k_1 = 1 + 1{,}4\tan\alpha_{ap} + 5{,}4\tan^2\alpha_{ap}$	$f_{m,y,d} = \mathbf{24{,}19 \text{ N/mm}^2}$ $f_{v,d} = \mathbf{2{,}76 \text{ N/mm}^2}$ $f_{c,90,d} = \mathbf{2{,}59 \text{ N/mm}^2}$ **The beam is adequate with respect to bending.**

References	Calculations	Output
Contract : Glulam Beams **Job Ref. No. : Example 5.7** **Part of Structure :** **Double-tapered Roof Beam** **Calculation Sheet No. : 4 of 7**		**Calculated by : W.McK.** **Checked by : B.Z.** **Date :**

References	Calculations	Output
Equation (6.43)	$k_1 = [1 + (1,4 \times 0,108) + (5,4 \times 0,108^2)] = 1,214$ $k_l = 1,214$	
Clause 6.4.3(4)	$\sigma_{m,d} = k_l \dfrac{6M_{ap,d}}{bh_{ap}^2} = 1,214 \times \dfrac{6 \times 329,1 \times 10^6}{115 \times 1350^2} = 11,44 \text{ N/mm}^2$	
Clause 6.4.3(5)	For double tapered beams $k_r = 1,0$ $k_r\, f_{m,y,d} = (1,0 \times 24,19) = 24,19 \text{ N/mm}^2 > 11,44 \text{ N/mm}^2$	**The beam is adequate with respect to the bending stress at the apex.**
Clause 6.4.3(6)	**Tensile stress perpendicular to the grain in the apex zone:**	
Equation (6.50)	$\sigma_{t,90,d} \leq k_{dis}\, k_{vol}\, f_{t,90,d}$	
Equation (6.54)	$\sigma_{t,90,d} = k_p \dfrac{6M_{ap,d}}{bh_{ap}^2}$ (**Note:** as indicated in the National Annex.)	
Equation (6.56)	$k_p = k_5 + k_6\left(\dfrac{h_{ap}}{r}\right) + k_7\left(\dfrac{h_{ap}}{r}\right)^2$ For double-tapered beams: $r = \infty$ $\therefore k_p = k_5$	
Equation (6.57)	$k_p = 0,2 \tan \alpha_{ap} = (0,2 \times 0,108) = 0,022$ $\sigma_{t,90,d} = 0,022 \times \dfrac{6 \times 329,1 \times 10^6}{115 \times 1350^2} = 0,21 \text{ N/mm}^2$	
Equation (6.51)	$k_{vol} = \left(\dfrac{V_0}{V}\right)^{0,2}$ for glued laminated timber	
Clause 6.4.3(6)	V_0 is the reference volume $= 0,01 \text{ m}^3$ V is the stressed volume of the apex zone as indicated in Figure 6.9 of the code $\leq 2V_b/3$ where V_b is the volume of the beam. $V = bh_{ap}^2\left(1 - \dfrac{\tan\alpha}{4}\right) = 0,115 \times 1,35^2\left(1 - \dfrac{0,108}{4}\right) = 0,204 \text{ m}^3$ $V_b = bL\dfrac{(h_e + h_{ap})}{2} = 0,115 \times 15,0 \times \dfrac{(0,54 + 1,35)}{2} = 1,63 \text{ m}^3$ $2V_b/3 = (2 \times 1,63)/3 = 1,09 \text{ m}^3 > 0,204 \text{ m}^3$ $\therefore V = 0,204 \text{ m}^3$ $k_{vol} = \left(\dfrac{V_0}{V}\right)^{0,2} = \left(\dfrac{0,01}{0,204}\right)^{0,2} = 0,547$	

Contract : Glulam Beams Job Ref. No. : Example 5.7	Calculated by : W.McK.
Part of Structure : Double-tapered Roof Beam	Checked by : B.Z.
Calculation Sheet No. : 5 of 7	Date :

References	Calculations	Output
Equation (6.52)	k_{dis} = 1,4 for double tapered and curved beams. Design tensile strength $f_{t,90,d}$ $= (f_{t,90,k} \times k_{mod} \times k_{sys})/\gamma_M$ (**Note:** k_h does not apply to $f_{t,90,k}$.) $f_{t,90,d} = (0,45 \times 0,9 \times 1,2)/1,25 = 0,39$ N/mm^2	$f_{t,90,d}$ = 0,39 N/mm^2
Equation (6.50)	$\sigma_{t,90,d} = 0,21$ N/mm$^2 < k_{dis} k_{vol} f_{t,90,d}$ $= (1,4 \times 0,547 \times 0,39) = 0,30$ N/mm^2	The beam is adequate with respect to tensile stresses at the apex.
Clause 6.4.3(7)	**Combined tension perpendicular to the grain and shear:** In this case the shear force at the apex is equal to zero.	
Clause 6.1.7	**Shear at the end:** Design shear force $F_{v,d} = (F_d \times 15,0)/2,0 = (11,7 \times 15,0)/2,0$ $= 87,75$ kN	$F_{v,d}$ = 87,75 kN
Clause 6.5.2	Design shear strength $f_{v,d} = (f_{v,k} \times k_{mod} \times k_v \times k_{sys})/\gamma_M$ k_v applies to notched beams only $f_{v,d} = (3,2 \times 0,9 \times 1,2)/1,25 = 2,76$ N/mm^2 Design shear stress $\tau_d = \dfrac{1,5 \times F_{v,d}}{A} = \dfrac{1,5 \times 87,75 \times 10^3}{62,1 \times 10^3}$ $= 2,12$ N/mm$^2 < 2,76$ N/mm^2	$f_{v,d}$ = 2,76 N/mm^2 The beam is adequate with respect to shear.
Clause 6.1.5	**Bearing at the end:** Design bearing force $F_{c,90,d} = 87,75$ kN Design bearing strength $f_{c,90,d} = (f_{c,90,k} \times k_{mod} \times k_{sys})/\gamma_M$ $= (3,0 \times 0,9 \times 1,2)/1,25$ $= 2,59$ N/mm^2 Assume a contact length of bearing $l = 125$ mm Design bearing stress $\sigma_{c,90,d} = \dfrac{F_{c,90,d}}{b \times l} = \dfrac{87,75 \times 10^3}{115 \times 125}$ $= 6,10$ N/mm^2	$F_{c,90,d}$ = 87,75 kN $f_{c,90,d}$ = 2,59 N/mm^2
Equation (6.3) Clause 6.1.5(2) Clause 6.1.5(3)	$\sigma_{c,90,d} \leq (k_{c,90} \times f_{c,90,d})$ Assume 'a' in Figure 6.2 of EC5 = 0 $k_{c,90} = \left(2,38 - \dfrac{l}{250}\right)\left(1 + \dfrac{h}{12l}\right)$ and $\leq 4,0$ $k_{c,90} = \left(2,38 - \dfrac{125}{250}\right)\left(1 + \dfrac{540}{12 \times 125}\right) = 2,56 \leq 4,0$ $(k_{c,90} \times f_{c,90,d}) = (2,56 \times 2,59) = 6,63$ N/mm$^2 \geq 6,10$ N/mm^2	$k_{c,90}$ = 2,56 Provide a minimum contact length l = 125 mm.

	Contract : **Glulam Beams** Job Ref. No. : **Example 5.7**	Calculated by : **W.McK.**
	Part of Structure : **Double-tapered Roof Beam**	Checked by : **B.Z.**
	Calculation Sheet No. : **6 of 7**	Date :

References	Calculations	Output
Clause 6.3.3 Equation (6.33)	**Lateral Torsional Stability:** $\sigma_{m,d} \le (k_{crit} \times f_{m,d})$	
Clause 6.3.3(5)	Since the compression flange is fully restrained by the decking and torsional restraint is provided at the supports the value of $k_{crit} = 1,0$.	**The section is adequate with respect to lateral torsional stability.**
Clause 7.2	**Serviceability Limit States:** **Deflection:** Limiting values for deflection of beams: $w_{net,fin} = w_{inst} + w_{creep} - w_c = w_{fin} - w_c$ Since there is no camber $w_c = 0$ and $w_{net,fin} = w_{inst} + w_{creep} = w_{fin}$	
Table 3.2	For glulam timber Service Class 2 $k_{def} = 0,8$ $w_{inst,G} (1 + k_{def}) = (w_{inst,m} + w_{inst,v})_G (1 + k_{def})$	$k_{def} = 0,8$
See Section 5.6.3.7	The equivalent member depth $h_{eq} \approx (C_{eq} \times h_e)$ where: $C_{eq} = (1 + 0,66C_y)$ when $0 < C_y < 1$ or $C_{eq} = (1 + 0,62C_y)$ when $1 < C_y < 3$ and $C_y = \dfrac{h_{ap} - h_e}{h_e} = \dfrac{(1350 - 540)}{540} = 1,5$ $C_{eq} = (1 + 0,62C_y) = [1 + (0,62 \times 1,5)] = 1,93$ $h_{eq} = (C_{eq} \times h_e) = (1,93 \times 540) = 1042,2$ mm Cross-sectional properties of the equivalent section: $I_{eq} = \dfrac{bh_{eq}^3}{12} = \dfrac{115 \times 1042,2^3}{12} = 10,85 \times 10^9$ mm^4 Permanent load deformation: $W = (2,0 \times 15,0) = 30,0$ kN $L = 15000$ mm $E_d = E_{0,g,mean} = 12,6$ kN/mm^2 $G_d = G_{g,mean} = 0,78$ kN/mm^2 $I_{eq} = 10,85 \times 10^9$ mm^4 $w_{inst,G,m} = \dfrac{5WL^3}{384E_d I_{eq}} = \dfrac{5 \times 30,0 \times 15000^3}{384 \times 12,6 \times 10,85 \times 10^9} = 9,64$ mm $w_{inst,G,v} = \dfrac{3WL}{20bh_{eq}G_d} = \dfrac{3 \times 30,0 \times 15000}{20 \times 115 \times 1042,2 \times 0,78} = 0,72$ mm	

References	Calculations	Output
	Contract : Glulam Beams Job Ref. No. : Example 5.7 **Part of Structure : Double-tapered Roof Beam** **Calculation Sheet No. : 7 of 7**	**Calculated by : W.McK.** **Checked by : B.Z.** **Date :**

References	Calculations	Output
	$w_{\text{inst,G}} = (w_{\text{inst,m}} + w_{\text{inst,v}})_G = (9{,}64 + 0{,}72) = 10{,}36 \text{ mm}$ $w_{\text{fin,G}} = w_{\text{inst,G}}(1 + k_{\text{def}}) = 10{,}36 \times (1 + 0{,}8) = 18{,}65 \text{ mm}$ Variable load deformation: $W = (6{,}0 \times 15{,}0) = 90{,}0 \text{ kN}$ $L = 15000 \text{ mm}$ $E_d = E_{0,\text{mean}} = 12{,}6 \text{ kN/mm}^2$ $G_d = G_{g,\text{mean}} = 0{,}78 \text{ kN/mm}^2$ $I_{\text{eq}} = 10{,}85 \times 10^9 \text{ mm}^4$ Using proportion: $w_{\text{inst,Q1,m}} = \dfrac{5WL^3}{384 E_d I_{\text{eq}}} = 9{,}64 \times \dfrac{90}{30} = 28{,}92 \text{ mm}$ $w_{\text{inst,Q1,v}} = \dfrac{3WL}{20 b h_{\text{eq}} G_d} = 0{,}72 \times \dfrac{90}{30} = 2{,}16 \text{ mm}$	
UK NA to BS EN 1990 Table A1.1	$\psi_{2,1} = 0{,}0$ for wind load $w_{\text{inst,Q1}} = (w_{\text{inst,m}} + w_{\text{inst,v}})_{Q1} = (28{,}92 + 2{,}16) = 31{,}08 \text{ mm}$ $w_{\text{fin,Q1}} = w_{\text{inst,Q1}}(1 + \psi_{2,1} k_{\text{def}}) = 31{,}08 \times [1 + (0{,}0 \times 0{,}8)]$ $\qquad = 31{,}08 \text{ mm}$ $w_{\text{fin}} = w_{\text{fin,G}} + w_{\text{fin,Q1}} = (18{,}65 + 31{,}08) = 49{,}73 \text{ mm}$	$\boldsymbol{\psi_{2,1} = 0{,}0}$
National Annex Table NA.4	Limiting ratio $= l/150 = 15000/150 = 100 \text{ mm} > 49{,}73 \text{ mm}$	$\boldsymbol{w_{\text{fin}} = 49{,}73 \text{ mm}}$ **The beam is adequate** **with respect to** **deflection.**

6. Axially Loaded Members

Objective: *to illustrate the design of axially loaded members considering single tension and compression frame elements, single and spaced columns.*

6.1 Introduction

The design of axially loaded members considers any member where the applied loading induces either axial tension or axial compression. Members subject to axial forces frequently occur in bracing systems, roof trusses or lattice girders.

Frequently, in structural frames, sections are subjected to combined axial and bending effects, which may be caused by eccentric connections, wind loading or rigid-frame action. The design of such members is discussed and illustrated in Chapter 7. In EC5 the ultimate limit state design requirements for members subjected to stress in one principal direction are given in Section 6.1. This includes tension/compression stresses which are parallel/perpendicular to the grain.

In the case of compression stresses which are at an angle to the grain, reference should be made to Section 6.2.2. The stability of compression members must also be considered as given in Section 6.3 of the code. The relevant modification factors which apply to axially loaded members are summarized in Table 6.1.

Factors	Application	Clause Number	Table/Equation
k_{mod}	Relates to all strength properties for service and load-duration classes.	3.1.3	Table 3.1
k_h	Relates to the bending strength/tension strength parallel to the grain and to member depth/width.	3.2(3) 3.3(3) 3.4(3)	Equation (3.1) Equation (3.2) Equation (3.3)
$k_{c,90}$	Relates to compression strength perpendicular to the grain.	6.1.5	Equations (6.4), (6.5), (6.6) , (6.10)
k_y	Relates to stability and interaction equations - y axis	6.3.2	Equation (6.27)
k_z	Relates to stability and interaction equations - z axis	6.3.2	Equation (6.28)
$k_{c,y}$	Relates to stability and interaction equations - y axis	6.3.2	Equation (6.25)
$k_{c,z}$	Relates to stability and interaction equations - z axis	6.3.2	Equation (6.26)
β_c	Relates to stability and interaction equations	6.3.2	Equation (6.29)
k_{sys}	Relates to lateral load distribution	6.6	Generally 1,1

Table 6.1 Modification Factors – axially loaded members

In the case of trusses, guidance is given in Sections 9.2.1 and 9.2.2 with respect to the effective buckling lengths of compression members.

6.2 Design of Tension Members (Clauses 6.1.2 and 6.1.3)

The design of tension members is based on the effective area of the cross-section allowing for a reduction due to notches, bolts, dowels, screw holes or any mechanical fastener inserted in the member as indicated in Clauses 5.2(2) and 5.2(3). The structural detailing and control of connections (and hence allowances for holes in sections) is given in Section 10 of EC5; e.g. in the case of bolts and washers Clause 10.4.3(1) states that "Bolt holes in timber should have a diameter not more than 1mm larger than the bolt …." The use of glued joints, nails, dowels, connectors and screws is also considered.

In the case of tension parallel to the grain the following equation should be satisfied:

$$\sigma_{t,0,d} \leq f_{t,0,d} \qquad \text{(Equation (6.1) in EC5)}$$

where:

$\sigma_{t,0,d}$ is the design tensile stress $= \dfrac{F_{t,0d}}{A_{net}}$,

$F_{t,0,d}$ is the design tensile force parallel to the grain,

A_{net} is the net cross-sectional area allowing for fasteners as indicated in Clause 5.2(2). In the case of multiple fasteners all holes within a distance of half the minimum fastener spacings measured parallel to the grain (see Section 8.2 of EC5) from a given cross-section should be considered as occurring at that cross-section.

The design strength is given by:

$$f_{t,0,d} = (f_{t,0,k} \times k_{mod} \times k_h \times k_{sys})$$

The k_h factor applies to $f_{t,0,k}$ in rectangular solid timber sections with a characteristic density $\rho_k \leq 700$ kg/m^3 in which the width in tension is less than 150 mm. It is given by the following expression:

$$k_h = \min\begin{cases} \left(\dfrac{150}{h}\right)^{0,2} & \text{(Equation (3.1) in EC5)} \\ 1,3 \end{cases}$$

Tension stresses perpendicular to the grain occur in elements such as tapered beams, curved beams and connections. Timber has a relatively low $f_{t,90,k}$ value which is dependent on the stressed volume of material. When used in EC5 relating to the apex zone of double-tapered, curved and pitch cambered beams the characteristic value is modified by two factors k_{dis} and k_{vol}, i.e.

$$f_{t,90,d} = k_{dis} \times k_{vol} \times (f_{t,90,k} \times k_{mod} \times k_{sys}). \qquad \text{(Equation (6.50) in EC5)}$$

The k_h value does not apply to tension strength perpendicular to the grain.

6.2.1 Example 6.1: Collar Tie Member

A collar tie roof construction comprises two rafters and a collar tie (100 mm x 50 mm), connected to the rafters by 8 mm diameter steel bolts as shown in Figure 6.1. Assuming the wall does not provide any lateral restraint to the toe of the rafter, check the suitability of the collar tie.

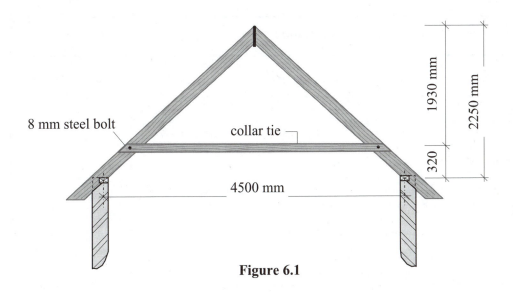

8 mm steel bolt

collar tie

1930 mm

2250 mm

320

4500 mm

Figure 6.1

Design data:

Spacing of rafter frames	450 mm
Distance between the centre-lines of the wall plates	4500,0 m
Timber species and grade	UK Douglas fir - GS
Steel bolts	8 mm diameter
Timber Service Class	2
Characteristic self-weight	0,71 kN/m^2
Characteristic imposed load due to maintenance access	0,36 kN/m^2
Characteristic snow load	0,30 kN/m^2

Assume:
  (i)   all loads are based on plan areas,
  (ii) the altitude of the structure is less than 1000 m above mean sea level,
  (iii) the roof construction is continuous over at least two spans.

6.2.1.1 Solution to Example 6.1

References	Calculations	Output
Contract : Roof **Job Ref. No. : Example 6.1** **Calculated by : W.McK.** **Part of Structure :** **Collar tie** **Checked by : B.Z.** **Calculation Sheet No. : 1 of 3** **Date :**		

References	Calculations	Output
BS EN 1990 BS EN 1991 BS EN 338:2003 BS EN 1995-1-1: 2004	Eurocode: Basis of Structural Design Eurocode 1: Actions on Structures Structural Timber - Strength Classes Eurocode 5: Design of Timber Structures Part 1.1: General - Common Rules and Rules for Buildings	
National Annex	UK National Annex to BS EN 1995-1-1	
BS EN 1912 Table 1	Assignment of timber type and grade to strength class: UK, Douglas fir, Grade GS – Strength Class C14	
BS EN 338:2003 Clause 5.0	Structural Timber - Strength Classes Characteristic values for C14 Timber	
Table 1	**Strength and Density Properties:** Tension parallel to grain $f_{t,0,k} = 8,0 \text{ N/mm}^2$ Mean density $\rho_{mean} = 350,0 \text{ kg/m}^3$	

BS EN 1995-1-1
 Clause 2.3.1.2
 Table NA.1
 (EC5: Table 2.3)
	Load-duration Classes: use Table NA.1 from National Annex
	 Self-weight - permanent
	 Roof loading (snow and maintenance) - short-term
	
	 Permanent action = $(0,71 \times 0,45)$ $= 0,32 \text{ kN/m}$
	 Variable action 1 due to imposed load $= (0,36 \times 0,45)$
	                                    $= 0,16 \text{ kN/m}$
	 Variable action 2 due to snow load $= (0,30 \times 0,45)$
	                                    $= 0,14 \text{ kN/m}$	

Clause 2.3.2.1	Load-duration and moisture influence on strength: k_{mod}	
Table 3.1	Solid timber: Service Class 2 Permanent actions $k_{mod} = 0,6$ Variable actions (short-term) $k_{mod} = 0,9$	
Clause 3.1.3(2)	Use k_{mod} corresponding to action with the shortest duration.	$k_{mod} = 0,9$
Table NA.3 EC5 (Table 2.3)	For solid timber – untreated and preservative treated	$\gamma_M = 1,3$
Clause 3.2.(3) Equation (3.1)	Member size: For timber in bending with $\rho_k \leq 700 \text{ kg/m}^3$ where the depth is less than 150 mm $f_{m,k}$ can be multiplied by k_h where:	

Contract : Roof Job Ref. No. : Example 6.1 Part of Structure : Collar tie Calculation Sheet No. : 2 of 3	Calculated by : W.McK. Checked by : B.Z. Date :

References	Calculations	Output
	$$k_h = \min \left\{ \begin{array}{l} \left(\dfrac{150}{h}\right)^{0,2} \\[2mm] 1,3 \end{array} \right.$$	
	$$k_h = \min \left\{ \begin{array}{l} \left(\dfrac{150}{100}\right)^{0,2} \\[2mm] 1,3 \end{array} \right. = \min \left\{ \begin{array}{l} 1,08 \\[1mm] 1,3 \end{array} \right.$$	$k_h = 1,08$
Clause 6.6	System strength: k_{sys} Since the roofing has adequate provision for lateral distribution of loading and the spacing of the trusses is $\le 1,2$ m the strength properties can be multiplied by k_{sys}.	$k_{sys} = 1,1$
BS EN 1990 Equation (6.10)	Eurocode: Basis of Structural Design $$E_d = \sum_{j\geq 1} \gamma_{G,j} G_{k,j} + \gamma_{Q,1} Q_{k,1} + \sum_{i>1} \gamma_{Q,i} \psi_{0,i} Q_{k,i} \quad j = 1 \ \text{ and } \ i = 2$$	
UK NA to BS EN 1990 Table NA.A1.2(B)	$\gamma_G = 1,35 \qquad \gamma_{Q,1} = 1,50 \qquad \gamma_{Q,2} = 1,50$	
NA to EN 1990 Table NA.A1.1	For imposed loads due to maintenance on roofs: $\psi_0 = 0,7$ For snow loads at H < 1000 m $\psi_0 = 0,5$	
BS EN 1990 Equation (6.10)	Assuming the imposed load to be the leading variable: Design load $= \gamma_G G_k + \gamma_{Q,1} Q_{k,1} + \gamma_{Q,2} \psi_{0,2} Q_{k,2}$ $= (1,35 \times 0,32) + (1,5 \times 0,16) + (1,5 \times 0,5 \times 0,14) = 0,78$ kN/m	
Equation (6.10)	Assuming the snow load to be the leading variable: Design load $= \gamma_G G_k + \gamma_{Q,1} Q_{k,1} + \gamma_{Q,2} \psi_{0,2} Q_{k,2}$ $= (1,35 \times 0,32) + (1,5 \times 0,14) + (1,5 \times 0,7 \times 0,16) = 0,81$ kN/m	
	Design load $F_{t,0d}$ = 0,81 kN/m	$F_{t,0,d} = 0,81$ kN/m

	0,81 kN/m 0,81 kN/m
	
	O O
	               1930 mm
	          $F_{t,0,d}$
	      2250 mm
	1,82 kN 1,82 kN 1,82 kN	

Contract : Roof Job Ref. No. : Example 6.1	Calculated by : W.McK.
Part of Structure : Collar tie	Checked by : B.Z.
Calculation Sheet No. : 3 of 3	Date :

References	Calculations	Output
	Σ Moments about the apex $= 0$ $(1,82 \times 2,25) - (0,81 \times 2,25 \times 1,125) - (F_{t,0,d} \times 1,93) = 0$ $\qquad\qquad\qquad\qquad\qquad F_{t,0,d} = 1,06$ kN	$F_{t,0,d} = 1,06$ kN
BS EN 1995-1-1 Clause 6.1.2 Equation (6.1)	Tension parallel to the grain: $\sigma_{t,0,d} \leq f_{t,0,d}$ Design tensile strength parallel to the grain: $f_{t,0,d} = (f_{t,0,k} \times k_{mod} \times k_h \times k_{sys} / \gamma_M)$ $\qquad = (8,0 \times 0,9 \times 1,08 \times 1,1)/1,3 = 6,58$ N/mm^2	$f_{t,0,d} = 6,58$ N/mm^2
Clause 5.2(2)	The cross-sectional area should be reduced to allow for the 8 mm steel bolts.	
Clause 10.4.3(1)	The bolt hole diameter should not be greater than 1 mm larger than the bolt diameter. Assume 9 mm diameter bolt holes. Net cross-sectional area $A_{net} = [(100 \times 50) - (9 \times 50)]$ $\qquad\qquad\qquad = 4550,0$ mm^2 Design tensile stress parallel to the grain: $\sigma_{t,0,d} = \dfrac{F_{t,0,d}}{A_{net}} = \dfrac{1,06 \times 10^3}{4550,0} = 0,23$ N/mm$^2 \ll 6,58$ N/mm^2 There is considerable reserve of strength in the collar tie.	The 100 mm x 50 mm section is adequate.

6.3 Design of Compression Members (Clauses 6.1.4, 6.1.5 and 6.3.2)

The design of compression members is more complex than that of tension members and encompasses the design of structural elements referred to as columns, stanchions or struts. The term struts is usually used when referring to members in lattice/truss frameworks, whilst the other two generally refer to vertical or inclined members supporting floors and/or roofs in structural frames.

As with tension members in many cases they are subjected to both axial and bending effects. This chapter deals with those members, which are subjected to concentric axial loading.

The dominant mode of failure to be considered when designing struts is flexural buckling. Flexural buckling failure is caused by secondary bending effects induced by factors such as:

- ◆ The inherent eccentricity of applied loads due to asymmetric connection details,
- ◆ Imperfections present in the cross-section and/or profile of a member throughout its length. The allowable deviation from straightness when using either visual or machine grading (i.e. bow not greater than 20 mm lateral displacement over a length of 2,0 m, see BS EN 518:1995 - Table 2 and BS EN 519:1995 – Table 1) is inadequate when considering compression members. More severe restrictions such as $L/300$ for structural timber and $L/500$ for glulam sections should be considered,
- ◆ Non-uniformity of material properties throughout a member.

The effects of these characteristics are to introduce initial curvature, secondary bending and consequently premature failure by buckling before the stress in the material reaches the failure value. Since they cannot generally be quantified, their influence on the failure load of compressive members is implicit in the modification factors adopted in EC. Primarily these factors; $k_{c,y}$, $k_{c,z}$, k_y and k_z relate to slenderness values λ_y and λ_z and relative slenderness values $\lambda_{rel,y}$ and $\lambda_{rel,z}$.

6.3.1 Slenderness

Slenderness values are defined as: $\lambda = \dfrac{L_e}{i}$ where L_e is the effective buckling length and i is the radius of gyration.

6.3.2 Relative Slenderness

The relative slenderness can be defined as $\lambda_{rel} = \sqrt{\dfrac{f_{c,0,k}}{\sigma_{crit}}}$ where σ_{crit} is the Euler buckling

stress equal to $\dfrac{\pi^2 E}{\left(L_e/i\right)^2}$.

The relative slenderness values are given in Clause 6.3.2 of EC5 as follows:

$$\lambda_{rel,y} = \frac{\lambda_y}{\pi}\sqrt{\frac{f_{c,0,k}}{E_{0,05}}} \qquad \text{(Equation (6.21) in EC5)}$$

$$\lambda_{rel,z} = \frac{\lambda_z}{\pi}\sqrt{\frac{f_{c,0,k}}{E_{0,05}}} \qquad \text{(Equation (6.22) in EC5)}$$

where $E_{0,05}$ is the fifth percentile value of the modulus of elasticity parallel to the grain.

A practical and realistic assessment of the critical slenderness of a strut is the most important criterion in determining the compressive strength.

6.3.3 Effective Buckling Length

The effective length is considered to be the actual length of the member between points of restraint, multiplied by a coefficient to allow for effects such as stiffening due to end connections of the frame of which the member is a part. Typical effective lengths based on different end support conditions are shown in Figure 6.2.

When considering members in triangulated frameworks (other than trussed rafters which are considered in BS 5268 : Part 3 : 2006) with continuous compression members, e.g. the top chord of a lattice girder, reference should be made to Clause 9.2 of EC5 in which guidance is given relating to effective length for a variety of circumstances.

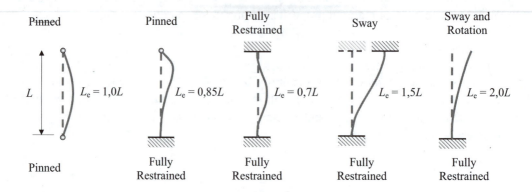

Figure 6.2   Effective Lengths (L_e)

6.3.3.1 Compression Parallel to the Grain (Clause 6.1.4)

In the case of stocky members which are not susceptible to flexural buckling, i.e. when both $\lambda_{rel,y}$ and $\lambda_{rel,z}$ are $\leq 0,3$ (see Clause 6.3.2(2)), the full compressive strength as given in Clause 6.1.4 can be utilized, i.e.

$\sigma_{c,0,d} \leq f_{c,0,d}$ (Equation (6.2) in EC5)

where
$f_{c,0,d} = (f_{c,0,k} \times k_{mod} \times k_{sys})/\gamma_m$

When flexural buckling must be considered reference should be made to Section 6.3 of the code. The governing Equations (6.23) and (6.24) relate to compression or combined compression and bending, i.e.

$$\frac{\sigma_{c,0,d}}{k_{c,y} f_{c,0,d}} + \frac{\sigma_{m,y,d}}{f_{m,y,d}} + k_m \frac{\sigma_{m,z,d}}{f_{m,z,d}} \leq 1,0 \qquad \text{(Equation (6.23) in EC5)}$$

and

$$\frac{\sigma_{c,0,d}}{k_{c,z} f_{c,0,d}} + k_m \frac{\sigma_{m,y,d}}{f_{m,y,d}} + \frac{\sigma_{m,z,d}}{f_{m,z,d}} \leq 1,0 \qquad \text{(Equation (6.24) in EC5)}$$

In the case of only axial compressive loading these equations reduce to:

$$\sigma_{c,0,d} \le k_{c,y} f_{c,0,d}$$

and

$$\sigma_{c,0,d} \le k_{c,z} f_{c,0,d}$$

where:

$$k_{c,y} = \frac{1}{k_y + \sqrt{k_y^2 - \lambda_{rel,y}^2}} \qquad \text{(Equation (6.25) in EC5)}$$

$$k_y = 0,5\left(1 + \beta_c\left(\lambda_{rel,y} - 0,3\right) + \lambda_{rel,y}^2\right) \qquad \text{(Equation (6.27) in EC5)}$$

$$k_{c,z} = \frac{1}{k_z + \sqrt{k_z^2 - \lambda_{rel,z}^2}} \qquad \text{(Equation (6.26) in EC5)}$$

$$k_z = 0,5\left(1 + \beta_c\left(\lambda_{rel,z} - 0,3\right) + \lambda_{rel,z}^2\right) \qquad \text{(Equation (6.28) in EC5)}$$

where:
β_c is a factor to allow for deviation from target dimensions and is given in Equation (6.29) in EC5. Generally the variation in solid timber is greater than that found in glued laminated and LVL members.

6.3.4 Example 6.2: Concentrically Loaded Column

A symmetrically loaded internal column is required to support four beams as shown in Figure 6.3. The top and bottom can be considered to be held in position but not in direction. A lateral restraint is provided at the mid-height as indicated. Using the design data given, check the suitability of a 75 mm x 155 mm, solid timber section for the column.

Design data:

Timber section size	75 mm x 155 mm
Strength class	C18
Timber Service Class	2
Characteristic long-term load	8,0 kN
Characteristic medium-term load	24,0 kN

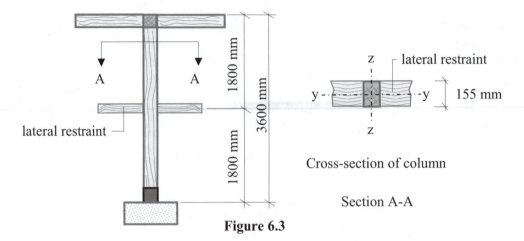

Figure 6.3

6.3.4.1 Solution to Example 6.2

Contract : Column Job Ref. No. : Example 6.2	Calculated by : W.McK.
Part of Structure : Concentrically loaded column	Checked by : B.Z.
Calculation Sheet No. : 1 of 3	Date :

References	Calculations	Output
BS EN 1990 BS EN 1991 BS EN 338:2003 BS EN 1995-1-1: 2004 National Annex	Eurocode: Basis of Structural Design Eurocode 1: Actions on Structures Structural Timber - Strength Classes Eurocode 5: Design of Timber Structures Part 1.1: General – Common Rules and Rules for Buildings UK National Annex to BS EN 1995-1-1	
BS EN 338:2003 Clause 5.0	Structural Timber - Strength Classes Characteristic values for C18 Timber	
Table 1	**Strength and Stiffness Properties:** Compression parallel to grain $f_{c,0,k} = 18,0 \text{ N/mm}^2$ 5% Modulus of elasticity parallel to the grain $E_{0,05} = 6,0 \text{ kN/m}^3$	
Clause 2.3.2.1	Load-duration and moisture influence on strength: k_{mod}	
Table 3.1	Solid timber: Service Class 2 Permanent actions $k_{mod} = 0,6$ Variable actions (medium-term) $k_{mod} = 0,8$	
Clause 3.1.3(2)	Use k_{mod} corresponding to action with the shortest duration.	$k_{mod} = 0,8$
Table NA.3 EC5 (Table 2.3) Clause 6.6	For solid timber – untreated and preservative treated System strength factor (k_{sys}) does not apply in this case.	$\gamma_M = 1,3$

References	Calculations	Output
Contract : Column Job Ref. No. : Example 6.2 **Part of Structure : Concentrically loaded column** **Calculation Sheet No. : 2 of 3**		**Calculated by : W.McK.** **Checked by : B.Z.** **Date :**

References	Calculations	Output
BS EN 1990 Equation (6.10)	Eurocode: Basis of Structural Design $E_d = \displaystyle\sum_{j \geq 1} \gamma_{G,j} G_{k,j} + \gamma_{Q,1} Q_{k,1}$	
UK NA to BS EN 1990 Table NA.A1.2(B)	$\gamma_G = 1{,}35 \qquad \gamma_{Q,1} = 1{,}50$	
BS EN 1990 Equation (6.10)	The design load $= \gamma_G G_k + \gamma_{Q,1} Q_{k,1}$ Design load $F_{c,0,d} \;\; = (1{,}35 \times 8{,}0) + (1{,}5 \times 24{,}0) = 46{,}8$ kN	$F_{c,0,d} = \textbf{46,8 kN}$
BS EN 1995-1-1 Clause 6.1.4 Equation (6.2)	Compression parallel to the grain: $\sigma_{c,0,d} \leq f_{c,0,d}$ Design compressive strength parallel to the grain: $f_{c,0,d} = (f_{c,0,k} \times k_{mod} \times k_{sys}) / \gamma_M$ $\qquad\;\; = (18{,}0 \times 0{,}8)/1{,}3 = 11{,}08$ N/mm^2 Design compressive stress parallel to the grain: $\sigma_{c,0,d} = \dfrac{F_{c,0,d}}{A} \;=\; \dfrac{46{,}8 \times 10^3}{(75 \times 155)} = 4{,}03 \text{ N/mm}^2 \; < \; 11{,}08 \text{ N/mm}^2$	$f_{c,0,d} = \textbf{11,08 N/mm}^2$
Clause 6.3 Clause 6.3.2	Stability of members: Columns subjected to either compression or combined compression and bending.	
Equation (6.23)	$\dfrac{\sigma_{c,0,d}}{k_{c,y} f_{c,0,d}} + \dfrac{\sigma_{m,y,d}}{f_{m,y,d}} + k_m \dfrac{\sigma_{m,z,d}}{f_{m,z,d}} \leq 1{,}0$	
Equation (6.24)	$\dfrac{\sigma_{c,0,d}}{k_{c,z} f_{c,0,d}} + k_m \dfrac{\sigma_{m,y,d}}{f_{m,y,d}} + \dfrac{\sigma_{m,z,d}}{f_{m,z,d}} \leq 1{,}0$	
	Since $\sigma_{m,y,d}$ and $\sigma_{m,z,d} = 0$, Equations (6.23) and (6.24) reduce to: $\dfrac{\sigma_{c,0,d}}{k_{c,y} f_{c,0,d}} \leq 1{,}0$ and $\dfrac{\sigma_{c,0,d}}{k_{c,z} f_{c,0,d}} \leq 1{,}0$	
Clause 6.3.2(1)	Relative slenderness	
Equation (6.21)	$\lambda_{rel,y} = \dfrac{\lambda_y}{\pi} \sqrt{\dfrac{f_{c,0,d}}{E_{0,05}}}$ where $\lambda_y = \dfrac{L_{y,ef}}{i_y}$ $L_{y,ef} = (1{,}0 \times 3600) = 3600$ mm	

Contract : Column Job Ref. No. : Example 6.2 Part of Structure : Concentrically loaded column Calculation Sheet No. : 3 of 3	Calculated by : W.McK. Checked by : B.Z. Date :

References	Calculations	Output
	$i_y = \dfrac{h}{2\sqrt{3}} = \dfrac{155}{2\sqrt{3}} = 44,75 \qquad \lambda_y = \dfrac{3600}{44,75} = 80,45$ $\lambda_{rel,y} = \dfrac{80,45}{\pi}\sqrt{\dfrac{18,0}{6,0\times10^3}} = 1,40$	
Equation (6.22)	$\lambda_{rel,z} = \dfrac{\lambda_z}{\pi}\sqrt{\dfrac{f_{c,0,d}}{E_{0,05}}} \;\text{ where }\; \lambda_z = \dfrac{L_{z,ef}}{i_z}$ $L_{z,ef} = (1,0 \times 1800) = 1800 \text{ mm}$ $i_z = \dfrac{b}{2\sqrt{3}} = \dfrac{75}{2\sqrt{3}} = 21,65 \qquad \lambda_z = \dfrac{1800}{21,65} = 83,14$ $\lambda_{rel,z} = \dfrac{83,14}{\pi}\sqrt{\dfrac{18,0}{6,0\times10^3}} = 1,45$	
Equation (6.27)	$k_y = 0,5\left[1 + \beta_c\left(\lambda_{rel,y} - 0,3\right) + \lambda_{rel,y}^2\right]$	
Equation (6.29)	$\beta_c = 0,2$ for solid timber $k_y = 0,5\times\left[1 + 0,2\times\left(1,4 - 0,3\right) + 1,4^2\right] = 1,59$	
Equation (6.25)	$k_{cy} = \dfrac{1}{k_y + \sqrt{k_y^2 - \lambda_{rel,y}^2}} = \dfrac{1}{1,59 + \sqrt{1,59^2 - 1,4^2}} = 0,43$	
Equation (6.28)	$k_z = 0,5\left[1 + \beta_c\left(\lambda_{rel,z} - 0,3\right) + \lambda_{rel,z}^2\right]$	
Equation (6.29)	$\beta_c = 0,2$ for solid timber $k_z = 0,5\times\left[1 + 0,2\times\left(1,45 - 0,3\right) + 1,45^2\right] = 1,67$	
Equation (6.25)	$k_{cz} = \dfrac{1}{k_z + \sqrt{k_z^2 - \lambda_{rel,z}^2}} = \dfrac{1}{1,67 + \sqrt{1,67^2 - 1,45^2}} = 0,40$	
Equation (6.23)	$\dfrac{\sigma_{c,0,d}}{k_{c,y}f_{c,0,d}} = \dfrac{4,03}{0,43\times11,08} = 0,85 \le 1,0$ and	
Equation (6.24)	$\dfrac{\sigma_{c,0,d}}{k_{c,z}f_{c,0,d}} = \dfrac{4,03}{0,40\times11,08} = 0,91 \le 1,0$	**The 75 mm x 155 mm section is adequate.**

6.3.5 Example 6.3: Covered Walkway

A covered walkway is to be constructed comprising a series of timber columns and timber beams with a glazed roof as shown in Figure 6.4. Check the suitability of the proposed timber section for a typical internal column.

Design data:

Timber species and grade	British grown Douglas fir Grade GS
Timber section size	100 mm x 50 mm
Timber Service Class	3
Characteristic long-term load	0,32 kN/m^2
Characteristic short-term load	0,75 kN/m^2

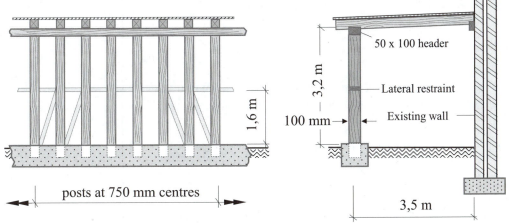

Figure 6.4

6.3.5.1 Solution to Example 6.3

<table>
<tr>
<td colspan="3">Contract : Walkway   Job Ref. No. : Example 6.3
Part of Structure :   Typical Internal Column
Calculation Sheet No. : 1 of 4</td>
<td colspan="2">Calculated by : W.McK.
Checked by : B.Z.
Date :</td>
</tr>
<tr>
<td>References</td>
<td colspan="3" align="center">Calculations</td>
<td align="center">Output</td>
</tr>
<tr>
<td>BS EN 1990
BS EN 1991
BS EN 338:2003
BS EN 1995-1-1:
2004

National Annex

BS EN 1912
Table 1</td>
<td colspan="3">Eurocode: Basis of Structural Design
Eurocode 1: Actions on Structures
Structural Timber - Strength Classes
Eurocode 5: Design of Timber Structures
Part 1.1: General – Common Rules and Rules for Buildings

UK - National Annex to BS EN 1995-1-1

Assignment of timber type and grade to strength class:
British grown Douglas fir, Grade GS – Strength Class C14</td>
<td></td>
</tr>
</table>

References	Calculations	Output
Contract : Walkway	**Job Ref. No. : Example 6.3**	**Calculated by : W.McK.**
Part of Structure :	**Typical Internal Column**	**Checked by : B.Z.**
Calculation Sheet No. : 2 of 4		**Date :**

References	Calculations	Output
BS EN 338:2003 Clause 5.0	Structural Timber - Strength Classes Characteristic values for C14 Timber	
Table 1	**Strength Properties:** Compression parallel to the grain $\qquad$ $f_{c,0,k} = 16,0$ N/mm^2 Compression perpendicular to the grain $\quad$ $f_{c,90,k} = 2,0$ N/mm^2 **Stiffness Properties:** 5% Modulus of elasticity parallel to the grain $\qquad\qquad\qquad\qquad E_{0,05} = 4,7$ kN/m^3 Permanent action/post $= 0,5 \times (0,32 \times 0,75 \times 3,5) = 0,42$ kN Variable action (short-term)/post $= 0,5 \times (0,75 \times 0,75 \times 3,5)$ $\qquad\qquad\qquad\qquad\qquad = 0,98$ kN	
Clause 2.3.2.1 Table 3.1	Load-duration and moisture influence on strength: k_{mod} Solid timber: $\qquad\qquad\qquad\qquad$ Service Class 3 Permanent actions $\qquad\qquad\qquad k_{mod} = 0,5$ Variable actions (short-term) $\qquad k_{mod} = 0,7$	
Clause 3.1.3(2)	Use k_{mod} corresponding to action with the shortest duration.	$k_{mod} = 0,7$
Table NA.3 EC5 (Table 2.3) Clause 6.6	For solid timber – untreated and preservative treated System strength factor (k_{sys}) does not apply in this case.	$\gamma_M = 1,3$
BS EN 1990 Equation (6.10)	Eurocode: Basis of Structural Design $E_d = \sum_{j \geq 1} \gamma_{G,j} G_{k,j} + \gamma_{Q,1} Q_{k,1}$	
UK NA to EC Table NA.A1.2(B) BS EN 1990 Equation (6.10)	$\gamma_G = 1,35 \qquad \gamma_{Q,1} = 1,50$ Design load $= \gamma_G G_k + \gamma_{Q,1} Q_{k,1}$ Design load $F_{c,0,d} = (1,35 \times 0,42) + (1,5 \times 0,98) = 2,04$ kN	$F_{c,0,d} = 2,04$ kN
BS EN 1995-1-1 Clause 6.1.4 Equation (6.2)	Compression parallel to the grain: $\sigma_{c,0,d} \leq f_{c,0,d}$ Design compressive strength parallel to the grain: $f_{c,0,d} = (f_{c,0,k} \times k_{mod} \times k_{sys})/\gamma_M$ $\qquad = (16,0 \times 0,7)/1,3 = 8,62$ N/mm^2 Design compressive stress parallel to the grain: $\sigma_{c,0,d} = \dfrac{F_{c,0,d}}{A} = \dfrac{2,04 \times 10^3}{(100 \times 50)} = 0,41$ N/mm$^2 \ll 8,62$ N/mm^2	$f_{c,0,d} = 8,62$ N/mm^2 $\sigma_{c,0,d} = 0,41$ N/mm^2

Contract : Walkway Job Ref. No. : Example 6.3 Part of Structure : Typical Internal Column Calculation Sheet No. : 3 of 4	Calculated by : W.McK. Checked by : B.Z. Date :

References	Calculations	Output
Clause 6.3 Clause 6.3.2	Stability of members: Columns subjected to either compression or combined compression and bending.	
Equation (6.23)	$\dfrac{\sigma_{c,0,d}}{k_{c,y}f_{c,0,d}} + \dfrac{\sigma_{m,y,d}}{f_{m,y,d}} + k_m\dfrac{\sigma_{m,z,d}}{f_{m,z,d}} \leq 1{,}0$	
Equation (6.24)	$\dfrac{\sigma_{c,0,d}}{k_{c,z}f_{c,0,d}} + k_m\dfrac{\sigma_{m,y,d}}{f_{m,y,d}} + \dfrac{\sigma_{m,z,d}}{f_{m,z,d}} \leq 1{,}0$	
	Since $\sigma_{m,y,d}$ and $\sigma_{m,z,d} = 0$, Equations (6.23) and (6.24) reduce to: $\dfrac{\sigma_{c,0,d}}{k_{c,y}f_{c,0,d}} \leq 1{,}0$ and $\dfrac{\sigma_{c,0,d}}{k_{c,z}f_{c,0,d}} \leq 1{,}0$	
Clause 6.3.2(1)	Relative slenderness	
Equation (6.21)	$\lambda_{rel,y} = \dfrac{\lambda_y}{\pi}\sqrt{\dfrac{f_{c,0,d}}{E_{0,05}}}$ where $\lambda_y = \dfrac{L_{y,ef}}{i_y}$ $L_{y,ef} = (0{,}85 \times 3150) = 2678$ mm $i_y = \dfrac{h}{2\sqrt{3}} = \dfrac{100}{2\sqrt{3}} = 28{,}87$ $\lambda_y = \dfrac{2678}{28{,}87} = 92{,}76$ $\lambda_{rel,y} = \dfrac{92{,}76}{\pi}\sqrt{\dfrac{16{,}0}{4{,}7\times10^3}} = 1{,}72$	
Equation (6.22)	$\lambda_{rel,z} = \dfrac{\lambda_z}{\pi}\sqrt{\dfrac{f_{c,0,d}}{E_{0,05}}}$ where $\lambda_z = \dfrac{L_{z,ef}}{i_z}$ $L_{z,ef} \geq (1{,}0 \times 1550) = 1550$ mm $L_{z,ef} \geq (0{,}85 \times 1600) = 1360$ mm $\therefore L_{z,ef} = 1550$ mm $i_z = \dfrac{b}{2\sqrt{3}} = \dfrac{50}{2\sqrt{3}} = 14{,}43$ $\lambda_z = \dfrac{1550}{14{,}43} = 107{,}42$ $\lambda_{rel,z} = \dfrac{107{,}42}{\pi}\sqrt{\dfrac{16{,}0}{4{,}7\times10^3}} = 2{,}0$	
Equation (6.27)	$k_y = 0{,}5\left[1 + \beta_c\left(\lambda_{rel,y} - 0{,}3\right) + \lambda_{rel,y}^2\right]$	
Equation (6.29)	$\beta_c = 0{,}2$ for solid timber $k_y = 0{,}5\times\left[1 + 0{,}2\times\left(1{,}72 - 0{,}3\right) + 1{,}72^2\right] = 2{,}12$	

Contract : Walkway Job Ref. No. : Example 6.3	Calculated by : W.McK.
Part of Structure : Typical Internal Column	Checked by : B.Z.
Calculation Sheet No. : 4 of 4	Date :

References	Calculations	Output
Equation (6.25)	$k_{cy} = \dfrac{1}{k_y + \sqrt{k_y^2 - \lambda_{rel,y}^2}} = \dfrac{1}{2,12 + \sqrt{2,12^2 - 1,72^2}} = 0,30$	
Equation (6.28)	$k_z = 0,5\left[1 + \beta_c\left(\lambda_{rel,z} - 0,3\right) + \lambda_{rel,z}^2\right]$	
Equation (6.29)	$\beta_c = 0,2$ for solid timber	
	$k_z = 0,5 \times \left[1 + 0,2 \times (2,0 - 0,3) + 2,0^2\right] = 2,67$	
Equation (6.25)	$k_{cz} = \dfrac{1}{k_z + \sqrt{k_z^2 - \lambda_{rel,z}^2}} = \dfrac{1}{2,67 + \sqrt{2,67^2 - 2,0^2}} = 0,23$	
Equation (6.23)	$\dfrac{\sigma_{c,0,d}}{k_{c,y} f_{c,0,d}} = \dfrac{0,41}{0,30 \times 8,62} = 0,16 \le 1,0$	The 100 mm x 50 mm section is adequate with respect to compressive stress.
	and	
Equation (6.24)	$\dfrac{\sigma_{c,0,d}}{k_{c,z} f_{c,0,d}} = \dfrac{0,41}{0,23 \times 8,62} = 0,20 \le 1,0$	

Check the bearing stress in the header at the top of the post.

Header: 50 mm x 100 mm

100 mm

Clause 6.1.5 Equation (6.3)	Compression perpendicular to the grain: $\sigma_{c,90,d} \le k_{c,90}\, f_{c,90,d}$	
	Design compressive strength perpendicular to the grain: $f_{c,0,d} = (f_{c,90,k} \times k_{mod} \times k_{sys})/\gamma_M$ $\qquad = (2,0 \times 0,7)/1,3 = 1,08 \text{ N/mm}^2$	$f_{c,90,d} = 1,08 \text{ N/mm}^2$
Clause 6.1.5.(3)	Beam member resting on internal support	
Equation (6.5)	$k_{c,90} = \left(2,38 - \dfrac{l}{250}\right)\left(1 + \dfrac{h}{6l}\right)$	
	$\qquad = \left(2,38 - \dfrac{50}{250}\right)\left(1 + \dfrac{50}{6 \times 50}\right) = 2,74$	
Equation (6.3)	$k_{c,90}\, f_{c,90,d} = (2,54 \times 1,08) = 2,74 \text{ N/mm}^2$	
	Design compressive stress perpendicular to the grain: $\sigma_{c,0,d} = \dfrac{F_{c,0,d}}{A} = \dfrac{2,04 \times 10^3}{(100 \times 50)} = 0,41 \text{ N/mm}^2 < 2,74 \text{ N/mm}^2$	Header is adequate with respect to bearing.

6.4 Built-up Columns (Annex C)

Built-up columns are fabricated by joining together several pieces of timber using mechanical fasteners or glue. The component parts are not connected continuously throughout their length but at discrete locations, either in contact or separated by using packs or gussets, using e.g. nails, bolts, glue or proprietary connectors as shown in Figure 6.5.

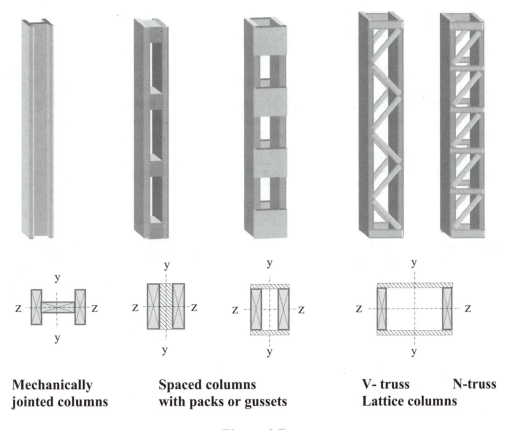

Mechanically jointed columns

Spaced columns with packs or gussets

V- truss N-truss Lattice columns

Figure 6.5

Guidance on the design requirements of built-up columns is given in Clause C.1.2. In all cases the load-carrying capacity should satisfy the requirements of, i.e.

(i)   for column deflection in the y-direction (i.e. buckling about the z-z axis) as indicated in Figures C.1 and C.3 (of EC5) the load-carrying capacity should be taken as the sum of the load-carrying capacities of the individual members,

(ii)   for column deflection in the z-direction (i.e. buckling about the y-y axis) as indicated in Figures C.1 and C.3 (of EC5) the following expression should be satisfied:

$\sigma_{c,0,d} \leq k_c f_{c,0,d}$ (Equation (C.1) in EC5)

where:

$$\sigma_{c,0,d} = \frac{F_{c,0,d}}{A_{tot}}$$

A_{tot} is the total cross-sectional area,

k_c is determined in accordance with 6.3.2 but with an effective slenderness ratio λ_{ef} determined in accordance with Sections C.2 to C.4 of Annex C.

The calculation of k_c requires a value for slenderness. For each of the cases indicated in Figure 6.5 an effective slenderness can be evaluated as indicated in Annex C and below in Table 6.2.

Mechanically jointed columns - Section C.2	Spaced columns with packs or gussets – Section C.3	Lattice columns with glued or nailed joints – Section C.4
$\lambda_{ef} = l\sqrt{\dfrac{A_{tot}}{I_{ef}}}$	$\lambda_{ef} = \sqrt{\lambda^2 + \eta\dfrac{n}{2}\lambda_1^2}$	$\lambda_{ef} = \max\begin{cases} \lambda_{tot}\sqrt{1+\mu} \\ 1{,}05\lambda_{tot} \end{cases}$
The variables in the above equations are defined in the relevant Sections of Annex C.		

Table 6.2

The fasteners transfer shear forces and their value should be determined in accordance with Annex B, Equation (B.10). The value of V in this equation should be determined using Equation (C.5) in Annex, i.e.

$$V_d = \begin{cases} \dfrac{F_{c,d}}{120k_c} & \text{for } \lambda_{ef} < 30 \\[2ex] \dfrac{F_{c,d}\lambda_{ef}}{3600k_c} & \text{for } 30 \leq \lambda_{ef} < 60 \\[2ex] \dfrac{F_{c,d}}{60k_c} & \text{for } 60 \leq \lambda_{ef} \end{cases} \qquad \text{(Equation (C.5) in EC5)}$$

In the case of spaced columns the shear forces on the gussets or packs should be calculated from Equation (C.13) in EC5:

$$T_d = \frac{V_d l_1}{a_1} \qquad \text{(Equation (C.5) in EC5)}$$

where:
l_1 is the centre-to-centre spacing between the packs or gussets,
a_1 is the centre to centre spacing of the individual members.

6.4.1 Example 6.4: Spaced Glulam Column

An internal spaced column in an exhibition centre is required to support permanent and medium-term loads of 40 kN and 70 kN respectively. The column comprises two shafts each of four laminations with glued packing pieces as shown in Figure 6.6. Check the suitability of the proposed section indicated.

Data:

Glulam strength class		GL 24h
Service Class		2
Length of column	l	4,2 m
Clear space between the shafts	a	120 mm
Centre-to-centre of shafts	a_1	185 mm
Centre-to-centre of packs	l_1	760 mm
Length of packs	l_2	250 mm

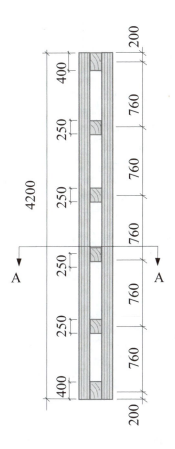
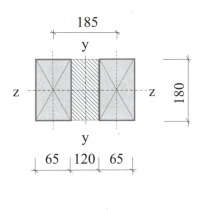

Figure 6.6

6.4.1.1 Solution to Example 6.4

References	Calculations	Output
Contract : Conf Centre Job Ref. No. : Example 6.4 **Part of Structure : Spaced Glulam Column** **Calculation Sheet No. : 1 of 4**		**Calculated by : W.McK.** **Checked by : B.Z.** **Date :**

References	Calculations	Output
BS EN 1990 BS EN 1991 BS EN 338:2003 BS EN 1995-1-1: 2004 National Annex	Eurocode: Basis of Structural Design Eurocode 1: Actions on Structures Structural Timber - Strength Classes Eurocode 5: Design of Timber Structures Part 1.1: General – Common Rules and Rules for Buildings. UK National Annex to BS EN 1995-1-1	
BS EN 1194 1999	Glued Laminated Timber – Strength Classes and Characteristic Values	
BS EN 1194 Table 1	Characteristic values for GL 24h glulam **Strength Properties:** Compression parallel to grain $f_{c,0,g,k} = 24{,}0$ N/mm^2 **Stiffness Properties:** 5% modulus of elasticity parallel to the grain $E_{0,g,05} = 9{,}4$ kN/mm^2	
Clause 2.3.2.1 Table 3.1	Load-duration and moisture influence on strength: k_{mod} Glued laminated timber: Service Class 2 Permanent actions $k_{mod} = 0{,}6$ Variable actions (medium-term) $k_{mod} = 0{,}8$	
Clause 3.1.3(2)	Use k_{mod} corresponding to action with the shortest duration.	**$k_{mod} = 0{,}8$**
Table NA.3 EC5 (Table 2.3) Clause 6.6	For glued laminated timber System strength factor (k_{sys}) does not apply in this case.	**$\gamma_M = 1{,}25$**
BS EN 1990 Equation (6.10) UK NA to Table NA.A1.2(B)	Eurocode: Basis of Structural Design $E_d = \sum_{j\geq1} \gamma_{G,j} G_{k,j} + \gamma_{Q,1} Q_{k,1}$ $\gamma_G = 1{,}35$ $\gamma_{Q,1} = 1{,}50$	
BS EN 1990 Equation (6.10)	Design load $= \gamma_G G_k + \gamma_{Q,1} Q_{k,1}$ Design load $F_{c,0,d} = (1{,}35 \times 40{,}0) + (1{,}5 \times 70) = 159{,}0$ kN	**$F_{c,0,d} = 159{,}0$ kN**
Equation (6.2)	$\sigma_{c,0,d} \leq f_{c,0,d}$ Design compressive strength parallel to the grain: $f_{c,0,d} = (f_{c,0,k} \times k_{mod})/\gamma_M$ $= (24{,}0 \times 0{,}8)/1{,}25 = 15{,}36$ N/mm^2	**$f_{c,0,d} = 15{,}36$ N/mm^2**

Contract : Conf Centre Job Ref. No. : Example 6.4	Calculated by : W.McK.
Part of Structure : Spaced Glulam Column	Checked by : B.Z.
Calculation Sheet No. : 2 of 4	Date :

References	Calculations	Output
	Design compressive stress parallel to the grain:	
	$$\sigma_{c,0,d} = \frac{F_{c,0,d}}{A} = \frac{159,0 \times 10^3}{2 \times (180 \times 65)} = 6,79 \text{ N/mm}^2 < 15,36 \text{ N/mm}^2$$	$\sigma_{c,0,d} = 6,79$ N/mm^2
Annex C	The assumptions given in Clause C.3.1(2) are satisfied, i.e.	
Clause C.3.1(2)	- the cross-section is composed of two, three or four identical shafts;	satisfied
	- the cross-sections are symmetrical about both axes;	satisfied
	- the number of restrained bays is at least three, i.e. the shafts are at least connected at the ends and at the third points;	satisfied
	- the free distance between the shafts is not greater than three times the shaft thickness h for columns with packs, i.e. $a = 120$ mm $< (3 \times 65) = 195$ mm;	satisfied
	- the pack length $l_2 \geq 1,5a$, 250 mm $\geq (1,5 \times 120) = 180$ mm;	satisfied
	- the columns are subjected to concentric axial loads;	satisfied
	- the joints and packs should be designed in accordance with Clause C.2.2.	
Annex C Clause C 3.2.(1)	Built-up columns: For deflection in the y direction (i.e. buckling about the z-z axis) the capacity is given by the sum of the capacities of the individual members.	
Clause 6.3.2	Effective length for an individual member $l_e = 4200$ mm $$i_z = \sqrt{\frac{I_z}{A}} = \sqrt{\frac{65 \times 180^3}{12 \times 65 \times 180}} = 51,96 \text{ mm}$$ $$\lambda_z = \frac{l_e}{i_z} = \frac{4200}{51,96} = 80,83$$ $$\lambda_{rel,z} = \frac{\lambda_z}{\pi}\sqrt{\frac{f_{c,0,k}}{E_{0,05}}} = \frac{80,83}{\pi}\sqrt{\frac{24,0}{9400,0}} = 1,3$$	
Equation (6.28)	$k_z = 0,5(1 + \beta_c(\lambda_{rel,z} - 0,3) + \lambda_{rel,z}^2)$	
Equation (6.29)	k_c is related to glued laminated timber use $\beta_c = 0,1$ $$k_z = 0,5 \times \left[1 + 0,1 \times (1,3 - 0,3) + 1,3^2\right] = 1,4$$	

Contract : Conf Centre Job Ref. No. : Example 6.4	Calculated by : W.McK.
Part of Structure : Spaced Glulam Column	Checked by : B.Z.
Calculation Sheet No. : 3 of 4	Date :

References	Calculations	Output
Equation (6.26)	$k_{cz} = \dfrac{1}{k_z + \sqrt{k_z^2 - \lambda_{rel,z}^2}} = \dfrac{1}{1,4 + \sqrt{1,4^2 - 1,3^2}} = 0,52$ $k_c f_{c,0,d} = (0,52 \times 15,36) = 7,99 \text{ N/mm}^2 > 6,79 \text{ N/mm}^2$	**Column is adequate with respect to z-z axis.**
Clause C 3.2.(2)	For deflection in the z direction (i.e. buckling about the y-y axis) the capacity should calculated in accordance with Clause C.1.2 and Equation (C.1).	
Equation (C.1)	$\sigma_{c,0,d} \leq k_c f_{c,0,d}$ where $\sigma_{c,0,d} = \dfrac{F_{c,0,d}}{A_{tot}}$ and k_c is determined in accordance with Clause 6.3.2 using $\lambda_y = \lambda_{ef}$ as defined in Clause C.3.2(2) and Equation (C.10) to (C.12).	
Equation (C.10)	$\lambda_{ef} = \sqrt{\lambda^2 + \eta \dfrac{n}{2} \lambda_1^2}$	
Equation (C.11)	$\lambda = l\sqrt{A_{tot}/I_{tot}}$ $l = 4200$ mm $A_{tot} = (2 \times 180 \times 65) = 23,4 \times 10^3 \text{ mm}^2$ $I_{tot} = \dfrac{b\left[(2h+a)^3 - a^3\right]}{12}$ where $b = 180$ mm, $a = 120$ mm, $h = 65$ mm $I_{tot} = \dfrac{180 \times \left[(2 \times 65 + 120)^3 - 120^3\right]}{12} = 208,46 \times 10^6 \text{ mm}^2$ $\lambda = 4200 \times \sqrt{23,4 \times 10^3 / 208,46 \times 10^6} = 44,5$	
Equation (C.12)	$\lambda_1 = \sqrt{12}\dfrac{l_1}{h} \geq 30$ $\lambda_1 = \left(\sqrt{12} \times \dfrac{760}{65}\right) = 40,5 \geq 30$ The number of shafts $n = 2$	
Table C.1	For glued packs with permanent/long-term and medium-term/short-term loads, $\eta = 1$	
Equation (C.10)	$\lambda_{ef} = \sqrt{44,5^2 + \left(1,0 \times \dfrac{2}{2} \times 40,5^2\right)} = 60,17$	

References	Calculations	Output
Contract : Conf Centre Job Ref. No. : Example 6.4 **Part of Structure : Spaced Glulam Column** **Calculation Sheet No. : 4 of 4**		**Calculated by : W.McK.** **Checked by : B.Z.** **Date :**

References	Calculations	Output
Clause 6.3.2		
Equation (6.22)	$\lambda_{rel,y} = \dfrac{\lambda_z}{\pi}\sqrt{\dfrac{f_{c,0,k}}{E_{0,05}}} = \dfrac{60,17}{\pi}\sqrt{\dfrac{24,0}{9400,0}} = 0,97$	
Equation (6.28)	$k_y = 0,5(1 + \beta_c(\lambda_{rel,y} - 0,3) + \lambda_{rel,y}^2)$	
Equation (6.29)	k_c is related to glued laminated timber use $\beta_c = 0,1$ $k_y = 0,5 \times \left[1 + 0,1 \times (0,97 - 0,3) + 0,97^2\right] = 1,0$	
Equation (6.26)	$k_{cy} = \dfrac{1}{k_y + \sqrt{k_y^2 - \lambda_{rel,y}^2}} = \dfrac{1}{1,0 + \sqrt{1,0^2 - 0,97^2}} = 0,80$ $k_c f_{c,0,d} = (0,8 \times 15,4) = 12,32 \text{ N/mm}^2 > 6,79 \text{ N/mm}^2$	**Column is adequate with respect to y-y axis.**

6.5 Parallel-chord Lattice Beams

For flat and mono-pitch roofs of 6 m to 10 m span, ply-web or glulam beams are popular but there are cases, particularly larger spans, in which parallel lattice beams would be the choice either for architectural, functional or manufacturing reasons. A lattice beam is ideally suited to accommodate large diameter service pipes/ducts within the depth of the beam. One possible disadvantage is that the large forces which are induced in the web members necessitate heavier joints at the nodes. The section sizes may need to be increased to accommodate the connections. Economically designed lattice beams are deeper than either glulam or ply-web beams having an effective depth (H) of approximately 1/10 to 1/8 of the span.

A typical construction consists of a top and bottom chord with internal members such as the Pratt (or N) Truss or Warren Truss shown in Figures 6.7 and 6.8.

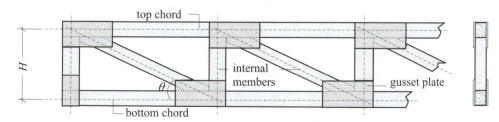

Pratt Girder

Figure 6.7

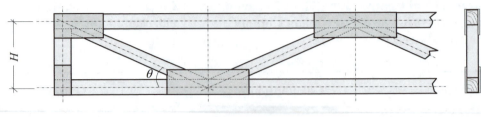

Warren Truss

Figure 6.8

There are numerous other possible member configurations which can be used. It is preferable to keep the number of joints to a minimum to reduce fabrication costs and the deflection due to joint slip, except in the case of glued girders. Generally the inclination of the sloping web members is greater than or equal to 30°.

The internal members are frequently the same size throughout the span and the same width as the main chords. This arrangement simplifies the manufacturing of the girder. They usually have a low slenderness ratio and compression members are unlikely to have any significant reduction in strength. A consequence of this is that the internal configuration is more dependent on manufacturing requirements or appearance than design strength.

Trusses in which all members are single and in the same plane (see Figures 6.7 and 6.8) are known as mono-chord trusses. The gusset plates can be manufactured from plywood (glued or nailed), thin steel plates (nailed, screwed or bolted), single-sided tooth plates or shear-plate connectors or pre-punched metal-plate fasteners, and normally occur on each side of the truss. All the members of the lattice girders are usually surfaced on all sides irrespective of the gussets used.

In double chord trusses the chords and/or some internal members are fabricated using two sections for individual members, as indicated in Figure 6.9.

Figure 6.9

The chords and internal members are normally connected using split-rings or tooth-plates. The use of single connector units rather than multiple dowel-type fasteners more closely reflects the assumption of pinned joints in the analysis to determine the member forces. The deformation of such trusses is influenced by joint slip and will be more significant than will occur when using glued joints.

The design of section sizes for truss members may need to be increased to accommodate the particular type of fastener used when detailed connection design is carried out.

Double compression members in triangulated frames can be designed with or without spacers, depending on the level of lateral restraint provided by additional structural members such as purlins. If spacers are not used each component part is considered to support a proportion of the load and designed accordingly as a strut; when spacers are used the combined member can be designed in the same manner as a spaced column.

6.5.1 Example 6.5: Lattice Girder

A series of Warren trusses as indicated in Figure 6.9 are spaced at 1.2 m centres and support the roof of a small workshop. Check the suitability of the proposed section for the diagonal members of the truss. (**Note:** the top chord is subject to secondary bending effects which are dealt with in Chapter 7.)

Data:
Timber: European whitewood   Grade SS
Gusset plates: Finnish, birch-faced plywood glued/nailed to each side of single members
Characteristic permanent load $1,2 \ kN/m^2$
Characteristic variable load $0,6 \ kN/m^2$

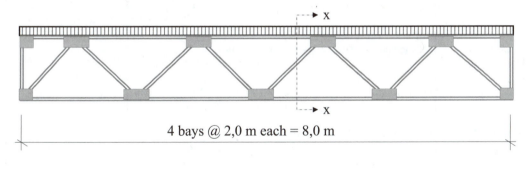

4 bays @ 2,0 m each = 8,0 m

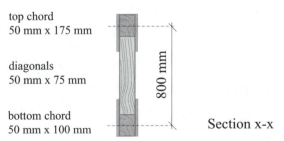

top chord
50 mm x 175 mm

diagonals
50 mm x 75 mm

800 mm

bottom chord
50 mm x 100 mm

Section x-x

Figure 6.9

312 Design of Structural Timber to EC5

6.5.1.1 Solution to Example 6.5

References	Calculations	Output
Contract : Workshop Job Ref. No. : Example 6.5 **Part of Structure : Lattice Girder Diagonals** **Calculation Sheet No. : 1 of 5**		**Calculated by : W.McK.** **Checked by : B.Z.** **Date :**

References	Calculations	Output
BS EN 1990 BS EN 1991 BS EN 338:2003 BS EN 1995-1-1: 2004	Eurocode: Basis of Structural Design Eurocode 1: Actions on Structures Structural Timber - Strength Classes Eurocode 5: Design of Timber Structures Part 1.1: General – Common Rules and Rules for Buildings.	
National Annex	UK National Annex to BS EN 1995-1-1	
BS EN 1912 Table 1	Assignment of timber type and grade to strength class: European whitewood, Grade SS - Strength Class C24	
BS EN 338:2003 Clause 5.0	Structural Timber - Strength Classes Characteristic values for C24 Timber	
Table 1	**Strength Properties:** Tension parallel to grain $f_{t,0,k}$ = 14,0 N/mm^2 Compression parallel to grain $f_{c,0,k}$ = 21,0 N/mm^2 **Stiffness Properties:** 5% modulus of elasticity parallel to the grain $E_{0,05}$ = 7,4 kN/mm^2 Permanent load = (1,2 × 1,2) = 1,44 kN/m Variable load (short-term) = (0,6 × 1,2) = 0,72 kN/m	
Clause 2.3.2.1 Table 3.1	Load-duration and moisture influence on strength: k_{mod} Solid timber: Service Class 2 Permanent actions k_{mod} = 0,6 Variable actions (medium-term) k_{mod} = 0,8	
Clause 3.1.3(2)	Use k_{mod} corresponding to action with the shortest duration.	k_{mod} = **0,8**
Table NA.3 EC5 (Table 2.3)	For solid timber – untreated and preservative treated	γ_M = **1,3**
Clause 6.6	System strength factor (k_{sys}) does not apply in this case.	
BS EN 1990 Equation (6.10)	Eurocode: Basis of Structural Design $E_d = \sum_{j\geq1}\gamma_{G,j}G_{k,j} + \gamma_{Q,1}Q_{k,1}$	
UK Annex to EC BS EN 1990 Table NA.A1.2(B) BS EN 1990 Equation (6.10)	γ_G = 1,35 $\gamma_{Q,1}$ = 1,50 Design load = $\gamma_G G_k + \gamma_{Q,1}Q_{k,1}$ Design load F_d = (1,35 × 1,44) + (1,5 × 0,72) = 3,02 kN	F_d = **3,02 kN/m**

Contract : Workshop Job Ref. No. : Example 6.5	Calculated by : W.McK.
Part of Structure : Lattice Girder Diagonals	Checked by : B.Z.
Calculation Sheet No. : 2 of 5	Date :

References	Calculations	Output
See Chapter 4 Section 4.7.6	Consider the load to be distributed to each internal panel point. Load/internal panel point = $(3,02 \times 2,0) = 6,04$ kN Total load $= (6,04 \times 4,0) = 24,16$ kN Support reactions $= (24,16/2,0) = 12,08$ kN	
	Consider Section x-x Length of the diagonal $= \sqrt{(0,8^2 + 1,0^2)}$ $= 1,281$ m $\sin\theta = \dfrac{0,8}{1,281} = 0,625$ Resolve the vertical forces $\Sigma F_y = 12,08 - 1,51 + F_1\sin\theta = 0$ $F_1 = -\dfrac{10,57}{0,625} = -16,91$ kN Compression	$F_{c,0,d} = \textbf{16,91 kN}$

References	Calculations	Output
	Contract : Workshop Job Ref. No. : Example 6.5 **Part of Structure : Lattice Girder Diagonals** **Calculation Sheet No. : 3 of 5**	**Calculated by : W.McK.** **Checked by : B.Z.** **Date :**

References	Calculations	Output
	Consider Section y-y	
	1,51 kN 4,53 kN	
	$\sin\theta = \dfrac{0,8}{1,281} = 0,625$	
	Resolve the vertical forces	
	$\Sigma F_y = 12,08 - 1,51 - 4,53 - F_2\sin\theta = 0$	
	$F_2 = +\dfrac{6,04}{0,625} = +9,66 \text{ kN}$ Tension	$F_{t,0,d} = 9,66$ kN
	Compression Diagonal:	
Clause 9.2.1(3)	Assume the effective length of the strut is equal to the actual distance between the assumed intersection points.	
BS EN 1995-1-1		
Clause 6.1.4	Compression parallel to the grain:	
Equation (6.2)	$\sigma_{c,0,d} \leq f_{c,0,d}$	
	Design compressive strength parallel to the grain:	
	$f_{c,0,d} = (f_{c,0,k} \times k_{mod} \times k_{sys})/\gamma_M$	
	$\quad\quad = (21,0 \times 0,8)/1,3 = 12,92 \text{ N/mm}^2$	$f_{c,0,d} = 12,92$ N/mm²
	Design compressive stress parallel to the grain:	
	$\sigma_{c,0,d} = \dfrac{F_{c,0,d}}{A} = \dfrac{16,91 \times 10^3}{(50 \times 75)} = 4,51 \text{ N/mm}^2 \ll 12,92 \text{ N/mm}^2$	$\sigma_{c,0,d} = 4,51$ N/mm²
Clause 6.3	Stability of members:	
Clause 6.3.2	Columns subjected to either compression or combined compression and bending.	
Equation (6.23)	$\dfrac{\sigma_{c,0,d}}{k_{c,y}f_{c,0,d}} + \dfrac{\sigma_{m,y,d}}{f_{m,y,d}} + k_m\dfrac{\sigma_{m,z,d}}{f_{m,z,d}} \leq 1,0$	
Equation (6.24)	$\dfrac{\sigma_{c,0,d}}{k_{c,z}f_{c,0,d}} + k_m\dfrac{\sigma_{m,y,d}}{f_{m,y,d}} + \dfrac{\sigma_{m,z,d}}{f_{m,z,d}} \leq 1,0$	
	Since $\sigma_{m,y,d}$ and $\sigma_{m,z,d} = 0$, Equations (6.23) and (6.24) reduce to:	

References	Calculations	Output
	Contract : Workshop Job Ref. No. : Example 6.5 **Part of Structure : Lattice Girder Diagonals** **Calculation Sheet No. : 4 of 5**	**Calculated by : W.McK.** **Checked by : B.Z.** **Date :**

References	Calculations	Output
	$$\dfrac{\sigma_{c,0,d}}{k_{c,y}f_{c,0,d}} \le 1,0 \quad \text{and} \quad \dfrac{\sigma_{c,0,d}}{k_{c,z}f_{c,0,d}} \le 1,0$$	
Clause 6.3.2(1)	Relative slenderness	
Equation (6.21)	$\lambda_{rel,y} = \dfrac{\lambda_y}{\pi}\sqrt{\dfrac{f_{c,0,d}}{E_{0,05}}}$ where $\lambda_y = \dfrac{L_{y,ef}}{i_y}$	
	$L_{y,ef} = 1281$ mm	
	$i_y = \dfrac{h}{2\sqrt{3}} = \dfrac{75}{2\sqrt{3}} = 21,65 \qquad \lambda_y = \dfrac{1281}{21,65} = 59,17$	
	$\lambda_{rel,y} = \dfrac{59,17}{\pi}\sqrt{\dfrac{12,92}{7,4\times10^3}} = 0,79$	
Equation (6.22)	$\lambda_{rel,z} = \dfrac{\lambda_z}{\pi}\sqrt{\dfrac{f_{c,0,d}}{E_{0,05}}}$ where $\lambda_z = \dfrac{L_{z,ef}}{i_z}$	
	$L_{z,ef} = 1281$ mm $i_z = \dfrac{b}{2\sqrt{3}} = \dfrac{50}{2\sqrt{3}} = 14,43 \qquad \lambda_z = \dfrac{1281}{14,43} = 88,77$	
	$\lambda_{rel,z} = \dfrac{88,77}{\pi}\sqrt{\dfrac{12,92}{7,4\times10^3}} = 1,18$	
Equation (6.27)	$k_y = 0,5\left[1 + \beta_c\left(\lambda_{rel,y} - 0,3\right) + \lambda_{rel,y}^2\right]$	
Equation (6.29)	$\beta_c = 0,2$ for solid timber	
	$k_y = 0,5\times\left[1 + 0,2\times\left(0,79 - 0,3\right) + 0,79^2\right] = 0,86$	
Equation (6.25)	$k_{cy} = \dfrac{1}{k_y + \sqrt{k_y^2 - \lambda_{rel,y}^2}} = \dfrac{1}{0,86 + \sqrt{0,86^2 - 0,79^2}} = 0,83$	
Equation (6.28)	$k_z = 0,5\left[1 + \beta_c\left(\lambda_{rel,z} - 0,3\right) + \lambda_{rel,z}^2\right]$	
Equation (6.29)	$\beta_c = 0,2$ for solid timber	

References	Calculations	Output
	Contract : Workshop Job Ref. No. : Example 6.5 **Part of Structure : Lattice Girder Diagonals** **Calculation Sheet No. : 5 of 5** **Calculated by : W.McK.** **Checked by : B.Z.** **Date :**	

References	Calculations	Output
	$k_z = 0,5 \times \left[1 + 0,2 \times (1,18 - 0,3) + 1,18^2 \right] = 1,28$	
Equation (6.26)	$k_{cz} = \dfrac{1}{k_z + \sqrt{k_z^2 - \lambda_{rel,z}^2}} = \dfrac{1}{1,28 + \sqrt{1,28^2 - 1,18^2}} = 0,56$	
Equation (6.23)	$\dfrac{\sigma_{c,0,d}}{k_{c,y} f_{c,0,d}} = \dfrac{4,51}{0,83 \times 12,92} = 0,42 < 1,0$ and	The 50 mm x 75 mm section is adequate with respect to compressive stress and stability.
Equation (6.24)	$\dfrac{\sigma_{c,0,d}}{k_{c,z} f_{c,0,d}} = \dfrac{4,51}{0,56 \times 12,92} = 0,62 < 1,0$	
Clause 3.2.(3) Equation (3.1)	**Tension Diagonal:** Member size: For timber in bending with $\rho_k \le 700$ kg/m^3 where the depth is less than 150 mm $f_{m,k}$ can be multiplied by k_h where: $k_h = \min\left\{\begin{array}{l}\left(\dfrac{150}{h}\right)^{0,2} \\ 1,3\end{array}\right. = \min\left\{\begin{array}{l}\left(\dfrac{150}{75}\right)^{0,2} \\ 1,3\end{array}\right. = \min\left\{\begin{array}{l}1,15 \\ 1,3\end{array}\right.$	$k_h = 1,15$
BS EN 1995-1-1 Clause 6.1.2 Equation (6.1)	Tension parallel to the grain: $\sigma_{t,0,d} \le f_{t,0,d}$ Design tensile strength parallel to the grain: $f_{t,0,d} = (f_{t,0,k} \times k_{mod} \times k_h \times k_{sys})/\gamma_M$ $= (14,0 \times 0,8 \times 1,15)/1,3 = 9,91$ N/mm^2 Design tensile stress parallel to the grain: $\sigma_{t,0,d} = \dfrac{F_{t,0,d}}{A} = \dfrac{9,66 \times 10^3}{(50 \times 75)} = 2,58$ N/mm$^2 \ll 9,91$ N/mm^2 There is considerable reserve of strength in the tie member	$f_{t,0,d} = 9,91$ N/mm^2 The 50 mm x 75 mm section is adequate.

6.5.1.2 Compression Perpendicular to the Grain (Clause 6.1.5)

Compression perpendicular to the grain is normally considered when checking the design bearing strength at supports and point of applied load, (see Chapter 5, Section 5.3.3).

6.5.1.3 Compression at an Angle to the Grain (Clause 6.1.5)

The compression stress in joints is frequently inclined at an angle to the grain direction in one or more members, e.g. in the case of joints as shown in Figure 6.10.

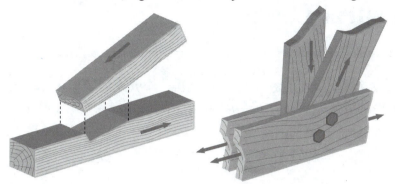

Figure 6.10

Experimental evidence (STEP 1 (58)) has demonstrated that compression strengths are very sensitive to change in the angle of inclination (α) when it is small as shown in Figure 6.11, i.e. the slope of the strength/α curve is relatively high at low angles and is virtually zero between 75° and 90°.

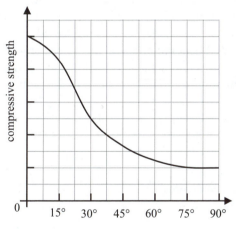

α – angle between load and grain directions

Figure 6.11

An approximation to the behaviour at failure of timber loaded at an angle to the grain is given in Equation (6.16) of EC5 and the design compressive stress should satisfy the following expression:

$$\sigma_{c,\alpha,d} \leq \frac{f_{c,0,d}}{\dfrac{f_{c,0,d}}{k_{c,90} f_{c,90,d}} \sin^2\alpha + \cos^2\alpha} \qquad \text{(Equation (6.26) in EC5)}$$

where:

$\sigma_{c,\alpha,d}$ is the compressive stress at an angle to the grain,

$k_{c,90}$ is a factor given in Clause 6.1.5 taking into account the effect of any stresses perpendicular to the grain.

7. Members Subject to Combined Axial and Flexural Loads

Objective: *to illustrate the design of members subject to combined bending and axial loads including lattice girder elements and portal frame members.*

7.1 Introduction

Many structural elements such as beams and truss members are subjected to a single dominant effect, i.e. applied bending or axial stresses. Secondary effects which also occur are often insignificant and can be neglected. There are, however, numerous elements in which the combined effects of bending and axial stresses must be considered, e.g. rigid-jointed frames such as portals, chords in lattice girders with applied loading between the node points, and columns with eccentrically applied loading. The behaviour of such members is dependent on the interaction characteristics of the individual components of load. Generally, members resisting combined bending and tension are easier to design than those resisting combined bending and compression. This is due to the susceptibility of the latter to associated buckling effects. Interaction equations are given for both cases in BS EN 1995-1-1::2004.

The relevant modification factors which apply to members subject to combined loading are those which apply to the individual types, i.e. axially loaded and flexural members, and are summarized in Table 7.1.

7.2 Combined Bending and Axial Tension (Clause 6.2.3)

The interaction equations for members subject to combined bi-axial bending and axial tension are given in Clause 6.2.3 as:

$$\frac{\sigma_{t,0,d}}{f_{t,0,d}} + \frac{\sigma_{m,y,d}}{f_{m,y,d}} + k_m \frac{\sigma_{m,z,d}}{f_{m,z,d}} \leq 1,0 \qquad \text{(Equation (6.17) in EC5)}$$

and

$$\frac{\sigma_{t,0,d}}{f_{t,0,d}} + k_m \frac{\sigma_{m,y,d}}{f_{m,y,d}} + \frac{\sigma_{m,z,d}}{f_{m,z,d}} \leq 1,0 \qquad \text{(Equation (6.18) in EC5)}$$

where:
$\sigma_{t,0,d}$ is the design tensile stress,
$f_{t,0,d}$ is the design tensile strength,
$\sigma_{m,y,d}$ is the design bending stress about the major y-y axis,
$f_{m,y,d}$ is the design bending strength about the major y-y axis,

$\sigma_{m,z,d}$ is the design bending stress about the minor z-z axis,
$f_{m,z,d}$ is the design bending strength about the minor z-z axis,

The values of $f_{t,0,d}$, $f_{m,y,d}$ and $f_{m,z,d}$ are evaluated using the modification factors where appropriate as described in Chapters 5 and 6. The value of k_m is given in Clause 6.1.6(2) and makes an allowance for the re-distribution of stresses and the effect of inhomogeneities of the material in a cross-section. The values adopted are as indicated in Table 7.2.

Factors	Application	Clause Number	Value/Location
k_{mod}	Relates to all strength properties for service and load-duration classes.	3.1.3	Table 3.1
k_h	Relates to the bending strength/tension strength parallel to the grain and to member depth/width.	3.2(3) 3.3(3) 3.4(3)	Equation (3.1) Equation (3.2) Equation (3.3)
$k_{c,90}$	Relates to compression strength perpendicular to the grain.	6.1.5	Equations (6.4), (6.5), (6.6), (6.10)
k_y	Relates to stability and interaction equations - y axis	6.3.2	Equation (6.27)
k_z	Relates to stability and interaction equations - z axis	6.3.2	Equation (6.28)
$k_{c,y}$	Relates to stability and interaction equations - y axis	6.3.2	Equation (6.25)
$k_{c,z}$	Relates to stability and interaction equations - z axis	6.3.2	Equation (6.26)
k_m	Relates to inhomogeneities in material.	6.1.6.2	0,7 – rectangular 1,0 – other sections
β_c	Relates to stability and interaction equations.	6.3.2	Equation (6.29)
k_{sys}	Relates to lateral load distribution.	6.6	Generally 1,1

Table 7.1   Modification Factors

Material	Cross-section shape	k_m
Solid timber, Glued laminated timber, Laminated veneer lumber	Rectangular	0,7
	Other	1,0
All other wood-based structural products	All	1,0

Table 7.2

7.2.1 Example 7.1: Pitched Roof Truss – Ceiling Tie

An existing pitched roof truss is required to support an additional load on the main tie, as indicated in Figure 7.1. Assuming all loading given to be medium-term, check the suitability of the section used for the main tie to resist the combined axial tension and bending moment.

Design data:

Timber Strength Class	C14
Timber Service Class	2
Main tie cross-section	50 mm x 125 mm
Design axial load in tie before additional load	6,0 kN
Design vertical load P applied to mid-span of tie	2,0 kN

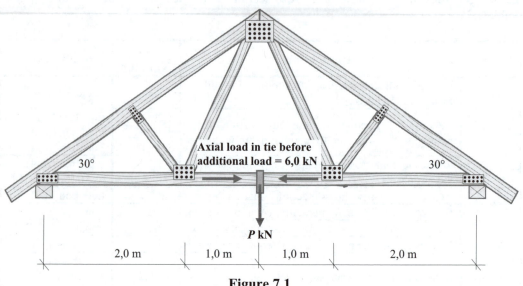

Figure 7.1

7.2.1.1 Solution to Example 7.1

Contract : Roof Truss Job Ref. No. : Example 7.1 Part of Structure : Main tie Calculation Sheet No. : 1 of 5	Calculations by : W.McK. Checked by : B.Z. Date :

References	Calculations	Output
BS EN 338:2003 BS EN 1995-1-1: 2004 National Annex BS EN 338:2003 Clause 5.0 Table 1	Structural Timber - Strength Classes Eurocode 5: Design of Timber Structures Part 1.1: General - Common Rules and Rules for Buildings. UK National Annex to BS EN 1995-1-1: Structural Timber - Strength Classes Characteristic values for C14 Timber **Strength Properties:** Bending strength parallel to the grain $f_{m,k} = 14,0 \text{ N/mm}^2$ Tension parallel to grain $f_{t,0,k} = 8,0 \text{ N/mm}^2$ **Stiffness Properties:** 5% Modulus of elasticity parallel to the grain $E_{0,05} = 4,7 \text{ kN/m}^3$	

Contract : Roof Truss Job Ref. No. : Example 7.1	Calculations by : W.McK.
Part of Structure : Main tie	Checked by : B.Z.
Calculation Sheet No. : 2 of 5	Date :

References	Calculations	Output
	Mean density $\qquad\qquad\qquad \rho_{mean} = 350{,}0 \text{ kg/m}^3$	
Clause 2.3.2.1	Load-duration and moisture influence on strength: k_{mod}	
Table 3.1	Solid timber: $\qquad\qquad\qquad$ Service Class 2 Medium-term actions $\qquad\quad k_{mod} = 0{,}8$	$k_{mod} = 0{,}8$
Table NA.3 EC5 (Table 2.3) Clause 3.2.(3) Equation (3.1)	For solid timber – untreated and preservative treated Member size: For timber in bending with $\rho_k \leq 700 \text{ kg/m}^3$ where the depth is less than 150 mm, $f_{m,k}$ can be multiplied by k_h where: $k_h = \min\begin{cases} \left(\dfrac{150}{h}\right)^{0{,}2} \\ 1{,}3 \end{cases}$ $k_h = \min\begin{cases} \left(\dfrac{150}{125}\right)^{0{,}2} \\ 1{,}3 \end{cases} = \min\begin{cases} 1{,}04 \\ 1{,}3 \end{cases}$	$\gamma_M = 1{,}3$ $k_h = 1{,}04$
Clause 6.6	System strength: k_{sys} Assume that lateral load distribution does not exist and k_{sys} does not apply. **Section Properties:** Cross-sectional area $A = 50 \times 125 = 6{,}25 \times 10^3 \text{ mm}^2$ Section modulus $\quad W_y = \dfrac{50 \times 125^2}{6} = 130{,}21 \times 10^3 \text{ mm}^3$ Existing axial load in the tie (tension) $= 6{,}0$ kN The additional load P induces both axial and bending load effects in the main tie. The additional **axial** load on the tie is determined assuming the load applied between nodes to be distributed statically to each of the adjacent nodes.	

O

1,0 kN   1,0 kN

2,0 kN

References	Calculations	Output
	Σ Moments about 'O' = 0	
	$(1,0 \times 3,0) - (1,0 \times 1,0) - (F_t \times 1,732) = 0$	
	$\qquad\qquad\qquad\qquad\qquad F_t = 1,16$ kN	
	Total axial load in the tie $F_{t,0,d} = (6,0 + 1,16) = 7,16$ kN	$F_{t,0,d} = 7,16$ kN
	The secondary bending moment can be estimated by assuming the main tie to be a three span beam as shown and evaluating the maximum bending moment.	
Appendix A	The bending moment can be determined using the coefficients given in Appendix A of this text.	

Contract : Roof Truss   Job Ref. No. : Example 7.1
Part of Structure :   Main tie
Calculation Sheet No. : 3 of 5

Calculations by : W.McK.
Checked by : B.Z.
Date :

References	Calculations	Output
	Contract : Roof Truss Job Ref. No. : Example 7.1 **Part of Structure : Main tie** **Calculation Sheet No. : 4 of 5**	**Calculations by : W.McK.** **Checked by : B.Z.** **Date :**

References	Calculations	Output
Clause 5.4.3(3)	The effect of deflection at the nodes and partial fixity at the connections should be taken into account by a reduction of 10% of the moments at the inner supports of the member. The inner support moments should be used to calculate the span bending moments.	
	Reduced support moments coefficients $= -(0,9 \times 0,075)$ $\qquad\qquad\qquad\qquad\qquad\qquad = -0,068$	
	Free bending moment coefficient $= +0,25$	
	Modified mid-span moment coefficient: $= (-0,068 + 0,25) = +0,182$	
	Mid-span bending moment $= +0,182PL$ $M_{y,d} = +(0,182 \times 2,0 \times 2,0) = +0,73$ kNm	$M_{y,d} = 0,73$ kNm
	Design tensile stress parallel to the grain $\sigma_{t,0,d} = \dfrac{F_{c,0,d}}{A}$	
	$\sigma_{t,0,d} = \dfrac{7,16 \times 10^3}{6,25 \times 10^3} = 1,15$ N/mm^2	$\sigma_{t,0,d} = 1,15$ N/mm^2
	Design bending stress parallel to the grain $\sigma_{m,y,d} = \dfrac{M_{y,d}}{W_y}$	
	$\sigma_{m,y,d} = \dfrac{0,73 \times 10^6}{130,21 \times 10^3} = 5,61$ N/mm^2	$\sigma_{m,y,d} = 5,61$ N/mm^2
	Design tensile strength parallel to the grain: $f_{t,0,d} = (f_{t,0,k} \times k_{mod} \times k_h \times k_{sys})/\gamma_M$ $\qquad = (8,0 \times 0,8 \times 1,04)/1,3 = 5,12$ N/mm^2	$f_{t,0,d} = 5,12$ N/mm^2
	Design bending strength parallel to the grain: $f_{m,y,d} = (f_{m,y,k} \times k_{mod} \times k_h \times k_{sys})/\gamma_M$ $\qquad = (14,0 \times 0,8 \times 1,04)/1,3 = 8,96$ N/mm^2	$f_{t,0,d} = 8,96$ N/mm^2
Clause 6.2.3(1) Equation (6.17)	Combined bending and axial tension: $\dfrac{\sigma_{t,0,d}}{f_{t,0,d}} + \dfrac{\sigma_{m,y,d}}{f_{m,y,d}} + k_m \dfrac{\sigma_{m,z,d}}{f_{m,z,d}} \le 1,0$ $\qquad\qquad\qquad\qquad\qquad\qquad \blacktriangle$ zero $\dfrac{1,15}{5,12} + \dfrac{5,61}{8,96} = (0,22 + 0,63) = 0,85 \le 1,0$	

References	Calculations	Output
Equation (6.18)	$\dfrac{\sigma_{t,0,d}}{f_{t,0,d}} + k_m \dfrac{\sigma_{m,y,d}}{f_{m,y,d}} + \dfrac{\sigma_{m,z,d}}{f_{m,z,d}} \leq 1,0$ zero	
Clause 6.1.6	For solid timber, rectangular section $k_m = 0,7$	**The section is adequate with respect to combined bending and axial tension.**
	Since k_m is less than 1,0 Equation (6.18) is less critical.	

Contract : Roof Truss Job Ref. No. : Example 7.1
Part of Structure : Main tie
Calculation Sheet No. : 5 of 5

Calculations by : W.McK.
Checked by : B.Z.
Date :

7.3 Combined Bending and Axial Compression (Clauses 6.2.4 and 6.3.2)

As indicated in Chapter 6, Section 6.3, in slender members subjected to axial compressive loads there is a tendency for lateral instability to occur. This type of failure is called flexural buckling and is reflected in the k_{cy} and k_{cz} modification factors when checking the suitability of members with respect to stability in accordance with Clause 6.3.

When combined bending and axial compressive stresses occur simultaneously in a section there is an increased tendency for buckling failure to occur. The axial load and its associated secondary bending effect, in addition to the primary bending effect induced by the applied lateral load, are shown in Figure 7.2.

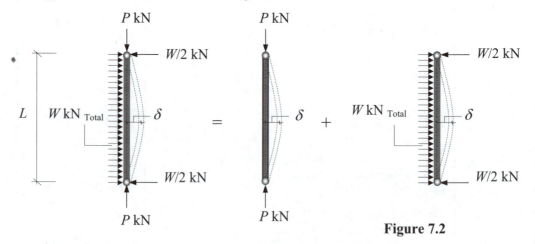

Figure 7.2

δ is the total deflection due to imperfections as described in Section 6.3 of Chapter 6, in addition to bending deflection due to the applied lateral load.

$$\text{Total bending moment} \quad = (P \times \delta) + \left(\frac{WL}{8}\right)$$

Typical interaction diagrams for combined bending and axial compression are shown in Figure 7.3, (see STEP 1 (58)).

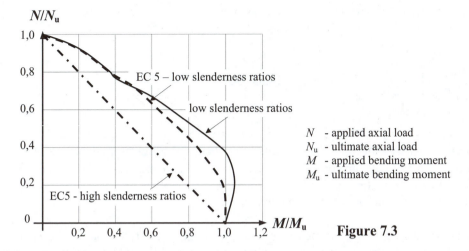

$N - $ applied axial load
$N_u - $ ultimate axial load
$M - $ applied bending moment
$M_u - $ ultimate bending moment

Figure 7.3

In stocky members where the slenderness ratio is low and buckling is not considered to be a problem (as indicated in Clause 6.3.2(2) of EC5, i.e. $\lambda_{rel,z} \leq 0,3$ and $\lambda_{rel,y} \leq 0,3$) the capability of timber to behave plastically in compression is reflected in Equations (6.19) and (6.20).

$$\left(\frac{\sigma_{c,0,d}}{f_{c,0,d}}\right)^2 + \frac{\sigma_{m,y,d}}{f_{m,y,d}} + k_m \frac{\sigma_{m,z,d}}{f_{m,z,d}} \leq 1,0 \qquad \text{(Equation (6.19) in EC5)}$$

and

$$\left(\frac{\sigma_{c,0,d}}{f_{c,0,d}}\right)^2 + k_m \frac{\sigma_{m,y,d}}{f_{m,y,d}} + \frac{\sigma_{m,z,d}}{f_{m,z,d}} \leq 1,0 \qquad \text{(Equation (6.20) in EC5)}$$

where:
$\sigma_{c,0,d}$ is the design compressive stress,
$f_{c,0,d}$ is the design compressive strength,
$\sigma_{m,y,d}$, $f_{m,y,d}$, $\sigma_{m,z,d}$, $f_{m,z,d}$ and k_m are as before.

In cases where $\lambda_{rel,z}$ and $\lambda_{rel,y}$ are greater than these limits, i.e. where the slenderness is considered to be sufficiently high to induce flexural buckling, the use of a linear interaction expression is more appropriate. In such cases, Equations (6.23) and (6.24) should be checked, i.e.

$$\frac{\sigma_{c,0,d}}{k_{c,y} f_{c,0,d}} + \frac{\sigma_{m,y,d}}{f_{m,y,d}} + k_m \frac{\sigma_{m,z,d}}{f_{m,z,d}} \leq 1,0 \qquad \text{(Equation (6.23) in EC5)}$$

and

$$\frac{\sigma_{c,0,d}}{k_{c,z} f_{c,0,d}} + k_m \frac{\sigma_{m,y,d}}{f_{m,y,d}} + \frac{\sigma_{m,z,d}}{f_{m,z,d}} \leq 1,0 \qquad \text{(Equation (6.24) in EC5)}$$

where the $k_{c,y}$ and $k_{c,z}$ can be determined as indicated in Chapter 6.

When lateral torsional bending is a consideration, i.e. in beams subjected to combined bending and compression, Equation (6.35) should be satisfied, i.e.

$$\left(\frac{\sigma_{m,d}}{k_{crit}\,f_{m,d}}\right)^2 + \frac{\sigma_{c,d}}{k_{c,z}\,f_{c,0,d}} \leq 1,0 \qquad \text{(Equation (6.19) in EC5)}$$

where:
$\sigma_{m,d}$ and $f_{m,d}$ are the design bending stress and design bending strength respectively,
$\sigma_{c,d}$ and $f_{c,d}$ are the design compressive stress and design compressive strength respectively,
k_{crit} is given by Equation (6.34) (see Chapter 5),
$k_{c,z}$ is given by Equation (6.26) (see Chapter 6).

The non-linear behaviour in members with low slenderness ratios is a consequence of the higher values of compressive stress required to induce failure. Unlike timber subjected to tension which exhibits a linear stress-strain curve until failure is reached, timber subjected to compressive stresses exhibits considerable plasticity (i.e. non-linear behaviour) after the initial elastic linear deformation. At high values of slenderness the stresses are relatively low when buckling occurs and hence a linear response is more appropriate, whilst at low values of slenderness relatively high values of stress occur at failure and hence a non-linear approximation is more realistic.

7.3.1 Example 7.2: Lattice Girder – Top Chord

A lattice girder supports a series of roof purlins as shown in Figure 7.4. Using the data provided, check the suitability of the proposed sections for:

i)   the top chord of the girder and
ii) the left hand column.

Design data:

Timber Strength Class	C24
Timber Service Class	2
Top chord to be a continuous member	2/50 mm x 225 mm double section
Column section	250 mm x 250 mm
Spacing of frames	4,0 m
Spacing of purlins	600 mm

All purlins may be assumed to be adequately fastened to the top chord and carried back to other bracing or support systems.

Design permanent (long-term) load on top chord	0,75 kN/m^2
Design variable (medium-term) load on top chord	1,12 kN/m^2
Design wind loading (short-term) on columns	as indicated in Figure 7.4

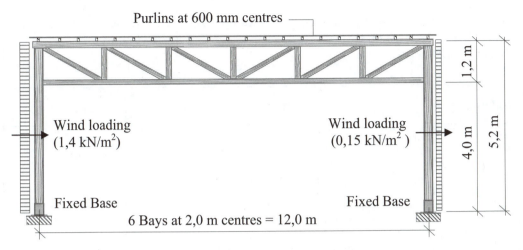

Purlins at 600 mm centres

Wind loading
(1,4 kN/m²)

Wind loading
(0,15 kN/m²)

1,2 m

4,0 m

5,2 m

Fixed Base

Fixed Base

6 Bays at 2,0 m centres = 12,0 m

Figure 7.4

7.3.1.1 Solution to Example 7.2

Contract : Lattice Girder Job Ref. No. : Example 7.2 Part of Structure : Top Chord Calculation Sheet No. : 1 of 11	Calculations by : W.McK. Checked by : B.Z. Date :

References	Calculations	Output
BS EN 338:2003 BS EN 1995-1-1: 2004 National Annex	Structural Timber - Strength Classes Eurocode 5: Design of Timber Structures Part 1.1: General – Common Rules and Rules for Buildings. UK National Annex to BS EN 1995-1-1:	
BS EN 338:2003 Clause 5.0	Structural Timber - Strength Classes Characteristic values for C24 Timber	
Table 1	**Strength Properties:** Bending strength parallel to the grain $f_{m,k} = 24,0$ N/mm² Compression parallel to grain $f_{c,0,k} = 21,0$ N/mm² **Stiffness Properties:** 5% Modulus of elasticity parallel to the grain $E_{0,05} = 7,4$ kN/m³ Mean density $\rho_{mean} = 420,0$ kg/m³	
Clause 2.3.2.1	Load-duration and moisture influence on strength: k_{mod}	
Table 3.1	Solid timber: Permanent actions $k_{mod} = 0,6$ Variable actions (medium-term) $k_{mod} = 0,8$ Variable actions (short-term) $k_{mod} = 0,9$	Service Class 2
Clause 3.1.3(2)	Use k_{mod} corresponding to action with the shortest duration.	

Contract : Lattice Girder Job Ref. No. : Example 7.2 Part of Structure : Top Chord Calculation Sheet No. : 2 of 11	Calculations by : W.McK. Checked by : B.Z. Date :

References	Calculations	Output
Table NA.3 EC5 (Table 2.3) Clause 6.6	For the top chord For the column For solid timber – untreated and preservative treated System strength factor (k_{sys}) does not apply in this case, i.e. the spacing of the trusses is greater than 1,2 m **Consider the top chord: 2/50 mm x 225 mm** Section properties/member: Cross-section $A = (50 \times 225) = 11{,}25 \times 10^3 \text{ mm}^2$ Section modulus $W_y = \dfrac{bh^2}{6} = \dfrac{50 \times 225^2}{6} = 421{,}88 \times 10^3 \text{ mm}^3$ Radius of gyration $i_y = \dfrac{d}{2\sqrt{3}} = \dfrac{225}{2\sqrt{3}} = 64{,}95 \text{ mm}$ $i_z = \dfrac{b}{2\sqrt{3}} = \dfrac{50}{2\sqrt{3}} = 14{,}43 \text{ mm}$ Area of roof supported by each girder $= (12{,}0 \times 4{,}0) = 48{,}0 \text{ m}^2$ Total load on girder $= (0{,}75 + 1{,}12) \times 48{,}0 = 89{,}76 \text{ kN}$ Load at each node point $= \dfrac{89{,}76}{6} = 14{,}96 \text{ kN}$ This load is applied through the purlins at 600 mm centres, and can be considered to be uniformly distributed along the continuous chord inducing both axial and bending effects. 89,76 kN Equivalent loading system: 7,48 kN 14,96 kN 14,96 kN 14,96 kN 14,96 kN 14,96 kN 7,48 kN X X Axial Effects 	$k_{mod} = 0{,}8$ $k_{mod} = 0{,}9$ $\gamma_M = 1{,}3$

References	Calculations	Output

| | **Contract : Lattice Girder Job Ref. No. : Example 7.2**
Part of Structure :   Top Chord
Calculation Sheet No. : 3 of 11 | **Calculations by : W.McK.**
Checked by : B.Z.
Date : |

Bending Effects

Appendix A
(this text)

Bending moment coefficients are approximately the same as those given for a four span beam.

$-0,107$ $-0,071$ $-0,107$

$+0,036$ $+0,036$

Clause 5.4.3(3)

The effect of deflection at the nodes and partial fixity at the connections should be taken into account by a reduction of 10% of the moments at the inner supports of the member. The inner support moments should be used to calculate the span bending moments.

Reduced support moments coefficients $\approx -(0,9 \times 0,071)$
$$= -0,064$$
Free bending moment coefficient $= +0,125$

Modified mid-span moment coefficient:
$$= (-0,064 + 0,125) = +0,061$$

Mid-span bending moment $= +0,061WL$
$$= +(0,061 \times 14,96 \times 2,0)   = +1,83 \text{ kNm}$$

(Alternatively simplify using $wL^2/8$)

The maximum axial load in the top chord occurs at mid-span at Section X-X as shown.

7,48 kN   14,96 kN   14,96 kN

F_c

1,2 m

44,88 kN   2,0 m   2,0 m   2,0 m

Σ Moments about 'O' $= 0$
$$(44,88 \times 6,0) - (7,48 \times 6,0) - (14,96 \times 4,0) - (14,96 \times 2,0)$$
$$+ (F_c \times 1,2) = 0$$
$$F_c = -112,20 \text{ kN (Compression)}$$

Contract : Lattice Girder Job Ref. No. : Example 7.2 Part of Structure : Top Chord Calculation Sheet No. : 4 of 11	Calculations by : W.McK. Checked by : B.Z. Date :

References	Calculations	Output

	The top chord can be designed as two separate members each supporting 50% of the axial load and moment or as a spaced column supporting the full load.
	
	Consider each section as a separate member.
	Axial load / member $= 0,5 \times 112,20 = 56,10$ kN
	Bending moment / member $= 0,5 \times 1,83 = 0,92$ kNm
	
	Design compressive strength parallel to the grain:
	$f_{c,0,d} = (f_{c,0,k} \times k_{mod} \times k_{sys})/\gamma_M$
	        $= (21,0 \times 0,8)/1,3 = 12,92$ N/mm^2
	
	Design compressive stress parallel to the grain:
	$\sigma_{c,0,d} = \dfrac{F_{c,0,d}}{A} = \dfrac{56,1 \times 10^3}{\left(11,25 \times 10^3\right)} = 4,99$ N/mm^2
	
	Design bending strength parallel to the grain:
	$f_{my,d} = (f_{m,y,k} \times k_{mod} \times k_{sys})/\gamma_M$ (Note: k_h does not apply).
	        $= (24,0 \times 0,8)/1,3 = 14,77$ N/mm^2
	
	Design bending stress parallel to the grain:
	$\sigma_{m,y,d} = \dfrac{M_{y,d}}{W_y} = \dfrac{0,92 \times 10^6}{\left(421,88 \times 10^3\right)} = 2,18$ N/mm^2
		$F_{c,0,d} = \mathbf{56,10}$ **kN**
		$M_{y,d} = \mathbf{0,92}$ **kNm**
		
		$f_{c,0,d} = \mathbf{12,92}$ **N/mm^2**
		
		$\sigma_{c,0,d} = \mathbf{4,99}$ **N/mm^2**
		
		$f_{m,y,d} = \mathbf{14,77}$ **N/mm^2**
		
		$\sigma_{m,y,d} = \mathbf{2,18}$ **N/mm^2**

Clause 6.3 Clause 6.3.2	Stability of members: Columns subjected to either compression or combined compression and bending.	
Clause 6.3.2(1) Equation (6.21)	Relative slenderness $\lambda_{rel,y} = \dfrac{\lambda_y}{\pi}\sqrt{\dfrac{f_{c,0,d}}{E_{0,05}}}$ where $\lambda_y = \dfrac{L_{y,ef}}{i_y}$ $L_{y,ef} = (1,0 \times 2000) = 2000$ mm; $i_y = 64,95$ $\lambda_y = \dfrac{2000}{64,95} = 30,79$ $\lambda_{rel,y} = \dfrac{30,79}{\pi}\sqrt{\dfrac{21,0}{7,4 \times 10^3}} = 0,52$	
Equation (6.22)	$\lambda_{rel,z} = \dfrac{\lambda_z}{\pi}\sqrt{\dfrac{f_{c,0,d}}{E_{0,05}}}$ where $\lambda_z = \dfrac{L_{z,ef}}{i_z}$ $L_{z,ef} = (1,0 \times 600) = 600$ mm; $i_z = 14,43$	

Contract : Lattice Girder Job Ref. No. : Example 7.2 Part of Structure : Top Chord Calculation Sheet No. : 5 of 11	Calculations by : W.McK. Checked by : B.Z. Date :

References	Calculations	Output
	$\lambda_z = \dfrac{600}{14,43} = 41,58$ $\lambda_{rel,z} = \dfrac{41,58}{\pi}\sqrt{\dfrac{21,0}{7,4\times10^3}} = 0,71$	
Clause 6.3.2(3)	Since $\lambda_{rel,y} \geq 0,3$ and $\lambda_{rel,z} \geq 0,3$ use Equations (6.23) and (6.24).	
Equation (6.23)	$\dfrac{\sigma_{c,0,d}}{k_{c,y}f_{c,0,d}} + \dfrac{\sigma_{m,y,d}}{f_{m,y,d}} + k_m \dfrac{\sigma_{m,z,d}}{f_{m,z,d}} \leq 1,0$ zero	
Equation (6.24)	$\dfrac{\sigma_{c,0,d}}{k_{c,z}f_{c,0,d}} + k_m \dfrac{\sigma_{m,y,d}}{f_{m,y,d}} + \dfrac{\sigma_{m,z,d}}{f_{m,z,d}} \leq 1,0$ zero	
Equation (6.27) Equation (6.29)	$k_y = 0,5\left[1+\beta_c\left(\lambda_{rel,y}-0,3\right)+\lambda^2_{rel,y}\right]$ $\beta_c = 0,2$ for solid timber $k_y = 0,5\times\left[1+0,2\times\left(0,52-0,3\right)+0,52^2\right] = 0,66$	
Equation (6.25)	$k_{cy} = \dfrac{1}{k_y+\sqrt{k_y^2-\lambda^2_{rel,y}}} = \dfrac{1}{0,66+\sqrt{0,66^2-0,52^2}} = 0,94$	
Equation (6.28) Equation (6.29)	$k_z = 0,5\left[1+\beta_c\left(\lambda_{rel,z}-0,3\right)+\lambda^2_{rel,z}\right]$ $\beta_c = 0,2$ for solid timber $k_z = 0,5\times\left[1+0,2\times\left(0,71-0,3\right)+0,71^2\right] = 0,79$	
Equation (6.26)	$k_{cz} = \dfrac{1}{k_z+\sqrt{k_z^2-\lambda^2_{rel,z}}} = \dfrac{1}{0,79+\sqrt{0,79^2-0,71^2}} = 0,88$	
Clause 6.1.6	$k_m = 0,7$ for rectangular sections	
Equation (6.23)	$\dfrac{\sigma_{c,0,d}}{k_{c,y}f_{c,0,d}} + \dfrac{\sigma_{m,y,d}}{f_{m,y,d}} \leq 1,0$ $\dfrac{4,99}{\left(0,94\times12,92\right)} + \dfrac{2,18}{14,77} = (0,39+0,15) = 0,54 \leq 1,0$	
Equation (6.24)	$\dfrac{\sigma_{c,0,d}}{k_{c,z}f_{c,0,d}} + k_m \dfrac{\sigma_{m,y,d}}{f_{m,y,d}} \leq 1,0$	

References	Calculations	Output
	Contract : **Lattice Girder** Job Ref. No. : **Example 7.2** Part of Structure : **Top Chord** Calculation Sheet No. : **6 of 11**	Calculations by : **W.McK.** Checked by : **B.Z.** Date :

References	Calculations	Output
	$$\frac{4,99}{\left(0,88\times12,92\right)} + \frac{0,7\times2,18}{14,77} = (0,44 + 0,10) = 0,54 \le 1,0$$ **Column Section: 250 mm x 250 mm** Section properties: Cross-section $A = (250 \times 250) = 62,50 \times 10^3 \text{ mm}^2$ Section modulus $W_y = \dfrac{bh^2}{6} = \dfrac{250\times250^2}{6} = 2,60 \times 10^6 \text{ mm}^3$ Radius of gyration $i_y = \dfrac{d}{2\sqrt3} = \dfrac{250}{2\sqrt3} = 72,17 \text{ mm}$ $i_z = \dfrac{b}{2\sqrt3} = \dfrac{250}{2\sqrt3} = 72,17 \text{ mm}$	The **2/50 mm x 225 mm section is adequate.**
Chapter 4 Section 4.8	Frames with fixed bases in which sway occurs have a point of contraflexure in the columns. $$h_c \approx \frac{h}{2}\left(\frac{h+2H}{2h+H}\right)$$ h_c is the assumed height of the point of contraflexure (i.e. zero moment) above the fixed base. This assumption is sufficiently accurate for design purposes even when the loading on each column is not equal. Spacing of frames = 4,0 m Horizontal loading on column AB = $(1,4 \times 4,0) = 5,6$ kN/m Horizontal loading on column DC = $(0,15 \times 4,0) = 0,6$ kN/m	

Contract : Lattice Girder Job Ref. No. : Example 7.2 Part of Structure : Top Chord Calculation Sheet No. : 7 of 11	Calculations by : W.McK. Checked by : B.Z. Date :

References	Calculations	Output

$$h_c = \frac{h}{2}\left(\frac{h+2H}{2h+H}\right) = \frac{4,0}{2}\left(\frac{4,0+(2\times5,2)}{8+5,2}\right) = 2,18 \text{ m}$$

Consider a horizontal section through the point of contraflexure

89,76 kN (total)

B   C

3,02 m

5,6 kN/m   1,82 m   0,6 kN/m

S P_1 S P_2

V_1 12,0 m V_2

$\Sigma F_x = 0$
$(5,6 \times 3,02) + (0,6 \times 3,02) - 2S = 0 \quad \therefore S = 9,36 \text{ kN}$
Σ Moments about $P_1 = 0$

$$\frac{5,6\times3,02\times3,02}{2} + (89,76 \times 6,0) + \frac{0,6\times3,02\times3,02}{2} -12V_2 = 0$$

$$\therefore V_2 = 47,24 \text{ kN}$$

$\Sigma F_y = 0$
$V_1 - 89,76 + 47,24 = 0 \qquad\qquad \therefore V_1 = 42,52 \text{ kN}$

Bending moment at C $= -(9,36 \times 1,82) + \dfrac{0,6\times1,82\times1,82}{2}$

$$= -16,04 \text{ kNm}$$

Bending moment at B $= +(9,36 \times 1,82) - \dfrac{5,6\times1,82\times1,82}{2}$

$$= +7,76 \text{ kNm}$$

Consider column AB below the point of contraflexure to
determine the value of the horizontal reaction and bending
moment at the base

9,36 kN

5,6 kN/m

2,18 m

M_A H_A

A

42,52 kN

Contract : Lattice Girder Job Ref. No. : Example 7.2 Part of Structure : Top Chord Calculation Sheet No. : 8 of 11	Calculations by : W.McK. Checked by : B.Z. Date :

References	Calculations	Output
	(see below)	

$\Sigma F_x = 0$

$(5{,}6 \times 2{,}18) + 9{,}36 - H_A = 0 \qquad \therefore H_A = 21{,}57 \text{ kN}$

$\Sigma M_A = 0$

$-M_A + \dfrac{5{,}6 \times 2{,}18 \times 2{,}18}{2} + (9{,}36 \times 2{,}18) = 0$

$\therefore M_A = 33{,}71 \text{ kNm}$

Consider column CD below the point of contraflexure to determine the value of the horizontal reaction and bending moment at the base

$\Sigma F_x = 0$

$(0{,}6 \times 2{,}18) + 9{,}36 - H_D = 0 \qquad \therefore H_D = 10{,}67 \text{ kN}$

$\Sigma M_D = 0$

$-M_D + \dfrac{0{,}6 \times 2{,}18 \times 2{,}18}{2} + (9{,}36 \times 2{,}18) = 0$

$\therefore M_D = 21{,}83 \text{ kNm}$

Bending Moment Diagrams for columns AB and CD

Vertical reaction at A = 42,52 kN
Vertical reaction at D = 47,24 kN

Column AB is the critical one.

References	Calculations	Output

Contract : Lattice Girder Job Ref. No. : Example 7.2
Part of Structure :   Top Chord
Calculation Sheet No. : 9 of 11

Calculations by : W.McK.
Checked by : B.Z.
Date :

References	Calculations	Output

Design axial loading   = 42,52 kN
Design bending moment   = 33,71 kNm

Output:
$F_{c,0,d} = $ **42,52 kNm**
$M_{y,d} = $ **33,71 kNm**

Design compressive stress parallel to the grain:

$$\sigma_{c,0,d} = \frac{F_{c,0,d}}{A} = \frac{42,52 \times 10^3}{\left(62,5 \times 10^3\right)} = 0,68 \text{ N/mm}^2$$

$\sigma_{c,0,d} = $ **0,68 N/mm²**

Design bending stress parallel to the grain:

$$\sigma_{m,y,d} = \frac{M_{y,d}}{W_y} = \frac{33,71 \times 10^6}{\left(2,60 \times 10^6\right)} = 12,97 \text{ N/mm}^2$$

$\sigma_{m,y,d} = $ **12,97 N/mm²**

P_1 1,82 m   4,0 m P_2

2,18 m

Clause 6.3
Clause 6.3.2

Stability of members:
Columns subjected to either compression or combined compression and bending.

Clause 6.3.2(1)

Relative slenderness

Equation (6.21)

$$\lambda_{rel,y} = \frac{\lambda_y}{\pi} \sqrt{\frac{f_{c,0,d}}{E_{0,05}}} \quad \text{where} \quad \lambda_y = \frac{L_{y,ef}}{i_y}$$

$L_{y,ef} = (2,0 \times 2180) = 4360$ mm; $i_y = 72,17$

$$\lambda_y = \frac{4360}{72,17} = 60,41$$

$$\lambda_{rel,y} = \frac{60,41}{\pi} \sqrt{\frac{21,0}{7,4 \times 10^3}} = 1,02$$

Equation (6.22)

$$\lambda_{rel,z} = \frac{\lambda_z}{\pi} \sqrt{\frac{f_{c,0,d}}{E_{0,05}}} \quad \text{where} \quad \lambda_z = \frac{L_{z,ef}}{i_z}$$

$L_{z,ef} = (0,85 \times 4000) = 3400$ mm; $i_z = 72,17$

$$\lambda_z = \frac{3400}{72,17} = 47,11$$

$$\lambda_{rel,z} = \frac{47,11}{\pi} \sqrt{\frac{21,0}{7,4 \times 10^3}} = 0,80$$

References	Calculations	Output
Contract : Lattice Girder Job Ref. No. : Example 7.2 **Part of Structure : Top Chord** **Calculation Sheet No. : 10 of 11**		**Calculations by : W.McK.** **Checked by : B.Z.** **Date :**

References	Calculations	Output
Clause 6.3.2(3)	Since $\lambda_{rel,y} \geq 0,3$ and $\lambda_{rel,z} \geq 0,3$ use Equations (6.23) and (6.24).	
Equation (6.23)	$\dfrac{\sigma_{c,0,d}}{k_{c,y}f_{c,0,d}} + \dfrac{\sigma_{m,y,d}}{f_{m,y,d}} + k_m \dfrac{\sigma_{m,z,d}}{f_{m,z,d}} \leq 1,0$ <div align="right">zero</div>	
Equation (6.24)	$\dfrac{\sigma_{c,0,d}}{k_{c,z}f_{c,0,d}} + k_m \dfrac{\sigma_{m,y,d}}{f_{m,y,d}} + \dfrac{\sigma_{m,z,d}}{f_{m,z,d}} \leq 1,0$ <div align="right">zero</div>	
Equation (6.27) Equation (6.29)	$k_y = 0,5\left[1 + \beta_c\left(\lambda_{rel,y} - 0,3\right) + \lambda_{rel,y}^2\right]$ $\beta_c = 0,2$ for solid timber $k_y = 0,5\times\left[1 + 0,2\times\left(1,02 - 0,3\right) + 1,02^2\right] = 1,09$	
Equation (6.25)	$k_{cy} = \dfrac{1}{k_y + \sqrt{k_y^2 - \lambda_{rel,y}^2}} = \dfrac{1}{1,09 + \sqrt{1,09^2 - 1,02^2}} = 0,68$	
Equation (6.28) Equation (6.29)	$k_z = 0,5\left[1 + \beta_c\left(\lambda_{rel,z} - 0,3\right) + \lambda_{rel,z}^2\right]$ $\beta_c = 0,2$ for solid timber $k_z = 0,5\times\left[1 + 0,2\times\left(0,80 - 0,3\right) + 0,80^2\right] = 0,87$	
Equation (6.26)	$k_{cz} = \dfrac{1}{k_z + \sqrt{k_z^2 - \lambda_{rel,z}^2}} = \dfrac{1}{0,87 + \sqrt{0,87^2 - 0,80^2}} = 0,83$	
Clause 6.1.6	$k_m = 0,7$ for rectangular sections	
	Design compressive strength parallel to the grain: $f_{c,0,d} = (f_{c,0,k} \times k_{mod} \times k_{sys})/\gamma_M$ $\quad = (21,0 \times 0,9)/1,3 = 14,54$ N/mm^2	$f_{c,0,d} = \mathbf{14,54}$ **N/mm^2**
	Design bending strength parallel to the grain: $f_{my,d} = (f_{m,y,k} \times k_{mod} \times k_{sys})/\gamma_M$ $\quad = (24,0 \times 0,9)/1,3 = 16,62$ N/mm^2	$f_{m,y,d} = \mathbf{16,62}$ **N/mm^2**
Equation (6.23)	$\dfrac{\sigma_{c,0,d}}{k_{c,y}f_{c,0,d}} + \dfrac{\sigma_{m,y,d}}{f_{m,y,d}} \leq 1,0$ $\dfrac{0,68}{\left(0,68\times14,54\right)} + \dfrac{12,97}{16,62} = (0,07 + 0,78) = 0,85 < 1,0$	

Contract : Lattice Girder Job Ref. No. : Example 7.2	Calculations by : W.McK.
Part of Structure : Top Chord	Checked by : B.Z.
Calculation Sheet No. : 11 of 11	Date :

References	Calculations	Output
Equation (6.24)	$\dfrac{\sigma_{c,0,d}}{k_{c,z}f_{c,0,d}} + k_m \dfrac{\sigma_{m,y,d}}{f_{m,y,d}} \leq 1,0$ $\dfrac{0,68}{(0,83 \times 14,54)} + \dfrac{0,7 \times 12,97}{16,62} = (0,06 + 0,55) = 0,61 < 1,0$	**A 250 mm x 250 mm section is adequate.**

7.3.2 Example 7.3: Laminated Portal Frame

A church building is to be constructed using a series of three-pinned, pitched roof portal frames as shown in Figure 7.5. Using the data given, check the suitability of the glued laminated frame members to resist the combined axial and bending effects at the knee.

Design data:

Glulam Strength Class	GL 28h
Timber Service Class	2
Centres of the frames	3,0 m
Design permanent load based on plan area	2,0 kN/m^2
Design variable (medium-term) load based on plan area	1,45 kN/m^2

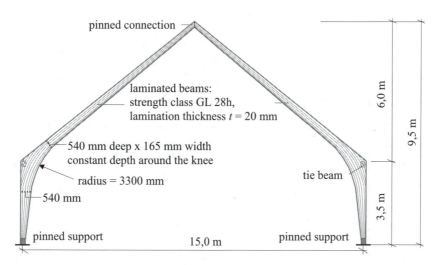

pinned connection

laminated beams:
strength class GL 28h,
lamination thickness t = 20 mm

540 mm deep x 165 mm width
constant depth around the knee

radius = 3300 mm

540 mm

tie beam

pinned support   pinned support

15,0 m

6,0 m

9,5 m

3,5 m

Figure 7.5

7.3.2.1 Solution to Example 7.3

Contract : Church Hall Job Ref. No. : Example 7.3	Calculations by : W.McK.
Part of Structure : Portal frame	Checked by : B.Z.
Calculation Sheet No. : 1 of 5	Date :

References	Calculations	Output
BS EN 1990 BS EN 1991 BS EN 338:2003 BS EN 1995-1-1: 2004 National Annex BS EN 1194 1999	Eurocode: Basis of Structural Design Eurocode 1: Actions on Structures Structural Timber - Strength Classes Eurocode 5: Design of Timber Structures Part 1-1: General – Common Rules and Rules for Buildings UK National Annex to BS EN 1995-1-1: Glued Laminated Timber – Strength Classes and Characteristic Values	

BS EN 1194
Table 1
	Characteristic values for GL 28h glulam
	Strength Properties:
	Bending $f_{m,g,k} = 28{,}0$ N/mm^2
	Tension perpendicular to grain $f_{t,90,g,k} = 0{,}45$ N/mm^2
	Compression parallel to grain $f_{c,0,g,k} = 26{,}5$ N/mm^2
	Compression perpendicular to grain $f_{c,90,g,k} = 3{,}0$ N/mm^2
	Shear $f_{v,g,k} = 3{,}2$ N/mm^2
	
	Stiffness Properties:
	Mean modulus of elasticity parallel to the grain
	                              $E_{0,mean} = 12{,}6$ kN/mm^2
	5% modulus of elasticity parallel to the grain
	                              $E_{0,g,05} = 10{,}2$ kN/mm^2
	Mean shear modulus $G_{g,mean} = 0{,}78$ kN/mm^2
	Mean density $\rho_{g,mean} = 410{,}0$ kg/m^3	

BS EN 1995 Clause 2.3.2.1	Load-duration and moisture influence on strength: k_{mod}	
Table 3.1	Glued laminated timber: Service Class 2 Permanent actions $k_{mod} = 0{,}6$ Medium-term actions $k_{mod} = 0{,}8$	
Clause 3.1.3(2)	Use k_{mod} corresponding to action with the shortest duration.	$k_{mod} = 0{,}8$
National Annex Table NA.3 EC5 (Table 2.3)	Partial factor for material properties and resistance: γ_M For glued laminated timber	$\gamma_M = 1{,}25$
Clause 3.3 Clause 3.3(3)	Member size: Since the depth is less than the reference depth for glulam beams (600 mm) the k_h value can be used:	
Equation (3.2)	$k_h = \min \left\{ \begin{array}{l} \left(\dfrac{600}{h}\right)^{0,1} \\ 1,1 \end{array} \right. = \min \left\{ \begin{array}{l} \left(\dfrac{600}{540}\right)^{0,1} \\ 1,1 \end{array} \right. = \min \left\{ \begin{array}{l} 1,01 \\ 1,1 \end{array} \right.$	$k_h = 1{,}01$

References	Calculations	Output
Contract : **Church Hall** Job Ref. No. : **Example 7.3**		Calculations by : **W.McK.**

Contract : Church Hall   Job Ref. No. : Example 7.3
Part of Structure :   Portal frame
Calculation Sheet No. : 2 of 5

Calculations by : W.McK.
Checked by : B.Z.
Date :

References	Calculations	Output
Clause 6.6(4) Figure 6.12	System strength: k_{sys} Lamination thickness = 20 mm thick: Number of laminations = 540/20 = 27 > 8	$k_{sys} = 1{,}2$

Design load = $(2{,}0 + 1{,}45) = 3{,}45$ kN/m^2
Portal frames are spaced at 3,0 m centres
Load / frame = $(3{,}45 \times 3{,}0) = 10{,}35$ kN/m

$\Sigma M_A = 0$
$(10{,}35 \times 15{,}0 \times 7{,}5) - (15{,}0 \times V_E) = 0$ $\therefore V_E = 77{,}63$ kN

$\Sigma F_y = 0$
$V_A - (10{,}35 \times 15{,}0) + V_E = 0$ $\therefore V_A = 77{,}63$ kN

Consider a section of the frame through the pin at the ridge.

References	Calculations	Output
Contract : Church Hall Job Ref. No. : Example 7.3 **Part of Structure : Portal frame** **Calculation Sheet No. : 3 of 5**		**Calculations by : W.McK.** **Checked by : B.Z.** **Date :**

$\Sigma M_C = 0$

$$(77{,}63 \times 7{,}5) - 9{,}5 H_A - \left(10{,}35 \times \frac{7{,}5^2}{2}\right) = 0 \quad H_A = 30{,}65 \text{ kN}$$

$\Sigma F_x = 0 \quad H_A - H_E \quad = \quad 0 \qquad\qquad \therefore \quad H_E = 30{,}65 \text{ kN}$

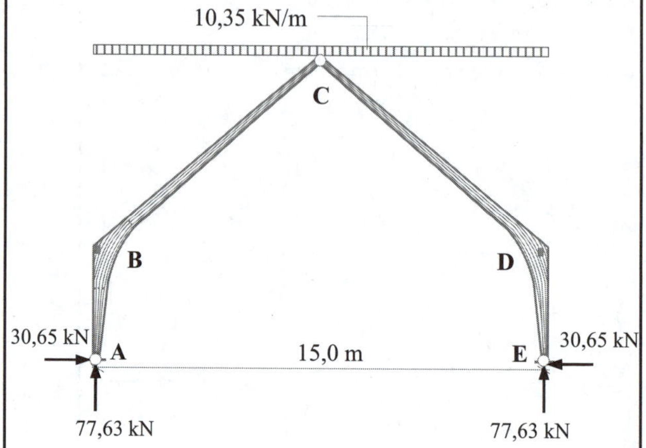

Bending moment at B $= - (30{,}65 \times 3{,}5) = - 107{,}28$ kNm

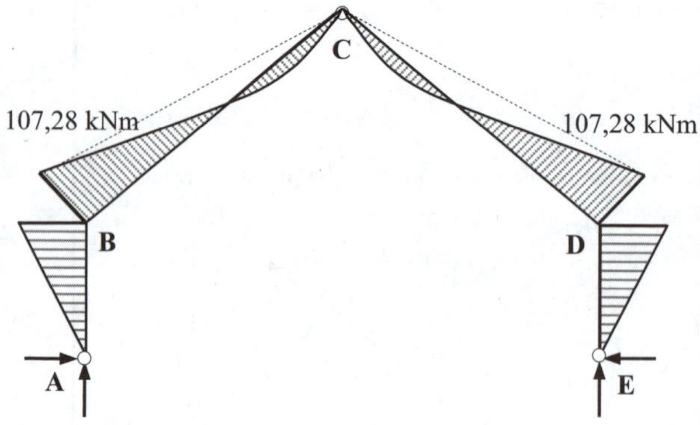

Design bending moment $M_{y,d} = 107{,}28$ kNm
Design axial loading $F_{c,0,d} = 77{,}63$ kN

Consider the cross-section at the knee: 540 mm x 165 mm
Section properties/member:
Cross-section $A = (540 \times 165) = 89{,}1 \times 10^3 \text{ mm}^2$

Output column:

$M_{y,d} = \mathbf{107{,}28}$ **kNm**
$F_{c,0,d} = \mathbf{77{,}63}$ **kN**

References	Calculations	Output

Contract : Church Hall   Job Ref. No. : Example 7.3
Part of Structure :   Portal frame
Calculation Sheet No. : 4 of 5

Calculations by : W.McK.
Checked by : B.Z.
Date :

Section modulus $W_y = \dfrac{bh^2}{6} = \dfrac{165 \times 540^2}{6} = 8{,}02 \times 10^6 \text{ mm}^3$

Radius of gyration $i_y = \dfrac{d}{2\sqrt{3}} = \dfrac{540}{2\sqrt{3}} = 155{,}88 \text{ mm}$

$i_z = \dfrac{b}{2\sqrt{3}} = \dfrac{165}{2\sqrt{3}} = 47{,}63 \text{ mm}$

Design compressive strength parallel to the grain:
$f_{c,0,d} = (f_{c,0,k} \times k_{mod} \times k_{sys})/\gamma_M$
$= (26{,}5 \times 0{,}8 \times 1{,}2)/1{,}25 = 20{,}35 \text{ N/mm}^2$

$f_{c,0,d} = 20{,}35 \text{ N/mm}^2$

Design compressive stress parallel to the grain:
$\sigma_{c,0,d} = \dfrac{F_{c,0,d}}{A} = \dfrac{77{,}63 \times 10^3}{(89{,}1 \times 10^3)} = 0{,}87 \text{ N/mm}^2$

$\sigma_{c,0,d} = 0{,}87 \text{ N/mm}^2$

Design bending strength parallel to the grain:
$f_{m,y,d} = (f_{m,y,k} \times k_{mod} \times k_h \times k_{sys})/\gamma_M$
$= (28{,}0 \times 0{,}8 \times 1{,}01 \times 1{,}2)/1{,}25 = 21{,}72 \text{ N/mm}^2$

$f_{m,y,d} = 21{,}72 \text{ N/mm}^2$

Design bending stress parallel to the grain:
$\sigma_{m,y,d} = \dfrac{M_{y,d}}{W_y} = \dfrac{107{,}28 \times 10^6}{(8{,}02 \times 10^6)} = 13{,}38 \text{ N/mm}^2$

$\sigma_{m,y,d} = 13{,}38 \text{ N/mm}^2$

Clause 6.3
Clause 6.3.2

Stability of members:
Columns subjected to either compression or combined compression and bending.

Clause 6.3.2(1)

Relative slenderness

Equation (6.21)

$\lambda_{rel,y} = \dfrac{\lambda_y}{\pi}\sqrt{\dfrac{f_{c,0,d}}{E_{0,05}}}$ where $\lambda_y = \dfrac{L_{y,ef}}{i_y}$

$L_{y,ef} \approx (1{,}0 \times 3500) = 3500 \text{ mm};\quad i_y = 155{,}88$

$\lambda_y = \dfrac{3500}{155{,}88} = 22{,}45$

$\lambda_{rel,y} = \dfrac{22{,}45}{\pi}\sqrt{\dfrac{20{,}35}{10{,}2 \times 10^3}} = 0{,}32$

Equation (6.22)

$\lambda_{rel,z} = \dfrac{\lambda_z}{\pi}\sqrt{\dfrac{f_{c,0,d}}{E_{0,05}}}$ where $\lambda_z = \dfrac{L_{z,ef}}{i_z}$

References	Calculations	Output
	Contract : Church Hall Job Ref. No. : Example 7.3 **Calculations by : W.McK.** **Part of Structure : Portal frame** **Checked by : B.Z.** **Calculation Sheet No. : 5 of 5** **Date :**	

References	Calculations	Output
	Assume that restraint to the z-z axis of the column is provided by the brickwork walls of the building. $L_{z,ef} = 0$ $\lambda_z = 0$ and $\lambda_{rel,z} = 0$	
Clause 6.3.2(3)	Since $\lambda_{rel,y} \geq 0,3$ use Equations (6.23) and (6.24).	
Equation (6.23)	$\dfrac{\sigma_{c,0,d}}{k_{c,y}f_{c,0,d}} + \dfrac{\sigma_{m,y,d}}{f_{m,y,d}} + k_m \dfrac{\sigma_{m,z,d}}{f_{m,z,d}} \leq 1,0$ zero	
Equation (6.24)	$\dfrac{\sigma_{c,0,d}}{k_{c,z}f_{c,0,d}} + k_m \dfrac{\sigma_{m,y,d}}{f_{m,y,d}} + \dfrac{\sigma_{m,z,d}}{f_{m,z,d}} \leq 1,0$ zero	
Equation (6.27)	$k_y = 0,5\left[1 + \beta_c\left(\lambda_{rel,y} - 0,3\right) + \lambda_{rel,y}^2\right]$	
Equation (6.29)	$\beta_c = 0,1$ for glued laminated timber $k_y = 0,5\times\left[1 + 0,1\times\left(0,32 - 0,3\right) + 0,32^2\right] = 0,55$	
Equation (6.25)	$k_{cy} = \dfrac{1}{k_y + \sqrt{k_y^2 - \lambda_{rel,y}^2}} = \dfrac{1}{0,55 + \sqrt{0,55^2 - 0,32^2}} = 1,0$	
	Assume $k_{cz} = 1,0$ since slenderness is zero.	
Clause 6.1.6	$k_m = 0,7$ for rectangular sections	
Equation (6.23)	$\dfrac{\sigma_{c,0,d}}{k_{c,y}f_{c,0,d}} + \dfrac{\sigma_{m,y,d}}{f_{m,y,d}} \leq 1,0$ $\dfrac{0,87}{\left(1,0\times 20,35\right)} + \dfrac{13,38}{21,72} = (0,04 + 0,62) = 0,66 < 1,0$	
Equation (6.24)	$\dfrac{\sigma_{c,0,d}}{k_{c,z}f_{c,0,d}} + k_m \dfrac{\sigma_{m,y,d}}{f_{m,y,d}} \leq 1,0$ $\dfrac{0,87}{\left(1,0\times 20,35\right)} + \dfrac{0,7\times 13,38}{21,72} = (0,04 + 0,43) = 0,47 < 1,0$	**The 554 mm x 165 mm section is adequate.**
	Note: Similar calculations are required to check the combined stresses at various sections along the column and rafter sections where the cross-section reduces. In addition radial stresses at the knee should be checked. In this particular case since the negative bending moment decreases the curvature these will be compressive; other load cases may induce tensile radial stresses, (see Section 6.4 of EC5).	

8. Roof Trusses

Objective: *to illustrate the commonly used structural forms for pitched roofs and introduce BS 5268-3:2006 for the design of trussed rafters.*

8.1 Introduction

Timber pitched roof structures are used for two main reasons: firstly to ensure adequate weather-proofing and secondly, if required, to provide roof-space utilisation such as additional storage. Relatively low pitch angles such as 10° are acceptable for roofs surfaced with multi-layered bituminous felts or similar sheet materials. In situations where a roof is tiled, slated or covered with overlapping profile sheeting, the pitch must be sufficient to permit rapid drainage of water to prevent the ingress of water by wind driven rain. The minimum slope necessary is dependent on the type of covering, particularly the degree of overlap between individual units. Typically, large slates require approximately 22½° slopes whilst small slates and single lap tiles may require nearer 35° slopes. Plain tiles require at least a 40° slope to inhibit water penetration. The construction of a timber pitched roof can take one of many structural forms such as:

- ♦ mono-pitch roof,
- ♦ coupled roof,
- ♦ close coupled roof,
- ♦ collar roof,
- ♦ purlin roof   or
- ♦ trussed rafter roof

as indicated in Figures 8.1 to 8.3:

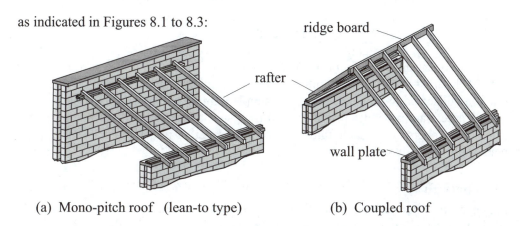

(a) Mono-pitch roof (lean-to type)   (b) Coupled roof

Figure 8.1

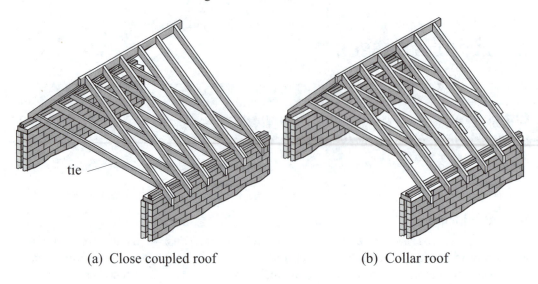

(a) Close coupled roof   (b) Collar roof

Figure 8.2

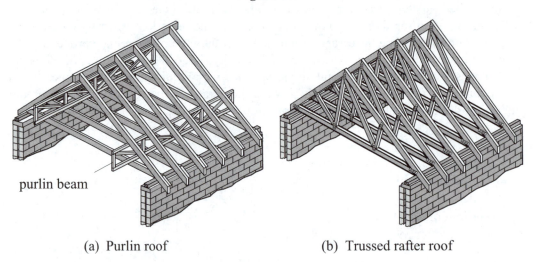

(a) Purlin roof   (b) Trussed rafter roof

Figure 8.3

All of the roof types shown in Figures 8.1 to 8.3, with the exception of the purlin roof and elements of the trussed rafter roofs, are essentially assembled on site at roof level. Their construction utilises basic timber sections and simple jointing methods, usually nailing.

Trussed rafters are normally pre-fabricated by a specialist manufacturer and are made using timber members of the same thickness fastened together in one plane by metal plate fasteners or plywood gussets.

8.2 Monopitch Roof

Monopitch roofs comprise a series of single rafters supported on wall plates directly carried on two walls, or on a wall-plate directly carried on a wall at the lower end and another wall-plate fastened to, or corbelled from, a wall at the upper level as shown in

Figure 8.1(a). The latter method is often referred to as a *lean-to construction*.

Since there is a tendency for an inclined rafter to induce horizontal thrust at the support, provision must be made to transfer this force to the wall and prevent sliding. ***Clearly the wall plate and the supporting wall must be capable of resisting the outward thrust from the rafter.***

The transfer of thrust is achieved by creating a notch (birdsmouth) in the rafter at the location where it meets the wall plate, as shown in Figure 8.4.

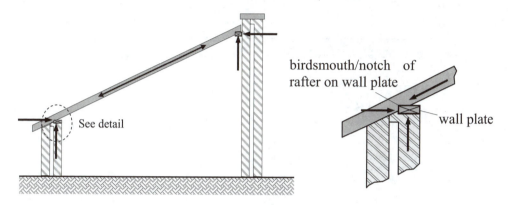

birdsmouth/notch of rafter on wall plate

wall plate

See detail

Figure 8.4

The wall plate is located at the top of the wall in a mortar bed to provide a level bearing for the rafters and distribute the end reaction evenly over the wall. The rafters are normally skew-nailed to the plate. A similar detail is used at the upper level. The covering material can be sheeted or tiled. When using sheeted materials care must be taken to ensure the appropriate rafter spacing is provided accurately to accommodate the standard sheet sizes.

In tiled roofs softwood battens are provided, the size of which depends upon the spacing of the rafters, normally ranging from 400 mm to 600 mm centres. When using larger spacings care must be taken to ensure that the battens have adequate stiffness to enable tiling or slating nails to be driven in.

The economic spans of monopitch roofs of this type are limited to approximately 3,0 m to 4,0 m.

8.3 Coupled Roof

A coupled roof is assembled by nailing rafters to a ridge board at the upper level and to wall plates at the lower level as shown in Figure 8.1(a).

The horizontal thrust which is produced at the wall plate normally limits the economic clear span to approximately 3,0 m. Spans above this level require a significant increase in the depth of the rafter to control deflection, and in addition buttressing of the support walls may be necessary to resist the lateral thrust. The ridge board locates the rafters in opposite alignment on each side preventing lateral movement.

This form of construction is sometimes used with very steeply pitched roofs in church buildings utilizing solid or glued laminated members and relatively low support walls. This structural arrangement, as indicated in Figure 8.5(b), significantly reduces the effects of horizontal thrust.

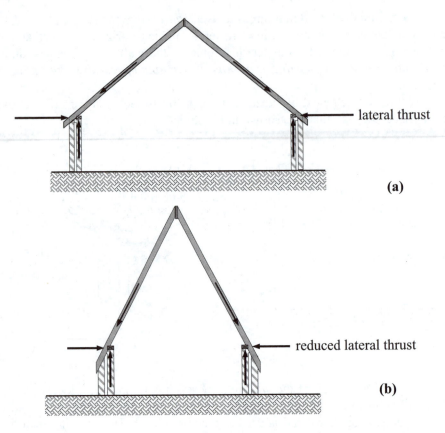

(a)

(b)

Figure 8.5

8.4 Close Coupled Roof

Close coupled roofs are a development of coupled roofs, in which the horizontal thrust at the lower rafter ends is contained internally within a triangulated framework as shown in Figure 8.6.

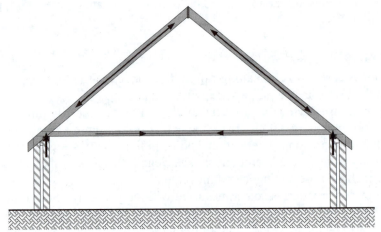

Figure 8.6

The additional tie member, which is securely nailed at its junctions with the rafters, may also be used as a ceiling joist. In this instance a heavier section than is required simply for tying is usually required. Economic spans for this type of construction are approximately 5,0 m to 6,0 m. In spans of this value, the section size adopted for the ceiling joist can be minimised by the introduction of additional supports such as binders and hangers as shown in Figure 8.7.

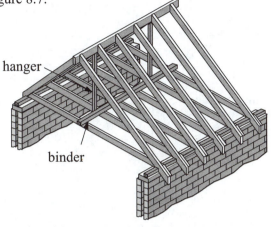

Figure 8.7

8.5 Collar Roof

Close couple roofs in which the tie is placed at a level above the supporting walls, typically one third to one half of the rise as shown in Figure 8.8, are referred to as collar roofs. This form of construction is less efficient in resisting the spread of the main rafters.

A disadvantage in this type of roof is that the section of rafter between the wall plate and the collar is subject to considerable bending. In addition, a bolted/connected joint between the collar and the rafter is normally required to adequately resist the rafter thrust. Collar roofs are economic for relatively short spans not exceeding 5,0 m.

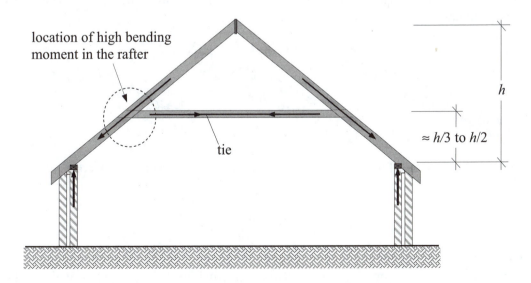

Figure 8.8

8.6 Purlin Roof

In coupled roofs as the span becomes larger, the required increase in rafter depth produces a less efficient and economic structural solution. This can be overcome by providing an additional support system to the rafters in the form of purlin beams as shown in Figure 8.3(a) and Figure 8.9.

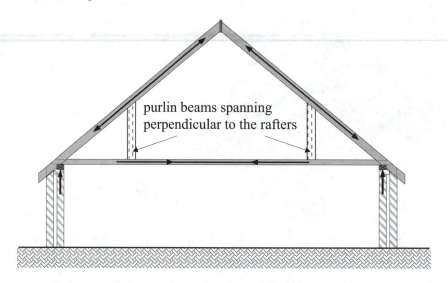

purlin beams spanning
perpendicular to the rafters

Figure 8.9

The purlins are beams spanning in a direction perpendicular to the rafters and providing intermediate support and hence reducing the effective span of the rafters. There are numerous possible structural configurations for the purlins; the most commonly used ones are illustrated in Figures 8.10(a) to 8.10(c). This form of construction is suited to spans exceeding 6,0 m.

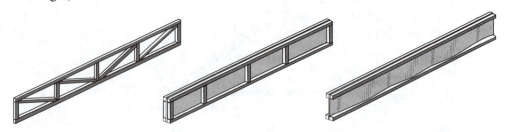

Trussed purlin beam   Ply-web beam with stiffeners   Ply-web beam with corrugated web
       (a)   (b)   (c)

Figure 8.10

8.7 Trussed Rafters

Light trussed rafters normally span between external load-bearing walls without the requirement for intermediate supports. They are fabricated using glued plywood gussets or metal plate fasteners as shown in Figure 8.11.

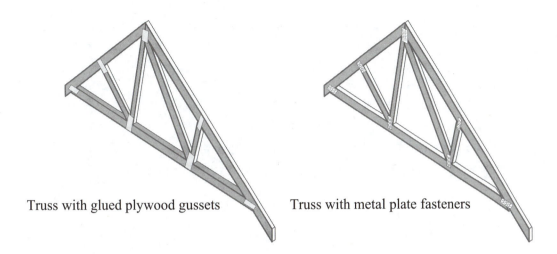

Truss with glued plywood gussets   Truss with metal plate fasteners

Figure 8.11

Ridge boards, hangers or binders are not required but *it is essential that adequate longitudinal and diagonal bracing of the whole roof and connection to the supporting structure are provided.* In BS 5268-3:2006 guidance is given for both the bracing requirements and the design of trussed rafters. In the majority of situations involving domestic housing the permissible span tables given in Annex B of this Part of the code can be used to determine appropriate rafter and ceiling tie section sizes. The permissible span tables give maximum permissible spans for the two truss configurations shown in Figure 8.12 and Figure 8.13 for a range of member sizes and roof pitches in selected classes of timber.

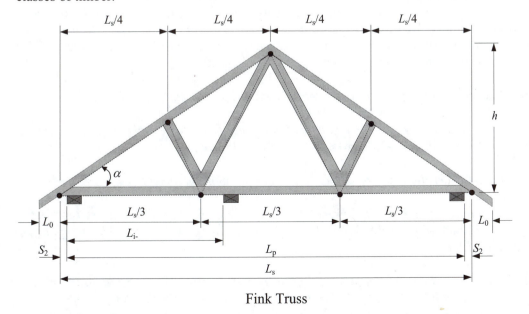

Fink Truss

Figure 8.12

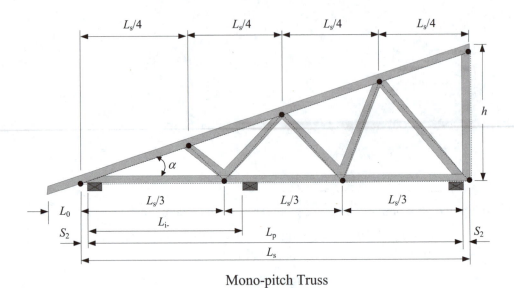

Mono-pitch Truss

Figure 8.13

The information in the tables is limited to spans not exceeding 12,0 m and has been obtained from extensive programs of testing trussed rafters. A number of conditions must be satisfied before the tables can be used without requiring any additional calculation. They are:

a) *"The dead load on the rafter members does not exceed 0,41 kN/m measured along the slope. For lightweight roofs subject to wind uplift, checks should be made for possible stress reversal in members, the effects of such reversals on joint design, and the need to provide additional lateral restraint to members normally in tension. The adequacy of the holding down restraints should also be checked in accordance with 7.2.3."* (Clause 7.2.3 in the code refers to restraint of a roof against wind uplift).

b) *"The dead load on the ceiling tie does not exceed 0,15 N/m."*

c) *"The method of support and capacity of any water tank carried by the trussed rafters is in accordance with the details given in Figure 7"* (of the code).

d) *"The trussed rafters do not support any other plant or equipment".*

e) *"The imposed load does not exceed those recommended in 6.4.2"* (Clause 6.4.2 refers to imposed and wind loading), *"with the exception of snow loading, which should not exceed 0,75 kN/m².*"

f) *"Each trussed rafter supports load from a section of roof not more than 0,6 m wide and the spacing between centres of trussed rafters is not more than 0,6 m, except where it is necessary to accommodate chimneys and other openings as described in 7,6"* (Clause 7.6 refers to limitation for hatch, chimney and other openings).

g) *"The joints between members have adequate strength in service and can resist normal handling forces without damage (see 6.5.6)"* (Clause 6.5.6 refers to joint design).

h) *"The cross-section dimensions, tolerances, moisture content and fabrication are in accordance with the recommendations of this part of BS 5268"* (i.e. BS 5268-3:2006).

i) *"The trussed rafter members are of one or more of the species listed in Table 1,"* (Part 3), *"and strength graded in accordance with 5.1.2"* (Clause 5.1.2 specifies grading requirements).

j) *"The strength class of timber used is not inferior to that recommended in Tables B.1 or B.2"* (Maximum permissible span Tables), *"and within each trussed rafter the rafter and ceiling tie members are of the same strength grade or strength class."*

k) *"The roof pitch is no greater than 35° and the supports are located in accordance with Figure 3"* (Figure 3 indicates the wall plate location in relation to truss members if a simplified analysis is being carried out as shown in Figure 8.12 and Figure 8.13 of this text), *"so that not less than half the width of each bearing is vertically below the eaves joint fastener and the distance S_2 is no greater than $S_1/3$ or 50 mm, whichever is the greater."*

l) *"Any overhang of the rafter at the eaves (dimension L_o in Figure 11) does not exceed 0,6 mm unless justified by calculation"* (Figure 11 indicates the two truss configurations, as indicated in Figures 8.12 and 8.13 of this text, to which the tables in Annex B apply).

m) *"Proper consideration is given to the stability of both the complete roof and the individual trussed rafters in accordance with Clause 7"* (Clause 7 refers to design responsibilities and overall stability).

BS 5268-3:2006   Annex B   Clause B.2   Internal members

"The internal members used for Fink trussed rafters in accordance with Table B.1 and Table B.2 should be of the same size and strength class as the smaller of either the rafter members or the ceiling tie. Smaller sizes and other strength classes may be used where they can be justified by test or calculation. In no case however should the internal members be less than 60 mm in depth (see 5.1.3).

For monopitch trussed rafters, the internal members and the vertical end member should be justified by test or calculation, but in no case should the depth be less than 60 mm (see 5.1.3), and the length should not exceed the appropriate value given in Table 4."

Note: Clause 5.1.3 in Part 3 of the code relates to timber sizes in accordance with BS EN 336 and Table 4 in Part 3 provides maximum lengths of internal members for given depths and thicknesses of timber.

9. Mechanical Fasteners

Objective: *to discuss, and illustrate the design of, the various types of mechanical fasteners, i.e. nails, staples, bolts, dowels, screws, punched metal-plates, split-ring, shear-plate and tooth-plate connectors and glue which are used in structural timber.*

9.1 Introduction

As a structural material, timber has been used for many hundreds of years. Traditionally the transfer of forces from one structural member to another was achieved by the construction of carpentry joints such as lap joints, cogging joints, tenon joints and framed joints as indicated in Figure 9.1. In many instances the physical contact or friction between members was relied upon to transfer the forces between them.

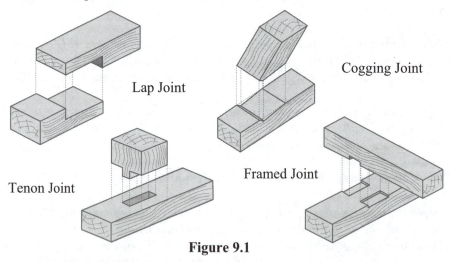

Figure 9.1

The use of mechanical fasteners is now firmly established as an essential part of modern economic design in timber. There are numerous types of fastener available ranging from dowel type fasteners such as:

- ◆ nails,
- ◆ staples,
- ◆ bolts,
- ◆ dowels and
- ◆ screws,

to individual types such as:

♦ punched, metal plate fasteners,
♦ split-ring connectors,
♦ shear-plate connectors and
♦ toothed-plate connectors.

Load transfer in the former group is achieved by bearing and shear stresses in the timber whilst in the latter group, it is achieved by bearing stresses at the surfaces of the members.

The design requirements for each of these types of fastener are given in Clauses and Tables of BS EN 1995-1-1:2004-1-1:2000 as indicated in Table 9.1 for dowel type connections and Table 9.2 for the individual types.

Type of Fastener	Type of connection	Clause Number	Tables
Nails: Section 8.3 Detailing: Clause 10.4.2	Laterally loaded: *timber-to-timber, panel-to-timber, steel-to-timber.* Axially loaded Combined laterally and axially loaded.	8.3.1 8.3.1.2 8.3.1.3 8.3.1.4 8.3.2 8.3.3	Table 8.1: k_{ef} related to nail spacing. Table 8.2: Minimum spacing and edge and end distances for nails.
Staples: Section 8.4	Round or nearly round or rectangular staples with bevelled or symmetrical pointed legs.		Table 8.3: Minimum spacing and edge and end distances for staples.
Bolts: Section 8.5 Detailing: Clause 10.4.3	Laterally loaded: *timber-to-timber, panel-to-timber, steel-to-timber.* Axially loaded.	8.5.1 8.5.1.1 8.5.1.2 8.5.1.3 8.5.2	Table 8.4: Minimum spacing and edge and end distances for bolts.
Dowels: Section 8.6 Detailing: Clause 10.4.4	Laterally loaded: *timber-to-timber, panel-to-timber, steel-to-timber.*	8.5.1 8.5.1.1 8.5.1.2 8.5.1.3	Table 8.5: Minimum spacing and edge and end distances for dowels.
Screws: Section 8.7 Detailing: Clause 10.4.5	Laterally loaded. Axially loaded. Combined laterally and axially loaded.	8.7.1 8.7.2 8.7.3	Table 8.6: Minimum spacing and edge and end distances for screws.

Table 9.1

Type of Fastener	Type of connection	Clause Number	Tables
Punched metal plate: Section 8.8	Punched metal plate fasteners with two orthogonal directions.	8.8.1 to 8.8.5	
Split-ring and shear-plate connectors: Section 8.9 Detailing: Table 10.1	Ring connectors of Type A or shear-plate connectors of type B according to BS EN 912 with diameter not greater than 200 mm.		Table 8.7: Minimum spacing and edge and end distances for ring and shear-plate connectors.
Tooth-plate connectors: Section 8.10 Detailing: Table 10.1	Toothed-plate connectors of Type C according to BS EN 912.	8.5.1 8.5.1.1 8.5.1.2 8.5.1.3 8.5.2	Tables 8.8 and 8.9: Minimum spacing and edge and end distances for tooth-plate connector types C1 to C9 and types C10/C11 respectively.
Glue Detailing: Section 10.3 Section C.4	Glued thin-webbed beams. Glued thin-flanged beams. Mechanically jointed and glued columns.	9.1.1 9.1.2 9.1.4	

Table 9.2

There are many proprietary types of fastener, such as punched metal plate fasteners with or without integral teeth, and splice plates (Figure 9.2), which are also used but are not included in the code.

Punched metal plate

Pre-formed splice plate

Figure 9.2

Many factors may influence the use of a particular type of fastener, e.g.

- ◆ method of assembly,
- ◆ connection details,
- ◆ purpose of connection,
- ◆ loading,
- ◆ permissible stresses,
- ◆ aesthetics.

Normally there are several methods of connection for any given joint which could provide an efficient structural solution. The choice will generally be dictated by consideration of cost, availability of skills, suitable fabrication equipment and desired finish required by the client. Glued joints generally require more rigorous conditions of application and control than mechanical fasteners.

The load carrying capacity of fasteners is defined in Clause 8.1.2(4) as:

$$F_{v,ef,Rk} = n_{ef}\,F_{v,Rk} \qquad \text{(Equation (8.1) in EC5)}$$

where:
$F_{v,ef,Rk}$ is the effective characteristic load-carrying capacity of one row of fasteners parallel to the grain,
n_{ef} is the effective number of fasteners in line parallel to the grain (see Clauses 8.3.1.1(8) and 8.5.1.1(4)),
$F_{v,Rk}$ is the characteristic load-carrying capacity of each fastener parallel to the grain (see Section 8.2).

Equation (8.1) relates to one row of fasteners parallel to the grain direction. In the case of a force acting at an angle to the direction of the row, $F_{v,ef,Rk}$ should be greater than or equal to the component of the force parallel to the row as shown in Figure 9.3.

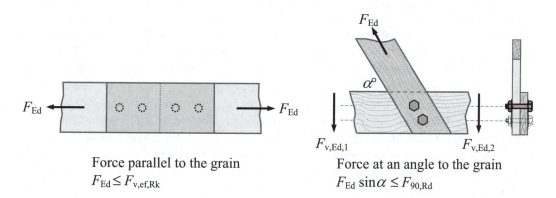

Force parallel to the grain
$F_{Ed} \le F_{v,ef,Rk}$

Force at an angle to the grain
$F_{Ed}\sin\alpha \le F_{90,Rd}$

Figure 9.3

Similarly, the possibility of splitting caused by the component of force perpendicular to the grain should be considered in accordance with Clause 8.1.4(2)P and Figure 8.1 of the code, i.e.

$$F_{v,\,Ed} \leq F_{90.Ed} \qquad \text{(Equation (8.2) in EC5)}$$

with $\quad F_{v,Ed} = \max \begin{cases} F_{v,Ed1} \\ F_{v,Ed2} \end{cases} \qquad \text{(Equation (8.3) in EC5)}$

where:

$F_{90,Ed}$ is the design splitting capacity in accordance with Clause 2.4.3, i.e. $k_{mod} \times \dfrac{F_{90,Rk}}{\gamma_m}$

The value of $F_{90,Rk}$ for softwoods is given in Clause 8.1.4(3) and Equation (8.4) as

$$14bw \sqrt{\frac{h_c}{\left(1 - \dfrac{h_c}{h}\right)}}$$

where:

$F_{90,Rk}$ is the characteristic splitting capacity in N,

w is a modification factor defined in Equation (8.5) of the code,

h_c is the loaded edge distance to the centre of the most distant fastener or to the edge of the punched metal plate fastener in mm,

h is the timber member height in mm,

b is the member thickness in mm,

In multiple shear plane connections the resistance of each shear plane should be determined by assuming that each shear plane is part of a series of three-member connections. This is illustrated in Examples 9.8, 9.10 and 9.11.

9.2 Lateral Load-carrying Capacity of Dowel-type Fasteners

The design capacity of dowel-type fasteners is based on the analysis carried out by Johansen (69) and is dependent on a number of factors such as the embedment strength of the fastener in the timber, the fastener yield moment and the axial withdrawal capacity of the fastener. Johansen (69) considered a number of possible failure modes relating to single- shear, double shear, timber-to-timber, timber-to-panel and steel-to-timber connections. The results of this research work are reflected in the code in terms of equations from which the characteristic load-carrying capacity for nails, staples, bolts dowels and screws per shear plane per fastener can be determined for each possible mode of failure; the minimum value being used for design. These Equations and variables for each of the fasteners considered are indicated in Table 9.3, Table 9.4 and Table 9.5.

The Equations incorporate an additional term to the Johansen values to allow for the 'rope effect' which enhances the strength. This effect is limited to a percentage (dependent on the type of fastener), of the value obtained using the Johansen equations. The 'rope effect' occurs in modes of failure where the fastener deforms and a component of the axial force induced by this deformation contributes to the joint shear strength.

Timber-to-timber and panel-to-timber connections - (Clause 8.2.2)		
Failure Mode	**Characteristic load-carrying capacity $F_{v,Rk}$**	**Equation No.**
	$f_{h,1,k}\, t_1\, d$	(8.6(a)) single shear
	$f_{h,2,k}\, t_2\, d$	(8.6(b)) single shear
	$\dfrac{f_{h,1,k}t_1 d}{1+\beta}\left[\sqrt{\beta+2\beta^2\left[1+\dfrac{t_2}{t_1}+\left(\dfrac{t_2}{t_1}\right)^2\right]+\beta^3\left(\dfrac{t_2}{t_1}\right)^2}-\beta\left(1+\dfrac{t_2}{t_1}\right)\right]+\dfrac{F_{ax,Rk}}{4}$	(8.6(c)) single shear
	$1{,}05\dfrac{f_{h,1,k}t_1 d}{2+\beta}\left[\sqrt{2\beta(1+\beta)+\dfrac{4\beta(2+\beta)M_{y,Rk}}{f_{h,1,k}d\,t_1^2}}-\beta\right]+\dfrac{F_{ax,Rk}}{4}$	(8.6(d)) single shear
	$1{,}05\dfrac{f_{h,1,k}t_2 d}{1+2\beta}\left[\sqrt{2\beta^2(1+\beta)+\dfrac{4\beta(1+2\beta)M_{y,Rk}}{f_{h,1,k}d\,t_2^2}}-\beta\right]+\dfrac{F_{ax,Rk}}{4}$	(8.6(e)) single shear
	$1{,}15\sqrt{\dfrac{2\beta}{1+\beta}}\sqrt{2M_{y,Rk}f_{h,1,k}d}+\dfrac{F_{ax,Rk}}{4}$	(8.6(f)) single shear
	$f_{h,1,k}\, t_1\, d$	(8.7(g)) double shear
	$0{,}5 f_{h,2,k}\, t_2\, d$	(8.7(h)) double shear
	$1{,}05\dfrac{f_{h,1,k}t_1 d}{2+\beta}\left[\sqrt{2\beta(1+\beta)+\dfrac{4\beta(2+\beta)M_{y,Rk}}{f_{h,1,k}d\,t_1^2}}-\beta\right]+\dfrac{F_{ax,Rk}}{4}$	(8.7(j)) double shear
	$1{,}15\sqrt{\dfrac{2\beta}{1+\beta}}\sqrt{2M_{y,Rk}f_{h,1,k}d}+\dfrac{F_{ax,Rk}}{4}$	(8.7(k)) double shear

Table 9.3

Steel-to-timber connections - (Clause 8.2.3)		
Failure Mode	**Characteristic load-carrying capacity $F_{v,Rk}$**	**Equation No.**
	$0{,}4 f_{h,k}\, t_1\, d$	(8.9(a)) single shear
	$1{,}15\sqrt{2 M_{y,Rk} f_{h,k} d} + \dfrac{F_{ax,Rk}}{4}$	(8.9(b)) single shear
	$f_{h,k} t_1 d\left[\sqrt{2+\dfrac{4 M_{y,Rk}}{f_{h,k} d\, t_1^2}} -1\right] + \dfrac{F_{ax,Rk}}{4}$	(8.10(c)) single shear
	$2{,}3\sqrt{M_{y,Rk} f_{h,k} d} + \dfrac{F_{ax,Rk}}{4}$	(8.10(d)) single shear
	$f_{h,k}\, t_1\, d$	(8.10(e)) single shear
	$f_{h,k}\, t_1\, d$	(8.11(f)) double shear
	$f_{h,k} t_1 d\left[\sqrt{2+\dfrac{4 M_{y,Rk}}{f_{h,1,k} d\, t_1^2}} -1\right] + \dfrac{F_{ax,Rk}}{4}$	(8.11(g)) double shear
	$2{,}3\sqrt{M_{y,Rk} f_{h,1,k} d} + \dfrac{F_{ax,Rk}}{4}$	(8.11(h)) double shear
	$0{,}5 f_{h,2,k}\, t_2\, d$	(8.12(j)) double shear
	$1{,}15\sqrt{2 M_{y,Rk} f_{h,2,k} d} + \dfrac{F_{ax,Rk}}{4}$	(8.12(k)) double shear
	$0{,}5 f_{h,2,k}\, t_2\, d$	(8.13(l)) double shear
	$2{,}3\sqrt{M_{y,Rk} f_{h,2,k} d} + \dfrac{F_{ax,Rk}}{4}$	(8.13(m)) double shear

Table 9.4

Symbol	Definition
$F_{v,Rk}$	the characteristic load-carrying capacity per shear plane per fastener
$M_{y,Rk}$	the characteristic yield fastener moment
$F_{ax,Rk}$	the characteristic axial withdrawal capacity of the fastener *
β	the ratio between the embedment strength of the members
$f_{h,i,k}$	the characteristic embedment strength in timber member i
d	the fastener diameter
Table 9.3 $- t_i$	the timber or board thickness or penetration depth, with i either 1 or 2 (see Clauses 8.3 to 8.7 in the code)
Table 9.4 $- t_1$	the smaller of the thickness of the timber side member or the penetration depth
Table 9.4 $- t_2$	the thickness of the timber middle member

* In Table 9.3 the contribution due to the 'rope effect' i.e. $(F_{ax,Rk}/4)$ should be limited to the following percentages of the capacity determined using only the Johansen equations, as follows:
Round nails: 15%, Square nails: 25%, Other nails: 50%, Screws: 100%,
Bolts: 25% and Dowels: 0%.
If the value of $F_{ax,Rk}$ is not known it should be neglected. In the case of single shear fasteners, $F_{ax,Rk}$ is taken as the lower of the capacities of the two members.

Table 9.5

9.3 Nailed Connections (Section 8.3)

A selection of nails is indicated in Figure 9.4. The most commonly used type of nails are (a) 'round plain wire' nails. Two variations of this type are (b) 'clout nails' (often referred to as slate, felt or plasterboard nails), which are simply large versions with a larger diameter head, and (c) 'lost head nails' in which the head is very small. The introduction of 'improved nails' such as (d) 'square twisted', and (e) 'helically threaded and annular ringed shank nails' with increased lateral and withdrawal resistance has proved very useful; particularly for fixing sheet materials to roofs and floors where 'popping' is often a problem with plain round nails.

Pneumatically driven nails, (f), when manufactured from suitably hardened and tempered steel, enable relatively straightforward fixing of timber to materials such as concrete, masonry and stone.

Pre-drilling of holes may be required to avoid splitting and to enable use with dense hardwoods such as greenheart and keruing timbers. In Clause 8.3.1.1(2) the code indicates that pre-drilling should be carried out when:

(i)   the characteristic density of the timber is greater than 500 kg/m^3 or
(ii) the diameter 'd' of the nail exceeds 8 mm.

As indicated in Clause 10.4.2(3) of the code, the pre-drilled holes should not be greater than $0,8 \times$ nail diameter. In timber-to-timber connections timber should be pre-drilled when the thickness of the timber members is smaller than that determined using Equation (8.18), i.e.

$$t = \max \begin{cases} 7d \\ (13d - 30)\dfrac{\rho_k}{400} \end{cases} \qquad \text{(Equation (8.18) in EC5)}$$

where t is the minimum thickness of timber member to avoid pre-drilling in mm.

In the UK Annex it is indicated that Clause 8.3.1.2(7) relating to pre-drilling of species which are especially sensitive to species does not apply to nailed joints.

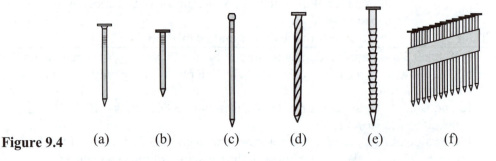

Figure 9.4   (a)   (b)   (c)   (d)   (e)   (f)

Nails are used for either locating timber e.g. a stud to a wall plate as in Figure 9.5(a), or for transferring forces such as the shear force and bending moment at a knee joint in a portal frame joint as shown in Figure 9.5(b).

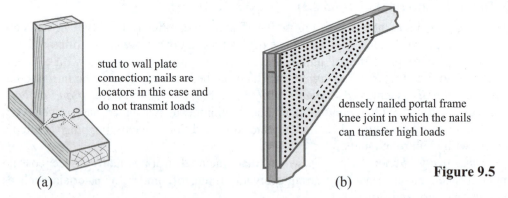

stud to wall plate connection; nails are locators in this case and do not transmit loads

densely nailed portal frame knee joint in which the nails can transfer high loads

Figure 9.5

(a)   (b)

The design of nailed joints is dependent on a number of factors, two of which are the *headside thickness* and the *pointside penetration* as shown in Figure 9.6:

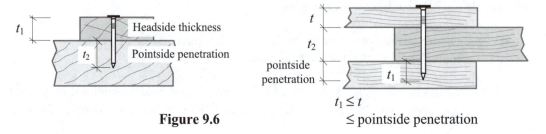

t_1 Headside thickness

t_2 Pointside penetration

t

t_2

pointside penetration

t_1

$t_1 \le t$

$\le$ pointside penetration

Figure 9.6

The lateral load-carrying capacity is dependent on the yield moment of the nails as indicated in Tables 9.3 and 9.4 above. This is defined in Clause 8.3.1.1(4) and Equation (8.14) of the code for nails with a minimum tensile strength of $f_u = 600$ N/mm^2 as:

$M_{y,Rk} = 0,3\,f_u\,d^{\,2,6}$ for round nails   and $M_{y,Rk} = 0,45\,f_u\,d^{\,2,6}$ for square nails.

9.3.1 Embedment Strength ($f_{h,k}$)

The embedment strength of timber is determined from an '*embedment strength test*' in which a sample of timber is subjected to a load imposed by a dowel bar as shown in Figure 9.10. The embedment strength is defined as the maximum load or the load at a specified limiting deformation divided by the projected area of the dowel in the specimen (see STEP 1 (58)).

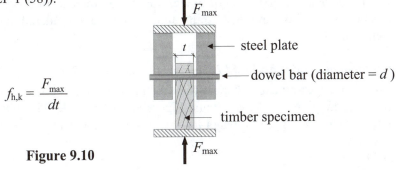

$$f_{h,k} = \frac{F_{max}}{dt}$$

Figure 9.10

The characteristic embedment strength for nails in various materials as given in the code is summarized in Table 9.6.

Material	Characteristic Embedment Strength - $f_{h,k}$	Equation No.
Timber and LVL diameter $d \leq 8$ mm	Without pre-drilling: $f_{h,k} = 0{,}082\,\rho_k\,d^{\,-0,3}$ N/mm^2	(8.15)
	With pre-drilling: $f_{h,k} = 0{,}082\,(1 - 0{,}01d)\,\rho_k$ N/mm^2	(8.16)
diameter $d > 8$ mm	See Clause 8.5.1 (i.e. as for bolts)	
Plywood head diameter $\geq 2d$ mm	$f_{h,k} = 0{,}11\rho_k\,d^{\,-0,3}$ N/mm^2	(8.20)
Hardwood (BS EN 622-2)	$f_{h,k} = 30\,\rho_k\,d^{\,-0,3}\,t^{\,0,6}$ N/mm^2	(8.21)
Particle board and OSB	$f_{h,k} = 65\,d^{\,-0,7}\,t^{\,0,1}$ N/mm^2	(8.22)
ρ_k is the characteristic density in kg/m^3, d is nail diameter in mm, t is the panel thickness in mm.		

Table 9.6

9.3.2 Pointside Penetration Lengths (Clause8.3.1.1 and 8.3.1.2)

The minimum pointside penetration lengths (see Figure 9.6) for nails are defined as follows:

For smooth nails: pointside penetration $\geq 8 \times$ nail diameter (d)

All other nails: pointside penetration $\geq 6 \times$ nail diameter (d)

9.3.3 End Grain (Clause 8.3.1.2)

The UK Annex permits the use of Clause 8.3.1.2(4) when considering the lateral load-carrying capacity for nails in the end grain of timber. If the requirements of this Clause are not met, then smooth nails on the end grain should not be considered capable of transmitting lateral forces.

9.3.4 Minimum Spacings, Edge and End Distances (Clause 8.3.1.2(5))

The minimum spacing, edge and end distances required to avoid splitting in nailed timber-to-timber connections are given in Table 8.2 and Figure 8.7 of the code. The values in the table are modified when applied to panel-to-timber connections and steel-to-timber connections as follows:
Clause (8.3.1.3(1)):
Panel-to-timber connections: minimum spacings = (Table 8.2 values $\times$ 0,85)
                         minimum end/edge distances = Table 8.2 values
Clause (8.3.1.4(1):
Steel-to-timber connections: minimum spacings = (Table 8.2 values $\times$ 0,70)
                         minimum end/edge distances = Table 8.2 values.

9.3.5 Overlapping Nails (Clause 8.3.1.1(7))

It is acceptable to overlap nails in the central member of three member connections provided that $(t - t_2)$ is greater than $4d$ as indicated in Figure 9.11.

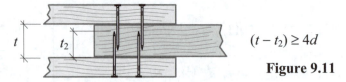

$$(t - t_2) \geq 4d$$

Figure 9.11

9.3.6 Minimum Number of Nails (Clause 8.3.1.1(9))

The minimum number of nails in a connection is two.

9.3.7 Effective Number of Fasteners (n_{ef})

If a single row of nails parallel to the grain is used, then unless the nails are staggered perpendicular to the grain by at least $(1,0 \times$ nail diameter $d)$, the load-carrying capacity parallel to the grain should be determined using an '*effective*' number of fasteners $\boldsymbol{n_{ef}}$ as given in Equation (8.17) of the code.

$n_{ef} = n^{k_{ef}}$ (Equation (8.17) in EC5)

where n is the number of nails and k_{ef} is given in Table 8.1 of the code.

Figure 9.12

The design of a selection of typical connections with laterally loaded nails is given in Examples 9.1 to 9.4. The selection covers a range of possible combinations of materials, subjected to single and double shear.

9.3.8 Example 9.1: Timber-to-timber Connection in Single Shear

A nailed timber-to-timber connection is required to transmit a lateral load as shown in Figure 9.13. Using the data given, design a suitable connection.

Design data:

Characteristic long-term variable load (Q_k)	800 N
Timber Class	C16
Moisture content	$\leq 20\%$
Type of nail	smooth, round
Pre-drilling	None
Nail diameter (d)	3,4 mm
Nail length (L)	70 mm
Tensile strength of wire (f_u)	600 N/mm^2

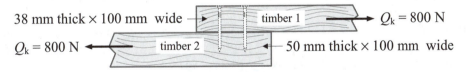

38 mm thick × 100 mm wide — timber 1 — $Q_k = 800$ N

$Q_k = 800$ N — timber 2 — 50 mm thick × 100 mm wide

Figure 9.13

Solution:

BS EN 338:2003

Clause 5.0	Characteristic values for C16 Timber
Table 1	Characteristic density: $\rho_k = 310,0$ kg/m^3
	Mean density: $\rho_{mean} = 370,0$ kg/m^3

BS EN 1995-1-1:2004

Clause 2.3.2.1	Load-duration and moisture influence on strength: k_{mod}	
Clause 2.3.1.3	Moisture content $\leq 20\%$	**Service Class 2**
Table 3.1	Long-term actions	$k_{mod} = 0,7$
NA: Table NA.3	Partial factor for material properties and resistance: γ_M	

For solid timber – untreated and preservative treated $\qquad$ $\gamma_M = 1,3$

For connections (except punched plate fasteners) $\qquad$ $\gamma_M = 1,3$

Clause 8.3.1.1(1)   Headside thickness $t_1 = 38$ mm   (single shear) $t_1 = 38$ mm

Pointside penetration $t_2 = (70 - 38) = 32$ mm $t_2 = 32$ mm

Clause 8.3.1.2(1)   Minimum pointside penetration for smooth nails $= 8d$

$8d = (8 \times 3,4) = 27,2$ mm $<$ actual penetration

Pointside penetration acceptable

Clause 8.3.1.1(4)   Characteristic yield moment for smooth nails with $f_u = 600$ N/mm^2

Equation (8.14) $M_{y,Rk} = 0,3 f_u d^{2,6} = (0,3 \times 600 \times 3,4^{2,6}) = 4336,28$ Nmm

$M_{y,Rk} = 4336,28$ Nmm

Clause 8.3.1.1(5)   Characteristic embedment strength: $f_{h,k}$

Equation (8.15) $f_{h,k} = 0,082\, \rho_k d^{-0,3}$

No pre-drilling for timber 1 or timber 2

Timber 1: $f_{h,1,k} = (0,082 \times 310 \times 3,4^{-0,3}) = 17,609$ N/mm^2

Timber 2: $f_{h,2,k} = (0,082 \times 310 \times 3,4^{-0,3}) = 17,609$ N/mm^2

Ratio of embedment strengths $\beta = \dfrac{f_{h,2,k}}{f_{h,1,k}} = 1,0$ **$\beta = 1,0$**

Clause 8.2.2(2) $F_{ax,Rk}$ is unknown and hence is the contribution from the rope effect in Equations (8.6(a)) to (8.6(f)) is taken as zero.

Note: the value $f_{ax,k}$ and hence $F_{ax,Rk}$ can be determined from Equations (8.23) to (8.26) provided that the pointside penetration is at least $12d$; in this it is less than $12d$ ($12 \times 3,4 = 40,8$ mm).

Clause 8.2.2(1)   Characteristic load-carrying capacity: $F_{v,Rk}$ – single shear

Equation (8.6(a)) $f_{h,1,k}\, t_1\, d = (17,609 \times 38 \times 3,4) = 2275,1$ N

$F_{v,Rk,a} = 2275,1$ N

Equation (8.6(b)) $f_{h,2,k}\, t_2\, d = (17,609 \times 32 \times 3,4) = 1915,9$ N

$F_{v,Rk,b} = 1915,9$ N

Equation (8.6(c)) $\dfrac{f_{h,1,k} t_1 d}{1+\beta} \left[\sqrt{\beta + 2\beta^2 \left[1 + \dfrac{t_2}{t_1} + \left(\dfrac{t_2}{t_1}\right)^2 \right] + \beta^3 \left(\dfrac{t_2}{t_1}\right)^2} - \beta \left(1 + \dfrac{t_2}{t_1}\right) \right]$

$= \dfrac{17,609 \times 38 \times 3,4}{1+1,0} \times \left[\sqrt{1,0 + 2 \times 1,0^2 \left[1 + \dfrac{32}{38} + \left(\dfrac{32}{38}\right)^2 \right] + 1,0^3 \left(\dfrac{32}{38}\right)^2} - 1,0 \left(1 + \dfrac{32}{38}\right) \right]$

$= 873,4$ kN $F_{v,Rk,c} = 873,4$ N

Equation (8.6(d)) $1,05\dfrac{f_{h,1,k}t_1 d}{2+\beta}\left[\sqrt{2\beta(1+\beta)+\dfrac{4\beta(2+\beta)M_{y,Rk}}{f_{h,1,k}d\,t_1^2}}-\beta\right]$

$= 1,05\times\dfrac{17,609\times38\times3,4}{2+1,0}\times\left[\sqrt{2\times1,0(1+1,0)+\dfrac{4\times1,0(2+1,0)\times4336,28}{17,609\times3,4\times38^2}}-1,0\right]$

$= 911,9$ kN $F_{v,Rk,d} = 911,9$ N

Equation (8.6(e)) $1,05\dfrac{f_{h,1,k}t_2 d}{1+2\beta}\left[\sqrt{2\beta^2(1+\beta)+\dfrac{4\beta(1+2\beta)M_{y,Rk}}{f_{h,1,k}d\,t_2^2}}-\beta\right]$

$= 1,05\times\dfrac{17,609\times32\times3,4}{1+2\times1,0}\times\left[\sqrt{2\times1,0^2(1+1,0)+\dfrac{4\times1,0(1+2\times1,0)\times4336,28}{17,609\times3,4\times32^2}}-1,0\right]$

$= 806,0$ kN $F_{v,Rk,e} = 806,0$ N

Equation (8.6(f)) $1,15\sqrt{\dfrac{2\beta}{1+\beta}}\sqrt{2M_{y,Rk}f_{h,1,k}d}$

$= 1,15\times\sqrt{\dfrac{2\times1,0}{1+1,0}}\times\sqrt{2\times4336,28\times17,609\times3,4} = 828,7$ N $F_{v,Rk,f} = 828,7$ N

The characteristic load-carrying capacity $F_{v,Rk}$ is minimum value determined using Equations (8.6(a)) to (8.6(f)).

$F_{v,Rk} = 806,0$ N corresponding with failure mode (e)

Equation (2.17)   Design load-carrying capacity of nails:

$$F_{v,Rd} = k_{mod}\dfrac{F_{v,Rk}}{\gamma_m} = 0,7\times\dfrac{806,0}{1,3} = 434,0\text{ N}$$

BS EN 1990:2002
Equation (6.10) $F_d = \sum_{j\geq1}\gamma_{G,j}G_{k,j}\ ''+''\ \gamma_{Q,1}Q_{k,1}\ ''+''\ \sum_{i>1}\gamma_{Q,i}\psi_{0,i}Q_{k,i}$

NA: Table NA.A1.2(B)
                   For variable actions: Leading actions $\gamma_{Q,1} = 1,5$
                   Design load $= F_{Q,d} = (1,5\times800) = 1200$ kN
BS EN 1995-1-1:2004

                   Number of nails required $n = \dfrac{F_{v,Rk}}{F_{Q,d}} = \dfrac{1200}{434} = 2,7$

Clause 8.3.1.1(9)   Minimum number of nails $= 2$

A nail pattern is selected to satisfy the minimum number of nails required, e.g.
(i) a single line of 3 nails or (ii) a double line of 4 nails.

Check the spacing requirements for (i):
Clause 8.3.1.1(8) Assume that the nails are staggered perpendicular to the grain.

Clause 8.3.1.2
(i) Single line of nails:
Figure 8.7

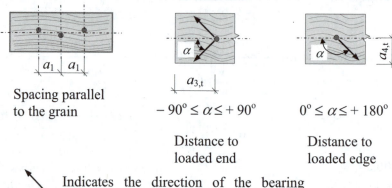

Spacing parallel
to the grain

$-90° \leq \alpha \leq +90°$

$0° \leq \alpha \leq +180°$

Distance to
loaded end

Distance to
loaded edge

Indicates the direction of the bearing
force imposed by the nail on the timber.

Figure 9.13(a)

timber 1

timber 2

Table 8.2

Nails without pre-drilling, $\rho_k = 310$ kg/m^3 (≤ 420 kg/m^3)
Angle between the force and grain direction $\alpha = 0°$, $d < 5$ mm
Minimum spacing parallel to the grain $a_1 = (5 + 5|\cos\alpha|)d$
$a_1 = [5 + (5 \times 1,0)] \times 3,4 = 34$ mm

Minimum distance to loaded end $a_{3,t} = (10 + 5|\cos\alpha|)d$
$a_{3,t} = [10 + (5 \times \cos0°)] \times 3,4 = 51$ mm

Minimum distance to loaded edge $a_{4,t} = (5 + 2|\sin\alpha|)d$
$a_{4,t} = [5 + (2 \times \sin0°)] \times 3,4 = 17$ mm

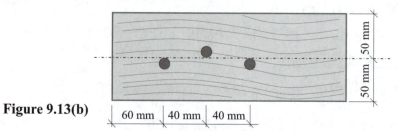

Figure 9.13(b)   60 mm   40 mm   40 mm

(ii) Double line of nails:
Figure 8.7

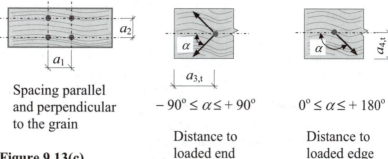

Spacing parallel
and perpendicular
to the grain

Figure 9.13(c)

$-90° \leq \alpha \leq +90°$

Distance to
loaded end

$0° \leq \alpha \leq +180°$

Distance to
loaded edge

Table 8.2

Nails without pre-drilling, $\rho_k = 310$ kg/m³ (≤ 420 kg/m³)
Angle between the force and grain direction $\alpha = 0°$, $d < 5$ mm
Minimum spacing parallel to the grain $a_1 = 34$ mm (as before)
Minimum spacing perpendicular to the grain $a_2 = 5d$
$a_2 = (5 \times 3,4) = 17$ mm
Minimum distance to loaded end $a_{3,t} = 51$ mm (as before)
Minimum distance to loaded edge $a_{4,t} = 17$ mm (as before)

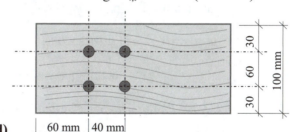

Figure 9.13(d)   60 mm 40 mm

Joint slip:
Clause 7.1(1)   Slip modulus K_{ser} per shear plane per fastener is given in Table 7.1.
Table 7.1   Nails without pre-drilling, $\rho_m = 370$ kg/m³, $d = 3,4$ mm,

$$K_{ser} = \rho_m^{1,5} d^{0,8}/30 = (370^{1,5} \times 3,4^{0,8})/30 = 631,49 \text{ N/mm}$$

Design load for the serviceability limit state $F_{Q,d} = 800$ N

BS EN 1990:2002
Equation (6.16b) $F_d = \sum_{j \geq 1} G_{k,j} \text{ "+" } \sum_{i \geq 1} \psi_{2,i} Q_{k,i}$

NA to BS EN 1990:2002
Table NA.A1.1   Quasi-permanent value for variable actions on domestic buildings
$\psi_{2,1} = 0,3$

BS EN 1995-1-1:2004
Table 3.2   Deformation factor for solid timber,   Service class 2
                       $k_{\text{def}} = 0{,}8$

(i) Single row of nails:

Equation (2.4) $u_{\text{fin,Q,1}} = u_{\text{inst,Q,1}}(1 + \psi_{2,1} k_{\text{def}})$

Load per nail $= \dfrac{F_{\text{Q,k}}}{n} = \dfrac{800}{3} = 266{,}67$ N

Instantaneous slip/side $u_{\text{inst}} = \dfrac{\text{Load per nail}}{K_{\text{ser}}} = \dfrac{266{,}67}{631{,}49} = 0{,}42$ mm/side

$u_{\text{fin,Q,1}} = 0{,}42 \times [1 + (0{,}3 \times 0{,}8)] = 0{,}52$ mm/side

Final joint opening $= (2 \times u_{\text{fin,Q,1}}) = (2 \times 0{,}52) = 1{,}04$ mm

(ii) Double row of nails:

Load per nail $= \dfrac{F_{\text{Q,k}}}{n} = \dfrac{800}{4} = 200{,}0$ N

Instantaneous slip/side $u_{\text{inst}} = \dfrac{\text{Load per nail}}{K_{\text{ser}}} = \dfrac{200{,}0}{631{,}49} = 0{,}32$ mm/side

$u_{\text{fin,Q,1}} = 0{,}32 \times [1 + (0{,}3 \times 0{,}8)] = 0{,}40$ mm/side

Final joint opening $= (2 \times u_{\text{fin,Q,1}}) = (2 \times 0{,}40) = 0{,}8$ mm

9.3.9 Example 9.2: Steel Plate-to-timber Connection in Single Shear

A nailed thin steel plate-to-timber connection is required to transmit a lateral load as shown in Figure 9.14. Using the data given, design a suitable connection.

Design data:

Characteristic long-term variable load (Q_k)	800 N
Timber Class	C16
Moisture content	$\leq 20\%$
Type of nail	smooth, round
Pre-drilling	None
Nail diameter (d)	3,4 mm
Nail length (L)	40 mm
Tensile strength of wire (f_u)	600 N/mm^2

Figure 9.14

Solution:
BS EN 338:2003
Clause 5.0   Characteristic values for C16 Timber
Table 1   Characteristic density: $\rho_k = 310,0$ kg/m^3
                    Mean density: $\rho_{mean} = 370,0$ kg/m^3

BS EN 1995-1-1:2004
Clause 2.3.2.1   Load-duration and moisture influence on strength: k_{mod}
Clause 2.3.1.3   Moisture content $\leq 20\%$ **Service Class 2**
Table 3.1   Long-term actions   **$k_{mod} = 0,7$**

NA: Table NA.3   Partial factor for material properties and resistance: γ_M
                   For solid timber – untreated and preservative treated   **$\gamma_M = 1,3$**
                   For connections (except punched plate fasteners)   **$\gamma_M = 1,3$**

Clause 8.2.3(1)   Steel plate thickness = 1,5 mm $\leq 0,5d$ = $(0,5 \times 3,4 = 1,7$ mm)
                                               Steel plate is classified as thin

                   t_1 is the smaller of the thickness of the timber side member or the
                   penetration depth.
                   Thickness of the timber side = 50 mm
                   Pointside penetration = $(40 - 1,5) = 38,5$ mm
                                                               $t_1 = 38,5$ mm

Clause 8.3.1.1(4) Characteristic yield moment for smooth nails with $f_u = 600$ N/mm^2
Equation (8.14) $M_{y,Rk} = 0,3f_u d^{2,6} = (0,3 \times 600 \times 3,4^{2,6}) = 4336,28$ Nmm
                                                      $M_{y,Rk} = 4336,28$ Nmm

Clause 8.3.1.1(5) Characteristic embedment strength: $f_{h,k}$
Equation (8.15) $f_{h,k} = 0,082\, \rho_k d^{-0,3}$
                   No pre-drilling for timber 1
                   Timber 1: $f_{h,k} = (0,082 \times 310 \times 3,4^{-0,3}) = 17,609$ N/mm^2

Clause 8.2.2(2) $F_{ax,Rk}$ is unknown and hence is the contribution from the rope effect in
                   Equation (8.9(b)) is taken as zero.
                   Note: the value $f_{ax,k}$ and hence $F_{ax,Rk}$ can be determined from
                   Equations (8.23) to (8.26) provided that the pointside penetration is at
                   least 12d; in this it is less than 12d ($12 \times 3,4 = 40,8$ mm).

Clause 8.2.3(3)   Characteristic load-carrying capacity: $F_{v,Rk}$ - thin plate in single shear

Equation (8.9(a)) $0,4\, f_{h,k}\, t_1\, d = (0,4 \times 17,609 \times 38,5 \times 3,4) = 922,0$ N
                                                      $F_{v,Rk,a} = 922,0$ N

Equation (8.9(b)) $1{,}15\sqrt{2M_{y,Rk}f_{h,k}d} = 1{,}15 \times \sqrt{2 \times 4336{,}28 \times 17{,}609 \times 3{,}4} = 828{,}7$ N

$$F_{v,Rk,b} = 828{,}7 \text{ N}$$

The characteristic load-carrying capacity $F_{v,Rk}$ is minimum value determined using Equations (8.9(a)) and (8.9(b)).

$F_{v,Rk} = 828{,}7$ N corresponding with failure mode (b)

Equation (2.17)   Design load-carrying capacity of nails:

$$F_{v,Rd} = k_{mod} \frac{F_{v,Rk}}{\gamma_{mt}} = 0{,}7 \times \frac{828{,}7}{1{,}3} = 446{,}2 \text{ N}$$

BS EN 1990:2002

Equation (6.10) $F_d = \sum_{j\geq1} \gamma_{G,j} G_{k,j} \; "+" \; \gamma_{Q,1} Q_{k,1} \; "+" \; \sum_{i>1} \gamma_{Q,i} \psi_{0,i} Q_{k,i}$

NA: Table NA.A1.2(B)

For variable actions: Leading actions $\gamma_{Q,1} = 1{,}5$

Design load $= F_{Q,d} = (1{,}5 \times 800) = 1200$ kN

Number of nails required $n = \dfrac{F_{v,Rk}}{F_{Q,d}} = \dfrac{1200}{446{,}2} = 2{,}7$

BS EN 1995-1-1:2004-1

Clause 8.3.1.1(9)   Minimum number of nails $= 2$

A nail pattern is selected to satisfy the minimum number of nails required, e.g.
(i) a single line of 3 nails or (ii) a double line of 4 nails.

Check the spacing requirements for (i):
Clause 8.3.1.2
(i) Single line of nails: assume that the nails are in line, no pre-drilled holes
Figure 8.7

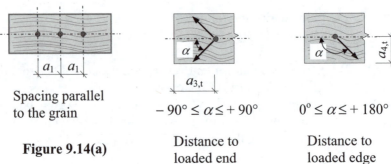

Spacing parallel to the grain

Figure 9.14(a)

$-90° \leq \alpha \leq +90°$

Distance to loaded end

$0° \leq \alpha \leq +180°$

Distance to loaded edge

Clause 8.3.1.4   Minimum edge and end distances are as given in Table 8.2. The minimum spacings are equal to $(0{,}7 \times$ Table 8.2 values).

Table 8.2   Nails without pre-drilling: $\rho_k = 310$ kg/m³ (≤ 420 kg/m³)
Angle between the force and grain direction $\alpha = 0°$, $d < 5$ mm

Minimum spacing parallel to the grain $a_1 = 0,7 \times \left(5 + 5|\cos\alpha|\right)d$

$a_1 = 0,7 \times [5 + (5 \times 1,0)] \times 3,4 = 23,8$ mm

Minimum distance to loaded end $a_{3,t} = \left(10 + 5|\cos\alpha|\right)d$

$a_{3,t} = [10 + (5 \times \cos0°)] \times 3,4 = 51$ mm

Minimum distance to loaded edge $a_{4,t} = \left(5 + 2|\sin\alpha|\right)d$

$a_{4,t} = [5 + (2 \times \sin0°)] \times 3,4 = 17$ mm

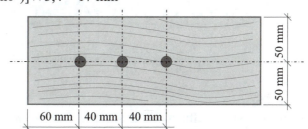

Figure 9.14(b)   60 mm | 40 mm | 40 mm

Clause 8.3.1.1(8)   Assume that the nails are in line, no pre-drilled holes.
Effective number of fasteners $n_{ef} = n^{k_{ef}}$

Table 8.1 $a_1 = 40$ mm $14d = (14 \times 3,4) = 47,6$ mm
Use interpolation to determine k_{ef} for $a_1/d = 40/3,4 = 11,76$
$k_{ef} = 0,85 + [(1,0 - 0,85)(4,0 - 1,76)/4,0] = 0,93$

Effective number of fasteners $n_{ef} = 3^{0,93} = 2,8 > 2,7$ required

(ii) Double line of nails:
Figure 8.7

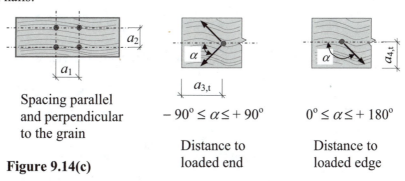

Spacing parallel
and perpendicular
to the grain $-90° \leq \alpha \leq +90°$ $0° \leq \alpha \leq +180°$

                                 Distance to   Distance to
Figure 9.14(c)   loaded end   loaded edge

Table 8.2   Nails without pre-drilling, $\rho_k = 310$ kg/m^3 (≤ 420 kg/m^3)

Angle between the force and grain direction $\alpha = 0°$, $d < 5$ mm

Minimum spacing parallel to the grain $a_1 = 34$ mm (as before)

Minimum spacing perpendicular to the grain $a_2 = 5d$
$a_2 = (5 \times 3,4) = 17$ mm

Minimum distance to loaded end $a_{3,t} = 51$ mm (as before)

Minimum distance to loaded edge $a_{4,t} = 17$ mm (as before)

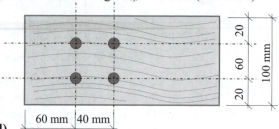

Figure 9.14(d)

Joint slip:

Clause 7.1(3)
Table 7.1

Slip modulus K_{ser} per shear plane per fastener is given in Table 7.1
Nails without pre-drilling, $\rho_m = 370$ kg/m^3, $d = 3,4$ mm,

$$K_{ser} = 2,0\,\rho_m^{1,5}d^{0,8}/30 = 2,0 \times (370^{1,5} \times 3,4^{0,8})/30 = 1262,97 \text{ N/mm}$$

Design load for the serviceability limit state $F_{Q,d} = 800$ N

BS EN 1990:2002
Equation (6.16b) $F_d = \sum_{j \geq 1} G_{k,j} \;"+"\; \sum_{i \geq 1} \psi_{2,i} Q_{k,i}$

NA. to EC
Table NA.A1.1

Quasi-permanent value for variable actions on domestic buildings
$\psi_{2,1} = 0,3$

BS EN 1995-1-1:2004
Table 3.2

Deformation factor for solid timber,   Service class 2
$k_{def} = 0,8$

(i) Single row of nails:

Equation (2.4) $u_{fin,Q,1} = u_{inst,Q,1}(1 + \psi_{2,1}k_{def})$

Load per nail $= \dfrac{F_{Q,k}}{n} = \dfrac{800}{3} = 266,67$ N

Instantaneous slip/side $u_{inst} = \dfrac{\text{Load per nail}}{K_{ser}} = \dfrac{266,67}{1262,97} = 0,21$ mm/side

$u_{fin,Q,1} = 0,21 \times [1 + (0,3 \times 0,8)] = 0,26$ mm/side

Final joint opening $= (2 \times u_{fin,Q,1}) = (2 \times 0,26) = 0,52$ mm

(ii) Double row of nails:

$$\text{Load per nail} = \frac{F_{Q,k}}{n} = \frac{800}{4} = 200{,}0 \text{ N}$$

$$\text{Instantaneous slip/side } u_{\text{inst}} = \frac{\text{Load per nail}}{K_{\text{ser}}} = \frac{200{,}0}{1262{,}97} = 0{,}16 \text{ mm/side}$$

$$u_{\text{fin,Q,1}} = 0{,}16 \times [1 + (0{,}3 \times 0{,}8)] = 0{,}20 \text{ mm/side}$$

$$\text{Final joint opening} = (2 \times u_{\text{fin,Q,1}}) = (2 \times 0{,}20) = 0{,}40 \text{ mm}$$

9.3.10 Example 9.3: Panel-to-timber Connection in Double Shear

A nailed panel-to-timber connection is required to transmit a lateral load as shown in Figure 9.15. Using the data given, design a suitable connection.

Design data:

Characteristic long-term variable load (Q_k)	2000 N
Timber Class	C16
Characteristic density of plywood (manufacturer's data) (ρ_k)	630 kg/m³
Mean density of plywood (manufacturer's data) (ρ_m)	680 kg/m³
Moisture content	$\leq 20\%$
Type of nail	smooth, round
Pre-drilling	plywood only
Nail diameter (d)	4,0 mm
Nail head diameter	$\geq 8{,}0$ mm
Nail length (L)	75 mm
Tensile strength of wire (f_u)	600 N/mm²

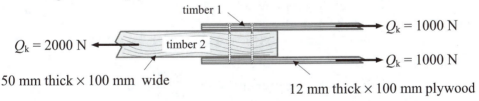

timber 1 $Q_k = 1000$ N

$Q_k = 2000$ N   timber 2

$Q_k = 1000$ N

50 mm thick × 100 mm wide

12 mm thick × 100 mm plywood

Figure 9.15

Solution:

BS EN 338:2003

Clause 5.0	Characteristic values for C16 Timber
Table 1	Characteristic density: $\rho_k = 310{,}0$ kg/m³
	Mean density: $\rho_{\text{mean}} = 370{,}0$ kg/m³

Manufacturer's Catalogue: Characteristic density of plywood $\rho_k = 630{,}0$ kg/m³

Mean density of plywood $\rho_{\text{mean}} = 680{,}0$ kg/m³

BS EN 1995-1-1:2004

Clause 2.3.2.1	Load-duration and moisture influence on strength: k_{mod}
Clause 2.3.1.3	Moisture content $\leq 20\%$ **Service Class 2**

Table 3.1   Long-term actions $k_{\text{mod,timber}} = k_{\text{mod,plywood}} = 0,7$

Equation (2.6) $k_{\text{mod}} = \sqrt{k_{\text{mod,1}} \times k_{\text{mod,2}}} = \sqrt{0,7 \times 0,7} = 0,7$ **$k_{\text{mod}} = 0,7$**

NA: Table NA.3   Partial factor for material properties and resistance: γ_M

For solid timber – untreated and preservative treated   **$\gamma_M = 1,3$**

For plywood – untreated and preservative treated   **$\gamma_M = 1,2$**

For connections (except punched plate fasteners)   **$\gamma_M = 1,3$**

Clause 8.3.1.1(1) Headside thickness t_1 is the minimum of the headside timber thickness and the pointside penetration.

Headside thickness = 12 mm

Pointside penetration = 12 mm   **$t_1 = 12$ mm**

The central member thickness   **$t_2 = 50$ mm**

Clause 8.3.1.1(4) Characteristic yield moment for smooth nails with $f_u = 600$ N/mm^2

Equation (8.14) $M_{\text{y,Rk}} = 0,3\, f_u\, d^{2,6} = (0,3 \times 600 \times 4,0^{2,6}) = 6616,5$ Nmm

$M_{\text{y,Rk}} = 6616,5$ Nmm

Clause 8.3.1.1(5) Characteristic embedment strength: $f_{h,k}$

Timber 1(plywood): characteristic density $\rho_k = 630$ kg/m^3

nail head diameter $\geq 2d$ (= 8,0 mm)

No pre-drilling for timber 1

Equation (8.20) $f_{h,k} = 0,11\, \rho_k\, d^{-0,3}$

Timber 1: $f_{h,1,k} = (0,11 \times 630 \times 4,0^{-0,3}) = 45,721$ N/mm^2

Timber 2 (solid timber): characteristic density $\rho_k = 310$ kg/m^3

No pre-drilling for timber 2

Equation (8.15) $f_{h,k} = 0,082\, \rho_k\, d^{-0,3}$

Timber 2: $f_{h,2,k} = (0,082 \times 310 \times 4,0^{-0,3}) = 16,771$ N/mm^2

Ratio of embedment strengths $\beta = \dfrac{16,771}{45,721} = 0,367$ **$\beta = 0,367$**

Clause 8.2.2(2) $F_{\text{ax,Rk}}$ is unknown and hence is the contribution from the rope effect in Equations (8.6(g)) to (8.6(k)) is taken as zero.

Note: the value $f_{\text{ax,k}}$ and hence $F_{\text{ax,Rk}}$ can be determined from Equations (8.23), to (8.26) provided that the pointside penetration is at least 12d: in this it is less than 12d (12 × 4,0 = 48,0 mm).

Clause 8.2.2(1)   Characteristic load-carrying capacity: $F_{\text{v,Rk}}$ - fasteners in double shear:

Equation (8.6(g)) $f_{h,1,k}\, t_1\, d = (45,721 \times 12 \times 4,0) = 2194, 6$ N

$F_{\text{v,Rk,g}} = 2194,6$ N

Equation (8.6(h)) $= 0.5\, f_{h,2,k}\, t_2\, d = (0.5 \times 16{,}771 \times 50 \times 4{,}0) = 1677{,}1$ N

$$F_{v,Rk,h} = 1677{,}1 \text{ N}$$

Equation (8.6(j)) $\quad 1{,}05 \dfrac{f_{h,1,k} t_1 d}{2 + \beta} \left[\sqrt{2\beta(1+\beta) + \dfrac{4\beta(2+\beta)M_{y,Rk}}{f_{h,1,k} d\, t_1^2}} - \beta \right] =$

$$1{,}05 \times \frac{45{,}720 \times 12 \times 4{,}0}{2 + 0{,}367} \left[\sqrt{2 \times 0{,}367 \times (1 + 0{,}367) + \frac{4 \times 0{,}367 (2 + 0{,}367) \times 6616{,}5}{45{,}721 \times 4{,}0 \times 12^2}} - 0{,}367 \right]$$

$$F_{v,Rk,j} = 976{,}1 \text{ N}$$

Equation (8.6(k)) $\quad 1{,}15 \sqrt{\dfrac{2\beta}{1+\beta}} \sqrt{2 M_{y,Rk} f_{h,1,k} d}$

$$= 1{,}15 \times \sqrt{\frac{2 \times 0{,}367}{1 + 0{,}367}} \times \sqrt{2 \times 6616{,}50 \times 45{,}721 \times 4{,}0} \qquad F_{v,Rk,k} = 1310{,}7 \text{ N}$$

The characteristic load-carrying capacity $F_{v,Rk}$ is minimum value determined using Equations (8.6(g)) to (8.6(k)).

$F_{v,Rk} = 976{,}1$ N corresponding with failure mode (j)

Equation (2.17) Design load-carrying capacity of nails:

$$F_{v,Rd} = k_{mod} \frac{F_{v,Rk}}{\gamma_{mt}} = 0{,}7 \times \frac{976{,}1}{1{,}3} = 525{,}6 \text{ N/shear plane}$$

For nails in double shear $F_{v,Rd} = (2 \times 525{,}6) = 1051{,}2$ N

BS EN 1990:2002
Equation (6.10) $\quad F_d = \sum_{j \geq 1} \gamma_{G,j} G_{k,j} \ "+" \ \gamma_{Q,1} Q_{k,1} \ "+" \ \sum_{i>1} \gamma_{Q,i} \psi_{0,i} Q_{k,i}$

NA: Table NA.A1.2(B)

For variable actions: Leading actions $\quad \gamma_{Q,1} = 1{,}5$

Design load $= F_{Q,d} = (1{,}5 \times 2000) = 3000$ kN

Number of nails required $n = \dfrac{F_{v,Rk}}{F_{Q,d}} = \dfrac{3000}{1051{,}2} = 2{,}9$

BS EN 1995-1-1:2004
Clause 8.3.1.1(9) Minimum number of nails = 2

A nail pattern is selected to satisfy the minimum number of nails required, e.g.

(i) a single line of 3 nails or (ii) a double line of 4 nails – the reader should complete this.

Check the spacing requirements for (i):

Clause 8.3.1.2
(i) Single line of nails: assume that the nails are in line
Figure 8.7

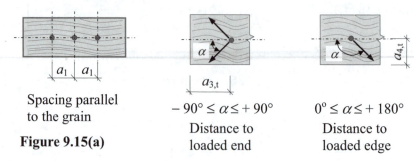

Spacing parallel
to the grain

Figure 9.15(a)

$-90° \le \alpha \le +90°$

Distance to
loaded end

$0° \le \alpha \le +180°$

Distance to
loaded edge

Spacings, end and edge distances:
Angle between the force and grain direction $\alpha = 0°$, $d < 5$ mm

Clause 8.3.1.3(1)   Timber 1 (plywood):
                    Nails without pre-drilling,

Table 8.2   Minimum spacing parallel to the grain $a_1 = 0,85 \times \left(4 + |\cos\alpha|\right)d$

$a_1 = 0,85 \times (4 + 1,0) \times 4,0 = 17,0$ mm

Clause 8.3.1.3(2)   Minimum distance to loaded end $a_{3,t} = \left(3 + 4|\sin\alpha|\right)d$

$a_{3,t} = [3 + (4 \times \sin 0°)] \times 4,0 = 12$ mm

Clause 8.3.1.3(2)   Minimum distance to loaded edge $a_{4,t} = \left(3 + 4|\sin\alpha|\right)d$

$a_{4,t} = [3 + (4 \times \sin 0°)] \times 4,0 = 12$ mm

Timber 2 (softwood):

Table 8.2   Nails without pre-drilling, $\rho_k = 310$ kg/m^3 (≤ 420 kg/m^3)

Clause 8.3.1.3(1)   Minimum spacing parallel to the grain $a_1 = \left(5 + 5|\cos\alpha|\right)d$

$a_1 = [5 + (5 \times 1,0)] \times 4,0 = 40,0$ mm

Minimum distance to loaded end $a_{3,t} = \left(10 + 5|\cos\alpha|\right)d$

$a_{3,t} = [10 + (5 \times \cos 0°)] \times 4,0 = 60$ mm

Minimum distance to loaded edge $a_{4,t} = \left(5 + 2|\sin\alpha|\right)d$

$a_{4,t} = [5 + (2 \times \sin 0°)] \times 4,0 = 20$ mm

Assume $a_1 = 40$ mm, $a_{3,t} = 60$ mm, $a_{4,t} = 50$ mm

Clause 8.3.1.1(8)   Assume that the nails are in line, no pre-drilled holes.

Table 8.1

Effective number of fasteners $n_{ef} = n^{k_{ef}}$
$a_1 = 40$ mm $\quad d = 4,0$ mm
$a_1/d = 40/4,0 = 10 \quad \therefore k_{ef} = 0,85$
Effective number of fasteners $n_{ef} = 3^{0,85} = 2,54 < 2,9$ required

Increase a_1 to 60 mm, $\quad a_1/d = 60/4,0 = 15 \quad \therefore k_{ef} = 1,0$
Effective number of fasteners $n_{ef} = 3^{1,0} = 3,0 > 2,9$ required

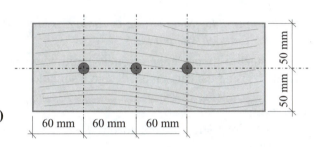

Figure 9.15(b)

Joint slip:

Clause 7.1(1)

Clause 7.1(2)

Slip modulus K_{ser} per shear plane per fastener is given in Table 7.1
Timber 1: plywood - $\rho_{m,1} = 680$ kg/m³
Timber 2: softwood - $\rho_{m,2} = 370$ kg/m³

Equation (7.1)

$$\rho_m = \sqrt{\rho_{m,1}\rho_{m,2}} = \sqrt{(680 \times 370)} = 501,60 \text{ kg/m}^3$$

Table 7.1

Nails without pre-drilling, $\quad \rho_m = 501,6$ kg/m³, $\quad d = 3,4$ mm

$$K_{ser} = \rho_m^{1,5} d^{0,8} / 30 = (501,6^{1,5} \times 4,0^{0,8})/30 = 1135,18 \text{ N/mm}$$

Design load for the serviceability limit state $F_{Q,d} = 2000$ N

BS EN 1990:2002

Equation (6.16b) $\quad F_d = \sum_{j \geq 1} G_{k,j} \; "+" \; \sum_{i \geq 1} \psi_{2,i} Q_{k,i}$

NA.

Table NA.A1.1

Quasi-permanent value for variable actions on domestic buildings
$\psi_{2,1} = 0,3$

BE EN 1995-1-1:2004

Table 3.2

Deformation factor for solid timber, Service class 2
$k_{def} = 0,8$

(i) Single row of nails:

Equation (2.4) $\quad u_{fin,Q,1} = u_{inst,Q,1}(1 + \psi_{2,1} k_{def})$

$$\text{Load per nail} = \frac{F_{Q,k}}{n} = \frac{2000}{2 \times 3} = 333{,}33 \text{ N}$$

$$\text{Instantaneous slip/side } u_{\text{inst}} = \frac{\text{Load per nail}}{K_{\text{ser}}} = \frac{333{,}33}{1135{,}18} = 0{,}29 \text{ mm/side}$$

$$u_{\text{fin,Q,1}} = 0{,}29 \times [1 + (0{,}3 \times 0{,}8)] = 0{,}36 \text{ mm/side}$$

$$\text{Final joint opening} = (2 \times u_{\text{fin,Q,1}}) = (2 \times 0{,}36) = 0{,}72 \text{ mm}$$

9.3.11 Example 9.4: Steel Plate-to-timber Three Member Connection with Single Shear for each Shear Plane

A nailed steel-to-timber connection is required to transmit a lateral load as shown in Figure 9.16. Using the data given, design a suitable connection.

Design data:

Characteristic long-term variable load (Q_k)	1600 N
Timber Class	C16
Moisture content	$\leq 20\%$
Type of nail	smooth, round
Pre-drilling	none
Nail diameter (d)	4,0 mm
Nail length (L)	35 mm
Tensile strength of wire (f_u)	600 N/mm^2

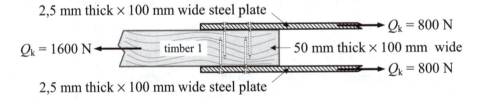

Figure 9.16

Solution:

BS EN 338:2003
Clause 5.0   Characteristic values for C16 Timber
Table 1   Characteristic density: $\rho_k = 310{,}0$ kg/m^3
                       Mean density: $\rho_{\text{mean}} = 370{,}0$ kg/m^3

BS EN 1995-1-1:2004
Clause 2.3.2.1   Load-duration and moisture influence on strength: k_{mod}
Clause 2.3.1.3   Moisture content $\leq 20\%$ **Service Class 2**

Table 3.1   Long-term actions $k_{\text{mod}} = 0{,}7$

NA: Table NA.3   Partial factor for material properties and resistance: γ_M
For solid timber – untreated and preservative treated $\gamma_M = \mathbf{1,3}$
For connections (except punched plate fasteners) $\gamma_M = \mathbf{1,3}$

Clause 8.2.3(1)   Steel plate thickness = 2,5 mm $> 0,5d = (0,5 \times 4,0) = 2,0$ mm
$< d = 4,0$ mm
Steel plate is classified as between thick and thin

Load-carrying capacity is calculated by linear interpolation between the limiting thin and thick plate values.

Clause 8.3.1.1(7)   Nail overlap:

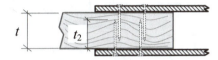

Figure 9.16(a)

$t = 50$ mm, $t_2 = (35 - 2,5) = 32,5$ mm
$(t - t_2) = (50 - 32,5) = 17,5$ mm $> 4d = (4 \times 4,0 = 16,0$ mm)
Overlap is satisfactory

Clause 8.3.1.1(1)   Headside thickness t_1 is the minimum of the headside plate thickness and the pointside penetration, (single shear).
Headside thickness = 2,5 mm
Effective pointside penetration = 50/2 = 25,0 mm
$t_1 = \mathbf{2,5}$ **mm**

The central member thickness $t_2 = \mathbf{50,0}$ **mm**

Clause 8.3.1.1(4)   Characteristic yield moment for smooth nails with $f_u = 600$ N/mm^2
Equation (8.14) $M_{y,Rk} = 0,3\, f_u d^{2,6} = (0,3 \times 600 \times 4,0^{2,6}) = 6616,5$ Nmm
$M_{y,Rk} = \mathbf{6616,5}$ **Nmm**

Clause 8.3.1.1(5)   Characteristic embedment strength: $f_{h,k}$
Equation (8.15) $f_{h,k} = 0,082\, \rho_k d^{-0,3}$
No pre-drilling for timber 1
Timber 1: $f_{h,k} = (0,082 \times 310 \times 4,0^{-0,3}) = 16,771$ N/mm^2

Clause 8.2.2(2) $F_{ax,Rk}$ is unknown and hence is the contribution from the rope effect in Equations (8.9) and (8.10) is taken as zero.

Note: the value $f_{ax,k}$ and hence $F_{ax,Rk}$ can be determined from Equations (8.23) to (8.26) provided that the pointside penetration is at least 12d; in this it is less than 12d ($12 \times 4,0 = 48,0$ mm).

Clause 8.2.3(3)   Characteristic load-carrying capacity: $F_{v,Rk}$ - thin plate in single shear

Equation (8.9(a)) $0,4\,f_{h,k}\,t_1\,d = (0,4 \times 16,771 \times 25,0 \times 4,0) = 670,8$ N

$$F_{v,Rk,a} = 670,8 \text{ N}$$

Equation (8.9(b)) $1,15\sqrt{2M_{y,Rk}\,f_{h,k}d} = 1,15 \times \sqrt{2 \times 6616,50 \times 16,771 \times 4,0} = 1083,5$ N

$$F_{v,Rk,b} = 1083,5 \text{ N}$$

The characteristic load-carrying capacity $F_{v,Rk}$ for a thin plate is the minimum value determined using Equations (8.9(a)) and (8.9(b)).

$$F_{v,Rk} = \mathbf{670,8 \text{ N} \text{ corresponding with failure mode (a)}}$$

Clause 8.2.3(3)   Characteristic load-carrying capacity: $F_{v,Rk}$ - thick plate in single shear

Equation (8.10(c)) $f_{h,k}t_1 d\left[\sqrt{2 + \dfrac{4M_{y,Rk}}{f_{h,k}d\,t_1^2}} - 1\right]$

$$= 16,771 \times 25,0 \times 4,0 \times \left[\sqrt{2 + \frac{4 \times 6616,5}{16,771 \times 4,0 \times 25,0^2}} - 1\right] \qquad F_{v,Rk,c} = 1043,3 \text{ N}$$

Equation (8.10(d)) $2,3\sqrt{M_{y,Rk}\,f_{h,k}d} = 2,3 \times \sqrt{6616,50 \times 16,771 \times 4,0} = 1532,3$ N

$$F_{v,Rk,d} = 1532,3 \text{ N}$$

Equation (8.10(e)) $f_{h,k}\,t_1\,d = (16,771 \times 25,0 \times 4,0) = 1677,1$ N

$$F_{v,Rk,e} = 1677,1 \text{ N}$$

The characteristic load-carrying capacity $F_{v,Rk}$ for a thick plate is the minimum value determined using Equations (8.10(c)), (8.10(d)) and (8.10(e)).

$$F_{v,Rk} = \mathbf{1043,3 \text{ N} \text{ corresponding with failure mode (c)}}$$

The characteristic load-carrying capacity $F_{v,Rk}$ for a plate with intermediate thickness is determined using interpolation Equations (8.9(a)) and (8.10(c)).

Clause 8.2.3(1) $F_{v,Rk} = F_{v,Rk,thin} + \dfrac{(t - t_{thin})}{(t_{thick} - t_{thin})} \times \left(F_{v,Rk,thick} - F_{v,Rk,thin}\right)$

$$= 670,84 + \frac{\left[2,5 - (0,5 \times 4,0)\right]}{\left[4,0 - (0,5 \times 4,0)\right]} \times (1043,3 - 670,8) = 764,0 \text{ N} \qquad \mathbf{F_{v,Rk} = 764,0 \text{ N}}$$

Equation (2.17)   Design load-carrying capacity of nails:

$$F_{v,Rd} = k_{mod}\frac{F_{v,Rk}}{\gamma_{mt}} = 0,7 \times \frac{763,97}{1,3} = 411,4 \text{ N}$$

BS EN 1990:2002
Equation (6.10) $F_d = \sum_{j \geq 1}\gamma_{G,j}G_{k,j} \text{ "+" } \gamma_{Q,1}Q_{k,1} \text{ "+" } \sum_{i>1}\gamma_{Q,i}\psi_{0,i}Q_{k,i}$

NA: Table NA.A1.2(B)

For variable actions: Leading actions $\gamma_{Q,1} = 1,5$
Design load $= F_{Q,d} = (1,5 \times 1600) = 2400$ kN

$$\text{Number of nails required } n = \frac{F_{v,Rk}}{F_{Q,d}} = \frac{2400}{411,4} = 5,8 \quad (= 2,9 \text{ /side})$$

BS EN 1995-1-1:2004
Clause 8.3.1.1(9)   Minimum number of nails $= 2$

A nail pattern is selected to satisfy the minimum number of nails required, e.g.

(i) a single line of 3 nails or (ii) a double line of 4 nails in each plate.

Check the spacing requirements for (i):
Clause 8.3.1.2
(i) Single line of nails: assume that the nails are in line, no pre-drilled holes
Figure 8.7

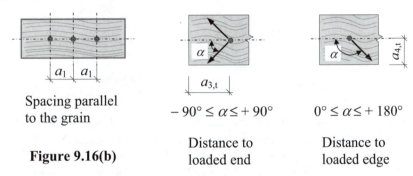

Spacing parallel
to the grain

Figure 9.16(b)

$-90° \leq \alpha \leq +90°$

Distance to
loaded end

$0° \leq \alpha \leq +180°$

Distance to
loaded edge

Clause 8.3.1.4   Minimum edge and end distances are as given in Table 8.2. The minimum spacings are equal to $(0,7 \times$ Table 8.2 values$)$.

Table 8.2   Nails without pre-drilling, $\rho_k = 310$ kg/m³ $(\leq 420$ kg/m³$)$

Angle between the force and grain direction $\alpha = 0°$, $d < 5$ mm

Minimum spacing parallel to the grain $a_1 = 0,7 \times (5 + 5|\cos\alpha|)d$

$a_1 = 0,7 \times [5 + (5 \times 1,0)] \times 4,0 = 28$ mm

Minimum distance to loaded end $a_{3,t} = \left(10 + 5\left|\cos\alpha\right|\right)d$

$a_{3,t} = [10 + (5 \times \cos0°)] \times 4{,}0 = 60$ mm

Minimum distance to loaded edge $a_{4,t} = \left(5 + 2\left|\sin\alpha\right|\right)d$

$a_{4,t} = [5 + (2 \times \sin0°)] \times 4{,}0 = 20$ mm

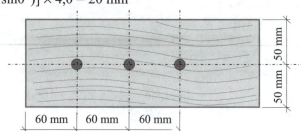

Figure 9.16(c)   60 mm | 60 mm | 60 mm

Clause 8.3.1.1(8)   Assume that the nails are in line, no pre-drilled holes.

Effective number of fasteners $n_{ef} = n^{k_{ef}}$

Table 8.1 $a_1 = 60$ mm $d = 4{,}0$ mm

$a_1/d = 60/4{,}0 = 15$ $\therefore k_{ef} = 1{,}0$

Effective number of fasteners $n_{ef} = 3^{1,0} = 3{,}0 > 2{,}9$ required

(ii) Double line of nails:

Figure 8.7

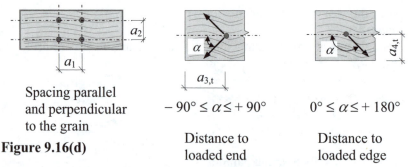

Spacing parallel
and perpendicular
to the grain

$-90° \leq \alpha \leq +90°$ $0° \leq \alpha \leq +180°$

Figure 9.16(d)

Distance to   Distance to
loaded end   loaded edge

Table 8.2   Nails without pre-drilling, $\rho_k = 310$ kg/m³ (≤ 420 kg/m³)

Angle between the force and grain direction $\alpha = 0°$, $d < 5$ mm

Minimum spacing parallel to the grain $a_1 = 28$ mm (as before)

Minimum spacing perpendicular to the grain $a_2 = (0{,}7 \times 5d)$

$a_2 = (0{,}7 \times 5 \times 4{,}0) = 14$ mm

Minimum distance to loaded end $a_{3,t} = 51$ mm (as before)

Minimum distance to loaded edge $a_{4,t} = 17$ mm (as before)

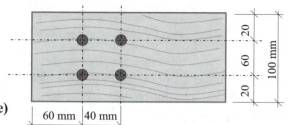

Figure 9.16(e)

Joint slip:

Clause 7.1(3) Slip modulus K_{ser} per shear plane per fastener is given in Table 7.1

Table 7.1 Nails without pre-drilling, $\rho_{\text{m}} = 370$ kg/m³, $d = 4,0$ mm,

$$K_{\text{ser}} = 2,0\,\rho_{\text{m}}^{1,5}d^{0,8}/30 = 2,0 \times (370^{1,5} \times 4,0^{0,8})/30 = 1438,33 \text{ N/mm}$$

Design load for the serviceability limit state $F_{Q,d} = 1600$ N

BS EN 1990:2002

Equation (6.16b) $F_{\text{d}} = \displaystyle\sum_{j\geq1} G_{\text{k,j}} \;''+'' \sum_{i\geq1} \psi_{2,i}Q_{\text{k,i}}$

NA.

Table NA.A1.1 Quasi-permanent value for variable actions on domestic buildings
$\psi_{2,1} = 0,3$

BS EN 1995-1-1:2004

Table 3.2 Deformation factor for solid timber, Service class 2
$k_{\text{def}} = 0,8$

(i) Single row of nails:

Equation (2.4) $u_{\text{fin,Q,1}} = u_{\text{inst,Q,1}}(1 + \psi_{2,1}k_{\text{def}})$

Load per nail $= \dfrac{F_{\text{Q,k}}}{n} = \dfrac{1600}{6} = 266,67$ N

Instantaneous slip/side $u_{\text{inst}} = \dfrac{\text{Load per nail}}{K_{\text{ser}}} = \dfrac{266,67}{1438,33} = 0,19$ mm/side

$u_{\text{fin,Q,1}} = 0,19 \times [1 + (0,3 \times 0,8)] = 0,24$ mm/side

Final joint opening $= (2 \times u_{\text{fin,Q,1}}) = (2 \times 0,24) = 0,48$ mm

(ii) Double row of nails:

Load per nail $= \dfrac{F_{\text{Q,k}}}{n} = \dfrac{1600}{8} = 200,0$ N

Instantaneous slip/side $u_{\text{inst}} = \dfrac{\text{Load per nail}}{K_{\text{ser}}} = \dfrac{200,0}{1438,33} = 0,14$ mm/side

$u_{\text{fin,Q,1}} = 0,14 \times [1 + (0,3 \times 0,8)] = 0,17$ mm/side

Final joint opening $= (2 \times u_{\text{fin,Q,1}}) = (2 \times 0,17) = 0,34$ mm

9.4 Stapled Connections (Section 8.4)

Stapled connection should be designed in accordance with the requirements of Section 8.4 in the code. The lateral design load-carrying capacity/staple is determined assuming the legs to be two nails with the same diameter as the staple, provided that the angle between the crown and the direction of the grain of the timber under the crown is greater than 30°. The definitions and staple dimensions are defined in Figure 8.9 and Figure 8.10 of the code and indicated in Figure 9.17. The minimum spacing, end and edge distances for staples are given in Table 8.3.

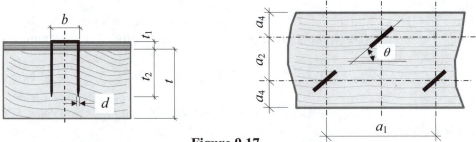

Figure 9.17

9.4.1 Example 9.5: Panel-to-timber Stapled Connection in Single Shear

A stapled panel-to-timber connection is required to transmit a lateral load as shown in Figure 9.18. Using the data given, design a suitable connection.

Design data:

Characteristic long-term variable load (Q_k)	1500 N
Timber Class	C18
Moisture content	$\leq 20\%$
Type of staples	smooth, round cross-section
Pre-drilling	none
Staple diameter (d)	3,5 mm
Staple length (L)	65 mm
Staple crown width (b)	25 mm
Angle between the crown and the direction of the grain under the crown (θ)	45°
Tensile strength of wire (f_u)	800 N/mm^2

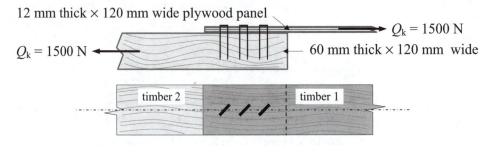

Figure 9.18

Solution:
BS EN 338:2003
Clause 5.0   Characteristic values for C18 Timber
Table 1   Characteristic density: $\rho_k = 320{,}0$ kg/m^3
                    Mean density: $\rho_{mean} = 380{,}0$ kg/m^3
Manufacturer's Catalogue: Characteristic density of plywood $\rho_k = 630{,}0$ kg/m^3

BS EN 1995-1-1:2004
Clause 2.3.2.1   Load-duration and moisture influence on strength: k_{mod}
Clause 2.3.1.3   Moisture content $\leq 20\%$ **Service Class 2**
Table 3.1   Long-term actions $k_{mod,timber} = k_{mod,plywood} = 0{,}7$
Equation (2.6) $k_{mod} = \sqrt{k_{mod,1} \times k_{mod,2}} = \sqrt{0{,}7 \times 0{,}7} = 0{,}7$ **$k_{mod} = 0{,}7$**

NA: Table NA.3   Partial factor for material properties and resistance: γ_M
                    For solid timber – untreated and preservative treated   **$\gamma_M = 1{,}3$**
                    For plywood – untreated and preservative treated   **$\gamma_M = 1{,}2$**
                    For connections (except punched plate fasteners)   **$\gamma_M = 1{,}3$**

Clause 8.3.1.1(1)   Headside thickness t_1 is the minimum of the headside plate thickness
                    and the pointside penetration, (single shear).
                    Headside thickness = 12,0 mm
                    Pointside penetration = $(65 - 12) = 53{,}0$ mm
                                                                                      $t_1 = 12{,}0$ mm
                                                                                      $t_2 = 53{,}0$ mm

Clause 8.4.3   The width b of the crown should be at least $6d$ and the pointside
                    penetration length t_2 should be at least $14d$.
                    Crown width:
                    $b = 25$ mm $> 6d = 6 \times 3{,}5 = 21$ mm
                    $t_2 = (65 - 12) = 53$ mm $> 14d = 14 \times 3{,}5 = 49$ mm
                                      Crown width and penetration length are adequate

Clause 8.3.1.1(4)   Characteristic yield moment for smooth staples with $f_u = 800$ N/mm^2
Equation (8.29) $M_{y,Rk} = 240d^{2,6} = (240 \times 3{,}5^{2,6}) = 6234{,}31$ Nmm
                                                                            $M_{y,Rk} = 6234{,}31$ Nmm

Clause 8.3.1.1(5)   Characteristic embedment strength: $f_{h,k}$
                    Timber 1 (plywood): characteristic density $\rho_k = 630$ kg/m^3
                                        nail head diameter $\geq 2d\, (= 8{,}0$ mm)
                    No pre-drilling for timber 1
Equation (8.20) $f_{h,k} = 0{,}11\, \rho_k\, d^{-0,3}$
                    Timber 1: $f_{h,1,k} = (0{,}11 \times 630 \times 3{,}5^{-0,3}) = 47{,}590$ N/mm^2

Timber 2 (solid timber): characteristic density $\rho_k = 320$ kg/m^3

No pre-drilling for timber 2

Equation (8.15) $f_{h,k} = 0,082\, \rho_k\, d^{-0,3}$

Timber 2: $f_{h,2,k} = (0,082 \times 320 \times 3,5^{-0,3}) = 18,020$ N/mm^2

Ratio of embedment strengths $\beta = \dfrac{18,020}{47,590} = 0,379$ $\boldsymbol{\beta = 0,379}$

Clause 8.2.2(2) $F_{ax,Rk}$ is unknown and hence is the contribution from the rope effect in Equations (8.6(a)) to (8.6(f)) is taken as zero.

Clause 8.2.2(1)   Characteristic load-carrying capacity: $F_{v,Rk}$ - fasteners in single shear:

Equation (8.6(a)) $f_{h,1,k}\, t_1\, d = (47,590 \times 12 \times 3,5) = 1998,8$ N $F_{v,Rk,a} = 1998,8$ N

Equation (8.6(b)) $f_{h,2,k}\, t_2\, d = (18,020 \times 53 \times 3,5) = 3342,7$ N $F_{v,Rk,b} = 3342,7$ N

Equation (8.6(c)) $\dfrac{f_{h,1,k}\, t_1\, d}{1+\beta}\left[\sqrt{\beta + 2\beta^2\left[1 + \dfrac{t_2}{t_1} + \left(\dfrac{t_2}{t_1}\right)^2\right] + \beta^3\left(\dfrac{t_2}{t_1}\right)^2} - \beta\left(1 + \dfrac{t_2}{t_1}\right) \right] =$

$\dfrac{47,590\times12\times3,5}{1+0,379}\times\left[\sqrt{0,379 + 2\times0,379^2\left[1 + \dfrac{53}{12} + \left(\dfrac{53}{12}\right)^2\right] + 0,379^3\left(\dfrac{53}{12}\right)^2} - 0,379\left(1 + \dfrac{53}{12}\right) \right]$

$= 1274,3$ kN $F_{v,Rk,c} = 1274,3$ N

Equation (8.6(d)) $1,05\dfrac{f_{h,1,k}\, t_1\, d}{2+\beta}\left[\sqrt{2\beta(1+\beta) + \dfrac{4\beta(2+\beta)M_{y,Rk}}{f_{h,1,k}\, d\, t_1^2}} - \beta \right] =$

$1,05\times\dfrac{47,590\times12\times3,5}{2+0,379}\times\left[\sqrt{2\times0,379(1+0,379) + \dfrac{4\times0,379(2+0,379)\times6234,31}{47,590\times3,5\times12^2}} - 0,379 \right]$

$= 907,6$ kN $F_{v,Rk,d} = 907,6$ N

Equation (8.6(e)) $1,05\dfrac{f_{h,1,k}\, t_2\, d}{1+2\beta}\left[\sqrt{2\beta^2(1+\beta) + \dfrac{4\beta(1+2\beta)M_{y,Rk}}{f_{h,1,k}\, d\, t_2^2}} - \beta \right] =$

$1,05\times\dfrac{47,590\times53\times3,5}{1+2\times0,379}\times\left[\sqrt{2\times0,379^2(1+0,379) + \dfrac{4\times0,379(1+2\times0,379)\times6234,31}{47,590\times3,5\times53^2}} - 0,379 \right]$

$= 1464,8$ kN $F_{v,Rk,e} = 1464,8$ N

Equation (8.6(f)) $1,15\sqrt{\dfrac{2\beta}{1+\beta}}\sqrt{2M_{y,Rk}f_{h,1,k}d}$

$= 1,15\times\sqrt{\dfrac{2\times0,379}{1+0,379}}\times\sqrt{2\times6234,31\times47,590\times3,5} = 1228,3 \text{ N}$ $F_{v,Rk,f} = 1228,3 \text{ N}$

The characteristic load-carrying capacity $F_{v,Rk}$ is minimum value determined using Equations (8.6(a)) to (8.6(f)).

$$F_{v,Rk} = \textbf{907,6 N corresponding with failure mode (d)}$$

Equation (2.17)   Design load-carrying capacity of staples:

$$F_{v,Rd} = k_{mod}\frac{F_{v,Rk}}{\gamma_m} = 0,7\times\frac{907,6}{1,3} = 488,7 \text{ N/staple leg}$$

For each staple $F_{v,Rd} = (2\times488,7) = 977,4 \text{ N}$

Clause 8.4(5)   Angle between the crown and the direction of the grain in the timber under the crown $\theta = 45°$

No reduction in capacity is required.

BS EN 1990:2002

Equation (6.10) $F_d = \sum\limits_{j\geq1}\gamma_{G,j}G_{k,j}\;"+"\;\gamma_{Q,1}Q_{k,1}\;"+"\;\sum\limits_{i>1}\gamma_{Q,i}\psi_{0,i}Q_{k,i}$

NA: Table NA.A1.2(B)

For variable actions: Leading actions $\gamma_{Q,1} = 1,5$

Design load = $F_{Q,d} = (1,5\times1500) = 2250 \text{ kN}$

Number of staples required $n = \dfrac{F_{v,Rk}}{F_{Q,d}} = \dfrac{2250}{977,4} = 2,3$

BS EN 1995-1-1:2004

Clause 8.3.1.1(9)   Minimum number of staples = 2

Clause 8.3.1.2

Single line of staples: assume that the staples are in line
Figure 8.7

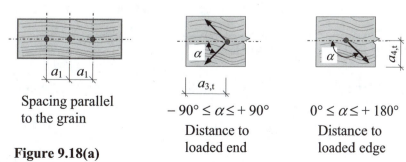

Spacing parallel
to the grain

Figure 9.18(a)

$-90° \leq \alpha \leq +90°$

Distance to
loaded end

$0° \leq \alpha \leq +180°$

Distance to
loaded edge

Spacings, end and edge distances:

Table 8.3   Angle between the force and grain direction $\alpha = 0°$
                            Angle between the crown and the direction of the grain in the timber
                            under the crown $\theta = 45°$

Clause 8.4(8)   Minimum spacing parallel to the grain $a_1 = \left(10 + 5|\cos\alpha|\right)d$
                            $a_1 = [10 + (5 \times 1,0)] \times 3,5 = 52,5$ mm

                            Minimum distance to loaded end $a_{3,t} = \left(15 + 5|\cos\alpha|\right)d$
                            $a_{3,t} = [15 + (5 \times \cos 0°)] \times 3,5 = 70$ mm

                            Minimum distance to loaded edge $a_{4,t} = \left(15 + 5|\sin\alpha|\right)d$
                            $a_{4,t} = [15 + (5 \times \sin 0°)] \times 3,5 = 52,5$ mm

                            Assume $a_1 = 60$ mm, $a_{3,t} = 70$ mm, $a_{4,t} = 60$ mm

Clause 8.3.1.1(8)   Assume that the staples are in line.
                            Effective number of fasteners $n_{ef} = n^{k_{ef}}$
Table 8.1 $a_1 = 60$ mm $d = 3,5$ mm
                            $a_1/d = 60/3,5 = 17,14$ $\therefore k_{ef} = 1,0$
                            Effective number of fasteners $n_{ef} = 3^{1,0} = 3,0 > 2,3$ required

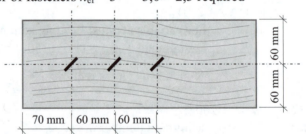

Figure 9.18(b)   70 mm | 60 mm | 60 mm

Joint slip:
Clause 7.1(1)   Slip modulus K_{ser} per shear plane per fastener is given in Table 7.1
Clause 7.1(2)   Timber 1: plywood - $\rho_{m,1} = 680$ kg/m^3
                            Timber 2: softwood - $\rho_{m,2} = 380$ kg/m^3

Equation (7.1) $\rho_m = \sqrt{\rho_{m,1}\rho_{m,2}} = \sqrt{(680 \times 380)} = 508,33$ kg/m^3

Table 7.1   Staples (without pre-drilling) $\rho_m = 508,33$ kg/m^3, $d = 3,5$ mm,

                            $K_{ser} = \rho_m^{1,5}d^{0,8}/30 = (508,33^{1,5} \times 3,5^{0,8})/30 = 1040,76$ N/mm

                            Design load for the serviceability limit state $F_{Q,d} = 1500$ N

BS EN 1990:2002

Equation (6.16b) $F_d = \sum_{j\geq1} G_{k,j} \; "+" \sum_{i\geq1} \psi_{2,i} Q_{k,i}$

NA.

Table NA.A1.1   Quasi-permanent value for variable actions on domestic buildings
$\psi_{2,1} = 0{,}3$

BS EN 1995-1-1:2004

Table 3.2   Deformation factor for solid timber, Service class 2
$k_{def} = 0{,}8$

Single row of staples:

Equation (2.4) $u_{fin,Q,1} = u_{inst,Q,1}(1 + \psi_{2,1} k_{def})$

Load per nail $= \dfrac{F_{Q,k}}{n} = \dfrac{1500}{2\times3} = 250{,}0$ N

Instantaneous slip/side $u_{inst} = \dfrac{\text{Load per nail}}{K_{ser}} = \dfrac{250{,}0}{1040{,}76} = 0{,}24$ mm/side

$u_{fin,Q,1} = 0{,}24 \times [1 + (0{,}3 \times 0{,}8)] = 0{,}30$ mm/side

Final joint opening $= (2 \times u_{fin,Q,1}) = (2 \times 0{,}30) = 0{,}60$ mm

9.5 Bolted Connections (Section 8.5)

Bolts are normally made from ordinary mild steel with hexagonal or square heads and nuts, see Figure 9.19:

Hexagonal head bolt

Square head bolt

Cup-square head bolt

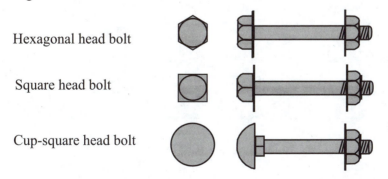

Figure 9.19

Generally bolts are used in single, double or multiple shear joints as shown in Figures 9.21 to 9.22.

The recommendations for the design of bolted joints are given in Section 8.5 of BS EN 1995-1-1:2004-1-1:2004. The threaded length of bolts is often inadequate (i.e. usually 2 × bolt diameter), and it may be necessary to order bolts with an extended threaded length. This is particularly so if they are to be used in conjunction with imbedded connectors (see Section 9.9).

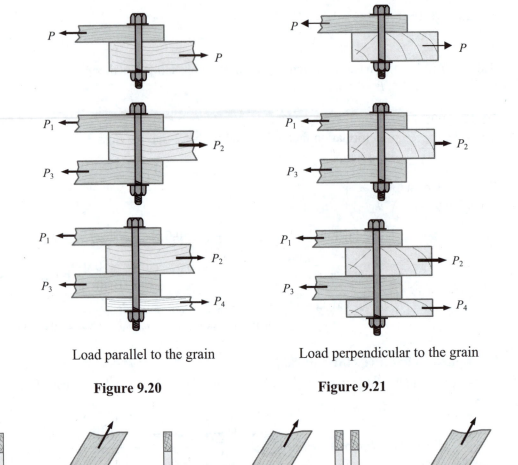

Load parallel to the grain   Load perpendicular to the grain

Figure 9.20   **Figure 9.21**

Load inclined at an angle α to the grain

Figure 9.22

9.5.1 Washers (Clause 10.4.3.(2))

Bolts should always be used with appropriate washers fitted under the head of each bolt and under each nut unless an equivalent bearing area is provided by e.g. a steel plate. When the nut is tightened at least one complete turn of the thread of the bolt should protrude through. The side length or diameter of a washer should be at least equal to (3 × bolt diameter) and the thickness under the head and nut at least equal to (0,3 × bolt diameter) in accordance with Clause 10.4.3(2).

Washers should have a full bearing area and in the case of axially loaded bolts, the bearing capacity is calculated in accordance with Clause 8.5.2(2), i.e. assuming a characteristic compressive strength of **3,0 $f_{c,90,k}$** on the bearing area ($A_{c,90}$). The bearing capacity per bolt of a steel plate should not exceed that of a circular washer with a diameter which is the minimum of (12 × plate thickness) or (4 × bolt diameter) as indicated in Clause 8.5.2(3).

9.5.2 Bolt Holes (Clause 10.4.3(1))

The bolt hole should be drilled to a diameter as close as practicable to the nominal bolt diameter, and no more than 1 mm larger than the bolt diameter in the case of timber and no more than the greater value of 2 mm or (0,1 × bolt diameter) in steel plates.

If necessary the bolts should be re-tightened after the equilibrium content has been reached to ensure that the load-carrying capacity and stiffness of a structure is maintained.

9.5.3 Spacings, End and Edge Distances (Clause 8.5.1.1(3))

The minimum requirements for bolt spacings, end and edge distances and are given in Table 8.4 and in Figure 8.7 of the code.

9.5.4 Example 9.6: Bolted Two-member Connection with the Applied Load Parallel to the Grain Direction

Using the design data given, determine the load-carrying capacity for the two-member joint shown in Figure 9.23.

Design data:

Timber Class	C24
Moisture content	≤ 20%
Load duration	medium-term
Bolt type	M12, 4.6 grade bolts
Number of bolts	1
Bolt diameter (d)	12 mm
Effective area of bolts (A_{eff})	84,3 mm^2
Tensile strength of bolts (f_u)	400 N/mm^2
Washer outer diameter (d_{wo})	36 mm
Washer inner diameter (d_{wi})	14 mm
Washer thickness (t_w)	3,6 mm

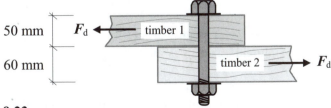

Figure 9.23

Solution:
BS EN 338:2003
Clause 5.0   Characteristic values for C24 Timber
Table 1   Characteristic density: $\rho_k = 350,0$ kg/m^3
                        Mean density: $\rho_{mean} = 420,0$ kg/m^3
                        Characteristic compressive strength perpendicular to the grain:
                        $f_{c,90,k} = 2,5$ N/mm^2

BS EN 1995-1-1:2004
Clause 2.3.2.1   Load-duration and moisture influence on strength: k_{mod}
Clause 2.3.1.3   Moisture content $\leq$ 20%   **Service Class 2**
Table 3.1   Medium-term actions $\boldsymbol{k_{mod} = 0,8}$
NA: Table NA.3   Partial factor for material properties and resistance: γ_M
                        For solid timber – untreated and preservative treated $\boldsymbol{\gamma_M = 1,3}$
                        For connections (except punched plate fasteners) $\boldsymbol{\gamma_M = 1,3}$

                        Thickness of timber 1 $\boldsymbol{t_1 = 50}$ **mm**
                        Thickness of timber 2 $\boldsymbol{t_2 = 60}$ **mm**

Clause 8.5.1.1(1)   Characteristic yield moment for bolts with $f_u = 400$ N/mm^2
Equation (8.30) $M_{y,Rk} = 0,3\,f_u\,d^{2,6} = (0,3 \times 400 \times 12,0^{2,6}) = 76745,4$ Nmm
                                                            $\boldsymbol{M_{y,Rk} = 76745,4}$ **Nmm**

Clause 8.5.1.1(2)   Characteristic embedment strength: $f_{h,k}$
                        Angle between the applied force and the grain $\alpha = 0°$
                        Timber 1 and Timber 2
Equation (8.32) $f_{h,0,k} = 0,082\,(1 - 0,01d)\rho_k = 0,082 \times [1 - (0,01 \times 12,0)] \times 350$
                                   $= 25,26$ N/mm^2

                        Ratio of embedment strengths $\beta = \dfrac{f_{h,2,k}}{f_{h,1,k}} = 1,0$ $\boldsymbol{\beta = 1,0}$

Clause 10.4.3(2)   Washer diameter $\geq 3d = (3 \times 12) = 36$ mm
                        Actual diameter = 36 mm = $3d$
                        Washer thickness $\geq 0,3d = (0,3 \times 12) = 3,6$ mm
                        Actual thickness = 3,6 mm = $0,3d$
                                          Washer diameter and thickness are adequate

Clause 8.5.2(1)   The axial load-bearing capacity and withdrawal capacity of a bolt
                        should be taken as the lower of:

                        (i)   the bolt tensile capacity,
                        (ii) the load-bearing capacity of either the washer or (for steel-to-
                              timber connections) the steel plate.

Clause 8.5.2(2) Load-bearing capacity of the washer $= (3,0 f_{c,90,k} \times A_{c,90})$
Bearing area:
$A_{c,90} = \pi (d_{wo}^2 - d_{wi}^2)/4 = [\pi \times (36^2 - 14^2)]/4 = 863,94 \text{ mm}^2$
$(3,0 f_{c,90,k} \times A_{c,90}) = (3,0 \times 2,5 \times 863,94) = 6479,55 \text{ N}$
Bolt tensile capacity $= (f_u \times A_{eff}) = (400 \times 84,3) = 33720 \text{ N}$

Clause 8.5.2(1) $F_{ax,Rk} =$ the minimum value between $(3,0 f_{c,90,k} \times A_{c,90})$ and $(f_u \times A_{eff})$
Characteristic axial withdrawal capacity: $F_{ax,Rk} = 6479,55 \text{ kN}$

Clause 8.2.2(2) For bolts, the rope effect is equal to $F_{ax,Rk}/4$ but no more than 25% of the Johansen part in Equations (8.6(c)) to (8.6(f)).
$$F_{ax,Rk}/4 = 6479,55/4 = 1619,9 \text{ N}$$

Clause 8.2.2(1) Characteristic load-carrying capacity: $F_{v,Rk}$ – single shear
Equation (8.6(a)) $f_{h,1,k} \, t_1 \, d = (25,26 \times 50 \times 12,0) = 15156,0 \text{ N}$
$$F_{v,Rk,a} = 15156,0 \text{ N}$$

Equation (8.6(b)) $f_{h,2,k} \, t_2 \, d = (25,26 \times 60 \times 12,0) = 18187,2 \text{ N}$
$$F_{v,Rk,b} = 18187,2 \text{ N}$$

Equation (8.6(c))
$$\frac{f_{h,1,k} t_1 d}{1+\beta}\left[\sqrt{\beta + 2\beta^2\left[1 + \frac{t_2}{t_1} + \left(\frac{t_2}{t_1}\right)^2\right] + \beta^3\left(\frac{t_2}{t_1}\right)^2} - \beta\left(1 + \frac{t_2}{t_1}\right)\right] + \frac{F_{ax,Rk}}{4}$$

$$\frac{25,26 \times 50 \times 12,0}{1+1,0}\left[\sqrt{1,0 + 2 \times 1,0^2\left[1 + \frac{60}{50} + \left(\frac{60}{50}\right)^2\right] + 1,0^3\left(\frac{60}{50}\right)^2} - 1,0\left(1 + \frac{60}{50}\right)\right] = 6954,3 \text{ N}$$

$F_{v,Rk,c} = \{6954,3 + \min[(0,25 \times 6954,3) \quad 1619,9)]\}$ $F_{v,Rk,c} = 8574,2 \text{ N}$

Equation (8.6(d)) $1,05\dfrac{f_{h,1,k} t_1 d}{2+\beta}\left[\sqrt{2\beta(1+\beta) + \dfrac{4\beta(2+\beta)M_{y,Rk}}{f_{h,1,k} d \, t_1^2}} - \beta\right] + \dfrac{F_{ax,Rk}}{4}$

$$1,05 \times \frac{25,26 \times 50 \times 12,0}{2+1,0}\left[\sqrt{2 \times 1,0(1+1,0) + \frac{4 \times 1,0(2+1,0) \times 76745,4}{25,26 \times 12,0 \times 50^2}} - 1,0\right] = 6809,5 \text{ N}$$

$F_{v,Rk,d} = \{6809,5 + \min[(0,25 \times 6809,5) \quad 1619,9)]\}$ $F_{v,Rk,d} = 8429,4 \text{ N}$

Equation (8.6(e)) $1,05\dfrac{f_{h,1,k} t_2 d}{1+2\beta}\left[\sqrt{2\beta^2(1+\beta) + \dfrac{4\beta(1+2\beta)M_{y,Rk}}{f_{h,1,k} d \, t_2^2}} - \beta\right] + \dfrac{F_{ax,Rk}}{4}$

$$1,05 \times \frac{25,26 \times 60 \times 12,0}{1+2 \times 1,0} \left[\sqrt{2 \times 1,0^2 \left(1+1,0\right) + \frac{4 \times 1,0 \left(1+2 \times 1,0\right) \times 76745,4}{25,26 \times 12,0 \times 60^2}} -1,0 \right] = 7644,3 \text{ N}$$

$F_{v,Rk,e} = \{7644,3 + \min \left[(0,25 \times 7644,3) \quad 1619,9) \right] \}$ $F_{v,Rk,e} = 9264,2 \text{ N}$

Equation (8.6(f)) $1,15 \sqrt{\dfrac{2\beta}{1+\beta}} \sqrt{2M_{y,Rk} f_{h,1,k} d} + \dfrac{F_{ax,Rk}}{4}$

$$1,15 \times \sqrt{\frac{2 \times 1,0}{1+1,0}} \times \sqrt{2 \times 76745,4 \times 25,26 \times 12,0} = 7844,2 \text{ N}$$

$F_{v,Rk,f} = \{7844,2 + \min \left[(0,25 \times 7844,2) \quad 1619,9) \right] \}$ $F_{v,Rk,f} = 9464,1 \text{ N}$

The characteristic load-carrying capacity $F_{v,Rk}$ is minimum value determined using Equations (8.6(a)) to (8.6(f)).

$F_{v,Rk}$ = 8429,4 N corresponding with failure mode (d)

Equation (2.17)   Design load-carrying capacity of bolts:

$$F_{v,Rd} = k_{mod} \frac{F_{v,Rk}}{\gamma_{mt}} = 0,8 \times \frac{8429,4}{1,3} = 5187,3 \text{ N}$$

The lateral load-carrying capacity of the joint = 5187,3 N

Spacing Requirements: (Clause 8.5.1.1(3))
The spacing requirements are similar to those required for nails but using Table 8.4 instead of Table 8.2 and Figure 8.7 – the reader should complete this.

Joint slip:
Joint slip is calculated in a similar manner to that for nails with the addition of + 1,0 mm to allow for the increased bolt hole diameter, as indicated in Table 7.1.

Clause 7.1(3)   Slip modulus K_{ser} per shear plane per fastener is given in Table 7.1
Table 7.1   Bolts, $\rho_m = 420 \text{ kg/m}^3$, $d = 12,0 \text{ mm}$,
                $K_{ser} = \rho_m^{1,5} d / 23 = (420^{1,5} \times 12,0)/23 = 4490,8 \text{ N/mm}$

                Assume $\gamma_{Q,1} = 1,5$
                Characteristic design load for the serviceability limit state $= F_{v,Rd}/\gamma_{Q,1}$
                $F_{Q,k} = 5187,3/1,5 = 3458,2 \text{ N}$

BS EN 1990:2002
Equation (6.16b) $F_d = \displaystyle\sum_{j\geq1} G_{k,j} \text{ "+" } \sum_{i\geq1} \psi_{2,i} Q_{k,i}$

NA.
Table NA.A1.1   Quasi-permanent value for variable actions on domestic buildings
$\psi_{2,1} = 0,3$

BS EN 1995-1-1:2004
Table 3.2   Deformation factor for solid timber,   Service class 2
$k_{def} = 0,8$
Equation (2.4) $u_{fin,Q,1} = u_{inst,Q,1}(1 + \psi_{2,1}k_{def})$

Load per bolt $= \dfrac{F_{Q,k}}{n} = 3458,2$ N

Instantaneous slip/side $u_{inst} = \dfrac{\text{Load per bolt}}{K_{ser}} = \dfrac{3458,2}{4490,8} = 0,77$ mm/side

$u_{fin,Q,1} = 0,77 \times [1 + (0,3 \times 0,8)] + 1,0 = 1,96$ mm/side

Final joint opening $= (2 \times u_{fin,Q,1}) = (2 \times 1,96) = 3,92$ mm

9.5.5 Example 9.7: Bolted Three-member Connection in Double Shear

Using the design data given, determine the load-carrying capacity for the three-member
joint shown in Figure 9.24.

Design data:

Timber Class	C24
Moisture content	$\leq 20\%$
Load duration	medium-term
Bolt type	M12, 4.6 grade bolts
Number of bolts	1
Bolt diameter (d)	12 mm
Effective area of bolts (A_{eff})	84,3 mm^2
Tensile strength of bolts (f_u)	400 N/mm^2
Washer outer diameter (d_{wo})	36 mm
Washer inner diameter (d_{wi})	14 mm
Washer thickness (t_w)	3,6 mm

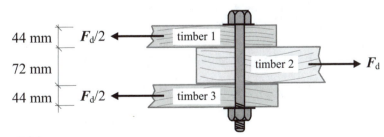

Figure 9.24

Solution:

The following variables are the same as in Example 9.6:

BS EN 338:2003

Clause 5.0   Characteristic values for C24 Timber

Table 1   Characteristic density: $\rho_k = 350{,}0 \text{ kg/m}^3$

                     Mean density: $\rho_{mean} = 420{,}0 \text{ kg/m}^3$

                     Characteristic compressive strength perpendicular to the grain:

                                                               $f_{c,90,k} = 2{,}5 \text{ N/mm}^2$

BS EN 1995-1-1:2004

Clause 2.3.2.1   Load-duration and moisture influence on strength: $k_{mod} = 0{,}8$

Clause 2.3.1.3   Moisture content $\leq 20\%$:   **Service Class 2**

NA: Table NA.3   Partial factor for material properties and resistance:

                     For solid timber – untreated and preservative treated $\gamma_M = 1{,}3$

                     For connections (except punched plate fasteners) $\gamma_M = 1{,}3$

Clause 8.5.1.1(1)   Characteristic yield moment for bolts: $M_{y,Rk} = 76745{,}4 \text{ Nmm}$

Clause 8.5.1.1(2)   Characteristic embedment strengths:

                     Angle between the applied force and the grain: $\alpha = 0°$

                     Timbers 1, 2 and 3 $f_{h,1,k} = 25{,}26 \text{ N/mm}^2$

                     Ratio of embedment strengths: $\beta = 1{,}0$

Clause 10.4.3(2)   **Washer diameter and thickness are adequate**

Clause 8.5.2(1)   The axial withdrawal capacity:

                                                        $F_{ax,Rk} = 6479{,}55 \text{ N}$

Clause 8.2.2(2)   For bolts, the rope effect is equal to $F_{ax,Rk}/4$ but no more than 25% of the Johansen part in Equations (8.6(j)) to (8.6(k)).

                                       $F_{ax,Rk}/4 = 6479{,}55/4 = 1619{,}9 \text{ N}$

                     Thickness of timbers 1 and 3 $t_1 = 44 \text{ mm}$

                     Thickness of timber 2 $t_2 = 72 \text{ mm}$

Clause 8.2.2(1)   Characteristic load-carrying capacity: $F_{v,Rk}$ - fasteners in double shear:

Equation (8.6(g)) $f_{h,1,k} \, t_1 \, d = (25{,}26 \times 44 \times 12{,}0) = 13337{,}3 \text{ N}$

                                                      $F_{v,Rk,g} = 13337{,}3 \text{ N}$

Equation (8.6(h)) $0{,}5 \, f_{h,2,k} \, t_2 \, d = (0{,}5 \times 25{,}26 \times 72 \times 12{,}0) = 10912{,}3 \text{ N}$

                                                      $F_{v,Rk,h} = 10912{,}3 \text{ N}$

Equation (8.6(j)) $1,05\dfrac{f_{h,1,k}t_1 d}{2+\beta}\left[\sqrt{2\beta(1+\beta)+\dfrac{4\beta(2+\beta)M_{y,Rk}}{f_{h,1,k}d\,t_1^2}}-\beta\right]+\dfrac{F_{axRk}}{4}$

$1,05\times\dfrac{25,26\times44\times12,0}{2+1,0}\left[\sqrt{2\times1,0\times(1+1,0)+\dfrac{4\times1,0(2+1,0)\times76745,4}{25,26\times12,0\times44^2}}-1,0\right]=6348,3\ N$

$F_{v,Rk,d}=\{6348,3+\min\,[(0,25\times6348,3)\quad1619,9)]\}$ $\hspace{2cm}$ $F_{v,Rk,j}=7935,4\ N$

Equation (8.6(k)) $1,15\sqrt{\dfrac{2\beta}{1+\beta}}\sqrt{2M_{y,Rk}f_{h,1,k}d}+\dfrac{F_{axRk}}{4}$

$1,15\sqrt{\dfrac{2\times1,0}{1+1,0}}\sqrt{2\times76745,4\times25,26\times12,0}=7844,2\ N$

$F_{v,Rk,k}=\{7844,2+\min\,[(0,25\times7844,2)\quad1619,9)]\}$ $\hspace{2cm}$ $F_{v,Rk,k}=9464,1\ N$

The characteristic load-carrying capacity $F_{v,Rk}$ is minimum value determined using Equations (8.6(g)) to (8.6(k)).

$$F_{v,Rk}=7935,4\ N\ \textbf{corresponding with failure mode (j)}$$

Equation (2.17)   Design load-carrying capacity of bolts:

$$F_{v,Rd}=k_{mod}\dfrac{F_{v,Rk}}{\gamma_{mt}}=0,8\times\dfrac{7935,4}{1,3}=4883,3\ N$$

Bolts in double shear: $(2\times4883,3)=9766,6\ N$

The lateral load-carrying capacity of the joint = 8311,7 N

Spacing Requirements and Joint slip can be determined as in Example 9.6.

9.5.6 Example 9.8: Bolted Four-member Connection in Multiple Shear

Using the design data given, determine the load-carrying capacity for the four-member joint shown in Figure 9.25.

Design data:

Timber Class	C24
Moisture content	$\leq 20\%$
Load duration	medium-term
Bolt type	M12, 4.6 grade bolts
Number of bolts	1
Bolt diameter (d)	12 mm
Effective area of bolts (A_{eff})	84,3 mm^2
Tensile strength of bolts (f_u)	400 N/mm^2

Washer outer diameter (d_{wo})   36 mm
Washer inner diameter (d_{wi})   14 mm
Washer thickness (t_w)   3,6 mm

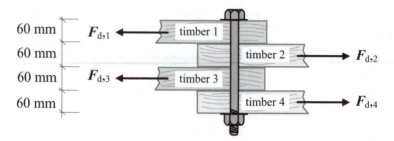

Figure 9.25

Solution:
The following variables are the same as in Example 9.6:
BS EN 338:2003
Clause 5.0   Characteristic values for C24 Timber
Table 1   Characteristic density: $\rho_k = 350,0$ **kg/m³**
                      Mean density: $\rho_{mean} = 420,0$ **kg/m³**
                      Characteristic compressive strength perpendicular to the grain:
                                                                             $f_{c,90,k} = 2,5$ **N/mm²**

BS EN 1995-1-1:2004
Clause 2.3.2.1   Load-duration and moisture influence on strength: $k_{mod} = 0,8$

Clause 2.3.1.3   Moisture content ≤ 20%:   **Service Class 2**

NA: Table NA.3   Partial factor for material properties and resistance:
                      For solid timber – untreated and preservative treated $\gamma_M = 1,3$
                      For connections (except punched plate fasteners) $\gamma_M = 1,3$

Clause 8.5.1.1(1)   Characteristic yield moment for bolts: $M_{y,Rk} = 76745,4$ **Nmm**

Clause 8.5.1.1(2)   Characteristic embedment strengths:
                      Angle between the applied force and the grain: $\alpha = 0°$
                      Timbers 1, 2, 3 and 4 $f_{h,1,k} = 25,26$ **N/mm²**

                      Ratio of embedment strengths: $\beta = 1,0$

Clause 10.4.3(2)   **Washer diameter and thickness are adequate**

Clause 8.5.2(1)   The axial withdrawal capacity:
                                                                             $F_{ax,Rk} = 6479,55$ N
Clause 8.2.2(2)   For bolts, the rope effect is equal to $F_{ax,Rk}/4$ but no more than 25% of
                      the Johansen part in Equations (8.6(j)) to (8.6(k)).
                                                                $F_{ax,Rk}/4 = 6479,55/4 = 1619,9$ N

Thickness of timbers 1, 2, 3 and 4   **$t = 60$ mm**

Clause 8.1.3 Multiple shear plane connections:

Clause 8.1.3(1) The resistance of each shear plane is determined assuming that each shear plane is part of a series of three-member connections.

Clause 8.1.3(2) This Clause indicates that in order to combine the resistance from individual shear planes in a multiple shear plane connection, the governing failure mode of the fasteners in the respective shear planes should be compatible with each other. They should not consist of a combination of failure modes (g) and (h) from Figure 8.2 for double shear, timber-to-timber connections with the other failure modes, **i.e. only consider failure modes (j) and (k) and ignore the possibility of embedment failure in individual timbers.**

There are two three-member systems:
(timber 1 + timber 2 + timber 3) and (timber 2 + timber 3 + timber 4)

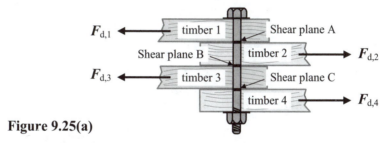

Figure 9.25(a)

Consider $(F_{d,1} + F_{d,3})$:
For timber 1, the loading capacity arises only from the capacity of shear plane A.
For timber 3, the loading capacity arises from the capacity of shear planes B and C.
$(F_{d,1} + F_{d,3})$ = shear capacity of the three shear planes (A + B + C).

Similarly for $(F_{d,2} + F_{d,4})$.

Due to the symmetry of the joint, only one side need be considered.

In the three-member system (timber 1 + timber 2 + timber 3), timber 3 is shared by shear planes B and C; as a consequence of this, the effective thickness for timber 3, as an outer member, should be reduced by 50%.

Consider the three-member system (timber 1 + timber 2 + timber 3):

Thickness of timbers 1, 2		$t = 60$ mm
Effective thickness of timber 3 = $(0,5 \times 60 = 30$ mm)	$t_{ef,3} = 30$ mm
Figure 8.2	t_1 is the smaller of t and $t_{ef,3}$	$t_1 = 30$ mm
		$t_2 = 60$ mm

Clause 8.2.2(1)   Characteristic load-carrying capacity: $F_{v,Rk}$ - fasteners in double shear:

Equation (8.6(g)) $f_{h,1,k}\, t_1\, d$ This mode of failure is not considered.

Equation (8.6(h)) $0,5\, f_{h,2,k}\, t_2\, d$ This mode of failure is not considered.

Equation (8.6(j)) $1,05\dfrac{f_{h,1,k} t_1 d}{2+\beta}\left[\sqrt{2\beta(1+\beta)+\dfrac{4\beta(2+\beta)M_{y,Rk}}{f_{h,1,k}d\, t_1^2}}-\beta\right]+\dfrac{F_{axRk}}{4}$

$1,05\times\dfrac{25,26\times30\times12,0}{2+1,0}\left[\sqrt{2\times1,0\times(1+1,0)+\dfrac{4\times1,0(2+1,0)\times76745,4}{25,26\times12,0\times30^2}}-1,0\right]=5461,1\ \text{N}$

$F_{v,Rk,j}$ = {5461,1+ min [(0,25 × 5461,1) 1619,9)]} $F_{v,Rk,j}$ = 6826,5 N

Equation (8.6(k)) $1,15\sqrt{\dfrac{2\beta}{1+\beta}}\sqrt{2M_{y,Rk}f_{h,1,k}d}+\dfrac{F_{axRk}}{4}$

$1,15\sqrt{\dfrac{2\times1,0}{1+1,0}}\sqrt{2\times76745,4\times25,26\times12,0}=7844,2\ \text{N}$

$F_{v,Rk,k}$ = {7844,2 + min [(0,25 × 7844,2) 1619,9)]} $F_{v,Rk,k}$ = 9464,1 N

The characteristic load-carrying capacity $F_{v,Rk}$ is minimum value determined using Equations (8.6(j)) and (8.6(k)).

$F_{v,Rk}$ = 6826,5 N corresponding with failure mode (j)

Equation (2.17)   Design load-carrying capacity of bolts:

$$F_{v,Rd}=k_{mod}\dfrac{F_{v,Rk}}{\gamma_{mt}}=0,8\times\dfrac{6826,5}{1,3}=4200,9\ \text{N/shear plane}$$

Capacity based on 3 shear planes: (3 × 4200,9) = 12602,7 N

The lateral load-carrying capacity of the joint = 12602,7 N

Spacing Requirements and Joint slip can be determined as in Example 9.6.

9.5.7 Example 9.9: Bolted Two-member Connection in Single Shear – Applied Load Perpendicular to the Grain.

Using the design data given, determine the load-carrying capacity for the two-member, joint shown in Figure 9.26.

Design data:

Timber Class	C22
Moisture content	≤ 20%
Load duration	long-term
Bolt type	M16, 4.6 grade bolts
Number of bolts	1
Bolt diameter (d)	16 mm
Effective area of bolts (A_{eff})	157,0 mm^2
Tensile strength of bolts (f_u)	400 N/mm^2
Washer outer diameter (d_{wo})	48 mm
Washer inner diameter (d_{wi})	18 mm
Washer thickness (t_w)	4,8 mm

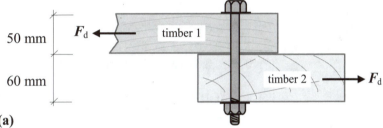

Figure 9.26(a)

Solution:
BS EN 338:2003

Clause 5.0	Characteristic values for C22 Timber	
Table 1	Characteristic density:	$\rho_k = \mathbf{340,0}$ **kg/m^3**
	Mean density:	$\rho_{\text{mean}} = \mathbf{410,0}$ **kg/m^3**
	Characteristic compressive strength perpendicular to the grain:	
		$f_{\text{c,90,k}} = \mathbf{2,4}$ **N/mm^2**

BS EN 1995-1-1:2004

Clause 2.3.2.1	Load-duration and moisture influence on strength: k_{mod}	
Clause 2.3.1.3	Moisture content ≤ 20%	**Service Class 2**
Table 3.1	Medium-term actions	$k_{\text{mod}} = \mathbf{0,7}$
NA: Table NA.3	Partial factor for material properties and resistance: γ_M	
	For solid timber – untreated and preservative treated	$\gamma_M = \mathbf{1,3}$
	For connections (except punched plate fasteners)	$\gamma_M = \mathbf{1,3}$
	Thickness of timber 1	$t_1 = \mathbf{50}$ **mm**
	Thickness of timber 2	$t_2 = \mathbf{60}$ **mm**

Clause 8.5.1.1(1)   Characteristic yield moment for bolts with $f_u = 400$ N/mm^2
Equation (8.30) $M_{\text{y,Rk}} = 0.3\, f_u\, d^{\,2,6} = (0.3 \times 400 \times 16.0^{2,6}) = 162141{,}1$ Nmm

$$M_{\text{y,Rk}} = \mathbf{162141{,}1} \text{ Nmm}$$

Clause 8.5.1.1(2)   Characteristic embedment strength: $f_{h,k}$

Equation (8.32)

Timber 1: Angle between the applied force and the grain $\alpha = 0°$

$f_{h,0,k} = 0,082 \ (1 - 0,01d)\rho_k = 0,082 \times [1 - (0,01 \times 16,0)] \times 340$
$= 23,42 \ \text{N/mm}^2$

Timber 2: Angle between the applied force and the grain $\alpha = 90°$
For bolts in timber and LVL and with diameters up to 30 mm and with an angle α to the grain:

Equation (8.31) $f_{h,\alpha,k} = \dfrac{f_{h,0,k}}{k_{90}\sin^2\alpha + \cos^2\alpha} \ \text{N/mm}^2$ where k_{90} is given in Equation (8.33)

Equation (8.33)

$k_{90} = \begin{cases} 1,35 + 0,015d & \text{for softwoods} \\ 1,30 + 0,015d & \text{for LVL} \\ 0,90 + 0,015d & \text{for hardwoods} \end{cases}$

$k_{90} = (1,35 + 0,015d) \quad = [1,35 + (0,015 \times 16)] = 1,59$

$f_{h,\alpha,k} = \dfrac{f_{h,0,k}}{k_{90}\sin^2\alpha + \cos^2\alpha} = \dfrac{23,42}{\left(1,59 \times \sin^2 90°\right) + \cos^2 90°} = 14,73 \ \text{N/mm}^2$

Ratio of embedment strengths $\beta = \dfrac{f_{h,2,k}}{f_{h,1,k}} = \dfrac{14,73}{23,42} = 0,63$ $\boldsymbol{\beta = 0,63}$

Clause 10.4.3(2)   Washer diameter $\geq 3d = (3 \times 16) = 48$ mm
Actual diameter $= 48$ mm $= 3d$
Washer thickness $\geq 0,3d = (0,3 \times 16) = 4,8$ mm
Actual thickness $= 4,8$ mm $= 0,3d$
                    Washer diameter and thickness are adequate

Clause 8.5.2(1)   The axial load-bearing capacity and withdrawal capacity of a bolt should be taken as the lower of:

(i) the bolt tensile capacity,
(ii) the load-bearing capacity of either the washer or (for steel-to-timber connections) the steel plate.

Clause 8.5.2(2)   Load-bearing capacity of the washer $= (3,0 \ f_{c,90,k} \times A_{c,90})$
Bearing area:
$A_{c,90} = \pi \ (d_{wo}^2 - d_{wi}^2)/4 = [\pi \times (48^2 - 18^2)]/4 = 1555,09 \ \text{mm}^2$
$(3,0 \ f_{c,90,k} \times A_{c,90}) = (3,0 \times 2,5 \times 1555,09) = 11196,65$ N
Bolt tensile capacity $= (f_u \times A_{eff}) = (400 \times 157,0) = 62800$ N

Clause 8.5.2(1) $F_{ax,Rk}$ = the minimum value between $(3,0 f_{c,90,k} \times A_{c,90})$ and $(f_u \times A_{eff})$
Characteristic axial withdrawal capacity: $\qquad F_{ax,Rk} = 11196,65$ kN

Clause 8.2.2(2) For bolts, the rope effect is equal to $F_{ax,Rk}/4$ but no more than 25% of the Johansen part in Equations (8.6(c)) to (8.6(f)).
$$F_{ax,Rk}/4 = 11196,65/4 = 2799,2 \text{ N}$$

Clause 8.2.2(1) Characteristic load-carrying capacity: $F_{v,Rk}$ – single shear
Equation (8.6(a)) $f_{h,1,k}\, t_1\, d = (23,42 \times 50 \times 16,0) = 18736,0$ N
$$F_{v,Rk,a} = 18736,0 \text{ N}$$

Equation (8.6(b)) $f_{h,2,k}\, t_2\, d = (14,73 \times 60 \times 16,0) = 14140,8$ N
$$F_{v,Rk,b} = 14140,8 \text{ N}$$

Equation (8.6(c))
$$\frac{f_{h,1,k} t_1 d}{1+\beta}\left[\sqrt{\beta + 2\beta^2\left[1 + \frac{t_2}{t_1} + \left(\frac{t_2}{t_1}\right)^2\right] + \beta^3\left(\frac{t_2}{t_1}\right)^2} - \beta\left(1 + \frac{t_2}{t_1}\right)\right] + \frac{F_{ax,Rk}}{4}$$

$$\frac{23,42\times50\times16,0}{1+0,63}\left[\sqrt{0,63 + 2\times0,63^2\left[1+\frac{60}{50}+\left(\frac{60}{50}\right)^2\right] + 0,63^3\left(\frac{60}{50}\right)^2} - 0,63\left(1+\frac{60}{50}\right)\right] = 6708,7 \text{ N}$$

$F_{v,Rk,c} = \{6708,7 + \min\,[(0,25 \times 6708,7)\ \ 2799,2)]\}$ $\qquad\qquad F_{v,Rk,c} = 8385,9$ N

Equation (8.6(d)) $1,05\dfrac{f_{h,1,k} t_1 d}{2+\beta}\left[\sqrt{2\beta(1+\beta) + \dfrac{4\beta(2+\beta)M_{y,Rk}}{f_{h,1,k}d\, t_1^2}} - \beta\right] + \dfrac{F_{ax,Rk}}{4}$

$$1,05\times\frac{23,42\times50\times16,0}{2+0,63}\left[\sqrt{2\times0,63(1+0,63) + \frac{4\times0,63(2+0,63)\times162141,1}{23,42\times16,0\times50^2}} - 0,63\right] = 8670,3 \text{ N}$$

$F_{v,Rk,d} = \{8670,3 + \min\,[(0,25 \times 8670,3)\ \ 2799,2)]\}$ $\qquad\qquad F_{v,Rk,d} = 10837,9$ N

Equation (8.6(e)) $1,05\dfrac{f_{h,1,k} t_2 d}{1+2\beta}\left[\sqrt{2\beta^2(1+\beta) + \dfrac{4\beta(1+2\beta)M_{y,Rk}}{f_{h,1,k}d\, t_2^2}} - \beta\right] + \dfrac{F_{ax,Rk}}{4}$

$$1,05\times\frac{23,42\times60\times16,0}{1+2\times0,63}\left[\sqrt{2\times0,63^2(1+0,63) + \frac{4\times0,63(1+2\times0,63)\times162141,1}{23,42\times16,0\times60^2}} - 0,63\right] = 8111,8 \text{ N}$$

$F_{v,Rk,e} = \{8111,8 + \min [(0,25 \times 8111,8)\ \ 2799,2)]\}$ $F_{v,Rk,e} = 10139,8N$

Equation (8.6(f)) $\quad = 1,15\sqrt{\dfrac{2\beta}{1+\beta}}\sqrt{2M_{y,Rk}f_{h,1,k}d} + \dfrac{F_{ax,Rk}}{4}$

$$1,15 \times \sqrt{\frac{2 \times 0,63}{1 + 0,63}} \times \sqrt{2 \times 162141,1 \times 23,42 \times 16,0} = 11145,6\ N$$

$F_{v,Rk,f} = \{11145,6 + \min [(0,25 \times 11145,6)\ \ 2799,2)]\}$ $F_{v,Rk,f} = 13932,0\ N$

The characteristic load-carrying capacity $F_{v,Rk}$ is minimum value determined using Equations (8.6(a)) to (8.6(f)).

$$\textbf{\textit{F}}_{\textbf{v,Rk}} = \textbf{8385,9 N corresponding with failure mode (c)}$$

Equation (2.17)   Design load-carrying capacity of bolts:

$$F_{v,Rd} = k_{mod}\frac{F_{v,Rk}}{\gamma_{mt}} = 0,7 \times \frac{8385,9}{1,3} = 4515,5\ N$$

$$\textbf{The lateral load-carrying capacity of the joint = 4515,5 N}$$

Note: Clause 8.1.4 specifies that when a force in a connection acts at an angle α to the grain (see Figure 8.1 of EC5), the possibility of splitting caused by the tension force component $F_{ED}\sin\alpha$, perpendicular to the grain, shall be taken into account. In this case, the angle of the force with the grain of timber 2 is $\alpha = 90°$.

9.5.8 Example 9.10: Bolted Four-member Connection in Multiple Shear – Applied Load Perpendicular to the Grain.

Using the design data given, determine the load-carrying capacity for the four-member joint shown in Figure 9.27.

Design data:

Timber Class	C22
Moisture content	$\leq 20\%$
Load duration	long-term
Bolt type	M16, 4.6 grade bolts
Number of bolts	1
Bolt diameter (d)	16 mm
Effective area of bolts (A_{eff})	157,0 mm^2
Tensile strength of bolts (f_u)	400 N/mm^2
Washer outer diameter (d_{wo})	48 mm
Washer inner diameter (d_{wi})	18 mm
Washer thickness (t_w)	4,8 mm

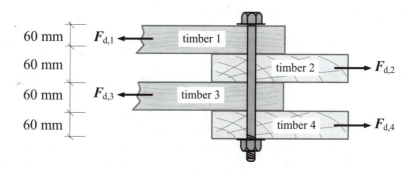

Figure 9.27

Solution:
The following variables are the same as in Example 9.9:
BS EN 338:2003

Clause 5.0	Characteristic values for C22 Timber	
Table 1	Characteristic density:	$\rho_k = 340{,}0 \text{ kg/m}^3$
	Mean density:	$\rho_{mean} = 410{,}0 \text{ kg/m}^3$

Characteristic compressive strength perpendicular to the grain:
$$f_{c,90,k} = 2{,}4 \text{ N/mm}^2$$

BS EN 19950-1-1:2004

Clause 2.3.2.1	Load-duration and moisture influence on strength:	$k_{mod} = 0{,}7$
Clause 2.3.1.3	Moisture content $\leq 20\%$:	**Service Class 2**
NA: Table NA.3	Partial factor for material properties and resistance:	
	For solid timber – untreated and preservative treated	$\gamma_M = 1{,}3$
	For connections (except punched plate fasteners)	$\gamma_M = 1{,}3$

Clause 8.5.1.1(1)   Characteristic yield moment for bolts: $M_{y,Rk} = 162141{,}1 \text{ Nmm}$

Clause 8.5.1.1(2)   Characteristic embedment strength: $f_{h,k}$
Timber 1: Angle between the applied force and the grain $\alpha = 0°$
$$f_{h,0,k} = 23{,}42 \text{ N/mm}^2$$

Timber 2: Angle between the applied force and the grain $\alpha = 90°$
$$f_{h,\alpha,k} = 14{,}73 \text{ N/mm}^2$$

Clause 10.4.3(2)   **Washer diameter and thickness are adequate**

Clause 8.5.2(1)   The axial withdrawal capacity:
$$F_{ax,Rk} = 11196{,}65 \text{ N}$$

Clause 8.2.2(2)   For bolts, the rope effect is equal to $F_{ax,Rk}/4$ but no more than 25% of
the Johansen part in Equations (8.6(j)) and (8.6(k)).
$$F_{ax,Rk}/4 = 11196{,}65/4 = 2799{,}2 \text{ N}$$

Thickness of timbers 1, 2, 3 and 4 $t = 60 \text{ mm}$

Clause 8.1.3   Multiple shear plane connections:
Clause 8.1.3(1)   The resistance of each shear plane is determined assuming that each
                      shear plane is part of a series of three-member connections.

Clause 8.1.3(2)   This Clause indicates that in order to combine the resistance from
                      individual shear planes in a multiple shear plane connection, the
                      governing failure mode of the fasteners in the respective shear planes
                      should be compatible with each other.
                      They should not consist of a combination of failure modes (g) and (h)
                      from Figure 8.2 for double shear, timber-to-timber connections with
                      the other failure modes,

                      **i.e. only consider failure modes (j) and (k) and ignore the
                      possibility of embedment failure in individual timbers.**

                      There are two three-member systems:
                      (timber 1 + timber 2 + timber 3) and (timber 2 + timber 3 + timber 4)

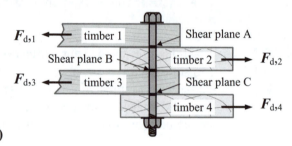

Figure 9.27(a)

Due to the asymmetry of the joint, both sides need to be considered separately and the
lowest capacity adopted for design.

Consider (timber 1 + timber 2 + timber 3):
Two members, timber 1 and timber 3, are loaded parallel to the grain and one member
timber 2, loaded perpendicular to the grain.

For timber 1, the loading capacity arises only from the capacity of shear plane A.
For timber 3, the loading capacity arises from the capacity of shear planes B and C.
$(F_{d,1} + F_{d,3})$ = **shear capacity of the three shear planes (A + B + C).**

In the three-member system timber 3 is shared by shear planes B and C; as a consequence
of this, the effective thickness for timber 3, as an outer member, should be reduced by
50%

Thickness of timbers 1, 2 $t = 60$ mm
Effective thickness of timber 3 = $(0,5 \times 60 = 30$ mm) $t_{ef,3} = 30$ mm

Figure 8.2 t_1 is the smaller of t and $t_{ef,3}$ **$t_1 = 30$ mm**
$t_2 = 60$ mm

$$\text{Ratio of embedment strengths } \beta = \frac{f_{h,2,k}}{f_{h,1,k}} = \frac{14,73}{23,42} = 0,63 \qquad \boldsymbol{\beta = 0,63}$$

Clause 8.2.2(1)   Characteristic load-carrying capacity: $F_{v,Rk}$ - fasteners in double shear:

Equation (8.6(g)) $f_{h,1,k}\, t_1\, d$ This mode of failure is not considered.

Equation (8.6(h)) $0,5\, f_{h,2,k}\, t_2\, d$ This mode of failure is not considered.

Equation (8.6(j)) $1,05\dfrac{f_{h,1,k}t_1 d}{2+\beta}\left[\sqrt{2\beta(1+\beta)+\dfrac{4\beta(2+\beta)M_{y,Rk}}{f_{h,1,k}d\,t_1^2}}-\beta\right]+\dfrac{F_{axRk}}{4}$

$$1,05\times\frac{23,42\times30\times16,0}{2+0,63}\left[\sqrt{2\times0,63\times(1+0,63)+\frac{4\times0,63(2+0,63)\times162141,1}{23,42\times16,0\times30^2}}-0,63\right]=7446,4\ \text{N}$$

$F_{v,Rk,j} = \{7446,4 + \min\,[(0,25 \times 7446,4)\ \ 2799,2)]\}$ $F_{v,Rk,j} = 9308,0$ N

Equation (8.6(k)) $1,15\sqrt{\dfrac{2\beta}{1+\beta}}\sqrt{2M_{y,Rk}f_{h,1,k}d}+\dfrac{F_{axRk}}{4}$

$$1,15\sqrt{\frac{2\times0,63}{1+0,63}}\sqrt{2\times162141,1\times23,42\times16,0}=11145,6\ \text{N}$$

$F_{v,Rk,k} = \{11145,6 + \min\,[(0,25 \times 11145,6)\ \ 2799,2)]\}$ $F_{v,Rk,k} = 13932,0$ N

The characteristic load-carrying capacity $F_{v,Rk}$ is minimum value determined using Equations (8.6(j)) and (8.6(k)).

$$\boldsymbol{F_{v,Rk} = 9308,0 \text{ N corresponding with failure mode (j)}}$$

Equation (2.17)   Design load-carrying capacity of bolts for shear planes A and B:

$$F_{v,Rd} = k_{mod}\frac{F_{v,Rk}}{\gamma_{mt}} = 0,7\times\frac{9308,0}{1,3} = 5012,0 \text{ N/shear plane}$$

Consider (timber 2 + timber 3 + timber 4):
One member, timber 3 is loaded parallel to the grain and two members, timber 2 and timber 4, are loaded perpendicular to the grain.
For timber 4, the loading capacity arises only from the capacity of shear plane C.
For timber 3, the loading capacity arises from the capacity of shear planes B and C.
$(F_{d,2} + F_{d,4}) =$ **shear capacity of the three shear planes (A + B + C).**

In the three-member system timber 2 is shared by shear planes A and B; as a consequence of this, the effective thickness for timber 2, as an outer member, should be reduced by 50%.

	Thickness of timbers 3, 4	$t = 60$ mm
	Effective thickness of timber 2 = $(0,5 \times 60 = 30$ mm)	$t_{ef,2} = 30$ mm
Figure 8.2	t_1 is the smaller of t and $t_{ef,2}$	$t_1 = \mathbf{30}$ **mm**
		$t_2 = \mathbf{60}$ **mm**

Ratio of embedment strengths $\beta = \dfrac{f_{h,2,k}}{f_{h,1,k}} = \dfrac{23,42}{14,73} = 1,59$ $\qquad$ **$\beta = 1,59$**

Clause 8.2.2(1)   Characteristic load-carrying capacity: $F_{v,Rk}$ - fasteners in double shear:

Equation (8.6(g)) $f_{h,1,k}\, t_1\, d$ $\qquad\qquad$ This mode of failure is not considered.

Equation (8.6(h)) $0,5\, f_{h,2,k}\, t_2\, d$ $\qquad\qquad$ This mode of failure is not considered.

Equation (8.6(j)) $1,05 \dfrac{f_{h,1,k} t_1 d}{2+\beta}\left[\sqrt{2\beta(1+\beta)+\dfrac{4\beta(2+\beta)M_{y,Rk}}{f_{h,1,k}\, d\, t_1^2}} - \beta\right] + \dfrac{F_{axRk}}{4}$

$1,05 \times \dfrac{14,73 \times 30 \times 16,0}{2+1,59}\left[\sqrt{2 \times 1,59 \times (1+1,59)+\dfrac{4 \times 1,59(2+1,59) \times 162141,1}{14,73 \times 16,0 \times 30^2}} - 1,59\right] = 7193,3$ N

$F_{v,Rk,j} = \{7193,3 + \min\,[(0,25 \times 7193,3) \quad 2799,2)]\}$ $\qquad\qquad F_{v,Rk,j} = 8991,7$ N

Equation (8.6(k)) $1,15\sqrt{\dfrac{2\beta}{1+\beta}}\sqrt{2M_{y,Rk}\, f_{h,1,k}\, d} + \dfrac{F_{axRk}}{4}$

$1,15\sqrt{\dfrac{2 \times 1,59}{1+1,59}}\sqrt{2 \times 162141,1 \times 14,73 \times 16,0} = 11140,0$ N

$F_{v,Rk,k} = \{11140,0 + \min\,[(0,25 \times 11140,0) \quad 2799,2)]\}$ $\qquad\qquad F_{v,Rk,k} = 13925,0$ N

The characteristic load-carrying capacity $F_{v,Rk}$ is minimum value determined using Equations (8.6(j)) and (8.6(k)).

$$F_{v,Rk} = \mathbf{8991,7}\ \textbf{N corresponding with failure mode (j)}$$

Equation (2.17)   Design load-carrying capacity of bolts for shear planes A and B:

$$F_{v,Rd} = k_{mod}\dfrac{F_{v,Rk}}{\gamma_{mt}} = 0,7 \times \dfrac{8911,7}{1,3} = 4841,7\ \text{N/shear plane}$$

The capacity of the laterally loaded bolts for each of the shear planes:

Shear plane A: $F_{v,Rd} = 5012,0$ N

Shear plane B: min of $\begin{cases} 5012,0 \\ 4841,7 \end{cases}$ $F_{v,Rd} = 4841,7$ N

Shear plane C: $F_{v,Rd} = 4841,7$ N

The capacity of the joint for both systems $= (5012,0 + 4841,7 + 4841,7) = 14695,4$ N

Conservatively, the lateral load-carrying capacity of the joint can be estimated as:
(3 × lowest shear plane strength) = (3 × 4841,7) = 14524,1 N

Spacing Requirements and Joint slip can be determined as for previous Examples.

9.5.9 *Example 9.11: Bolted Four-member Connection in Multiple Shear – Applied Load at α = 30° to the Grain.*

Using the design data given, determine the load-carrying capacity for the four-member, joint shown in Figure 9.28.

Design data:

Timber Class	C16
Moisture content	$\leq 20\%$
Load duration	short-term
Bolt type	M12, 4.6 grade bolts
Number of bolts	1
Bolt diameter (d)	12 mm
Effective area of bolts (A_{eff})	84,3 mm^2
Tensile strength of bolts (f_u)	400 N/mm^2
Washer outer diameter (d_{wo})	36 mm
Washer inner diameter (d_{wi})	14 mm
Washer thickness (t_w)	3,6 mm

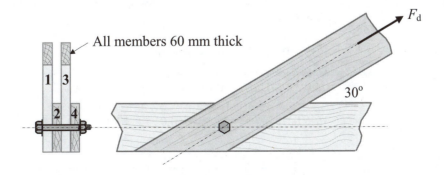

Figure 9.28

Solution:
BS EN 338:2003
Clause 5.0   Characteristic values for C16 Timber
Table 1   Characteristic density: ρ_k = **310,0 kg/m³**
                    Mean density: ρ_{mean} = **370,0 kg/m³**
                    Characteristic compressive strength perpendicular to the grain:
                                                                               $f_{c,90,k}$ = **2,2 N/mm²**

BS EN 1995-1-1:2004
Clause 2.3.2.1   Load-duration and moisture influence on strength: k_{mod}
Clause 2.3.1.3   Moisture content ≤ 20%   **Service Class 2**
Table 3.1   Short term actions k_{mod} = **0,9**

NA: Table NA.3   Partial factor for material properties and resistance: γ_M
                    For solid timber – untreated and preservative treated γ_M = **1,3**
                    For connections (except punched plate fasteners) γ_M = **1,3**

Clause 8.5.1.1(1)   Characteristic yield moment for bolts with f_u = 400 N/mm²
Equation (8.30) $M_{y,Rk} = 0,3\, f_u\, d^{2,6} = (0,3 \times 400 \times 12,0^{2,6}) = 76745,4$ Nmm
                                                                $M_{y,Rk}$ = 76745,4 Nmm

Clause 8.5.1.1(2)   Characteristic embedment strength: $f_{h,k}$
                    Timbers 1 and 3: Angle between the applied force and the grain $\alpha = 0°$

Equation (8.32) $f_{h,0,k} = 0,082\,(1 - 0,01d)\rho_k = 0,082 \times [1 - (0,01 \times 12,0)] \times 310$
                    $= 22,37$ N/mm²

                    Timbers 2 and 4: Angle between the applied force and the grain $\alpha = 90°$
                    For bolts in timber and LVL and with diameters up to 30 mm and with
                    an angle α to the grain:

Equation (8.31) $f_{h,\alpha,k} = \dfrac{f_{h,0,k}}{k_{90}\sin^2\alpha + \cos^2\alpha}$ N/mm² where k_{90} is given in Equation (8.33)

Equation (8.33) $k_{90} = \begin{cases} 1,35 + 0,015d & \text{for softwoods} \\ 1,30 + 0,015d & \text{for LVL} \\ 0,90 + 0,015d & \text{for hardwoods} \end{cases}$

                    $k_{90} = (1,35 + 0,015d)\quad = [1,35 + (0,015 \times 12)] = 1,53$

                    $f_{h,\alpha,k} = \dfrac{f_{h,0,k}}{k_{90}\sin^2\alpha + \cos^2\alpha} = \dfrac{22,37}{(1,53 \times \sin^2 90°) + \cos^2 90°} = 14,62$ N/mm²

Clause 10.4.3(2) Washer diameter $\geq 3d = (3 \times 12) = 36$ mm
Actual diameter = 36 mm = $3d$
Washer thickness $\geq 0,3d = (0,3 \times 12) = 3,6$ mm
Actual thickness = 3,6 mm = $0,3d$
Washer diameter and thickness are adequate

Clause 8.5.2(1) The axial load-bearing capacity and withdrawal capacity of a bolt should be taken as the lower of:

(i)   the bolt tensile capacity,
(ii) the load-bearing capacity of either the washer or (for steel-to-timber connections) the steel plate.

Clause 8.5.2(2) Load-bearing capacity of the washer = $(3,0\, f_{c,90,k} \times A_{c,90})$
Bearing area:
$A_{c,90} = \pi\,(d_{wo}^2 - d_{wi}^2)/4 = [\pi \times (36^2 - 14^2)]/4 = 863,94$ mm^2
$(3,0\, f_{c,90,k} \times A_{c,90}) = (3,0 \times 2,2 \times 863,94) = 5702,0$ N

Bolt tensile capacity = $(f_u \times A_{eff}) = (400 \times 84,3) = 33720$ N

Clause 8.5.2(1) The axial withdrawal capacity:
$$F_{ax,Rk} = 5702,0 \text{ N}$$

Clause 8.2.2(2) For bolts, the rope effect is equal to $F_{ax,Rk}/4$ but no more than 25% of the Johansen part in Equations (8.6(j)) and (8.6(k)).
$$F_{ax,Rk}/4 = 5702,0/4 = 1425,5 \text{ N}$$

Thickness of timbers 1, 2, 3 and 4   **$t = 60$ mm**

Clause 8.1.3   Multiple shear plane connections:
Clause 8.1.3(1)   The resistance of each shear plane is determined assuming that each shear plane is part of a series of three-member connections.

Clause 8.1.3(2)   This Clause indicates that in order to combine the resistance from individual shear planes in a multiple shear plane connection, the governing failure mode of the fasteners in the respective shear planes should be compatible with each other.
They should not consist of a combination of failure modes (g) and (h) from Figure 8.2 for double shear, timber-to-timber connections with the other failure modes,
i.e. only consider failure modes (j) and (k) and ignore the possibility of embedment failure in individual timbers.

There are two three-member systems:
(timber 1 + timber 2 + timber 3) and (timber 2 + timber 3 + timber 4)

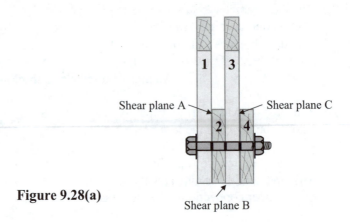

1   3

Shear plane A ⟍   ⟋ Shear plane C

2   4

Figure 9.28(a)   ↗
                                    Shear plane B

Due to the asymmetry of the joint, both sides need be considered separately and the lowest capacity adopted for design.

Consider (timber 1 + timber 2 + timber 3):
Two members, timber 1 and timber 3, are loaded parallel to the grain and one member timber 2, loaded at $\alpha = 30°$ to the grain.
For timber 1, the loading capacity arises only from the capacity of shear plane A.
For timber 3, the loading capacity arises from the capacity of shear planes B and C.

In the three-member system timber 3 is shared by shear planes B and C; as a consequence of this, the effective thickness for timber 3, as an outer member, should be reduced by 50%.

	Thickness of timbers 1, 2	$t = 60$ mm
	Effective thickness of timber 3 = $(0,5 \times 60 = 30$ mm)	$t_{ef,3} = 30$ mm
Figure 8.2	t_1 is the smaller of t and $t_{ef,3}$	$\mathbf{t_1 = 30}$ **mm**
		$t_2 = 60$ mm

$$\text{Ratio of embedment strengths } \beta = \frac{f_{h,2,k}}{f_{h,1,k}} = \frac{14,62}{22,37} = 0,65 \qquad \boldsymbol{\beta = 0,65}$$

Clause 8.2.2(1)   Characteristic load-carrying capacity: $F_{v,Rk}$ - fasteners in double shear:

Equation (8.6(g)) $f_{h,1,k}\, t_1\, d$ This mode of failure is not considered.

Equation (8.6(h)) $0,5\, f_{h,2,k}\, t_2\, d$ This mode of failure is not considered.

Equation (8.6(j)) $1,05\dfrac{f_{h,1,k}t_1 d}{2+\beta}\left[\sqrt{2\beta(1+\beta)+\dfrac{4\beta(2+\beta)M_{y,Rk}}{f_{h,1,k}d\, t_1^2}}-\beta\right]+\dfrac{F_{axRk}}{4}$

$$1,05\times\frac{22,37\times30\times12,0}{2+0,65}\left[\sqrt{2\times0,65\times(1+0,65)+\frac{4\times0,65(2+0,65)\times76745,4}{22,37\times12,0\times30^2}}-0,65\right]=4568,6 \text{ N}$$

$F_{\text{v,Rk,j}}$ = {4568,6 + min [(0,25 × 4568,6) 1425,5)]} $F_{\text{v,Rk,j}}$ = 5710,8 N

Equation (8.6(k)) $1,15\sqrt{\dfrac{2\beta}{1+\beta}}\sqrt{2M_{\text{y,Rk}}f_{\text{h,1,k}}d}+\dfrac{F_{\text{axRk}}}{4}$

$$1,15\sqrt{\frac{2\times0,65}{1+0,65}}\sqrt{2\times76745,4\times22,37\times12,0}=6552,3 \text{ N}$$

$F_{\text{v,Rk,k}}$ = {6552,3 + min [(0,25 × 6552,3) 1425,5)]} $F_{\text{v,Rk,k}}$ = 7977,8 N

The characteristic load-carrying capacity $F_{\text{v,Rk}}$ is minimum value determined using Equations (8.6(j)) and (8.6(k)).

$F_{\text{v,Rk}}$ = 5710,8 N corresponding with failure mode (j)

Equation (2.17)   Design load-carrying capacity of bolts for shear planes A and B:

$$F_{\text{v,Rd}} = k_{\text{mod}}\frac{F_{\text{v,Rk}}}{\gamma_{\text{m}}} = 0,9\times\frac{5710,8}{1,3} = 3953,6 \text{ N/shear plane}$$

Consider (timber 2 + timber 3 + timber 4):
One member, timber 3 is loaded parallel to the grain and two members, timber 2 and timber 4, are loaded at α = 30° to the grain.

For timber 4, the loading capacity arises only from the capacity of shear plane C.

For timber 3, the loading capacity arises from the capacity of shear planes B and C.

In the three-member system timber 2 is shared by shear planes A and B; as a consequence of this, the effective thickness for timber 2, as an outer member, should be reduced by 50%.

	Thickness of timbers 3, 4	t = 60 mm
	Effective thickness of timber 2 = (0,5 × 60 = 30 mm)	$t_{\text{ef,2}}$ = 30 mm
Figure 8.2	t_1 is the smaller of t and $t_{\text{ef,2}}$	**t_1 = 30 mm**
		t_2 = 60 mm

Ratio of embedment strengths $\beta = \dfrac{f_{\text{h,2,k}}}{f_{\text{h,1,k}}} = \dfrac{22,37}{14,62} = 1,53$ **β = 1,53**

Clause 8.2.2(1)   Characteristic load-carrying capacity: $F_{v,Rk}$ - fasteners in double shear:
Equation (8.6(g)) $f_{h,1,k}\, t_1\, d$ This mode of failure is not considered.

Equation (8.6(h)) $0{,}5\, f_{h,2,k}\, t_2\, d$ This mode of failure is not considered.

Equation (8.6(j)) $1{,}05\dfrac{f_{h,1,k}t_1 d}{2+\beta}\left[\sqrt{2\beta(1+\beta)+\dfrac{4\beta(2+\beta)M_{y,Rk}}{f_{h,1,k}d\, t_1^2}}-\beta\right]+\dfrac{F_{axRk}}{4}$

$1{,}05\times\dfrac{14{,}62\times30\times12{,}0}{2+1{,}53}\left[\sqrt{2\times1{,}53\times(1+1{,}53)+\dfrac{4\times1{,}53(2+1{,}53)\times76745{,}4}{14{,}62\times12{,}0\times30^2}}-1{,}53\right]=4291{,}3\ N$

$F_{v,Rk,k}=\{4291{,}3+\min[(0{,}25\times4291{,}3)\quad 1425{,}5)]\}$ $F_{v,Rk,k}=5364{,}1\ N$

Equation (8.6(k)) $1{,}15\sqrt{\dfrac{2\beta}{1+\beta}}\sqrt{2M_{y,Rk}f_{h,1,k}d}+\dfrac{F_{axRk}}{4}$

$1{,}15\sqrt{\dfrac{2\times1{,}53}{1+1{,}53}}\sqrt{2\times76745{,}4\times14{,}62\times12{,}0}=6563{,}0\ N$

$F_{v,Rk,k}=\{6563{,}0+\min[(0{,}25\times6563{,}0)\quad 1425{,}5)]\}$ $F_{v,Rk,k}=7988{,}5\ N$

The characteristic load-carrying capacity $F_{v,Rk}$ is minimum value determined using Equations (8.6(j)) and (8.6(k)).

$F_{v,Rk}=5364{,}1\ N$ corresponding with failure mode (j)

Equation (2.17)   Design load-carrying capacity of bolts for shear planes A and B:

$$F_{v,Rd}=k_{mod}\dfrac{F_{v,Rk}}{\gamma_m}=0{,}9\times\dfrac{5364{,}1}{1{,}3}=3713{,}6\ N/\text{shear plane}$$

The capacity of the laterally loaded bolts for each of the shear planes:

Shear plane A: $F_{v,Rd}=3953{,}6\ N$
Shear plane B: min of $\begin{cases}3953{,}6\\3713{,}6\end{cases}$ $F_{v,Rd}=3713{,}6\ N$
Shear plane C: $F_{v,Rd}=3713{,}6\ N$

The capacity of the joint for both systems $=(3953{,}6+3713{,}6+3713{,}6)=11380{,}8\ N$

Conservatively, the lateral load-carrying capacity of the joint can be estimated as:
$(3\times\text{lowest shear plane strength})=(3\times3713{,}6)=11140{,}8\ N$
Spacing Requirements and Joint slip can be determined as for previous Examples.

9.5.10 Example 9.12: Bolted Knee-brace

A structural frame comprising two columns, a roof truss and two knee-braces is shown in Figures 9.29(a), (b) and (c). Lateral wind loading induces a 30 kN force on the knee-brace as indicated in Figure 9.29(b).
Using the design data given, check the suitability of the bolt details to transmit the loading.

Design data:

Timber Class	C30
Moisture content	$\leq 20\%$
Load duration	short term
Bolt type	M20, 8.8 grade bolts
Number of bolts	2
Bolt diameter (d)	20 mm
Effective area of bolts (A_{eff})	245,0 mm^2
Tensile strength of bolts (f_u)	800 N/mm^2
Washer outer diameter (d_{wo})	60 mm
Washer inner diameter (d_{wi})	22 mm
Washer thickness (t_w)	6 mm

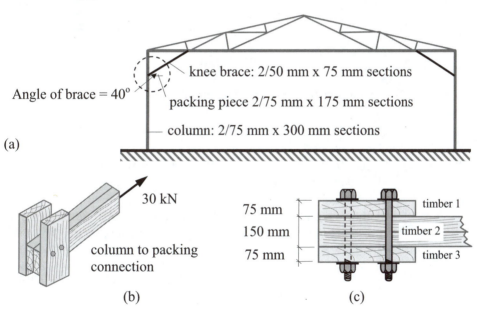

Figure 9.29

Solution:
BS EN 338:2003
Clause 5.0   Characteristic values for C22 Timber
Table 1   Characteristic density: $\rho_k = 380,0$ **kg/m**^{3}
                    Mean density: $\rho_{mean} = 460,0$ **kg/m**^{3}
                    Characteristic compressive strength perpendicular to the grain:
                                                          $f_{c,90,k} = 2,7$ **N/mm**^{2}

BS EN 1995-1-1:2004

Clause 2.3.2.1	Load-duration and moisture influence on strength: k_{mod}	
Clause 2.3.1.3	Moisture content $\leq 20\%$	**Service Class 2**
Table 3.1	Short term actions	$k_{mod} = 0{,}9$
NA: Table NA.3	Partial factor for material properties and resistance: γ_M	
	For solid timber – untreated and preservative treated	$\gamma_M = 1{,}3$
	For connections (except punched plate fasteners)	$\gamma_M = 1{,}3$

Clause 8.5.1.1(1)   Characteristic yield moment for bolts with $f_u = 400$ N/mm^2

Equation (8.30) $M_{y,Rk} = 0{,}3\, f_u\, d^{2,6} = (0{,}3 \times 800 \times 20{,}0^{2,6}) = 162141{,}1$ Nmm

$$M_{y,Rk} = 162141{,}1 \text{ Nmm}$$

(i) Consider the column to packing connection:

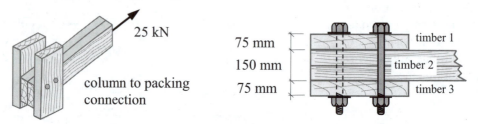

Figure 9.29(d)

Clause 8.5.1.1(2)   Characteristic embedment strength: $f_{h,k}$

Timber 2:Angle between the applied force and the grain $\alpha = 0°$

Equation (8.32) $f_{h,0,k} = 0{,}082\,(1 - 0{,}01d)\rho_k = 0{,}082 \times [1 - (0{,}01 \times 20{,}0)] \times 380$
$\qquad\qquad\qquad\quad = 24{,}93$ N/mm^2

Timbers 1 and 3:Angle between the applied force and the grain $\alpha = 50°$
For bolts in timber and LVL and with diameters up to 30 mm and with
an angle α to the grain:

Equation (8.31) $f_{h,\alpha,k} = \dfrac{f_{h,0,k}}{k_{90}\sin^2\alpha + \cos^2\alpha}$ N/mm^2 where k_{90} is given in Equation (8.33)

Equation (8.33) $k_{90} = \begin{cases} 1{,}35 + 0{,}015d & \text{for softwoods} \\ 1{,}30 + 0{,}015d & \text{for LVL} \\ 0{,}90 + 0{,}015d & \text{for hardwoods} \end{cases}$

$k_{90} = (1{,}35 + 0{,}015d) = [1{,}35 + (0{,}015 \times 20)] = 1{,}65$

$f_{h,\alpha,k} = \dfrac{f_{h,0,k}}{k_{90}\sin^2\alpha + \cos^2\alpha} = \dfrac{24{,}93}{(1{,}65 \times \sin^2 50°) + \cos^2 50°} = 18{,}05$ N/mm^2

Characteristic embedment strength: $f_{h,1,k} = 18{,}05$ N/mm^2
Characteristic embedment strength: $f_{h,2,k} = 24{,}93$N/mm^2

Ratio of embedment strengths $\beta = \dfrac{f_{h,2,k}}{f_{h,1,k}} = \dfrac{24{,}93}{18{,}05} = 1{,}38$ $\boldsymbol{\beta = 1{,}38}$

Thickness of timbers 1 and 3 $\boldsymbol{t_1 = 75}$ **mm**
Thickness of timber 2 $\boldsymbol{t_2 = 150}$ **mm**

Clause 10.4.3(2)   Washer diameter $\geq 3d = (3 \times 20) = 60$ mm
Actual diameter $= 60$ mm $= 3d$
Washer thickness $\geq 0{,}3d = (0{,}3 \times 20) = 6{,}0$ mm
Actual thickness $= 6{,}0$ mm $= 0{,}3d$
                    Washer diameter and thickness are adequate

Clause 8.5.2(1)   The axial load-bearing capacity and withdrawal capacity of a bolt should be taken as the lower of:

(i)   the bolt tensile capacity,
(ii) the load-bearing capacity of either the washer or (for steel-to-timber connections) the steel plate.

Clause 8.5.2(2)   Load-bearing capacity of the washer $= (3{,}0\,f_{c,90,k} \times A_{c,90})$
Bearing area:
$A_{c,90} = \pi\,(d_{wo}^2 - d_{wi}^2)/4 = [\pi \times (60^2 - 22^2)]/4 = 2447{,}30$ mm^2
$(3{,}0\,f_{c,90,k} \times A_{c,90}) = (3{,}0 \times 2{,}7 \times 2447{,}30) = 19823{,}1$ N

Bolt tensile capacity $= (f_u \times A_{eff}) = (800 \times 245{,}0) = 196000$ N

Clause 8.5.2(1)   The axial withdrawal capacity:
                                        $F_{ax,Rk} = 19823{,}1$ N

Clause 8.2.2(2)   For bolts, the rope effect is equal to $F_{ax,Rk}/4$ but no more than 25% of the Johansen part in Equations (8.6(j)) to (8.6(k)).
                            $F_{ax,Rk}/4 = 19823{,}1/4 = 4955{,}8$ N

Clause 8.2.2(1)   Characteristic load-carrying capacity: $F_{v,Rk}$ - fasteners in double shear:

Equation (8.6(g)) $f_{h,1,k}\,t_1\,d = (18{,}05 \times 75 \times 20{,}0) = 27075{,}0$ N
                                        $F_{v,Rk,g} = 27075{,}0$ N

Equation (8.6(h)) $0{,}5\,f_{h,2,k}\,t_2\,d = (0{,}5 \times 24{,}93 \times 150 \times 20{,}0) = 37395{,}0$ N
                                        $F_{v,Rk,h} = 37395{,}0$ N

Equation (8.6(j)) $1,05\dfrac{f_{h,1,k}t_1 d}{2+\beta}\left[\sqrt{2\beta(1+\beta)+\dfrac{4\beta(2+\beta)M_{y,Rk}}{f_{h,1,k}d\,t_1^2}}-\beta\right]+\dfrac{F_{axRk}}{4}$

$1,05\times\dfrac{18,05\times75\times20,0}{2+1,38}\left[\sqrt{2\times1,38\times(1+1,38)+\dfrac{4\times1,38(2+1,38)\times579280,93}{18,05\times20,0\times75^2}}-1,38\right]=17396,8\ N$

$F_{v,Rk,j}=\{17396,8+\min\,[(0,25\times17396,8)\quad4955,8)]\}$ $F_{v,Rk,j}=21746,0\ N$

Equation (8.6(k)) $1,15\sqrt{\dfrac{2\beta}{1+\beta}}\sqrt{2M_{y,Rk}f_{h,1,k}d}+\dfrac{F_{axRk}}{4}$

$1,15\sqrt{\dfrac{2\times1,38}{1+1,38}}\sqrt{2\times579280,93\times18,05\times20,0}=25326,6\ N$

$F_{v,Rk,k}=\{25326,6+\min\,[(0,25\times25326,6)\quad4955,8)]\}$ $F_{v,Rk,k}=30282,4\ N$

The characteristic load-carrying capacity $F_{v,Rk}$ is minimum value determined using Equations (8.6(j)) to (8.6(k)).

$F_{v,Rk}=21746,0\ N$ corresponding with failure mode (j)

Equation (2.17)   Design load-carrying capacity of bolts:

$$F_{v,Rd}=k_{mod}\dfrac{F_{v,Rk}}{\gamma_m}=0,9\times\dfrac{21746,0}{1,3}=15054,9\ N$$

Bolts in double shear: $(2\times15054,9)=30109,8\ N/bolt$

Assume two bolts which are parallel to the grain on the inner packing but neither parallel nor perpendicular to the grain of the columns.

$$F_{v,Rd.total}=(2\times30109,8)=60219,6\ N/connection$$

BS EN 1990:2002

Equation (6.10) $F_d=\sum_{j\geq1}\gamma_{G,j}G_{k,j}\ ''+''\ \gamma_{Q,1}Q_{k,1}\ ''+''\ \sum_{i>1}\gamma_{Q,i}\psi_{0,i}Q_{k,i}$

NA: Table NA.A1.2(B)

For variable actions: Leading actions $\gamma_{Q,1}=1,5$

Design load $=F_{Q,d}=(1,5\times30000)=45000\ N$

BS EN 1995-1-1:2004

Clause 8.5.1.1(3)   Spacing requirements using Table 8.4 and Figure 7.

Table 8.4   Timber 2: Angle between the force and grain direction $\alpha=0°$

Minimum spacing parallel to the grain $a_1=(4+|\cos\alpha|)d$

$a_1 = [4 + \cos 0°)] \times 20,0 = 100$ mm

Minimum spacing perpendicular to the grain $a_2 = 4d$
$a_2 = (4 \times 20,0) = 80$ mm

Minimum distance to loaded end $a_{3,t} = \max \begin{cases} 7d \\ 80 \text{ mm} \end{cases}$

$a_{3,t} = \max \begin{cases} (7 \times 20) \\ 80 \end{cases} = 140$ mm

Minimum distance to loaded edge $a_{4,t} = \max \begin{cases} (2 + 2\sin\alpha)d \\ 3d \end{cases}$

$a_{4,t} = \max \begin{cases} (2 + 2\sin 0°) \times 20 \\ (3 \times 20) \end{cases} = 60$ mm

Timbers 1 and 3: Angle between the force and grain direction $\alpha = 50°$
Minimum spacing parallel to the grain $a_1 = (4 + |\cos\alpha|)d$

$a_1 = [4 + (\cos 50°)] \times 20,0 = 92,9$ mm

Minimum spacing perpendicular to the grain $a_2 = 4d$
$a_1 = (4 \times 20,0) = 80$ mm

Minimum distance to loaded end $a_{3,t} = 140$ mm (as above)

Minimum distance to loaded edge $a_{4,t} = 60$ mm (as above)

Load capacity of column members = 60219,6 N/connection > $F_{Q,d}$
Design is satisfactory for the column members

Clause 8.5.1.1(4) Loading capacity of packing members = $n_{ef} \times F_{v,Rd,total}$
Assume $a_1 = 150$ mm

Equation (8.34) $n_{ef} = \min \begin{cases} n \\ n^{0,9} \sqrt[4]{\dfrac{a_1}{13d}} \end{cases} = \min \begin{cases} 2 \\ 2^{0,9} \times \sqrt[4]{\dfrac{150}{13 \times 20}} \end{cases} = 1,63$

$n_{ef} \times F_{v,Rd,total} = (1,63 \times 30109,8) = 49079,0$ N > $F_{Q,d}$
Design is satisfactory for the packing pieces

(ii) Consider the knee-brace to packing connection:

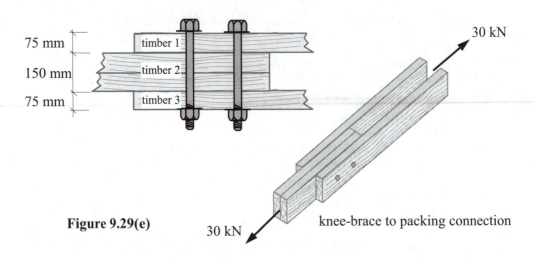

75 mm

150 mm

75 mm

Figure 9.29(e)   30 kN   knee-brace to packing connection

Thickness of timbers 1, and 3 $t_1 = 75$ mm
Thickness of timbers 2 $t_2 = 150$ mm

Angle between the applied force and the grain for all timbers $\alpha = 0°$

Equation (8.32) $f_{h,0,k} = 0,082\,(1 - 0,01d)\rho_k = 0,082 \times [1 - (0,01 \times 20,0)] \times 380$
                   $= 24,93$ N/mm^2

Characteristic embedment strength: $f_{h,1,k} = 24,93$ N/mm^2
Characteristic embedment strength: $f_{h,2,k} = 24,93$N/mm^2

Ratio of embedment strengths $\beta = \dfrac{f_{h,2,k}}{f_{h,1,k}} = 1,0$ $\beta = 1,0$

Clause 8.2.2(1)   Characteristic load-carrying capacity: $F_{v,Rk}$ - fasteners in double shear:

Equation (8.6(g)) $f_{h,1,k}\,t_1\,d = (24,93 \times 75 \times 20,0) = 37395,0$ N

                                                         $F_{v,Rk,g} = 37395,0$ N

Equation (8.6(h)) $0,5\,f_{h,2,k}\,t_2\,d = (0,5 \times 24,93 \times 150 \times 20,0) = 37395,0$ N

                                                         $F_{v,Rk,h} = 37395,0$ N

Equation (8.6(j)) $1,05\dfrac{f_{h,1,k}t_1 d}{2+\beta}\left[\sqrt{2\beta(1+\beta)+\dfrac{4\beta(2+\beta)M_{y,Rk}}{f_{h,1,k}d\,t_1^2}} - \beta\right]+\dfrac{F_{axRk}}{4}$

$$1,05 \times \frac{24,93 \times 75 \times 20,0}{2+1,0} \left[\sqrt{2 \times 1,0 \times (1+1,0) + \frac{4 \times 1,0 (2+1,0) \times 579280,93}{24,93 \times 20,0 \times 75^2}} - 1,0 \right] = 20225,0 \text{ N}$$

$$F_{v,Rk,j} = \{20225,0 + \min [(0,25 \times 20225,0) \quad 4955,8)]\} \qquad\qquad F_{v,Rk,j} = 25181,0 \text{ N}$$

Equation (8.6(k)) $1,15 \sqrt{\dfrac{2\beta}{1+\beta}} \sqrt{2M_{y,Rk} f_{h,1,k} d} + \dfrac{F_{axRk}}{4}$

$$1,15 \sqrt{\frac{2 \times 1,0}{1+1,0}} \sqrt{2 \times 579280,93 \times 24,93 \times 20,0} = 27639,7 \text{ N}$$

$$F_{v,Rk,k} = \{27639,7 + \min [(0,25 \times 27639,0) \quad 4955,8)]\} \qquad\qquad F_{v,Rk,k} = 32595,5 \text{ N}$$

The characteristic load-carrying capacity $F_{v,Rk}$ is minimum value determined using Equations (8.6(g)) to (8.6(k)).

$$F_{v,Rk} = 25181,0 \text{ N corresponding with failure mode (j)}$$

Equation (2.17)   Design load-carrying capacity of bolts:

$$F_{v,Rd} = k_{mod} \frac{F_{v,Rk}}{\gamma_{mt}} = 0,9 \times \frac{25181,0}{1,3} = 17433,0 \text{ N}$$

Bolts in double shear: $(2 \times 17433,0) = 34866,0$ N/bolt

Assume two bolts which are parallel to the grain on the inner packing and the knee-brace.

$$F_{v,Rd.total} = (2 \times 34866,0) = 69732,0 \text{ N/connection}$$

Clause 8.5.1.1(4)   Loading capacity of knee-brace members $= n_{ef} \times F_{v,Rd.total}$
Assume $a_1 = 125$ mm

Equation (8.34) $n_{ef} = \min \begin{Bmatrix} n \\ n^{0,9} \sqrt[4]{\dfrac{a_1}{13d}} \end{Bmatrix} = \min \begin{Bmatrix} 2 \\ 2^{0,9} \times \sqrt[4]{\dfrac{125}{13 \times 20}} \end{Bmatrix} = \begin{Bmatrix} 2 \\ 1,55 \end{Bmatrix} = 1,55$

$$n_{ef} \times F_{v,Rd.total} = (1,55 \times 34866,0) = 54042,3 \text{ N} > F_{Q,d}$$

Design is satisfactory for the packing pieces

See Figure 9.29(f) for bolt spacing.

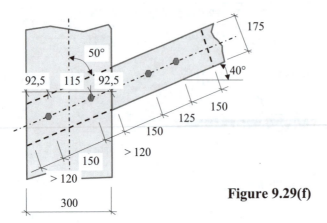

Figure 9.29(f)

9.6 Dowelled Connections (Section 8.6)

Dowels are generally pieces of steel rod fitting tightly into drilled holes and used in single, double or multiple shear joints. Unlike bolts, which require holes drilled as close as practicable to the nominal bolt diameter, and no more than 1 or 2 mm larger than the bolt diameter, dowels should be inserted in pre-drilled holes having a diameter not greater than the dowel itself (see Clause 10.4.4 of BS EN 1995-1-1:2004-1-1:2004). Since clearance holes reduce the capacity of bolted joints, there is a tendency to adopt the use of dowels. In addition to an increased structural efficiency, it is possible to insert dowels which do not fully penetrate the thickness of a member, hence producing a more acceptable visual appearance.

The dowel diameter should be greater than 6 mm and less than 30 mm as indicated in Clause 8.6(2) of the code. Tolerances on the dowel diameter of $-0/+0,1$ mm are specified in Clause 10.4.4.

The design procedure for laterally loaded dowels is the same as that for bolts with the exception of the minimum spacing, end and edge distances which are specified in Table 8.5 of the code.

9.7 Screwed Connections (Section 8.7)

A very wide variety of shapes, finishes and types of screw are available; the most commonly used type are shown in Figure 9.30. The increased cost and slower rate of application limit the use of screws in purely structural work. The withdrawal resistance of screws is considerably greater than that of round wire nails whilst the lateral load-carrying capacity is nominally less since the yield strength of the threaded portion is smaller than the yield strength of the shank. The insertion of screws should be into pre-drilled holes with a diameter equal to the shank diameter and no deeper than the length of the shank. The requirements for pre-drilling holes are given in Clause 10.4.5 of the code. It is important to note that screws should not be driven by hammering since this significantly reduces the load-carrying capacity, particularly the withdrawal loads; they should be turned in the hole using a lubricant if necessary.

Screws are used for timber-to-timber connections, for fixing metal plates to timber and for plywood-to-timber joints.

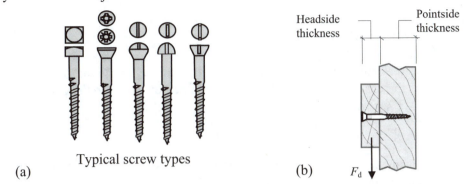

Typical screw types

(a)   (b) F_d

Figure 9.30

9.7.1 Laterally Loaded Screws (Clause 8.7.1)

Provided that the following conditions are satisfied, i.e.

  (i)   the effective diameter is taken as the smooth shank diameter, and
  (ii) the smooth shank penetrates into the member containing the point of the screw by not less than $4d$,

the load-carrying capacity of laterally-loaded screws is determined using an effective diameter d_{ef} and the appropriate equations from Section 8.2 as follows:

for smooth shank screws with a diameter $d \leq 6$ mm use rules given in Section 8.3.1- nails
for smooth shank screws with a diameter $d > 6$ mm use rules given in Section 8.3.2 - bolts

as indicated in Clauses 8.7.1(4) and 8.7.1(5).

9.7.2 Axially Loaded Screws (Clause 8.7.2)

The characteristic withdrawal capacity of axially loaded connections should be determined in accordance with Clause 8.7.2 and Equation (8.38).

$$F_{ax,\alpha,Rk} = n_{ef}\, (\pi\, d\, l_{ef})^{0,8}\, f_{ax,\alpha,k} \qquad \text{(Equation (8.38) in EC5)}$$

where:
$F_{ax,\alpha,Rk}$ is the characteristic withdrawal capacity of the connection at an angle α to the grain,
n_{ef} is the effective number of screws,
         Note: for a connection with a group of screws loaded by a force component parallel to the shank, the effective number of screws is given by $n_{ef} = n^{0,9}$ where n is the number of screws acting together in a connection.
d is the outer diameter measured on the threaded part,
l_{ef} is the pointside penetration length of the threaded part minus one screw diameter,

$f_{ax,\alpha,k}$ is the characteristic withdrawal strength at an angle α to the grain given by Equation (8.39).

$$f_{ax,\alpha,k} = \frac{f_{ax,k}}{\sin^2\alpha + 1,5\cos^2\alpha}$$ (Equation (8.39) in EC5)

where:

$f_{ax,\alpha,k}$ is the characteristic withdrawal strength at an angle α to the grain
$f_{ax,k}$ is the characteristic withdrawal strength perpendicular to the grain
$\quad\quad = 3,6 \times 10^{-3} \rho_k^{1,5}$ in N/mm^2
ρ_k is the characteristic density, in kg/m^3.

The minimum spacing and edge distances for axially loaded screws are given in Table 8.6. The pointside penetration length of the threaded part should be $\geq 6d$ (see Clause 8.7.2(3)).

9.7.3 Combined Laterally and Axially Loaded Screws (Clause 8.7.3)

Connections, which are subject to combined lateral and axial loading, should satisfy Equation (8.28) given in Clause 8.3.3(1):

$$\left(\frac{F_{ax,Ed}}{F_{ax,Rd}}\right)^2 + \left(\frac{F_{v,Ed}}{F_{v,Rd}}\right)^2 \leq 1$$ (Equation (8.28) in EC5)

where:
$F_{ax,Rd}$ and $F_{v,Rd}$ are the design load-carrying capacities of the connection loaded with axial load and lateral load respectively.

9.7.4 Example 9.13: Screwed Panel-to-timber, Three-member Connection in Single Shear

A tension splice is shown in Figure 9.31. Using the design data given determine the maximum long-term load which can be transmitted by the screws indicated.

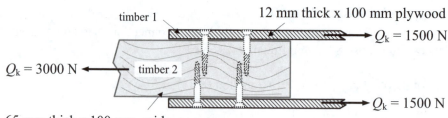

Figure 9.31

Design data:
Service Class 2
Characteristic long-term variable load (Q_k)   3000 N
Timber Class   C18
Moisture content $\leq 20\%$

Type of screw	No.8 slotted, countersunk head wood screws with smooth shank
Pre-drilling	Plywood only
Shank diameter (d)	4,17 mm
Screw length (L_s)	60 mm
Tensile strength of screw (f_u)	400 N/mm^2

Solution:

BS EN 338:2003

Clause 5.0   Characteristic values for C18 Timber

Table 1   Characteristic density: $\rho_k = 320,0$ kg/m^3

Mean density: $\rho_{mean} = 380,0$ kg/m^3

Manufacturer's Catalogue: Characteristic density of plywood $\rho_k = 630,0$ kg/m^3

Mean density $\rho_{mean} = 680,0$ kg/m^3

BS EN 1995-1-1:2004

Clause 2.3.2.1   Load-duration and moisture influence on strength: k_{mod}

Clause 2.3.1.3   Moisture content $\leq 20\%$ **Service Class 2**

Table 3.1   Long-term actions $k_{mod,timber} = k_{mod,plywood} = 0,7$

Equation (2.6) $k_{mod} = \sqrt{k_{mod,1} \times k_{mod,2}} = \sqrt{0,7 \times 0,7} = 0,7$ $\boldsymbol{k_{mod} = 0,7}$

NA: Table NA.3   Partial factor for material properties and resistance: γ_M

For solid timber – untreated and preservative treated $\boldsymbol{\gamma_M = 1,3}$

For plywood – untreated and preservative treated $\boldsymbol{\gamma_M = 1,2}$

For connections (except punched plate fasteners) $\boldsymbol{\gamma_M = 1,3}$

Clause 8.7.1(2)   Effective diameter $d_{ef} = d = 4,17$ mm

Manufacturer   Shank length $L_{sh} = 30,0$ mm

Smooth shank penetration $= (30,0 - 12,0) = 18,0$ mm

$\geq 4d = (4 \times 4,17) = 16,18$ mm

$\therefore$ Rules given in Clause 8.2 apply

Clause 8.7.1(5) $d < 6$ mm   **Rules given in Clause 8.3.1 for nails apply**

Clause 8.7.2(3)   Minimum point side penetration length of threaded part $L_{th} = 30$ mm

$> 6d = 6 \times 4,17 = 25,02$ mm

Clause 8.3.1.1(1) Headside thickness t_1 is the minimum of the headside timber thickness
and the pointside penetration.

Headside thickness $= 12,0$ mm

Pointside penetration $= 48,0$ mm $\boldsymbol{t_1 = 12}$ **mm**

The central member thickness (assume 50%) $\boldsymbol{t_2 = 32,5}$ **mm**

Clause 8.3.1.1(4) Characteristic yield moment for screws with $f_u = 400$ N/mm^2

Equation (8.14) $M_{y,Rk} = 0,3 f_u d^{2,6} = (0,3 \times 400 \times 4,17^{2,6}) = 4915,1$ Nmm

$\boldsymbol{M_{y,Rk} = 4915,1}$ **Nmm**

Clause 8.3.1.1(5) Characteristic embedment strength: $f_{h,k}$
                   Timber 1 (plywood): characteristic density $\rho_k = 630$ kg/m^3
                   Screw effective diameter = 4,17 mm
                   Pre-drilling for timber 1
Clause 10.4.5(2)   Assume hole diameter = 3,0 mm
Equation (8.20) $f_{h,k} = 0,11\,\rho_k\,d^{-0,3}$
                   Timber 1: $f_{h,1,k} = (0,11 \times 630 \times 4,17^{-0,3}) = 45,154$ N/mm^2

                   Timber 2 (solid timber): characteristic density $\rho_k = 320$ kg/m^3
                   No pre-drilling for timber 2
Equation (8.15) $f_{h,k} = 0,082\,\rho_k\,d^{-0,3}$
                   Timber 2: $f_{h,2,k} = (0,082 \times 320 \times 4,17^{-0,3}) = 17,097$ N/mm^2

                   Ratio of embedment strengths $\beta = \dfrac{17,097}{45,154} = 0,379$ **$\beta = 0,379$**

Clause 8.7.2(5)   Characteristic withdrawal strength perpendicular to the grain:
Equation (8.40) $f_{ax,k,1} = 3,6 \times 10^{-3}\,\rho_k^{1,5} = (3,6 \times 10^{-3} \times 630^{1,5}) = 56,926$ N/mm^2

                   $f_{ax,k,2} = 3,6 \times 10^{-3}\,\rho_k^{1,5} = (3,6 \times 10^{-3} \times 320^{1,5}) = 20,608$ N/mm^2

                   Angle between the withdrawal force and the grain $\alpha = 90°$
                   Characteristic withdrawal strength perpendicular to the grain:
Equation (8.39) $f_{ax,\alpha,k,1} = \dfrac{f_{ax,k,1}}{\sin^2\alpha + 1,5\cos^2\alpha} = \dfrac{59,926}{\sin^2 90° + 1,5\cos^2 90°} = 59,926$ N/mm^2

                   $f_{ax,\alpha,k,2} = \dfrac{f_{ax,k,2}}{\sin^2\alpha + 1,5\cos^2\alpha} = \dfrac{20,608}{\sin^2 90° + 1,5\cos^2 90°} = 20,608$ N/mm^2

                   The withdrawal strength is the smaller value $f_{ax,\alpha,k} = 20,608$ N/mm^2

Clause 8.7.2(4)   Characteristic withdrawal capacity:
Equation (8.38) $F_{ax,\alpha,Rk} = n_{ef}(\pi d\,l_{ef})^{0,8}\,f_{ax,\alpha,k} = (\pi d\,l_{ef})^{0,8}\,f_{ax,\alpha,k}/$screw
                   l_{ef} is the pointside penetration length of the threaded part minus one
                   screw diameter
                   $l_{ef} = (30,0 - 4,17) = 25,83$ mm
                   $(\pi d\,l_{ef})^{0,8}\,f_{ax,\alpha,k} = (\pi \times 4,17 \times 25,83)^{0,8} \times 20,608 = 2175,5$ N / screw
                                                                              $F_{ax,Rk} = 2175,5$ N/screw

Clause 8.2.2(2)   For screws, the rope effect is equal to $F_{ax,Rk}/4$ but no more than 100%
                   of the Johansen part in Equations (8.6(c)) to (8.6(f)).
                                                   $F_{ax,Rk}/4 = 2175,5/4 = 543,9$ N

Clause 8.2.2(1)   Characteristic load-carrying capacity: $F_{v,Rk}$ – single shear

Equation (8.6(a)) $f_{h,1,k} \, t_1 \, d = (45,154 \times 12 \times 4,17) = 2259,49$ N

$$F_{v,Rk,a} = 2259,5 \text{ N}$$

Equation (8.6(b)) $f_{h,2,k} \, t_2 \, d = (17,097 \times 32,5 \times 4,17) = 2317,1$ N

$$F_{v,Rk,b} = 2317,1 \text{ N}$$

Equation (8.6(c)) $\dfrac{f_{h,1,k} t_1 d}{1+\beta}\left[\sqrt{\beta + 2\beta^2\left[1 + \dfrac{t_2}{t_1} + \left(\dfrac{t_2}{t_1}\right)^2\right] + \beta^3\left(\dfrac{t_2}{t_1}\right)^2} - \beta\left(1 + \dfrac{t_2}{t_1}\right)\right] + \dfrac{F_{ax,Rk}}{4}$

$$\frac{45,154\times12\times4,17}{1+0,379}\left[\sqrt{0,379 + 2\times0,379^2\left[1 + \frac{32,5}{12} + \left(\frac{32,5}{12}\right)^2\right] + 0,379^3\left(\frac{32,5}{12}\right)^2} - 0,379\left(1 + \frac{32,5}{12}\right)\right]$$

$$= 954,0 \text{ N}$$

$F_{v,Rk,c} = \{954,0 + \min\,[(1,0 \times 954,0)\quad 543,9)]\}$ $F_{v,Rk,c} = 1497,9$ N

Equation (8.6(d)) $1,05\dfrac{f_{h,1,k} t_1 d}{2+\beta}\left[\sqrt{2\beta(1+\beta) + \dfrac{4\beta(2+\beta)M_{y,Rk}}{f_{h,1,k} d \, t_1^2}} - \beta\right] + \dfrac{F_{ax,Rk}}{4}$

$$1,05\times\frac{45,154\times12\times4,17}{2+0,379}\left[\sqrt{2\times0,379(1+0,379) + \frac{4\times0,379(2+0,379)\times4915,1}{45,154\times4,17\times12^2}} - 0,379\right]$$

$$= 921,9 \text{ N}$$

$F_{v,Rk,d} = \{921,9 + \min\,[(1,0 \times 921,9)\quad 543,9)]\}$ $F_{v,Rk,d} = 1465,8$ N

Equation (8.6(e)) $1,05\dfrac{f_{h,1,k} t_2 d}{1+2\beta}\left[\sqrt{2\beta^2(1+\beta) + \dfrac{4\beta(1+2\beta)M_{y,Rk}}{f_{h,1,k} d \, t_2^2}} - \beta\right] + \dfrac{F_{ax,Rk}}{4}$

$$1,05\times\frac{45,154\times32,5\times4,17}{1+2\times0,379}\left[\sqrt{2\times0,379^2(1+0,379) + \frac{4\times0,379(1+2\times0,379)\times4915,1}{45,154\times4,17\times32,5^2}} - 0,379\right]$$

$$= 1099,1 \text{ N}$$

$F_{v,Rk,e} = \{1099,1 + \min\,[(1,0 \times 1099,1)\quad 543,9)]\}$ $F_{v,Rk,e} = 1643,0$ N

Equation (8.6(f)) $1,15\sqrt{\dfrac{2\beta}{1+\beta}}\sqrt{2M_{y,Rk}f_{h,1,k}d}+\dfrac{F_{ax,Rk}}{4}$

$$= 1,15\times\sqrt{\dfrac{2\times0,379}{1+0,379}}\times\sqrt{2\times4915,1\times45,154\times4,17}=1160,0\text{ N}$$

$F_{v,Rk,fr}=\{1160,0+\min[(1,0\times1160,0)\quad543,9)]\}$ $F_{v,Rk,f}=1703,9\text{ N}$

The characteristic load-carrying capacity $F_{v,Rk}$ is minimum value determined using Equations (8.6(a)) to (8.6(f)).

$F_{v,Rk}=1465,8$N corresponding with failure mode (d)

Equation (2.17)   Design load-carrying capacity of screws:

$$F_{v,Rd}=k_{mod}\dfrac{F_{v,Rk}}{\gamma_{mt}}=0,7\times\dfrac{1465,8}{1,3}=789,3\text{ N/screw}$$

BS EN 1990:2002

Equation (6.10) $F_d=\displaystyle\sum_{j\geq1}\gamma_{G,j}G_{k,j}\ ''+''\ \gamma_{Q,1}Q_{k,1}\ ''+''\ \sum_{i>1}\gamma_{Q,i}\psi_{0,i}Q_{k,i}$

NA: Table NA.A1.2(B)

   For variable actions: Leading actions $\gamma_{Q,1}=1,5$

   Design load $=F_{Q,d}=(1,5\times3000)=4500$ kN

   Number of screws required $n=\dfrac{F_{v,Rk}}{F_{Q,d}}=\dfrac{4500}{789,3}=5,7$

Clause 8.3.1.1(9)   Minimum number of screws $=2$

A screw pattern is selected to satisfy the minimum number of nails required, e.g. 8 screws, 4 on each side in two rows.

Clause 8.7.2(7) There is no force component parallel to the shank, $\therefore n_{ef}=n=8$

Check the spacing requirements:
Clause 8.3.1.2
Double line of screws with shank passing through the panel:
Figure 8.7

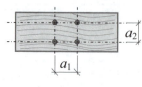

Spacing parallel and perpendicular to the grain	$-90°\leq\alpha\leq+90°$ Distance to loaded end	$0°\leq\alpha\leq+180°$ Distance to loaded edge

Figure 9.31(a)

Angle between the force and grain direction $\alpha = 0°$, $d < 5$ mm

| Table 8.2 | With pre-drilling for timber 1 |
| Clause 8.3.1.3(1) | Minimum spacing parallel to the grain $a_1 = 0{,}85(4 + |\cos\alpha|)d$ |

$a_1 = 0{,}85 \times [4 + \cos0°] \times 4{,}17 = 17{,}7$ mm

Minimum spacing perpendicular to the grain $a_2 = 0{,}85(3 + |\sin\alpha|)d$

$a_2 = 0{,}85 \times [3 + \sin0°] \times 4{,}17 = 10{,}6$ mm

Clause 8.3.1.3(2)   Minimum distance to loaded end $a_{3,t} = (3 + 4|\sin\alpha|)d$

$a_{3,t} = [3 + (4 \times \sin0°)] \times 4{,}17 = 12{,}5$ mm

Minimum distance to loaded edge $a_{4,t} = 3d$

$a_{4,t} = (3 \times 4{,}17) = 12{,}5$ mm

| Table 8.2 | Without pre-drilling for timber 2 $\rho_k = 320$ kg/m^3 (≤ 420 kg/m^3) |
| Clause 8.3.1.3(1) | Minimum spacing parallel to the grain $a_1 = (5 + 5|\cos\alpha|)d$ |

$a_1 = [5 + (5 \times 1{,}0)] \times 4{,}17 = 41{,}7$ mm

Minimum spacing perpendicular to the grain $a_2 = 5d$
$a_2 = (5 \times 4{,}17) = 20{,}9$ mm
Minimum distance to loaded end $a_{3,t} = (10 + 5|\cos\alpha|)d$

$a_{3,t} = [10 + (5 \times \cos0°)] \times 4{,}17 = 62{,}6$ mm

Minimum distance to loaded edge $a_{4,t} = (5 + 2|\sin\alpha|)d$

$a_{4,t} = [5 + (2 \times \sin0°)] \times 4{,}17 = 20{,}9$ mm

Selection of spacing:
$a_1 = 50$ mm $a_2 = 50$ mm

Selection of end and edge distances for timber 1:
$a_{3,t} = 30$ mm $a_{4,t} = 25$ mm

Selection of end and edge distances for timber 2:
$a_{3,t} = 65$ mm $a_{4,t} = 25$ mm

Joint slip:
Clause 7.1(1)	Slip modulus K_{ser} per shear plane per fastener is given in Table 7.1
Clause 7.1(2)	Timber 1: plywood - $\rho_{m,1} = 680$ kg/m^3
	Timber 2: softwood - $\rho_{m,2} = 380$ kg/m^3

Equation (7.1) $\rho_m = \sqrt{\rho_{m,1}\rho_{m,2}} = \sqrt{(680\times380)} = 508{,}33 \text{ kg/m}^3$

Table 7.1   Screws $\rho_m = 508{,}33 \text{ kg/m}^3, \quad d = 4{,}17 \text{ mm},$

$K_{ser} = \rho_m^{1,5}d/23 = (508{,}33^{1,5} \times 4{,}17)/23 = 2077{,}91 \text{ N/mm}$

Design load for the serviceability limit state $F_{Q,d} = 3000 \text{ N}$

BS EN 1990:2002
Equation (6.16b) $F_d = \sum_{j\geq1} G_{k,j} \; "+" \; \sum_{i\geq1} \psi_{2,i}Q_{k,i}$

NA.
Table NA.A1.1   Quasi-permanent value for variable actions on domestic buildings
$\psi_{2,1} = 0{,}3$
BS EN 1995-1-1:2004
Table 3.2   Deformation factor for solid timber,   Service class 2
$k_{def} = 0{,}8$

Equation (2.4) $u_{fin,Q,1} = u_{inst,Q,1}(1 + \psi_{2,1}k_{def})$

Load per screw $= \dfrac{F_{Q,k}}{n} = \dfrac{3000}{8} = 375{,}0 \text{ N}$

Instantaneous slip/side $u_{inst} = \dfrac{\text{Load per screw}}{K_{ser}}$

$= \dfrac{375{,}0}{2077{,}91} = 0{,}18 \text{ mm/side}$

$u_{fin,Q,1} = 0{,}18 \times [1 + (0{,}3 \times 0{,}8)] = 0{,}22 \text{ mm/side}$

Final joint opening $= (2 \times u_{fin,Q,1}) = (2 \times 0{,}22) = 0{,}44 \text{ mm}$

9.7.5 Example 9.14: Wall Cladding Subject to Wind Suction

A timber framed building has cladding screwed to timber studs as shown in Figure 9.32. Using the design data given, determine the number of screws required to resist the characteristic wind suction pressure indicated.

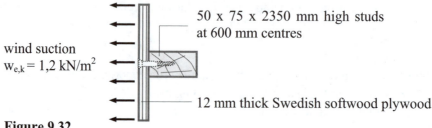

wind suction
$w_{e,k} = 1{,}2 \text{ kN/m}^2$

50 x 75 x 2350 mm high studs at 600 mm centres

12 mm thick Swedish softwood plywood

Figure 9.32

Design data:
Service Class 2
Characteristic short-term variable load ($w_{e,k}$)   1,2 kN/m^2
Timber Class   C14
Moisture content   > 20%
Type of screw   No.6 slotted, countersunk head wood screws with smooth shank
Pre-drilling   Plywood only
Shank diameter (d)   3,5 mm
Screw length (L_s)   44,5 mm
Tensile strength of screw (f_u)   400 N/mm^2

Solution:
BS EN 338:2003
Clause 5.0   Characteristic values for C14 Timber
Table 1   Characteristic density: ρ_k = 290,0 kg/m^3
BS EN 1995-1-1:2004
Clause 2.3.2.1   Load-duration and moisture influence on strength: k_{mod}
Clause 2.3.1.3   Moisture content > 20%   **Service Class 3**
Table 3.1   Short-term actions $k_{mod,timber}$ = $k_{mod,plywood}$ = 0,7

Equation (2.6) $k_{mod} = \sqrt{k_{mod,1} \times k_{mod,2}} = \sqrt{0,7 \times 0,7} = 0,7$ **k_{mod} = 0,7**

NA: Table NA.3   Partial factor for material properties and resistance: γ_M
                   For solid timber – untreated and preservative treated   **γ_M = 1,3**
                   For plywood – untreated and preservative treated   **γ_M = 1,2**
                   For connections (except punched plate fasteners)   **γ_M = 1,3**

Clause 8.7.2(3)   The minimum pointside penetration length of the threaded part = $6d$
                   $6d = (6 \times 3,5) = 21$ mm
BS1210:1963   Shank length $L_{sh} = (L_s/3) = (44,5/3) = 14,8$ mm
                   Pointside penetration length of the threaded part:
                   $(L_s - L_{sh}) = (44,5 - 14,8) = 29,7$ mm

Clause 8.7.2(3)   Effective pointside penetration length of the threaded part l_{ef}:
                   $(L_s - L_{sh}) - d = (29,7 - 3,5) = 26,2$ mm   **l_{ef} = 26,2 mm**

Clause 8.7.2(5)   Characteristic withdrawal strength perpendicular to the grain:
Equation (8.40) $f_{ax,k,2} = 3,6 \times 10^{-3}\, \rho_k^{1,5} = (3,6 \times 10^{-3} \times 290^{1,5}) = 17,78$ N/mm^2

                   Angle between the withdrawal force and the grain $\alpha = 90°$
                   Characteristic withdrawal strength perpendicular to the grain:

Equation (8.39) $f_{ax,\alpha,k,2} = \dfrac{f_{ax,k,2}}{\sin^2\alpha + 1,5\cos^2\alpha} = \dfrac{17,78}{\sin^2 90° + 1,5\cos^2 90°} = 17,78 \text{ N/mm}^2$

The withdrawal strength $f_{ax,\alpha,k} = 17,78 \text{ N/mm}^2$

Clause 8.7.2(4)   Characteristic withdrawal capacity:

Equation (8.38) $F_{ax,\alpha,Rk} = n_{ef}(\pi d\, l_{ef})^{0,8}\, f_{ax,\alpha,k} = (\pi d\, l_{ef})^{0,8}\, f_{ax,\alpha,k} \text{ N/screw}$

$(\pi d\, l_{ef})^{0,8}\, f_{ax,\alpha,k} = (\pi \times 3,5 \times 26,2)^{0,8} \times 17,78 = 1650,2 \text{ N / screw}$

Equation (2.17)   Design load-carrying capacity of screws:

$F_{v,Rd} = k_{mod}\dfrac{F_{v,Rk}}{\gamma_{mt}} = 0,7 \times \dfrac{1650,2}{1,3} = 888,6 \text{ N/screw}$

BS EN 1990:2002

Equation (6.10) $F_d = \sum_{j\geq 1}\gamma_{G,j}G_{k,j} \;''+''\; \gamma_{Q,1}Q_{k,1} \;''+''\; \sum_{i>1}\gamma_{Q,i}\psi_{0,i}Q_{k,i}$

NA: Table NA.A1.2(B)

For variable actions: Leading actions $\gamma_{Q,1} = 1,5$

Design load $= w_{ed} = (1,5 \times 1,2) = 1,8 \text{ kN/m}^2$

Design load/stud $= F_{w,d} = \dfrac{h \times s \times w_{ed}}{1000 \times 1000}$

where h is the stud height and s is the stud spacing.

Design load/stud $= F_{w,d} = \dfrac{2350 \times 600 \times 1,8}{1000 \times 1000} = 2,538 \text{ kN}$ (suction)

BS EN 1995-1-1:2004

Number of screws required $n = \dfrac{F_{w,d} \times 1000}{F_{ax,Rd}} = \dfrac{2,538 \times 1000}{888,6} = 2,9$

Assume screw spacing = 450 mm and distance to the ends = 50 mm

Number of screws provided $= \left(\dfrac{h-100}{450}+1\right) = \left(\dfrac{2350-100}{450}+1\right) = 6$

Clause 8.7.2(7)   Check the effective number of screws

The force component is parallel to the shank, $\therefore\; n_{ef} = n^{0,9} = 6,0^{0,9} = 5,0$

$> 2,9$ required

Provide 6/No.6 slotted, countersunk head wood screws with smooth shank at 450 mm centres in each stud.

Clause 8.7.2   Check the spacing and edge distances:
Table 8.6   Angle between the force and grain direction $\alpha = 90°$
                  Minimum spacing $a_1 = 4d = (4 \times 3,5) = 14,0$ mm
                  Minimum edge distance $a_{4,t} = 4d = 14,0$ mm
                  Selection of spacing:
                  $a_1 = 450$ mm $> 14,0$ mm
                  Selection of edge distances for timber 1:
                  $a_{4t} = 25,0$ mm $> 14,0$ mm

9.8 Split-ring, Shear-plate and Toothed-plate Connections

The design requirements for split-ring connectors of Type A and shear-plate connectors of type B according to BS EN 912 and with a diameter less than or equal to 200 mm, are given in Section 8.9 of EC5 whilst those for toothed-plate connectors are given in Section 8.10 of the code.

9.8.1 Split-ring Connectors (Section 8.9)

Split-ring connectors consist of circular bands of steel placed in pre-cut grooves in the contact faces of the members being joined. The connectors are available with either straight or bevelled sides as shown in Figure 9.33. Various manufacturers provide a range of diameters of ring to be used with standard diameter bolts. Both versions transfer the same load; the bevelled type however is easier to position and produces less slip than the straight-sided version. This is obviously advantageous when considering structural deflections. The groove-ring into which the split-ring is placed has a slightly larger diameter than the connector and consequently the connector is slightly sprung open when it is in position. The load carrying capacity of split-ring connectors is higher than double-sided tooth-plate connectors, and they can be used more easily in very dense timbers.

 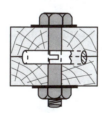   Split-ring connector

Figure 9.33

9.8.2 Shear-plate Connectors

Shear-plate connectors, as shown in Figure 9.34, are most frequently used for timber to structural steel joints or as in two single-sided tooth-plate connectors for demountable structures. The strength of smaller connectors is dependent on the strength of the central metal plate. The larger connectors have a reinforced centre plate resulting in its strength being governed by the shear strength of the bolt. A recess is cut into the timber into which the shear-plate connector is placed. Unlike the split-ring connector, the diameter of the recess is the same as that of the connector.

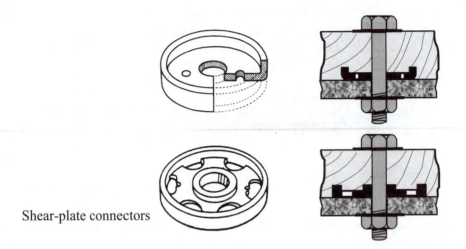

Shear-plate connectors

Figure 9.34

*Care must be taken when preparing and assembling a connectored joint. Since shrinkage is inherent in timber as it dries out and reaches an equilibrium moisture content, it is important to revisit timber joints which use bolts some time after erection and retighten any bolts which may have loosened.

The characteristic load-carrying capacity parallel to the grain per connector per shear plane for both types of connector is given in Clause 8.9(1) as follows:

$$F_{v,0,Rk} = \min \begin{cases} k_1\, k_2\, k_3\, k_4\left(35 d_c^{1,5}\right) & \text{(a)} \\ k_1\, k_3\, h_e\, \left(31{,}5 d_c\right) & \text{(b)} \end{cases} \qquad \text{(Equation (8.61 in EC5))}$$

where:
$F_{v,0,Rk}$ is the characteristic load-carrying capacity parallel to the grain, in N,
d_c is the connector diameter, in mm,
h_e is the embedment depth, in mm,
k_i are modification factors, with $i = 1$ to 4 defined by Equation (8.62 to 8.66) of EC5.

The characteristic load-carrying capacity at an angle α to the grain per connector per shear plane for both types of connector is given in Clause 8.9(8) as follows:

$$F_{v,\alpha,Rk} = \frac{F_{v,0,Rk}}{k_{90}\sin^2\alpha + \cos^2\alpha} \qquad \text{(Equation (8.67) in EC5)}$$

where:
$k_{90} = 1{,}3 + 0{,}001 d_c$
$F_{v,0,Rk}$ and d_c are as in Equation (8.61).

The minimum spacing, end and edge distances are given in Table 8.7, and illustrated in Figure 8.7.

9.8.3 Toothed-plate Connections (Section 8.10)

Toothed-plate connectors consist of a circular plate of steel with toothed edges as shown in Figures 9.35(a) and (b). Double-sided connectors are used for timber-to-timber connections and have teeth projecting alternately on either side. This type of connector is used in permanent joints and is imbedded into the timber by the compressive action of the nuts/washers when the bolts are tightened. A temporary high tensile steel screwed rod with plate washers larger than the connectors and nuts at the ends of the rod should be used to imbed the teeth of toothed-plate connectors before the insertion of the permanent ordinary mild steel bolt.

When using large connectors, some tropical hardwoods such as iroko, jarrah or teak, or in multiple member joints the forces necessary to imbed the connectors, may require the use of a ratchet spanner and ball-bearing washers. Toothed-plate connectors are not suitable for very dense timbers of strength class D50 and above. The permanent bolt used in double-sided toothed-plate connectors is assumed to stitch the units together and to be unloaded.

Single-sided connectors are used for wood-to-metal connections and have teeth projecting on one side only. The load is assumed to be transferred by shear from the timber to the connector, from the connector to the bolt and subsequently into the metal plate. Unlike the double-sided connector where the centre plate is unloaded, in this case, the connector centre plate is loaded and consequently is stiffened as shown in Figure 9.35(a). The single-sided connector is utilised in temporary or demountable structures where each member has a separate connector imbedded. The connectors are then placed back-to-back when the structure is assembled. The transfer of load is similar to that mentioned above, i.e. timber – connector – bolt – connector – timber.

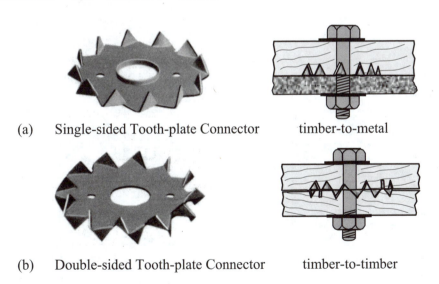

(a)   Single-sided Tooth-plate Connector   timber-to-metal

(b)   Double-sided Tooth-plate Connector   timber-to-timber

Figure 9.35

The characteristic load-carrying capacity is taken as the combined capacities of the toothed-plate connector and the associated bolt as given in Clause 8.5.

In the case of the connector the load-carrying capacity is given in Clause 8.10(2) as follows:

$$F_{v,Rk} = \min \begin{cases} 18\ k_1\ k_2\ k_3\ d_c^{1,5} & \text{for single-sided types} \\ 25\ k_1\ k_2\ k_3\ d_c^{1,5} & \text{for double sided types} \end{cases} \qquad \text{(Equation (8.72 in EC5))}$$

where:

$F_{v,Rk}$ is the characteristic load-carrying capacity of the toothed-plate connector in N,
d_c is dependent on the type of connector as defined in the code,
k_i are modification factors, with $i = 1$ to 3 defined by Equation (8.73) to (8.78) of EC5

The minimum spacing, end and edge distances are dependent on the type of connector and are given in Tables 8.8 and 8.9 and illustrated in Figure 8.7. In the case of connector types C1, C2, C6 and C7 with circular shape which are staggered the requirements of Clause 10.4.3 should be satisfied.

9.8.4 Example 9.15: Connectored Joints – Bolted Lattice Girder Node Connection

A node of a bolted, connectored lattice girder is shown in Figure 9.36. Using the variable loads and design data given, check the suitability of the joint considering:

    (i)   split-ring connectors   (A2 type with 64 mm diameter and M12 bolts),
    (ii)   shear-plate connectors   (B2 type with 67 mm diameter and M20 bolts),
    (iii)   tooth-plate connectors   (C6 type with 75 mm diameter and M12 bolts).

Design data:

Timber Class	C27
Load duration	medium-term
Service Class	2
Moisture content	$\leq 20\%$
Main ties – members 1 and 4	$2/50 \times 225$ mm
Diagonal – member 3	$1/50 \times 180$
Upright – member 2	$1/50 \times 180$

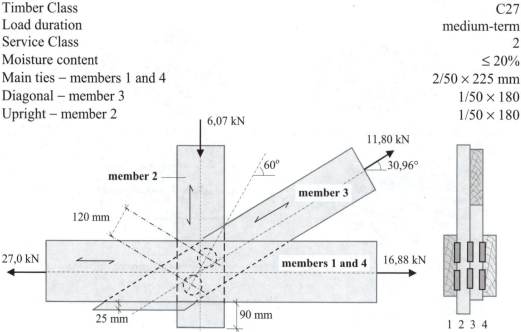

Figure 9.36

Cross-section

Solution:
The allowable connector loads are dependent upon the edge and end distances, spacing between the connectors and the direction of the load with respect to the grain. In this joint the connectors at each interface should be checked separately. Consider the exploded view of the joint shown in Figure 9.36(a) and the corresponding forces transferred through the connectors.

BS EN 338:2003
Clause 5.0   Characteristic values for C27 Timber
Table 1   Characteristic density: $\rho_k = 370,0 \text{ kg/m}^3$

BS EN 1995-1-1:2004
Clause 2.3.2.1   Load-duration and moisture influence on strength: k_{mod}
Clause 2.3.1.3   Moisture content ≤ 20%   **Service Class 2**
Table 3.1   Medium-term actions $k_{mod,timber} = 0,8$ $k_{mod} = 0,8$

NA: Table NA.3   Partial factor for material properties and resistance: γ_M
                 For solid timber – untreated and preservative treated $\gamma_M = 1,3$
                 For connections (except punched plate fasteners) $\gamma_M = 1,3$

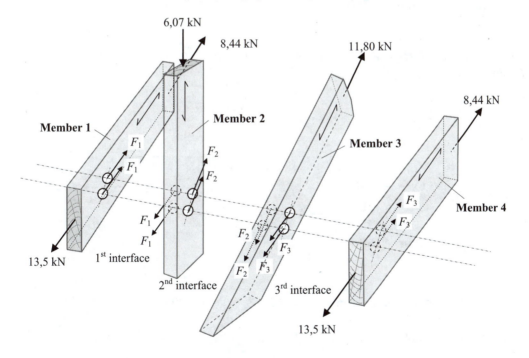

Figure 9.36(a)

BS EN 338:2003
Clause 5.0   Characteristic values for C14 Timber
Table 1   Characteristic density: $\rho_k = 290,0 \text{ kg/m}^3$

Consider the 1ˢᵗ Interface: *(consider the left-hand side first)*

Timber 1

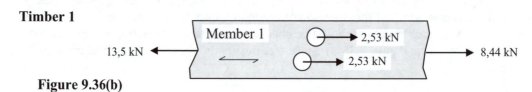

Figure 9.36(b)

Resultant member force = (13,5 − 8,44) = 5,06 kN ⟵ horizontal
Force on connectors = 5,06 kN (= 2,53 kN/connector) ⟶ horizontal
Grain direction is horizontal
Angle of connector force to the grain = 0°

Member 2

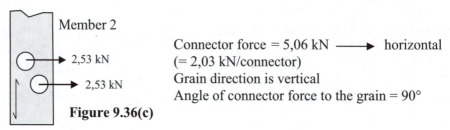

Connector force = 5,06 kN ⟶ horizontal
(= 2,03 kN/connector)
Grain direction is vertical
Angle of connector force to the grain = 90°

Figure 9.36(c)

Consider the 2ⁿᵈ Interface: *(consider the left-hand side first)*
Member 2

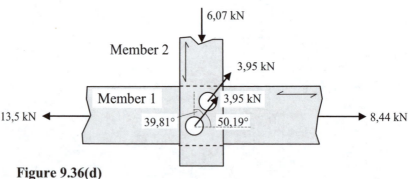

Figure 9.36(d)

Resultant member force = $\sqrt{6,07^2 + (13,5 - 8,44)^2}$ = 7,90 kN

Angle of resultant force to vertical $\beta = \tan^{-1}\left(\dfrac{5,06}{6,07}\right) = 39,81°$

Force on connectors = 7,90 kN (= 3,95 kN/connector)
Grain direction is vertical
Angle of connector force to vertical the grain = 39,81°

Member 3

Connector force = 7,90 kN
(= 3,95 kN/connector)
(39,81° to the vertical direction)

Grain direction is 30,96° to the horizontal
(59,04° to the vertical)

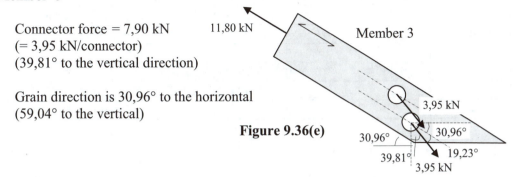

Figure 9.36(e)

Angle of connector force to the grain = (90° - 30,96° − 39,81°) = 19,23°

Consider the 3rd Interface: (consider the right-hand side)

Member 3

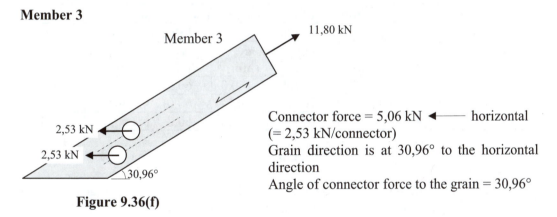

Figure 9.36(f)

Connector force = 5,06 kN ←——— horizontal
(= 2,53 kN/connector)
Grain direction is at 30,96° to the horizontal
direction
Angle of connector force to the grain = 30,96°

Member 4 (the same as member 1)

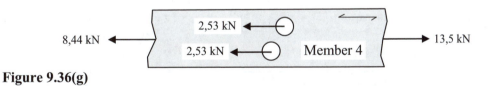

Figure 9.36(g)

Resultant member force = (13,5 − 8,44) = 5,06 kN ←——— horizontal
Force on connectors = 5,06 kN (= 2,53 kN/connector) ——→ horizontal
Grain direction is horizontal
Angle of connector force to the grain = 0°

The characteristic shear forces ($F_{v,k}$) on the connectors in each of the members and the corresponding angles to the grain (α) are summarised separately for each interface in Table 9.7.

Characteristic Applied Connector Forces and Angles to the Grain (2 Connectors)						
	1st Interface		2nd Interface		3rd Interface	
Member	$F_{v,k}$ (kN)	$\alpha°$	$F_{v,k}$ (kN)	$\alpha°$	$F_{v,k}$ (kN)	$\alpha°$
1	5,06	0°				
2	5,06	90°	7,90	39,81°		
3			7,90	19,23°	5,06	30,96°
4					5,06	0°

Table 9.7

The connector spacings parallel and perpendicular to the grain (a_1 and a_2), the distances to loaded and unloaded ends ($a_{3,t}$ and $a_{3,c}$), the distances at loaded and unloaded edges ($a_{4,t}$ and $a_{4,c}$), and the angles of the connector central line to the grain (θ), are summarised for all members in Table 9.8.

Summary of Spacings, Edge and End Distances (2 Connectors)							
	Spacings		End distances		Edge distances		Angle
Member	a_1 (mm)	a_2 (mm)	$a_{3,t}$ (mm)	$a_{3,c}$ (mm)	$a_{4,t}$ (mm)	$a_{4,c}$ (mm)	$\theta°$
1, 4	60,0	103,9	> 160,0	> 160,0	60,5	60,5	60°
2	103,9	60,0	> 160,0	150,5	60,0	60,0	30°
3	104,9	58,3	166,3	> 160,0	60,8	60,8	29,04°

Table 9.8

The capacity of connectors is dependent on the end and edge distance and the spacing of the units. In this joint, the details are given in Figures 9.36(h) to 9.36(j).

Members 1 and 4

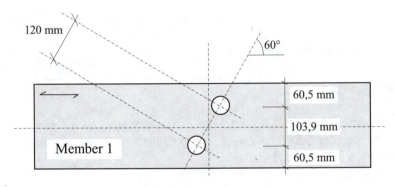

Figure 9.36(h)

Spacing = 120,0 mm
Edge distance = 60,5 mm

Member 2

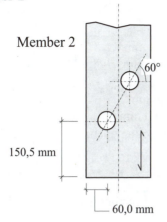

Member 2

150,5 mm

60°

60,0 mm

End distance = 150,5 mm
Edge distance = 60,0 mm

Figure 9.36(i)

Member 3

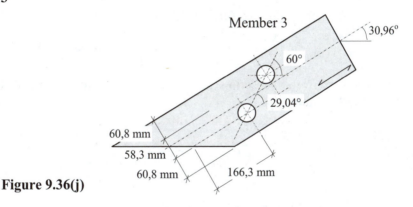

Member 3

30,96°

60°

29,04°

60,8 mm

58,3 mm

60,8 mm

166,3 mm

Figure 9.36(j)

End distance = 166,3 mm
Edge distance = 60,8 mm

Split-ring Connectors: (Section 8.9)

Timber members

Member thickness   All members = 50,0 mm
Characteristic density $\rho_k = 370,0$ kg/m^3

Split-ring connectors (Table A.2 of BS EN 912:2000)
Type: A2 connectors with parallel sides cut at one place on its circumference to form a tongue and a slot

Nominal diameter $d_{nom} = 64$ mm
Diameter $d_c = 72,0$ mm
Height $h_c = 19,0$ mm
Embedment depth $h_e = h_c / 2 = 19,0 / 2 = 9,5$ mm

Clause 8.9(2)   Minimum thickness of outer and inner members

Minimum outer member thickness:
$t_{outer,min} = 2{,}25\,h_e = (2{,}25 \times 9{,}5) = 21{,}4$ mm
$< 50{,}0$ mm
Minimum inner member thickness:
$t_{inner,min} = 3{,}75\,h_e = (3{,}75 \times 9{,}5) = 35{,}6$ mm
$< 50{,}0$ mm
(**Note:** members 2 and 3 also act as outer members in three-member systems.)

Design shear forces, $F_{v,d}$
The characteristic shear forces per connector of individual members on each interface can be obtained from Table 9.7, $F_{v,k}$, as

$F_{v,k\,(11)} = F_{v,k\,(21)} = 5{,}06 / 2 = 2{,}53$ kN
$F_{v,k\,(22)} = F_{v,k\,(32)} = 7{,}90 / 2 = 3{,}95$ kN
$F_{v,k\,(33)} = F_{v,k\,(43)} = 5{,}06 / 2 = 2{,}53$ kN

Note: the first numbers in the subscripts in the brackets are member numbers and the second ones are interface numbers.

BS EN 1990:2002
Equation (6.10) $F_d = \sum_{j\geq 1} \gamma_{G,j} G_{k,j}\ ''+''\ \gamma_{Q,1} Q_{k,1}\ ''+''\ \sum_{i>1} \gamma_{Q,i} \psi_{0,i} Q_{k,i}$

NA: Table NA.A1.2(B)
For variable actions: Leading actions $\gamma_Q = 1{,}5$
$F_{v,d\,(11)} = F_{v,d\,(21)} = \gamma_Q\,F_{v,k,11} = (1{,}5 \times 2{,}53) = 3{,}795$ kN
$F_{v,d\,(22)} = F_{v,d\,(32)} = \gamma_Q\,F_{v,k,22} = (1{,}5 \times 3{,}95) = 5{,}925$ kN
$F_{v,k\,(33)} = F_{v,k\,(43)} = \gamma_Q\,F_{v,k,33} = (1{,}5 \times 2{,}53) = 3{,}795$ kN

The design shear forces ($F_{v,d}$) on each of the connectors in each of the members and the corresponding angles to the grain (α) are summarised separately for each interface in Table 9.9.

Design Applied Connector Forces and Angles to the Grain (2 Connectors)						
	1st Interface		2nd Interface		3rd Interface	
Member	$F_{v,k}$ (kN)	$\alpha°$	$F_{v,k}$ (kN)	$\alpha°$	$F_{v,k}$ (kN)	$\alpha°$
1	3,795	0°				
2	3,795	90°	5,925	39,81°		
3			5,925	19,23°	3,795	30,96°
4					3,795	0°

Table 9.9

Clause 8.9.1   Characteristic load-carrying capacity parallel to the grain.
The characteristic load-carrying capacity parallel to the grain, $F_{v,0,Rk}$, per connector per shear plane for a diameter not bigger than 200 mm is given by Equation (8.61) in N as:

Equation (8.61) $$F_{v,0,Rk} = \min \begin{cases} k_1\, k_2\, k_3\, k_4 \left(35 d_c^{1,5}\right) & \text{(a)} \\ k_1\, k_3\, h_e\, (31,5 d_c) & \text{(b)} \end{cases}$$

where k_i are modification factors ($i = 1$ to 4) and are determined as follows:

Clause 8.9(3): k_1

Equation (8.62) $$k_1 = \min\left(1,0 \quad \frac{t_1}{3 h_e} \quad \frac{t_2}{5 h_e}\right) = \min\left(1,0 \quad \frac{50}{3 \times 9,5} \quad \frac{50}{5 \times 9,5}\right)$$
$$= \min\,(1,0 \quad 1,754, \quad 1,053) = 1,0 \qquad\qquad\qquad \mathbf{k_1 = 1,0}$$

Clause 8.9(4): k_2

Equation (8.63) $$k_2 = \begin{cases} \min\begin{cases} k_a \\ a_{3,t} \\ \dfrac{a_{3,t}}{2 d_c} \end{cases} & \text{for} -30° \leq \alpha \leq 30° \\[2mm] 1,0 & \text{for other values of } \alpha \end{cases}$$

The angles of the shear forces to the grain of individual members on each interface have been calculated and given in Tables 9.7 and 9.9 as:

$\alpha_{11} = 0°$	$\alpha_{21} = 90°$	$\alpha_{22} = 39,81°$
$\alpha_{32} = 19,23°$	$\alpha_{33} = 30,96°$	$\alpha_{43} = 0°$

The factor k_a in Equation (8.63) for connections with more than one connector per shear plane (here two connectors per shear plane) is given by Equation (8.64) as:

Equation (8.64) $k_a = 1,0$

Table 8.7   The minimum distance to the loaded end:
$a_{3,t,min} = 1,5 d_c = (1,5 \times 72) = 108,0$ mm

$a_{3,t}$ in Equation (8.63) is the actual distance to the loaded end which should be larger than $a_{3,t,min}$. The values of $a_{3,t}$ for members on each interface are given in Table 9.8 as follows:

$a_{3,t\,(11)} > 160,0$ mm	$a_{3,t\,(21)} > 160,0$ mm	$a_{3,t\,(22)} = 166,3$ mm
$a_{3,t\,(32)} = 166,3$ mm	$a_{3,t\,(33)} > 160,0$ mm	$a_{3,t\,(43)} > 160,0$ mm

All values of $a_{3,t}$ are larger than $a_{3,t,\,min} = 108,0$ mm.

The values of k_2 for individual members on each interface are calculated as

$$k_{2(11)} = k_{2(43)} = \min\left(1,0 \quad \frac{160}{2\times72}\right) = \min(1,0 \quad 1,111) = 1,0$$

$$k_{2(21)} = k_{2(22)} = k_{2(33)} = 1,0 \qquad (\alpha \text{ is out of the defined range})$$

$$k_{2(32)} = \min\left(1,0 \quad \frac{167,9}{2\times72}\right) = \min(1,0 \quad 1,166) = 1,0$$

$k_2 = 1,0$ can be used for all cases

Clause 8.9(5): k_3

Equation (8.65) $\quad k_3 = \min\left(1,75 \quad \dfrac{\rho_k}{350}\right) = \min\left(1,75 \quad \dfrac{370}{350}\right) = \min(1,75 \quad 1,057) = 1,057$

$k_3 = 1,057$

Clause 8.9(6): k_4

Equation (8.66) $\quad k_4 = 1,0$ for timber-to-timber connections $\qquad$ **$k_4 = 1,0$**

The characteristic load-carrying capacity parallel to the grain, $F_{v,0,Rk}$, per connector per shear plane for all members can now be determined from Equation (8.61).

Equation (8.61) $\quad F_{v,0,Rk} = \min \begin{cases} k_1\,k_2\,k_3\,k_4\left(35d_c^{1,5}\right) & \text{(a)} \\ k_1\,k_3\,h_e\left(31,5d_c\right) & \text{(b)} \end{cases}$

$$F_{v,0,Rk} = \min \begin{cases} \dfrac{1,0\times1,0\times1,057\times35\times72,0^{1,5}}{1000} \\ \dfrac{1,0\times1,057\times9,5\times31,5\times72,0}{1000} \end{cases} = \min \begin{cases} 22,602 \\ 22,774 \end{cases}$$

$$F_{v,0,Rk} = 22,602 \text{ kN}$$

Clause 8.9(8) $\qquad$ Characteristic load-carrying capacity at an angle α to the grain. The characteristic load-carrying capacity at angle α to the grain, $F_{v,\alpha,Rk}$, per connector per shear plane is given by Equation (8.67) as:

Equation (8.67) $\quad F_{v,\alpha,Rk} = \dfrac{F_{v,0,Rk}}{k_{90}\sin^2\alpha + \cos^2\alpha}$

where k_{90} is a factor which is given from Equation (8.68).

Equation (8.68) $\quad k_{90} = 1,3 + 0,001\,d_c = 1,3 + (0,001\times72,0) = 1,372$

The characteristic load-carrying capacity at angle α to the grain, $F_{v,\alpha,Rk}$, per connector per shear plane for individual members on each interface are calculated as:

$$F_{v,\alpha,Rk\,(11)} = F_{v,\alpha,Rk\,(43)} = \frac{22,602}{1,372 \times \sin^2 0° + \cos^2 0°} = 22,602 \text{ kN}$$

$$F_{v,\alpha,Rk\,(21)} = \frac{22,602}{1,372 \times \sin^2 90° + \cos^2 90°} = 16,474 \text{ kN}$$

$$F_{v,\alpha,Rk\,(22)} = \frac{22,602}{1,372 \times \sin^2 39,81° + \cos^2 39,81°} = 19,611 \text{ kN}$$

$$F_{v,\alpha,Rk\,(32)} = \frac{22,602}{1,372 \times \sin^2 19,23° + \cos^2 19,23°} = 21,725 \text{ kN}$$

$$F_{v,\alpha,Rk\,(33)} = \frac{22,602}{1,372 \times \sin^2 30,96° + \cos^2 30,96°} = 20,576 \text{ kN}$$

Clause 2.4.1(1)   Design load-carrying capacity at an angle α to the grain.
The design load-carrying capacity at angle α to the grain, $F_{v,\alpha,Rd}$, per connector per shear plane is determined using Equation (2.14).

Equation (2.14) $$F_{v,\alpha,Rd} = k_{mod} \frac{F_{v,\alpha,Rk}}{\gamma_M}$$

where:
k_{mod} is the modification factor taking into account the effect of the load duration and moisture content and is given from Table 3.1 for medium-term action and Service Class 2 as 0,80,
γ_M is the partial safety factor for connections and is given from Table NA.3 of the UK NA as 1,3.

The design load-carrying capacity at angle α to the grain, $F_{v,\alpha,Rd}$, per connector per shear plane for individual members on each interface are calculated as:

$$F_{v,\alpha,Rd\,(11)} = F_{v,\alpha,Rd\,(43)} = \frac{0,8 \times 22,602}{1,3} = 13.909 \text{ kN}$$

$$> F_{v,d\,(11)} = F_{v,d\,(43)} = 3,795 \text{ kN}$$

$$F_{v,\alpha,Rd\,(21)} = \frac{0,8 \times 16,474}{1,3} = 10,138 \text{ kN} > F_{v,d\,(21)} = 3,795 \text{ kN}$$

$$F_{v,\alpha,Rd\,(22)} = \frac{0,8 \times 19,611}{1,3} = 12,068 \text{ kN} > F_{v,d\,(22)} = 5,925 \text{ kN}$$

$$F_{v,\alpha,Rd\,(32)} = \frac{0,8 \times 21,725}{1,3} = 13,369 \text{ kN} > F_{v,d\,(32)} = 5,925 \text{ kN}$$

$$F_{v,\alpha,Rd\,(33)} = \frac{0,8 \times 20,576}{1,3} = 12,662 \text{ kN} > F_{v,d\,(33)} = 3,795 \text{ kN}$$

The design load-carrying capacity at angle α to the grain, $F_{v,\alpha,Rd}$, per connector per shear plane in each of the members for each interface together with the corresponding angles to the grain (α) are summarised in Table 9.10.

Design Load Carrying Capacities and Angles to the Grain (2 Connectors)						
	1st Interface		2nd Interface		3rd Interface	
Member	$F_{v,\alpha,Rd}$ (kN)	$\alpha°$	$F_{v,\alpha,Rd}$ (kN)	$\alpha°$	$F_{v,\alpha,Rd}$ (kN)	$\alpha°$
1	13,909	0°				
2	10,138	90°	12,068	39,81°		
3			13,369	19,23°	12,662	30,96°
4					13,909	0°

Table 9.10

The calculations clearly show that the split-ring connectors are adequate to sustain the applied shear forces.

Clause 8.9(9)   Minimum spacings and edge and end distances for split-ring connectors.
                   The minimum spacings and edge and end distances for split-ring connectors are specified in Table 8.7 of the code.

                   Member 1 on Interface 1 ($\alpha_{11} = 0°$)
Table 8.7   Minimum spacing parallel to the grain $a_{1,min}$
                   $$a_{1,min} = (1,2 + 0,8 \times |\cos 0°|) \times 72,0 = 144,0 \text{ mm}$$

                   Minimum spacing perpendicular to the grain $a_{2,min}$
                   $$a_{2,min} = 1,2 \times 72,0 = 86,4 \text{ mm}$$

Clause 8.9(10)   Reduction factors for the minimum spacings parallel and perpendicular to grain, k_{a1} and k_{a2}

The reduction factor for the minimum distances $a_{1,min}$ parallel to the grain, k_{a1}, is defined as:

$$k_{a1} = \min \left\{ \begin{array}{l} \dfrac{a_1}{a_{1,min}} \\ 1,0 \end{array} \right. = \min \left\{ \begin{array}{l} \dfrac{60,0}{144,0} \\ 1,0 \end{array} \right. = \min \left\{ \begin{array}{l} 0,417 \\ 1,0 \end{array} \right. = 0,417$$

The reduction factor for the minimum distance $a_{2,min}$ perpendicular to the grain, k_{a2}, is defined in Clause 8.9(10) as

$$k_{a2} = \min \left\{ \begin{array}{l} \dfrac{a_2}{a_{2,min}} \\ 1,0 \end{array} \right. = \min \left\{ \begin{array}{l} \dfrac{103,9}{86,4} \\ 1,0 \end{array} \right. = \min \left\{ \begin{array}{l} 1,203 \\ 1,0 \end{array} \right. = 1,0$$

Clause 8.9(13) Connectors are considered as positioned parallel where:
$k_{a2}\, a_2 < 0,5 k_{a1} a_1$ (see Table 9.8 for a_2 and a_1)
$k_{a2}\, a_2 = (1,0 \times 103,9 \text{ mm}) = 103,9 \text{ mm}$
$0,5 a_1 = (0,5 \times 60,0) = 30,0 \text{ mm}$

Since $k_{a2} a_2 > 0,5 k_{a1} a_1$ the two connectors used are regarded as staggered.

Clause 8.9(10) When the connectors are staggered, the minimum spacings parallel and perpendicular to the grain should comply with Equation (8.69).

Equation (8.69) $k_{a1}^{\,2} + k_{a2}^{\,2} \geq 1$ i.e. $0,417^2 + 1,0^2 = 1,174 > 1$

Equation (8.69) is satisfied

Clause 8.9(12) Effective number of connectors:
Since the above calculations confirm that the connectors are not positioned parallel to the grain, no reduction in the number of connectors need be applied to the load-carrying capacity.

Table 8.7 Minimum distance to the loaded end $a_{3,t,min}$
$a_{3,t,\,min} = 1,5 d_c = (1,5 \times 72,0) = 108,0 \text{ mm}$ $a_{3,t} = 160,0 \text{ mm}$

Minimum distance to the unloaded end $a_{3,c,min}$ $(\alpha = 180°)$
$a_{3,c,\,min} = 1,2 d_c = (1,2 \times 72,0) = 86,4 \text{ mm}$ $a_{3,c} = 160,0 \text{ mm}$

Minimum distance to the loaded edge $a_{4,t,min}$ $(\alpha = 0°)$
$a_{4,t,min} = \left(0,6 + 0,2 |\sin \alpha°|\right) d_c$
        $= \left(0,6 + 0,2 \times |\sin 0°|\right) \times 72,0 = 43,2 \text{ mm}$ $a_{4,t} = 60,5 \text{ mm}$

Minimum distance to the unloaded edge $a_{4,c,min}$

$a_{4,c} = 0,6d_c = (0,6 \times 72,0) = 43,2$ mm $a_{4,c} = 60,5$ mm

Table 8.7

Member 2 on Interface 1 ($\alpha_{21} = 90°$)
Minimum spacing parallel to the grain $a_{1,min}$

$a_{1,min} = (1,2 + 0,8 \times |\cos 90°|) \times 72,0 = 86,4$ mm

Minimum spacing perpendicular to the grain $a_{2,min}$

$a_{2,min} = 1,2 \times 72,0 = 86,4$ mm

Clause 8.9(10)

Reduction factors for the minimum spacings parallel and perpendicular to grain, k_{a1} and k_{a2}

The reduction factor for the minimum distances $a_{1,min}$ parallel to the grain, k_{a1}, is defined in Clause 8.9(10) as

$$k_{a1} = \min\begin{Bmatrix} \dfrac{a_1}{a_{1,min}} \\ 1,0 \end{Bmatrix} = \min\begin{Bmatrix} \dfrac{103,9}{86,4} \\ 1,0 \end{Bmatrix} = \min\begin{Bmatrix} 1,203 \\ 1,0 \end{Bmatrix} = 1,0$$

The reduction factor for the minimum distance $a_{2,min}$ perpendicular to the grain, k_{a2}, is defined in Clause 8.9(10) as

$$k_{a2} = \min\begin{Bmatrix} \dfrac{a_2}{a_{2,min}} \\ 1,0 \end{Bmatrix} = \min\begin{Bmatrix} \dfrac{60,0}{86,4} \\ 1,0 \end{Bmatrix} = \min\begin{Bmatrix} 0,694 \\ 1,0 \end{Bmatrix} = 0,694$$

Clause 8.9(13)

Connectors are considered as positioned parallel where:
$k_{a2}\, a_2 < 0,5k_{a1}a_1$ (see Table 9.8 for a_2 and a_1)
$k_{a2}\, a_2 = (1,0 \times 60,0$ mm $) = 60,0$ mm
$0,5a_1 = (0,5 \times 103,9) = 52,0$ mm

Since $k_{a2}a_2 > 0,5k_{a1}a_1$ the two connectors used are regarded as staggered.

Clause 8.9(10)

When the connectors are staggered, the minimum spacings parallel and perpendicular to the grain should comply with Equation (8.69).

Equation (8.69) $k_{a1}^2 + k_{a2}^2 \geq 1$ i.e. $1,0^2 + 0,694^2 = 1,482 > 1$

Equation (8.69) is satisfied

Clause 8.9(12)

Effective number of connectors:
Since the above calculations confirm that the connectors are not positioned parallel to the grain, no reduction in the number of connectors should be applied and so in the load-carrying capacity.

Table 8.7 Minimum distance to the loaded end $a_{3,t,min}$
$a_{3,t,\ min} = 1.5d_c = (1.5 \times 72.0) = 108.0$ mm $\qquad a_{3,t} = 160.0$ mm

Minimum distance to the unloaded end $a_{3,c,min}$ ($\alpha = 90°$)
$a_{3,c,\ min} = \left(0.4 + 1.6|\sin \alpha°|\right)d_c$
$\qquad = \left(0.4 + 1.6 \times |\sin 90°|\right) \times 72.0 = 144.0$ mm $\qquad a_{3,c} = 150.5$ mm

Minimum distance to the loaded edge $a_{4,t,min}$ ($\alpha = 90°$)
$a_{4,t,\ min} = \left(0.6 + 0.2|\sin \alpha°|\right)d_c$
$\qquad = \left(0.6 + 0.2 \times |\sin 90°|\right) \times 72.0 = 57.6$ mm $\qquad a_{4,t} = 60.0$ mm

Minimum distance to the unloaded edge $a_{4,c,min}$
$a_{4,c} = 0.6d_c = (0.6 \times 72.0) = 43.2$ mm $\qquad a_{4,c} = 60.0$ mm

Member 2 on Interface 2 ($\alpha_{22} = 39.81°$)

Table 8.7 Minimum spacing parallel to the grain $a_{1,min}$
$a_{1,min} = \left(1.2 + 0.8 \times |\cos 39.81°|\right) \times 72.0 = 130.6$ mm
Minimum spacing perpendicular to the grain $a_{2,min}$
$a_{2,min} = 1.2 \times 72.0 = 86.4$ mm

Clause 8.9(10) Reduction factors for the minimum spacings parallel and perpendicular to grain, k_{a1} and k_{a2}

The reduction factor for the minimum distances $a_{1,min}$ parallel to the grain, k_{a1}, is defined in Clause 8.9(10) as

$$k_{a1} = \min \begin{cases} \dfrac{a_1}{a_{1,min}} \\ 1.0 \end{cases} = \min \begin{cases} \dfrac{103.9}{130.6} \\ 1.0 \end{cases} = \min \begin{cases} 0.796 \\ 1.0 \end{cases} = 0.796$$

The reduction factor for the minimum distance $a_{2,min}$ perpendicular to the grain, k_{a2}, is defined in Clause 8.9(10) as

$$k_{a2} = \min \begin{cases} \dfrac{a_2}{a_{2,min}} \\ 1.0 \end{cases} = \min \begin{cases} \dfrac{60.0}{86.4} \\ 1.0 \end{cases} = \min \begin{cases} 0.694 \\ 1.0 \end{cases} = 0.694$$

Clause 8.9(13) Connectors are considered as positioned parallel where:
$k_{a2}\, a_2 < 0.5 k_{a1} a_1$ (see Table 9.8 for a_2 and a_1)
$k_{a2}\, a_2 = (1.0 \times 60.0 \text{ mm}) = 60.0$ mm
$0.5a_1 = (0.5 \times 103.9) = 52.0$ mm

Since $k_{a2}a_2 > 0,5k_{a1}a_1$ the two connectors used are regarded as staggered.

Clause 8.9(10)   When the connectors are staggered, the minimum spacings parallel and perpendicular to the grain should comply with Equation (8.69).

Equation (8.69) $k_{a1}^2 + k_{a2}^2 \geq 1$ i.e. $0,796^2 + 0,694^2 = 1,115 > 1$

Equation (8.69) is satisfied

Table 8.7   Minimum distance to the loaded end $a_{3,t,min}$
$a_{3,t,\,min} = 1,5d_c = (1,5 \times 72,0) = 108,0$ mm $a_{3,t} = 160,0$ mm
Minimum distance to the unloaded end $a_{3,c,min}$ ($\alpha = (180 - 39,81°)$)
$a_{3,c,\,min} = \left(0,4 + 1,6\left|\sin\alpha°\right|\right)d_c$

$= \left(0,4 + 1,6 \times \left|\sin(180 - 39,81)°\right|\right) \times 72,0 = 102,6$ mm

$a_{3,c} = 150,5$ mm

Minimum distance to the loaded edge $a_{4,t,min}$ ($\alpha = (180 - 39,81°)$)
$a_{4,t,\,min} = \left(0,6 + 0,2\left|\sin\alpha°\right|\right)d_c$

$= \left(0,6 + 0,2 \times \left|\sin(180 - 39,81)°\right|\right) \times 72,0 = 52,4$ mm

$a_{4,t} = 60,0$ mm

Minimum distance to the unloaded edge $a_{4,c,min}$
$a_{4,c} = 0,6d_c = (0,6 \times 72,0) = 43,2$ mm $a_{4,c} = 60,0$ mm

Member 3 on Interface 2 ($\alpha_{32} = 19,23°$)

Table 8.7   Minimum spacing parallel to the grain $a_{1,min}$
$a_{1,min} = \left(1,2 + 0,8 \times \left|\cos 19,23°\right|\right) \times 72,0 = 140,8$ mm
Minimum spacing perpendicular to the grain $a_{2,min}$
$a_{2,min} = 1,2 \times 72,0 = 86,4$ mm

Clause 8.9(10)   Reduction factors for the minimum spacings parallel and perpendicular to grain, k_{a1} and k_{a2}

The reduction factor for the minimum distances $a_{1,min}$ parallel to the grain, k_{a1}, is defined as:

$$k_{a1} = \min\begin{cases} \dfrac{a_1}{a_{1,min}} \\ 1,0 \end{cases} = \min\begin{cases} \dfrac{104,9}{140,8} \\ 1,0 \end{cases} = \min\begin{cases} 0,745 \\ 1,0 \end{cases} = 0,745$$

The reduction factor for the minimum distance $a_{2,min}$ perpendicular to the grain, k_{a2}, is defined in Clause 8.9(10) as

$$k_{a2} = \min \begin{cases} \dfrac{a_2}{a_{2,min}} \\ 1,0 \end{cases} = \min \begin{cases} \dfrac{58,3}{86,4} \\ 1,0 \end{cases} = \min \begin{cases} 0,675 \\ 1,0 \end{cases} = 0,675$$

Clause 8.9(13)   Connectors are considered as positioned parallel where:
$k_{a2}\, a_2 < 0,5 k_{a1} a_1$ (see Table 9.8 for a_2 and a_1)
$k_{a2}\, a_2 = (1,0 \times 58,3 \text{ mm}) = 58,3$ mm
$0,5 a_1 = (0,5 \times 104,9) = 52,5$ mm
Since $k_{a2} a_2 > 0,5 k_{a1} a_1$ the two connectors used are regarded as staggered.

Clause 8.9(10)   When the connectors are staggered, the minimum spacings parallel and perpendicular to the grain should comply with Equation (8.69).

Equation (8.69) $k_{a1}{}^2 + k_{a2}{}^2 \geq 1$ i.e. $0,745^2 + 0,675^2 = 1,011 > 1$

Equation (8.69) is satisfied

Clause 8.9(12)   Effective number of connectors:
Since the above calculations confirm that the connectors are not positioned parallel to the grain, no reduction in the number of connectors need be applied to the load-carrying capacity.

Table 8.7   Minimum distance to the loaded end $a_{3,t,min}$
$a_{3,t,\,min} = 1,5 d_c = (1,5 \times 72,0) = 108,0$ mm $a_{3,t} = 166,3$ mm

Minimum distance to the unloaded end $a_{3,c,min}$ ($\alpha = 19,23°$)
$a_{3,c,\,min} = \left(0,4 + 1,6 |\sin \alpha°|\right) d_c$

$\qquad = \left(0,4 + 1,6 \times |\sin(180 - 19,23)°|\right) \times 72,0 = 66,7$ mm

$\qquad\qquad\qquad\qquad\qquad\qquad\qquad\qquad\qquad a_{3,c} = 160,0$ mm
Minimum distance to the loaded edge $a_{4,t,min}$ ($\alpha = 19,23°$)
$a_{4,t,min} = \left(0,6 + 0,2 |\sin \alpha°|\right) d_c$

$\qquad = \left(0,6 + 0,2 \times |\sin(180 - 19,23)°|\right) \times 72,0 = 47,9$ mm

$\qquad\qquad\qquad\qquad\qquad\qquad\qquad\qquad\qquad a_{4,t} = 60,8$ mm
Minimum distance to the unloaded edge $a_{4,c,min}$
$a_{4,c} = 0,6 d_c = (0,6 \times 72,0) = 43,2$ mm $a_{4,c} = 60,8$ mm

Member 3 on Interface 3 ($\alpha_{33} = 30,96°$)
Table 8.7   Minimum spacing parallel to the grain $a_{1,min}$
$a_{1,min} = \left(1,2 + 0,8 \times |\cos 30,96°|\right) \times 72,0 = 135,8$ mm

Minimum spacing perpendicular to the grain $a_{2,min}$

$a_{2,min} = 1,2 \times 72,0 = 86,4$ mm

Clause 8.9(10)

Reduction factors for the minimum spacings parallel and perpendicular to grain, k_{a1} and k_{a2}

The reduction factor for the minimum distances $a_{1,min}$ parallel to the grain, k_{a1}, is defined as:

$$k_{a1} = \min \left\{ \begin{array}{c} \dfrac{a_1}{a_{1,min}} \\ 1,0 \end{array} \right. = \min \left\{ \begin{array}{c} \dfrac{104,9}{135,8} \\ 1,0 \end{array} \right. = \min \left\{ \begin{array}{c} 0,772 \\ 1,0 \end{array} \right. = 0,772$$

The reduction factor for the minimum distance $a_{2,min}$ perpendicular to the grain, k_{a2}, is defined in Clause 8.9(10) as

$$k_{a2} = \min \left\{ \begin{array}{c} \dfrac{a_2}{a_{2,min}} \\ 1,0 \end{array} \right. = \min \left\{ \begin{array}{c} \dfrac{58,3}{86,4} \\ 1,0 \end{array} \right. = \min \left\{ \begin{array}{c} 0,675 \\ 1,0 \end{array} \right. = 0,675$$

Clause 8.9(10)

When the connectors are staggered, the minimum spacings parallel and perpendicular to the grain should comply with Equation (8.69).

Equation (8.69) $k_{a1}^{2} + k_{a2}^{2} \geq 1$ i.e. $0,772^2 + 0,675^2 = 1,052 > 1$

Equation (8.69) is satisfied

Clause 8.9(12)

Effective number of connectors:
Since the above calculations confirm that the connectors are not positioned parallel to the grain, no reduction in the number of connectors need be applied to the load-carrying capacity.

Table 8.7

Minimum distance to the loaded end $a_{3,t,min}$
$a_{3,t,\,min} = 1,5d_c = (1,5 \times 72,0) = 108,0$ mm $a_{3,t} = 166,3$ mm

Minimum distance to the unloaded end $a_{3,c,min}$ ($\alpha = 30,96°$)
$a_{3,c,\,min} = \left(0,4 + 1,6 \left| \sin \alpha° \right| \right) d_c$

$\qquad = \left(0,4 + 1,6 \times \left| \sin (180 - 30,96)° \right| \right) \times 72,0 = 88,1$ mm

$a_{3,C} = 160,0$ mm

Minimum distance to the loaded edge $a_{4,t,min}$ ($\alpha = 30,96°$)
$a_{4,t,min} = \left(0,6 + 0,2 \left| \sin \alpha° \right| \right) d_c$

$\qquad = \left(0,6 + 0,2 \times \left| \sin (180 - 30,96)° \right| \right) \times 72,0 = 50,6$ mm

$a_{4,t} = 60,8$ mm

Minimum distance to the unloaded edge $a_{4,c,min}$

$a_{4,c} = 0,6d_c = (0,6 \times 72,0) = 43,2$ mm $a_{4,c} = 60,8$ mm

Member 4 on Interface 3 ($\alpha_{43} = 0°$)
The calculations are the same as those for Member 1 on Interface 1.

The minimum connector spacings parallel and perpendicular to the grain ($a_{1,min}$ and $a_{2,min}$), the minimum distances at loaded and unloaded ends ($a_{3,t,min}$ and $a_{3,c,min}$), and the minimum distances at loaded and unloaded edges ($a_{4,t,min}$ and $a_{4,c,min}$) are summarised for all members in Table 9.11, respectively.

Minimum Spacings, Edge and End Distances for Split-ring Connectors								
		Spacings			End distances		Edge distances	
Member	Interface	$a_{1,min}$ (mm)	$a_{2,min}$ (mm)	$k_{a1}{}^2+k_{a2}{}^2$ ($\geq 1,0$)	$a_{3,t,min}$ (mm)	$a_{3,c,min}$ (mm)	$a_{4,t,min}$ (mm)	$a_{4,c,min}$ (mm)
1	1	144,0		1,161		86,4	43,2	
2		86,4		1,466		144,0	57,6	
	2	130,6	86,4	1,080	108,0	102,6	52,4	43,2
3		140,8		1,011		66,7	47,9	
	3	135,8		1,049		88,1	50,6	
4		144,0		1,161		86,4	43,2	

Table 9.11

The comparison of the provided spacings and end/edge distances in Table 9.8 with the corresponding calculated minima in Table 9.11 shows that the requirements from Table 8.7 of the code are satisfied. Hence, the present design of split-ring connectors for the joint is adequate.

Shear-plate Connectors: (Clauses 8.9)
The same set of equations as used for split-ring connectors are used for shear-plate connectors.

Timber members
The timber members are the same as those for the design of split-ring connections.
Shear -plate connectors (Table B.2 of BS EN 912:2000)

Type: B2 connectors made of a circular flanged plate with a bolt-hole through the centre.
Nominal diameter $d_{nom} = 67$ mm
Diameter $d_c = 66,7$ mm
Height $h_c = 10,7$ mm
Embedment depth $h_e = h_c = 10,7$ mm
Bolt diameter $d = 20,0$ mm

Clause 8.9(2)   Minimum thickness of outer and inner members

Minimum outer member thickness:
$t_{outer,min} = 2{,}25\, h_e = (2{,}25 \times 10{,}7) = 24{,}08$ mm
$< 50{,}0$ mm
Minimum inner member thickness:
$t_{inner,min} = 3{,}75\, h_e = (3{,}75 \times 10{,}7) = 40{,}13$ mm
$< 50{,}0$ mm

Design shear forces, $F_{v,d}$

The design forces per connector of individual members on each interface can be obtained from Table 9.9.

Clause 8.9   **Design load-carrying capacity at an angle α to the grain**
Similar calculations for shear-plate connectors can be carried out as those for split-ring connectors except that the connector diameter d_c and embedment depth h_e are different.

The design load-carrying capacity at angle α to the grain, $F_{v,\alpha,Rd}$, per connector per shear plane in each of the members for each interface together with the corresponding angles to the grain (α) are summarised in Table 9.12.

Design Load Carrying Capacities and Angles to the Grain (2 Connectors)						
Member	1st Interface		2nd Interface		3rd Interface	
	$F_{v,\alpha,Rd}$ (kN)	$\alpha°$	$F_{v,\alpha,Rd}$ (kN)	$\alpha°$	$F_{v,\alpha,Rd}$ (kN)	$\alpha°$
1	11,596	0°				
2	8,482	90°	10,079	39,81°		
3			11,152	19,23°	10,569	30,96°
4					11,596	0°

Table 9.12

The calculations clearly show that the split-ring connectors are adequate to sustain the applied shear forces as illustrated in Table 9.9.

The calculations of the minimum spacings and edge and end distances for shear-plate connectors are also based on Table 8.7 of the code. The calculated minimum connector spacings parallel and perpendicular to the grain ($a_{1,min}$ and $a_{2,min}$), the minimum distances at loaded and unloaded ends ($a_{3,t,min}$ and $a_{3,c,min}$), and the minimum distances at loaded and unloaded edges ($a_{4,t,min}$ and $a_{4,c,min}$) are summarised for all members in Table 9.13.

The comparison of the provided spacings and end/edge distances in Table 9.8 with the corresponding calculated minima in Table 9.13 shows that the requirements from Table 8.7 of the code are satisfied. Hence, the present design of split-ring connectors for the joint is adequate.

Minimum Spacings, Edge and End Distances for Split-ring Connectors									
		Spacings			End distances		Edge distances		
Member	Interface	$a_{1,min}$ (mm)	$a_{2,min}$ (mm)	$k_{a1}^2+k_{a2}^2$ ($\geq 1,0$)	$a_{3,t,min}$ (mm)	$a_{3,c,min}$ (mm)	$a_{4,t,min}$ (mm)	$a_{4,c,min}$ (mm)	
1	1	133,4		1,202		80,0	40,0		
2		80,0		1,563		133,4	53,4		
	2	121,0	80,0	1,300	101,1	95,0	48,6	40,0	
3		130,4		1,178		61,8	44,4		
	3	125,8		1,226		81,6	46,9		
4		133,4		1,202		80,0	40,0		

Table 9.13

Toothed-plate Connectors: (Clauses 8.10)

Clause 8.10(1)   The characteristic load-carrying capacity of connections made using toothed-plate connectors should be taken as the summation of the characteristic load-carrying capacity of the connectors themselves and the connecting bolts according to Clause 8.5.

Timber members

Grade   C27
Member thickness   All members = 50,0 mm
Characteristic perpendicular compressive strength: $f_{c,90,k}$ = 2,6 N/mm^2
Characteristic density ρ_k = 370,0 kg/m^3

Toothed-plate connectors (Table C.6 of BS EN 912:2000)
Type: C6 double-sided connectors made of a circular plate with a bolt-hole through the centre.

Nominal diameter d_{nom} = 75 mm
Diameter d_c = 75,0 mm
Height h_c = 20,5 mm
Embedment depth h_e = h_c / 2 = 10,25 mm

Bolt connectors: (Clause 8.5.1 and Form G in BS 4320:1968)
Bolt grade   Grade 4.6
Bolt diameter d = 12,0 mm

Washer outer diameter: Clause 10.4.3(2) of EC5-1-1 and Form G in BS 4320:1968
$d_{w,outer}$ = 36,0 mm   ($3\,d$ = ($3 \times 12,0$) = 36,0 mm)

Washer inner diameter: (Form G in BS 4320:1968)
$d_{w,inner}$ = 14,0 mm

Washer thickness: Clause 10.4.3(2) of EC5-1-1 and Form G in BS 4320:1968)
$$t_w = 3,6 \text{ mm} \qquad (0,3\, d = 3 \times 12,0 = 3,6 \text{ mm})$$

Effective area of bolt: Manufacturer's data
$$A_{\text{eff}} = 84,3 \text{ mm}^2$$

Tensile strength: Manufacturer's data
$$f_u = 400 \text{ N/mm}^2$$

Clause 8.10(3)   The requirements of Clause 8.9(2) relating to minimum thicknesses applies to tooth-plate connectors.

Clause 8.9(2)   Minimum thickness of outer and inner members
Minimum outer member thickness $t_{\text{outer,min}}$
$$t_{\text{outer,min}} = 2,25\, h_e = (2,25 \times 10,25) = 23,1 \text{ mm} < 50,0 \text{ mm}$$

Minimum inner member thickness $t_{\text{inner,min}}$
$$t_{\text{inner,min}} = 3,75\, h_e = (3,75 \times 10,25) = 38,4 \text{ mm} < 50,0 \text{ mm}$$

(**Note:** members 2 and 3 also act as outer members in three-member systems.)

Design shear forces, $F_{v,d}$
The same design forces per connector of individual members on each interface as shown in Table 9.9 are used.

Clause 8.10(2)   Characteristic load-carrying capacity of toothed-plate connectors.
The characteristic load-carrying capacity, $F_{v,Rk(T)}$, per toothed-plate connector is given by Equation (8.72) in N as:

Equation (8.62) $F_{v,Rk\,(T)} = 25\, k_1\, k_2\, k_3\, d_c^{1,5}$

where k_i are modification factors ($i = 1$ to 3) and are determined as follows:
Clause 8.10(4): k_1

Equation (8.73) $k_1 = \min\left(1,0 \quad \dfrac{t_1}{3h_e} \quad \dfrac{t_2}{5h_e}\right) = \min\left(1,0 \quad \dfrac{50}{3\times 10,25} \quad \dfrac{50}{5\times 10,25}\right)$
$$= \min (1,0 \quad 1,626 \quad 1,976) = 1,0 \qquad\qquad \mathbf{k_1 = 0,976}$$

Clause 8.10(5): k_2 – Type C6

Equation (8.74) $k_2 = \min\begin{cases} 1,0 \\ a_{3,t} \\ 1,5 d_c \end{cases}$

The minimum distance to the loaded end $a_{3,t,min}$ is given from Equation (8.75) for type C6, i.e.

Equation (8.75) $a_{3,t,min} = \max \begin{cases} 1,1d_c \\ 7d \\ 80,0 \end{cases} = \max \begin{cases} 1,1\times75,0 \\ 7\times12,0 \\ 80,0 \end{cases} = \max \begin{cases} 82,5 \\ 84,0 \\ 80,0 \end{cases} = 84,0 = 84$ mm

$a_{3,t}$ in Equation (8.75) is the actual distance to the loaded end which should be larger than $a_{3,t,min}$. The values of $a_{3,t}$ for members on each interface are given in Table 9.8 as follows:

$a_{3,t(11)} > 160,0$ mm $a_{3,t(21)} > 160,0$ mm $a_{3,t(22)} = 166,3$ mm
$a_{3,t(32)} = 166,3$ mm $a_{3,t(33)} > 160,0$ mm $a_{3,t(43)} > 160,0$ mm
All values of $a_{3,t}$ are larger than $a_{3,t,\,min} = 84,0$ mm.

Equation (8.74) $k_2 = \min \begin{cases} 1,0 \\ \dfrac{a_{3,t}}{1,5d_c} \end{cases}$

$1,5d_c = (1,5 \times 75,0) = 112,5$ mm

Since all values of $a_{3,t}$ are larger than $1,5\ d_c$, $k_2 = 1,0$ can be used for all cases. $k_2 = 1,0$

Clause 8.10(6): k_3

Equation (8.78) $k_3 = \min\left(1,5 \quad \dfrac{\rho_k}{350}\right) = \min\left(1,5 \quad \dfrac{370}{350}\right) = \min(1,5 \quad 1,057) = 1,057$

$k_3 = 1,057$

The characteristic load-carrying capacity per toothed-plate connector for all members, $F_{v,Rk\,(T)}$, can be determined from Equation (8.72):

Equation (8.72) $F_{v,Rk\,(T)} = 25k_1\,k_2\,k_3\,d_c^{1,5} = \dfrac{25\times0,976\times1,0\times1,057\times75,0^{1,5}}{1000} = 16,75$ kN

The characteristic load-carrying capacity of bolt connectors to Clause 8.5 of the code.

Clause 8.5.1.1(2)   Characteristic embedment strength of timber parallel to the grain $f_{h,0,k}$

Equation (8.32) $f_{h,0,k} = 0,082\ (1 - 0,01d)\rho_k$
$= 0,082 \times [1 - (0,01 \times 12)] \times 370 = 26,699$ N/mm^2

The angles of the shear forces to the grain of individual members on each interface are:

$$\alpha_{11} = 0° \qquad \alpha_{21} = 90° \qquad \alpha_{22} = 39,81°$$
$$\alpha_{32} = 19,23° \qquad \alpha_{33} = 30,96° \qquad \alpha_{43} = 0°$$

Clause 8.5.1.1(2) Characteristic embedment strength of members on each interface at an angle α to the grain.

The characteristic embedment strength of members on each interface at an angle α to the grain for bolts up to 30 mm diameter, $f_{h,\alpha,k}$, is given by Equation (8.31):

Equation (8.31) $$f_{h,\alpha,k} = \frac{f_{h,0,k}}{k_{90} \sin^2 \alpha + \cos^2 \alpha}$$

where k_{90} is a factor and these value for softwood is calculated from Equation (8.33):

Equation (8.33) $$k_{90} = 1,35 + 0,05d = 1,35 + 0,015 \times 12 = 1,53$$

The characteristic embedment strengths of each timber member on each interface at an angle α to the grain can be determined from Equation (8.31) as follows:

$$f_{h,\alpha,k\,(11)} = \frac{f_{h,0,k}}{k_{90} \sin^2 \alpha_{11} + \cos^2 \alpha_{11}} = \frac{26,699}{1,53 \times \sin^2 0° + \cos^2 0°} = 26,70 \text{ N/mm}^2$$

$$f_{h,\alpha,k\,(21)} = \frac{f_{h,0,k}}{k_{90} \sin^2 \alpha_{21} + \cos^2 \alpha_{21}} = \frac{26,699}{1,53 \times \sin^2 90° + \cos^2 90°} = 17,45 \text{ N/mm}^2$$

$$f_{h,\alpha,k\,(22)} = \frac{f_{h,0,k}}{k_{90} \sin^2 \alpha_{22} + \cos^2 \alpha_{22}} = \frac{26,699}{1,53 \times \sin^2 39,81° + \cos^2 39,81°} = 21,93 \text{ N/mm}^2$$

$$f_{h,\alpha,k\,(32)} = \frac{f_{h,0,k}}{k_{90} \sin^2 \alpha_{32} + \cos^2 \alpha_{32}} = \frac{26,699}{1,53 \times \sin^2 19,23° + \cos^2 19,23°} = 25,25 \text{ N/mm}^2$$

$$f_{h,\alpha,k\,(33)} = \frac{f_{h,0,k}}{k_{90} \sin^2 \alpha_{33} + \cos^2 \alpha_{33}} = \frac{26,699}{1,53 \times \sin^2 30,96° + \cos^2 30,96°} = 23,42 \text{ N/mm}^2$$

$$f_{h,\alpha,k\,(43)} = \frac{f_{h,0,k}}{k_{90} \sin^2 \alpha_{43} + \cos^2 \alpha_{43}} = \frac{26,699}{1,53 \times \sin^2 0° + \cos^2 0°} = 26,70 \text{ N/mm}^2$$

The characteristic embedment strengths of each timber member at an angle α to the grain can be determined by adopting the lower value on both interfaces of each member:

$$f_{h,\alpha,k\,(1)} = f_{h,\alpha,k\,(11)} = 26,70 \text{ N/mm}^2$$

$$f_{h,\alpha,k\,(2)} = \min\left(f_{h,\alpha,k\,(21)} \quad f_{h,\alpha,k\,(22)}\right) = \min(17,45 \quad 21,93) = 17,45 \text{ N/mm}^2$$

$$f_{h,\alpha,k\,(3)} = \min\left(f_{h,\alpha,k\,(32)} \quad f_{h,\alpha,k\,(33)}\right) = \min(25,25 \quad 23,42) = 23,42 \text{ N/mm}^2$$

$$f_{h,\alpha,k\,(4)} = f_{h,\alpha,k\,(43)} = 26,70 \text{ N/mm}^2$$

Clause 8.5.1.1(1)   Characteristic value of yield moment $M_{y,Rk}$

Equation (8.30) $M_{y,Rk} = 0,3 f_u d^{2,6} = \left(0,3 \times 400 \times 12,0^{2,6}\right) = 76745,42 \text{ Nmm}$

Clause 8.5.2.1   Characteristic withdrawal capacity of bolts:

Clause 8.5.2(2)   Load-bearing capacity of the washer = $(3,0\, f_{c,90,k} \times A_{c,90})$
                    Bearing area:
                    $A_{c,90} = \pi\, (d_{wo}^2 - d_{wi}^2)/4 = [\pi \times (36^2 - 14^2)]/4 = 863,94 \text{ mm}^2$

                    $(3,0\, f_{c,90,k} \times A_{c,90}) = (3,0 \times 2,6 \times 863,94) = 6738,73 \text{ N}$

                    Bolt tensile capacity = $(f_u \times A_{eff}) = (400 \times 84,3) = 33720 \text{ N}$

                    Characteristic withdrawal capacity of bolts:
                    $F_{ax,Rk} = \min\left(F_{t,Rk} \quad F_{c,90,Rk}\right) = \min\left(33720 \quad 6738,732\right) = 6738,7 \text{ N}$

Clause 8.2.2(2)   For bolts, the rope effect is equal to $F_{ax,Rk}/4$ but no more than 25% of the Johansen part in Equations (8.6(j)) and (8.6(k)).
                    $$F_{ax,Rk}/4 = 6738,7/4 = 1684,7 \text{ kN}$$

Clause 8.5.1.1 Characteristic capacity of laterally loaded bolts
There are two three-member systems (members 1+2+3 and 2+3+4) in double shear and three shear interfaces. Due to asymmetry, both three-member systems need to be considered.

Characteristic capacity for the 1st three-member system
Consider (timber 1 + timber 2 + timber 3):
In the three-member system timber 3 is shared by shear planes B and C; as a consequence of this, the effective thickness for timber 3, as an outer member, should be reduced by 50%.

Thickness of timbers 1, 2	$t = 50$ mm
Effective thickness of timber 3 = $(0,5 \times 50 = 25$ mm$)$	$t_{ef,3} = 25$ mm

Figure 8.2 t_1 is the smaller of t and $t_{ef,3}$ $t_1 = 25$ mm

$t_2 = 50$ mm

Characteristic embedment strengths if timber members:

$$f_{h,1,k} = \min(f_{h,\alpha,k\,(1)}\quad f_{h,\alpha,k\,(3)}) = \min(26,70\quad 23,42) = 23,42 \text{ N/mm}^2$$

$$f_{h,2,k} = f_{h,\alpha,k\,(2)} = 17,450 \text{ N/mm}^2$$

Equation (8.6)   Ratio of embedment strengths $\beta = \dfrac{f_{h,2,k}}{f_{h,1,k}} = \dfrac{17,450}{23,415} = 0,75$ $\beta = 0,75$

Clause 8.2.2(1)   Characteristic load-carrying capacity of bolts :

Equation (8.6(g)) $f_{h,1,k}\, t_1\, d$ This mode of failure is not considered.

Equation (8.6(h)) $0,5\, f_{h,2,k}\, t_2\, d$ This mode of failure is not considered.

Equation (8.6(j)) $1,05\dfrac{f_{h,1,k}t_1 d}{2+\beta}\left[\sqrt{2\beta(1+\beta)+\dfrac{4\beta(2+\beta)M_{y,Rk}}{f_{h,1,k}d\,t_1^2}} - \beta\right] + \dfrac{F_{axRk}}{4}$

$$1,05\times\dfrac{23,42\times25,0\times12,0}{2+0,75}\left[\sqrt{2\times0,75\times(1+0,75)+\dfrac{4\times0,75\times(2+0,75)\times76745,42}{23,425\times12,0\times25,0^2}} - 0,75\right]$$

$$= 4683,3 \text{ N}$$

$F_{v,Rk,j} = \{4683,3 + \min[(0,25\times4683,3)\quad 1684,7)]\}$ $F_{v,Rk,j} = 5854,1$ N

Equation (8.7(k)) $1,15\sqrt{\dfrac{2\beta}{1+\beta}}\sqrt{2M_{y,Rk}f_{h,1,k}}\,d + \dfrac{F_{ax,Rk}}{4}$

$$=1,15\times\sqrt{\dfrac{2\times0,75}{1+0,75}}\sqrt{2\times76745,42\times23,42\times12,0} = 6992,8 \text{ N}$$

$F_{v,Rk,k} = \{6992,8 + \min[(0,25\times6992,8)\quad 1684,7)]\}$ $F_{v,Rk,k} = 8677,5$ N

The characteristic load-carrying capacity of bolts for system with members 1+2+3, $F_{v,Rk\,(1+2+3)}$, is the smaller of the above two capacities, i.e.

$$F_{v,Rk\,(1+2+3)} = \min\left(F_{v,Rk\,(j)}\quad F_{v,Rk\,(k)}\right) = \min(5854,1\quad 8677,5) = 5854,1 \text{ N}$$

Characteristic capacity for the 2nd three-member system

Consider (timber 2 + timber 3 + timber 4):
In the three-member system timber 2 is shared by two shear planes as a consequence of this, the effective thickness for timber 2, as an outer member, should be reduced by 50%.

	Thickness of timbers 3, 4	$t = 50$ mm
	Effective thickness of timber 2 = $(0,5 \times 50 = 25$ mm)	$t_{ef,2} = 25$ mm

Figure 8.2 t_1 is the smaller of t and $t_{ef,2}$ $t_1 = 25$ **mm**
                                                                       $t_2 = 50$ **mm**

Characteristic embedment strengths if timber members:

$$f_{h,1,k} = \min(f_{h,\alpha,k\,(2)}\ \ f_{h,\alpha,k\,(4)}) = \min(17,45,\ 26,70) = 17,45 \text{N/mm}^2$$

$$f_{h,2,k} = f_{h,\alpha,k\,(3)} = 23,42 \text{ N/mm}^2$$

Equation (8.6)   Ratio of embedment strengths $\beta = \dfrac{f_{h,2,k}}{f_{h,1,k}} = \dfrac{23,42}{17,45} = 1,34$ $\boldsymbol{\beta = 1,34}$

Clause 8.2.2(1)   Characteristic load-carrying capacity of bolts :

Equation (8.6(g)) $f_{h,1,k}\ t_1\ d$ This mode of failure is not considered.

Equation (8.6(h)) $0,5\ f_{h,2,k}\ t_2\ d$ This mode of failure is not considered.

Equation (8.6(j)) $1,05 \dfrac{f_{h,1,k}\ t_1\ d}{2+\beta}\left[\sqrt{2\beta(1+\beta)+\dfrac{4\beta(2+\beta)M_{y,Rk}}{f_{h,1,k}\ d\,t_1^2}}-\beta\right]+\dfrac{F_{ax,Rk}}{4}$

$$1,05\times\dfrac{17,45\times25,0\times12,0}{2+1,34}\times\left[\sqrt{2\times1,34\times(1+1,34)+\dfrac{4\times1,34\times(2+1,34)\times76745,42}{17,45\times12,0\times25,0^2}}-1,34\right]$$
$$= 4534,0 \text{ N}$$

$F_{v,Rk,j} = \{4534,0 + \min[(0,25 \times 4534,0)\ \ 1684,7)]\}$ $F_{v,Rk,j} = 5667,5$ N

Equation (8.7(k)) $1,15\sqrt{\dfrac{2\beta}{1+\beta}}\ \sqrt{2M_{y,Rk}f_{h,1,k}}\ d + \dfrac{F_{ax,Rk}}{4}$

$$= 1,15\times\sqrt{\dfrac{2\times1,34}{1+1,34}}\ \sqrt{2\times76745,42\times17,45\times12,0} = 6977,3 \text{ N}$$

$F_{v,Rk,k} = \{6977,3 + \min[(0,25 \times 6977,3)\ \ 1684,7)]\}$ $F_{v,Rk,k} = 8662,0$ N

The characteristic load-carrying capacity of bolts for system with members 2+3+4, $F_{v,Rk\,(2+3+4)}$, is the smaller of the above two capacities, i.e.

$$F_{v,Rk\,(2+3+4)} = \min\left(F_{v,Rk\,,(j)} \quad F_{v,Rk\,,(k)}\right) = \min\left(5667,5 \quad 8662,0\right) = 5667,5\ \text{N}$$

Characteristic capacity for the four-member system (members 1+2+3+4)
For conservative design, the characteristic load-carrying capacity of bolts for all four-members, $F_{v,Rk\,(B)}$ can be taken as the smaller of those for the two subsystems, i.e.

$$F_{v,Rk\,(B)} = \min\left(F_{v,Rk\,(1+2+3)} \quad F_{v,Rk\,(2+3+4)}\right)$$

$$= \min\left(5854,1 \quad 5667,5\right) = 5667,1\ \text{N} = 5,67\ \text{kN}$$

The difference between two capacities is small, i.e. 5,1%.

Clause 8.10(1)   Overall characteristic load-carrying capacity of toothed-plate connectors and bolts.
The overall characteristic load-carrying capacity, $F_{v,Rd}$, per connector per shear plane is determined as follows:
$F_{v,Rk}$ = capacity of the connector + capacity of the bolt
$$F_{v,Rk} = F_{v,Rk\,(T)} + F_{v,Rk\,(B)} = 16,75 + 5,67 = 22,42\ \text{kN}$$

Clause 2.4.1(1)   Overall design load-carrying capacity at an angle α to the grain, $F_{v,\,Rd}$
The overall design load-carrying capacity, $F_{v,Rd}$, per connector per shear plane is determined using Equation (2.14):

Equation (2.17) $$F_{v,Rd} = k_{mod}\,\frac{F_{v,Rk}}{\gamma_M} = 0,8 \times \frac{22,42}{1,3} = 13,80\ \text{kN}$$

The minimum spacings and edge and end distances for toothed-plate connectors can be determined in a similar manner to that for split-ring and shear-plate connectors, using Tables 8.8 and 8.9 depending on the type of connector.

The minimum connector spacings parallel and perpendicular to the grain ($a_{1,min}$ and $a_{2,min}$), the minimum distances at loaded and unloaded ends ($a_{3,t,min}$ and $a_{3,c,min}$), and the minimum distances at loaded and unloaded edges ($a_{4,t,min}$ and $a_{4,c,min}$) are summarised for all members in Table 9.14.

The comparison of the provided spacings and end/edge distances in Table 9.2 with the corresponding calculated minima in Table 9.10 shows that the requirements from Table 8.9 of the code are satisfied.

As shown in Section 9.53 of this text, Table 8.4 of the code specifies the minimum spacings and edge and end distances for bolt connectors. The minimum connector spacings parallel and perpendicular to the grain ($a_{1,min}$ and $a_{2,min}$), the minimum distances at loaded and unloaded ends ($a_{3,t,min}$ and $a_{3,c,min}$), and the minimum distances at loaded and unloaded edges ($a_{4,t,min}$ and $a_{4,c,min}$) are summarised for all members in Table 9.15.

Minimum Spacings, Edge and End Distances for Split-ring Connectors								
Member	Interface	Spacings			End distances		Edge distances	
		$a_{1,min}$ (mm)	$a_{2,min}$ (mm)	$k_{a1}^2+k_{a2}^2$ ($\geq 1,0$)	$a_{3,t,min}$ (mm)	$a_{3,c,min}$ (mm)	$a_{4,t,min}$ (mm)	$a_{4,c,min}$ (mm)
1	1	112,5	90,0	1,284	150,0	90,0	45,0	45,0
2		90,0		1,445		112,5	60,0	
	2	107,3		1,382		96,3	54,6	
3		111,2		1,309		90,0	49,9	
	3	109,3		1,342		90,6	52,7	
4		112,5		1,284		90,0	45,0	

Table 9.14

Minimum Spacings, Edge and End Distances for Bolt Connectors							
Member	Interface	Spacings		End distances		Edge distances	
		$a_{1,min}$ (mm)	$a_{2,min}$ (mm)	$a_{3,t,min}$ (mm)	$a_{3,c,min}$ (mm)	$a_{4,t,min}$ (mm)	$a_{4,c,min}$ (mm)
1	1	60,0	48,0	84,0	48,0	36,0	36,0
2		48,0			84,0	48,0	
	2	57,2			58,1	39,4	
3		59,3			48,0	36,0	
	3	58,3			49,0	36,3	
4		60,0			48,0	36,0	

Table 9.15

The comparison of the provided spacings and end/edge distances in Table 9.2 with the corresponding calculated minima in Table 9.11 shows that the requirements from Table 8.4 of the code are satisfied.

Requirements for bolts
Clause 8.10(10) specifies that for bolts used with toothed-plate connectors, Clause 10.4.3 applies. Some of the requirements by Clause 10.4.3 have been checked in the present calculations, e.g. the minimum diameter and thickness.

9.9 Glued Connections

The manufacture of glued joints, e.g. finger joints in glued-laminated beams, web-to-flange connections in ply-web beams and lap-joints including splices, should comply with the requirements of "BS EN 301:2006 Adhesives, phenolic and aminoplastic, for load-bearing timber structures — Classification and performance requirements". The structural requirements of adhesives are given in Section 3.6 of the code as follows:

"(1)P Adhesives for structural purposes shall produce joints of such strength and

durability that the integrity of the bond is maintained in the assigned service class throughout the expected life of the structure.

(2) Adhesives which comply with Type I specification as defined in EN 301 may be used in all service classes.

(3) Adhesives which comply with Type II specification as defined in EN 301 should only be used in service classes 1 or 2 and not under prolonged exposure to temperatures in excess of 50°C".

The structural detailing and control requirements are given in Section 10.3 as follows:

"(1) Where bond strength is a requirement for ultimate limit state design, the manufacture of glued joints should be subject to quality control, to ensure that the reliability and quality of the joint is in accordance with the technical specification.

(2) The adhesive manufacturer's recommendations with respect to mixing, environmental conditions for application and curing, moisture content of members and all factors relevant to the proper use of the adhesive should be followed.

(3) For adhesives which require a conditioning period after initial set, before attaining full strength, the application of load to the joint should be restricted for the necessary time".

A number of typical structural joints are shown in Figure 9.37

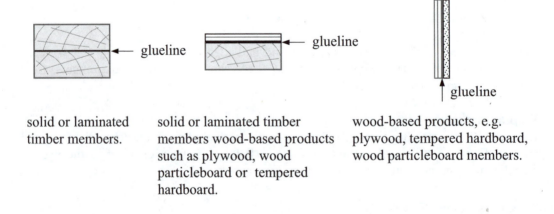

solid or laminated timber members.

solid or laminated timber members wood-based products such as plywood, wood particleboard or tempered hardboard.

wood-based products, e.g. plywood, tempered hardboard, wood particleboard members.

Figure 9.37

The possibility of differential shrinkage, distortion and stress concentrations at glued joints should also be considered. When mechanical fasteners are present in a glued joint, they are **not** considered to contribute to the strength of the joint.

Glued joints are designed on the basis of shear stresses. When the components of a joint are loaded parallel to the grain, the lesser of the shear strength parallel to the grain for the timbers being joined is used in the calculations.

Most gluing is carried out in controlled factory conditions involving manufacturers who conduct full scale testing of their products to ensure that an adequate factor of safety is achieved.

Since glue is always stronger than the timber being connected (assuming the recommended type of glue and manufacturing procedure have been adopted), the strength of the glueline is based on timber strengths. It is important to recognise that tensile components of stress perpendicular to the plane of the glueline are unacceptable.

9.9.1 Example 9.16: Glued Lap Joint – Timber-to-timber Connection in Single Shear

Using the design data given, determine the minimum contact area required to transmit the variable characteristic load indicated in the lap joint shown in Figure 9.38.

Figure 9.38   1400 N

Design data:

Timber Class	C16
Load duration	long-term
Service Class	2

Solution:

BS EN 338:2003
Clause 5.0   Characteristic values for C16 Timber
Table 1   Characteristic density: $\rho_k = 310{,}0$ kg/m³
                 Mean density: $\rho_{mean} = 370{,}0$ kg/m³
                 Characteristic shear strength $f_{v,Rk} = 1{,}8$ N/mm²

BS EN 1995-1-1:2004
Table 3.1   Long-term actions $k_{mod} = 0{,}7$
NA: Table NA.3   Partial factor for material properties and resistance: γ_M
                 For solid timber – untreated and preservative treated $\gamma_M = 1{,}3$

Equation (2.17)   Design shear strength $f_{v,Rd} = k_{mod} \dfrac{F_{v,Rk}}{\gamma_m} = 0{,}7 \times \dfrac{1{,}8}{1{,}3} = 0{,}97$ N

NA: Table NA.A1.2(B)
                 For variable actions: Leading actions $\gamma_{Q,1} = 1{,}5$
                 Design load $= F_{v,Q,d} = (1{,}5 \times 1400) = 2100$ N

Required glue area $= (F_{v,Q,d}/ f_{v,Rd}) = (2100/0{,}97) = 2165$ mm²

10. Overall Structural Stability

Objective: to introduce the principles of overall structural stability and robustness.

10.1 *Introduction*

In the previous chapters the requirements of strength, stiffness and stability of individual structural components have been considered in detail. It is also **essential** in any structural design to consider the requirements of **overall** structural stability.

The term **stability** has been defined in *Stability of Buildings* published by the Institution of Structural Engineers (68) in the following manner:

"Provided that displacements induced by normal loads are acceptable, then a building may be said to be stable if:
- *a minor change in its form, condition, normal loading or equipment would not cause partial or complete collapse and*
- *it is not unduly sensitive to change resulting from accidental or other actions.*
Normal loads include the permanent and variable actions for which the building has been designed.

The phrase 'is not unduly sensitive to change' should be broadly interpreted to mean that the building should be so designed that it will not be damaged by accidental or other actions to an extent disproportionate to the magnitudes of the original causes of damage."

This publication, and the inclusion of stability, robustness and accidental damage clauses in current design codes, is largely a consequence of the overall collapse or significant partial collapse of structures, e.g. the collapse of precast concrete buildings under erection at Aldershot in 1963 (66) and notably the Ronan Point Collapse due to a gas explosion in 1968 (67).

The Ronan Point failure occurred in May 1968 in a 23-storey precast building. A natural gas explosion in a kitchen triggered the progressive collapse of all of the units in one corner above and below the kitchen. The spectacular nature of the collapse had a major impact on the philosophy of structural design resulting in important revisions of design codes world-wide.

This case stands as one of the few landmark failures which have had a sustained impact on structural thinking.

The inclusion of such clauses in codes and building regulations is not new. The following is an extract from the 'CODE OF LAWS OF HAMMURABI (2200 BC), KING OF BABYLONIA' (the earliest building code yet discovered):

A. *If a builder builds a house for a man and do not make its construction firm and the house which he has built collapse and cause the death of the owner of the house – that builder shall be put to death.*

B. *If it cause the death of the son of the owner of the house – they shall put to death a son of that builder.*

C. *If it cause the death of a slave of the owner of the house – he shall give to the owner of the house a slave of equal value.*

D. *If it destroy property, he shall restore whatever it destroyed, and because he did not make the house which he built firm and it collapsed, he shall rebuild the house which collapsed at his own expense.*

E. *If a builder build a house for a man and do not make its construction meet the requirements and a wall fall in, that builder shall strengthen the wall at his own expense.*

Whilst this code is undoubtedly harsh it probably did concentrate the designer's mind on the importance of structural stability!

An American structural engineer, Dr Jacob Feld, spent many years investigating structural failure and suggested ten basic rules to consider when designing and/or constructing any structure (70):

1. *Gravity always works, so if you don't provide permanent support, something will fail.*

2. *A chain reaction will make a small fault into a large failure, unless you can afford a fail-safe design, where residual support is available when one component fails. In the competitive private construction industry, such design procedure is beyond consideration.*

3. *It only requires a small error or oversight – in design, in detail, in material strength, in assembly, or in protective measures – to cause a large failure.*

4. *Eternal vigilance is necessary to avoid small errors. If there are no capable crew or group leaders on the job and in the design office, then supervision must take over the chore of local control. Inspection service and construction management cannot be relied on as a secure substitute.*

5. *Just as a ship cannot be run by two captains, a construction job cannot be run by a committee. It must be run by one individual, with full authority to plan, direct, hire and fire, and full responsibility for production and safety.*

6. *Craftsmanship is needed on the part of the designer, the vendor, and the construction teams.*

7. *An unbuildable design is not buildable, and some recent attempts at producing striking architecture are approaching the limit of safe buildability, even with our most sophisticated equipment and techniques.*

8. *There is no foolproof design, there is no foolproof construction method, without guidance and proper and careful control.*

9. *The best way to generate a failure on your job is to disregard the lessons to be learnt from someone else's failures.*

10. *A little loving care can cure many ills. A little careful control of a job can avoid many accidents and failures.*

An appraisal of the overall stability of a complete structure during both the design and construction stages should be carried out by, and be the responsibility of, one individual. In many instances a number of engineers will be involved in designing various elements or sections of a structure but never the whole entity. It is **essential**, therefore, that one identified engineer carries out this vital appraisal function, including consideration of any temporary measures which may be required during the construction stage. It is stated in Clause 3.3(4)P of the BS EN 1990:2002:

"The following ultimate states shall be verified where they are relevant:
- *loss of equilibrium of the structure or any part of it, considered as a rigid body;*
- *failure by excessive deformation, transformation of the structure or any part of it into a mechanism, rupture, loss of stability of the structure or any part of it, including supports and foundations;*
- *failure caused by fatigue or other time-dependent effects."*

10.1.1 Structural Form

Generally, instability problems arise due to an inadequate provision to resist lateral loading (e.g. wind loading) on a structure. There are a number of well-established structural forms which, when used correctly, will ensure adequate stiffness, strength and stability. It is important to recognise that stiffness, strength and stability are three different characteristics of a structure. In simple terms:

- the *stiffness* determines the deflections which will be induced by the applied load system,
- the *strength* determines the maximum loads which can be applied before acceptable material stresses are exceeded and,
- the *stability* is an inherent property of the structural form which ensures that the building will remain stable.

The most common forms of structural arrangements which are used to transfer loads safely and maintain stability are:

- braced frames,
- unbraced frames,
- shear cores/walls,
- cross-wall construction,
- cellular construction,
- diaphragm action.

In many structures, a combination of one or more of the above arrangements is employed to ensure adequate load paths, stability and resistance to lateral loading. All buildings behave as complex three-dimensional structures with components frequently interacting compositely to resist the applied force system. Analysis and design processes are a simplification of this behaviour in which it is usual to analyse and design in two dimensions with wind loading considered separately in two mutually perpendicular directions.

10.1.2 Braced Frames

In braced frames lateral stability is provided in a structure by utilising systems of diagonal bracing in at least two vertical planes, preferably at right angles to each other. The bracing systems normally comprise a triangulated framework of members which are either in tension or compression. The horizontal floor or roof plane can be similarly braced at an appropriate level, as shown in Figure 10.1, or the floor/roof construction may be designed as a deep horizontal beam to transfer loads to the vertical, braced planes, as shown in Figure 10.2. There are a number of configurations of bracing which can be adopted to accommodate openings, services, etc. and are suitable for providing the required load transfer and stability.

In such systems the entire wind load on the building is transferred to the braced vertical planes and hence to the foundations at these locations.

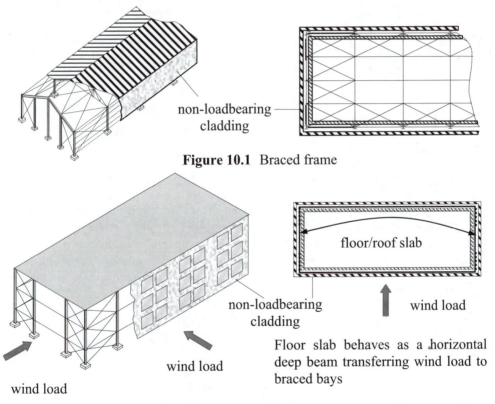

non-loadbearing cladding

Figure 10.1 Braced frame

floor/roof slab

non-loadbearing cladding

wind load

wind load

wind load

Floor slab behaves as a horizontal deep beam transferring wind load to braced bays

Figure 10.2 Braced frames

10.1.3 Unbraced Frames

Unbraced frames comprise structures in which the lateral stiffness and stability are achieved by providing an adequate number of rigid (moment-resisting) connections at appropriate locations. Unlike braced frames in which 'simple connections' only are required, the connections must be capable of transferring moments and shear forces. This is illustrated in the structure in Figure 10.3, in which stability is achieved in two mutually perpendicular directions using rigid connections. In wind direction A each typical transverse frame transfers its own share of the wind load to its own foundations through the moment connections and bending moments/shear forces/axial forces in the members. In wind direction B the wind load on either gable is transferred through the members and floors to stiffened bays (i.e. in the longitudinal section), and hence to the foundation at these locations. It is not necessary for *every* connection to be moment-resisting.

It is common for the portal frame action in a stiffened bay in wind direction B to be replaced by diagonal bracing whilst still maintaining the moment-resisting frame action to transfer the wind loads in direction A.

As with braced frames, in most cases the masonry cladding and partition walls are non-loadbearing.

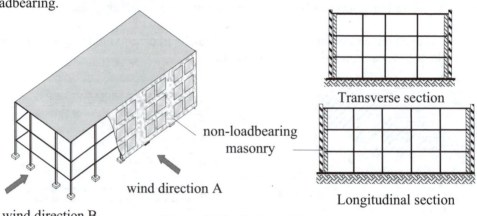

Figure 10.3   Unbraced frame

10.1.4 Shear Cores/Walls

The stability of modern high-rise buildings can be achieved using either braced or unbraced systems as described in Sections 4.5.2 and 4.5.3, or alternatively by the use of shear-cores and/or shear-walls. Such structures are generally considered as three-dimensional systems comprising horizontal floor plates and a number of strong-points provided by cores/walls enclosing stairs or lift shafts. A typical layout for such a building is shown in Figure 10.4.

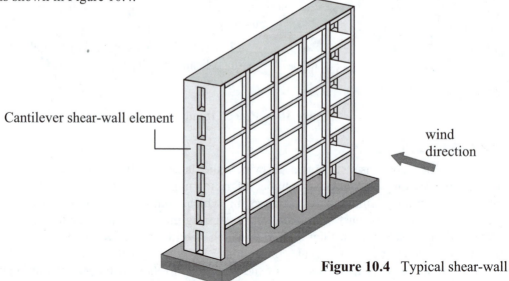

Figure 10.4   Typical shear-wall

In most cases the vertical loads are generally transferred to the foundations by a conventional skeleton of beams and columns whilst the wind loads are divided between several shear-core/wall elements according to their relative stiffness.

Where possible the plan arrangement of shear-cores and walls should be such that the centre-line of their combined stiffness is coincidental with the resultant of the applied wind load, as shown in Figure 10.5.

If this is not possible and the building is much stiffer at one end than the other, as in Figure 10.6, then torsion may be induced in the structure and must be considered. It is better at the planning stage to avoid this situation arising by selecting a judicious floor-plan layout. The floor construction must be designed to transfer the vertical loads (which are perpendicular to their plane) to the columns/wall elements in addition to the horizontal wind forces (in their own plane) to the shear-core/walls. In the horizontal plane they are designed as deep beams spanning between the strong-points.

There are many possible variations, including the use of concrete, steel, masonry and composite construction, which can be used to provide the necessary lateral stiffness, strength and stability.

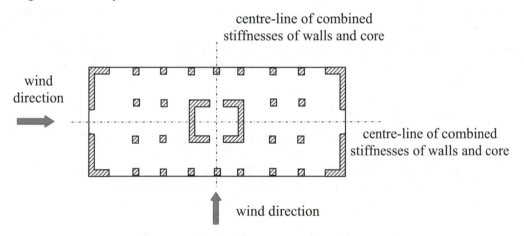

Figure 10.5 Efficient layout of shear-core/walls

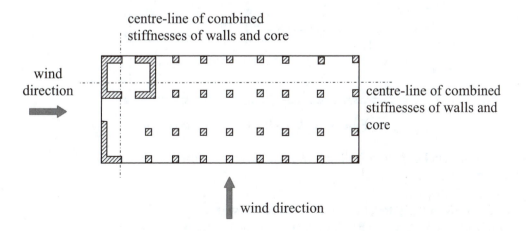

Figure 10.6 Inefficient layout of shear-core/walls

10.1.5 Cross-wall Construction

In long rectangular buildings which have repetitive, compartmental floor plans such as hotel bedroom units and classroom blocks, as shown in Figure 10.7, masonry cross-wall construction is often used.

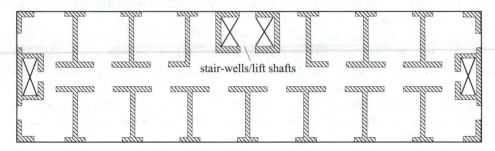

stair-wells/lift shafts

Figure 10.7 Cross-wall construction

Lateral stability parallel to the cross-walls is very high, with the walls acting as separate vertical cantilevers sharing the wind load in proportion to their stiffnesses. Longitudinal stability, i.e. perpendicular to the plane of the walls, must be provided by the other elements such as the box sections surrounding the stair-wells/lift-shafts, corridor and external walls.

10.1.6 Cellular Construction

It is common in masonry structures for the plan layout of walls to be irregular with a variety of exterior and interior walls, as shown in Figure 10.8.

The resulting structural form is known as 'cellular construction', and includes an inherent high degree of interaction between the intersecting walls. The provision of stair-wells and lift-shafts can also be integrated to contribute to the overall bracing of the structure.

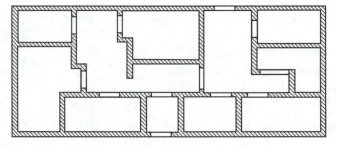

Figure 10.8 Cellular construction

It is important in both cross-wall and cellular masonry construction to ensure the inclusion of features such as:

♦ bonding or tying together of all intersecting walls,
♦ provision of returns where practicable at ends of load-bearing walls,
♦ provision of bracing walls to external walls,

♦ provision of internal bracing walls,
♦ provision of strapping of the floors and roof at their bearings to the load-bearing walls,

as indicated in *Stability of Buildings* (69).

10.1.7 Diaphragm Action

Floors, roofs, and in some cases cladding, behave as horizontal diaphragms which distribute lateral forces to the vertical wall elements. This form of structural action is shown in Figure 10.9.

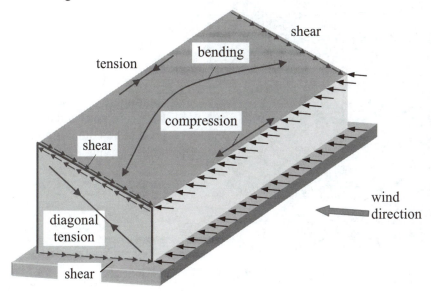

Figure 10.9 Diaphragm action

It is essential when utilising diaphragm action to ensure that each element and the connections between the various elements are capable of transferring the appropriate forces and providing adequate load-paths to the supports.

10.1.8 Accidental Damage and Robustness

It is inevitable that accidental loading such as vehicle impact or gas explosions will result in structural damage. A structure should be sufficiently robust to ensure that damage to small areas or failure of individual elements does not lead to progressive collapse or significant partial collapse. There are a number of strategies which can be adopted to achieve this, e.g.

♦ enhancement of continuity which includes increasing the resistance of connections between members and hence load-transfer capability,
♦ enhancement of overall structural strength including connections and members,
♦ provision of multiple load-paths to enable the load carried by any individual member to be transferred through adjacent elements in the event of local failure,

♦ the inclusion of load-shedding devices such as venting systems to allow the escape of gas following an explosion, or specifically designed weak elements/details to prevent transmission of load.

The robustness required in a building may be achieved by 'tying' the elements of a structure together using peripheral and internal ties at each floor and roof level, as indicated in Figure 10.10.

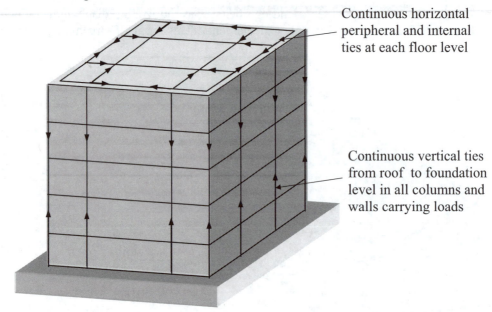

Continuous horizontal peripheral and internal ties at each floor level

Continuous vertical ties from roof to foundation level in all columns and walls carrying loads

Figure 10.10

An alternative to the 'fully tied' solution is one in which the consequences of the removal of each load-bearing member are considered in turn. If the removal of a member results in an unacceptable level of damage then this member must be strengthened to become a **protected member** (i.e. one which will remain intact after an accidental event), or the structural form must be improved to limit the extent of the predicted collapse. This process is carried out until all non-protected horizontal and vertical members have been removed one at a time.

Appendix A

Continuous Beam Coefficients

Uniformly Distributed Loads

It is assumed that all spans are equal and W is the **total uniformly distributed load/span**. Positive reactions are upwards and positive bending moments induce tension on the underside of the beams.

Two-span beams

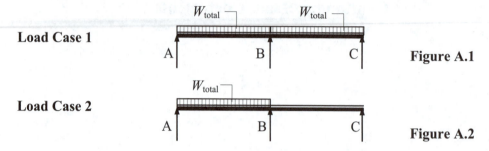

Load Case 1

Figure A.1

Load Case 2

Figure A.2

Load	Support	Support	Support	Span AB	Support B	Span BC
Case	V_A	V_B	V_C	M_{AB}	M_B	M_{BC}
1	0,375	1,250	0,375	0,070	− 0,125	0,070
2	0,438	0,625	− 0,063	0,096	− 0,063	-

Table A.1 - Coefficients for Reactions and Bending Moments

Vertical Reaction = coefficient × W
Bending Moment = coefficient × W × span (*L*)

Example: Determine the reaction at A and the bending moment at support B for the beam shown in Figure A.3.

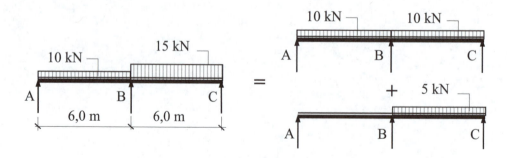

Figure A.3

V_A = (0,375× 10) − (0,063 × 5)   = + 3,345 kN
M_B = − (0,125 × 10 × 6,0) − (0,063 × 5 × 6,0)   = − 9,39 kNm

Three-span beams

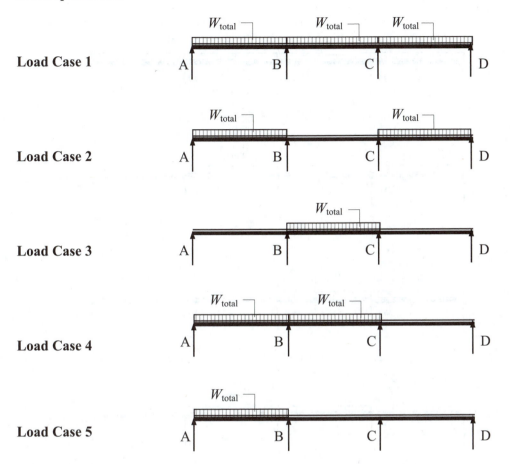

Figure A.4

Load Case	Support V_A	Support V_B	Support V_C	Support V_D	Span AB M_{AB}	Support B M_B	Span BC M_{BC}	Support C M_C	Span CD M_{CD}
1	0,4	1,1	1,1	0,4	0,080	− 0,100	0,025	− 0,100	0,080
2	0,45	0,55	0,55	0,45	0,101	− 0,050	-	− 0,050	0,101
3	− 0,05	0,55	0,55	− 0,05	-	− 0,05	0,075	− 0,05	-
4	0,383	1,2	0,45	− 0,033	0,073	− 0,117	0,054	−0,033	-
5	0,433	0,65	−0,10	0,017	0,094	− 0,067	-	−0,017	-

Table A.2 - Coefficients for Reactions and Bending Moments

Four-span beams

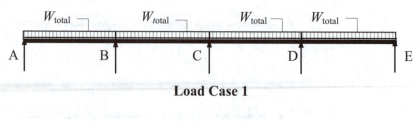

Load Case 1

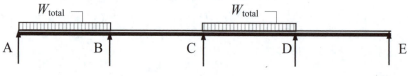

Load Case 2

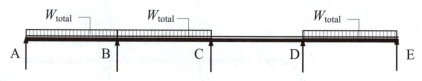

Load Case 3

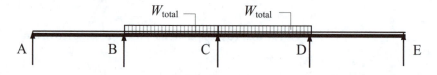

Load Case 4

Load Case 5

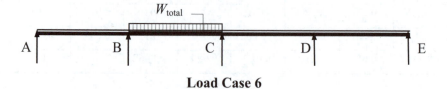

Load Case 6

Figure A.5

(Four-span beams cont,)

Load Case	Support V_A	Support V_B	Support V_C	Support V_D	Support V_D
1	0,393	1,143	0,928	1,143	0,393
2	0,446	0,572	0,464	0,572	− 0,054
3	0,38	1,223	0,357	0,598	0,442
4	− 0,036	0,464	1,144	0,464	− 0,036
5	0,433	0,652	− 0,107	0,027	− 0,005
6	− 0,049	0,545	0,571	− 0,080	0,014

Table A.3 - Coefficients for Reactions

Load Case	Span AB M_{AB}	Support B M_B	Span BC M_{BC}	Support C M_C	Span CD M_{CD}	Support D M_D	Span DE M_{DE}
1	0,077	− 0,107	0,036	− 0,071	0,036	− 0,107	0,077
2	0,100	− 0,054	-	− 0,036	0,081	− 0,054	-
3	0,072	− 0,121	0,061	− 0,018	-	− 0,058	0,098
4	-	− 0,036	0,056	− 0,107	0,056	− 0,036	-
5	0,094	− 0,067	-	+ 0,017	-	− 0,005	-
6	-	− 0,049	0,074	− 0,054	-	+ 0,014	-

Table A.4 - Coefficients for Bending Moments

Point Loads

It is assumed that all spans are equal and *P* is the **total central point-load/span**, Positive reactions are upwards and positive bending moments induce tension on the underside of the beams,

Two-span beams

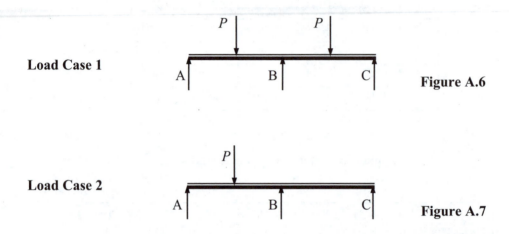

Load Case 1

Figure A.6

Load Case 2

Figure A.7

Load Case	Support V_A	Support V_B	Support V_C	Span AB M_{AB}	Support B M_B	Span BC M_{BC}
1	0,313	1,374	0,313	0,156	− 0,188	0,156
2	0,406	0,688	− 0,094	0,203	− 0,094	-

Table A.5 - Coefficients for Reactions and Bending Moments

Vertical Reaction = **coefficient** $\times P$
Bending Moment = **coefficient** $\times P \times$ **span** (*L*)

Three-span beams

Load Case 1

Load Case 2

Load Case 3

Load Case 4

Load Case 5

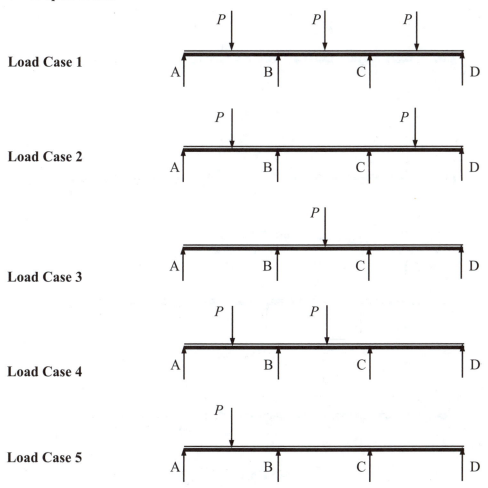

Figure A.8

Load Case	Support V_A	Support V_B	Support V_C	Support V_D	Span AB M_{AB}	Support B M_B	Span BC M_{BC}	Support C M_C	Span CD M_{CD}
1	0,350	1,150	1,150	0,350	0,175	− 0,150	0,100	− 0,150	0,175
2	0,425	0,575	0,575	0,425	0,213	− 0,075	-	− 0,075	0,213
3	− 0,075	0,575	0,575	− 0,075	-	− 0,075	0,175	− 0,075	-
4	0,325	1,300	0,425	− 0,050	0,163	− 0,175	0,138	− 0,050	-
5	0,400	0,725	− 0,15	0,025	0,200	− 0,100	-	+ 0,025	-

Table A.6 - Coefficients for Reactions and Bending Moments

Four-span beams

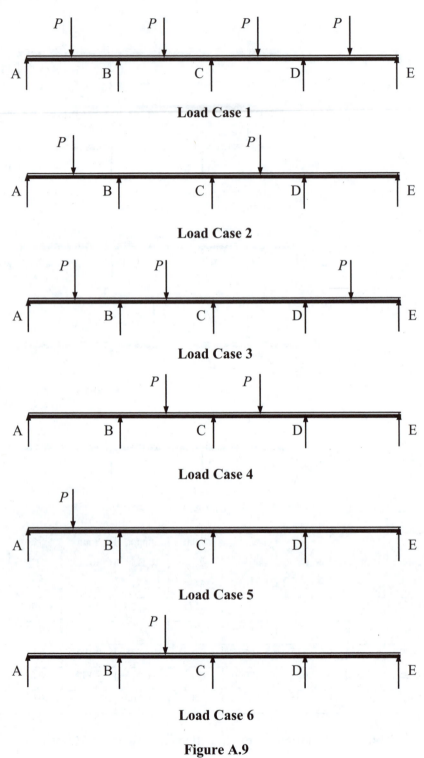

Load Case 1

Load Case 2

Load Case 3

Load Case 4

Load Case 5

Load Case 6

Figure A.9

(Four-span beams cont,)

Load Case	Support V_A	Support V_B	Support V_C	Support V_D	Support V_D
1	0,339	1,214	0,894	1,214	0,339
2	0,420	0,607	0,446	0,607	− 0,08
3	0,319	1,335	0,286	0,647	0,413
4	− 0,054	0,447	1,214	0,447	− 0,054
5	0,400	0,728	− 0,161	0,040	− 0,007
6	− 0,074	0,567	0,607	− 0,121	0,020

Table A.7 - Coefficients for Reactions

Load Case	Span AB M_{AB}	Support B M_B	Span BC M_{BC}	Support C M_C	Span CD M_{CD}	Support D M_D	Span DE M_{DE}
1	0,170	− 0,161	0,116	− 0,107	0,116	− 0,161	0,170
2	0,210	− 0,080	-	− 0,054	0,183	− 0,080	-
3	0,160	− 0,181	0,146	− 0,027	-	− 0,087	0,207
4	-	− 0,054	0,143	− 0,161	0,143	− 0,054	-
5	0,200	− 0,100	-	+ 0,027	-	− 0,007	-
6	-	− 0,074	0,173	− 0,080	-	+ 0,020	-

Table A.8 - Coefficients for Bending Moments

Appendix B

Shear Deformation of Beams

The shear deformation of beams can be estimated using the method indicated in reference (72) as follows.

Consider a beam comprising 'j' segments along its length and subjected to a general loading, then the shear deformation can be estimated from:

$$w_v = k \times \sum_{i=1}^{j} T_i v_i$$

where:

w_v is the shear deformation,

T_i is (the area of the shear force diagram for the ith segment due to the general loading) divided by $G_i A_i$,

G_i is the shear modulus of the material in the ith segment,

A_i is the cross-sectional area of the ith segment beam,

v_i is the shear force for the ith segment due to a unit load applied at the point of the desired deformation,

k is a constant for a given cross-sectional shape as follows:

Circular section ● $k = 1,11$ Rectangular section ▮ $k = 1,2$

I- sections and box-beams

$$k = \frac{4,5 \times \left[\dfrac{1}{p}(1-t)+t\right] \times \left\{\dfrac{1}{p^2}\left(\dfrac{t^5}{2}-t^3+\dfrac{t}{2}\right) + \dfrac{1}{p}\left[-t^5\left(\dfrac{3}{30\beta}+\dfrac{2}{3}\right)+t^3\left(\dfrac{1}{3\beta}+\dfrac{2}{3}\right)-\dfrac{t}{2\beta}+\dfrac{8}{30\beta}\right]+\dfrac{8t^5}{30}\right\}}{\left[\dfrac{1}{p}(1-t^3)+t^3\right]^2}$$

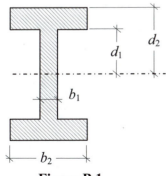

Figure B.1

The section dimensions relate to the *transformed values* in the case of cross-sections with different materials properties for the web and flanges.

$$\beta = \frac{\text{Shear moduls of the flange } (G_f)}{\text{Shear modulus of the web } (G_w)}$$

$$p = b_1/b_2 \quad \text{and} \quad t = d_1/d_2$$

Example B.1   Shear Deformation

Determine the shear deformations for a rectangular beam of breadth '*b*' and depth '*h*' when it is subjected to the loads indicated in Figures B.2(a), (b) and (c).

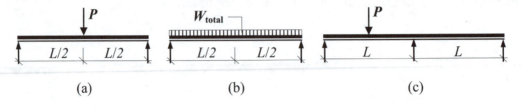

(a)   (b)   (c)

Figure B.2

(a)

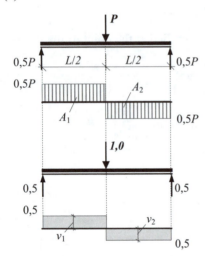

Applied load: Mid-span point-load = *P*

Applied Load Shear Force Diagram

Unit Load

Unit Load Shear Force Diagram

$$w_v = k \times \sum_{i=1}^{2} T_i v_i$$

$T_1 = (0,5P \times 0,5L)/bhG = 0,25PL/bhG; \quad v_1 = 0,5$

By symmetry:

$T_2 = 0,25PL/bhG; \quad v_2 = 0,5$

$$w_v = k \times \left[\frac{2 \times (0,25PL \times 0,5)}{bhG} \right] = k \times \frac{0,25PL}{bhG}$$

$k = 1,2$ for a rectangular cross-section

$$w_v = 1,2 \times \frac{0,25PL}{bhG} = \frac{0,3PL}{bhG}$$

(b)

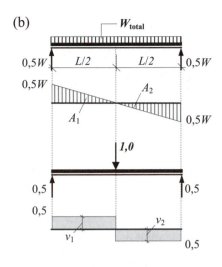

Applied load: Uniformly distributed load – W_{total}

Applied Load Shear Force Diagram

Unit Load

Unit Load Shear Force Diagram

$T_1 = (0.5W_{total} \times 0.25L)/bhG = 0.125W_{total}L/bhG;$ $v_1 = 0.5$
By symmetry:
$T_2 = 0.125W_{total}L/bhG;$ $v_2 = 0.5$

$$w_v = k \times \left[\frac{2 \times (0.125W_{total}L \times 0.5)}{bhG} \right] = k \times \frac{0.125W_{total}L}{bhG}$$

k = 1,2 for a rectangular cross-section

$$w_v = 1.2 \times \frac{0.125W_{total}L}{bhG} = \mathbf{\frac{0.15W_{total}L}{bhG}}$$

(c)

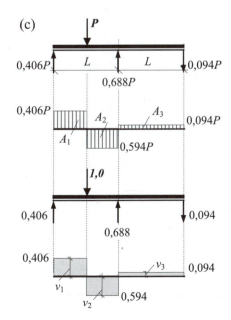

Applied load: Mid-span point-load = P
(Support reactions are given in Appendix A)

Applied Load Shear Force Diagram

Unit Load

Unit Load Shear Force Diagram

$T_1 = (0{,}406P \times 0{,}5L)/bhG = 0{,}203PL/bhG_1 \qquad v_1 = 0{,}406$

$T_2 = (0{,}594P \times 0{,}5L)/bhG = 0{,}297PL/bhG_1 \qquad v_2 = 0{,}594$

$T_3 = (0{,}094P \times 1{,}0L)/bhG = 0{,}094PL/bhG_1 \qquad v_3 = 0{,}094$

$$w_v = k \times \left[\frac{(0{,}203 \times 0{,}406 \times PL) + (0{,}297 \times 0{,}594 \times PL) + (0{,}094 \times 0{,}094 \times PL)}{bhG} \right]$$

$$= k \times \frac{0{,}268PL}{bhG}$$

k = 1,2 for a rectangular cross-section

$$w_v = 1{,}2 \times \frac{0{,}268PL}{bhG} = \frac{\mathbf{0{,}321PL}}{\mathbf{bhG}}$$

Note: the +ve and −ve signs of T_i and v_i should be taken into account during multiplication.

Example B.2   k-values

Determine the k-value for the ply-web beam shown in Figure B.3

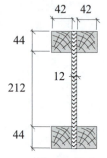

Flange:

Mean modulus of elasticity parallel to the grain

$E_{0,m}$ = 11,0 kN/mm^2

Modulus of rigidity G_v = 0,69 kN/mm^2

Web:

Mean modulus of elasticity in bending parallel to the grain

$E_{0,m}$ = 4,93 kN/mm^2

Modulus of rigidity – panel shear G_v = 1,08 kN/mm^2

Figure B.3

The cross-section must be transformed into one material (see Chapter 4, Section 4.13) Consider a transformed section in terms of the *web* material as follows:

Modular ratio $n = \dfrac{E_{\text{flange}}}{E_{\text{web}}} = \dfrac{11{,}0}{4{,}93} = 2{,}231; \qquad \beta = \dfrac{G_{\text{flange}}}{G_{\text{web}}} = \dfrac{0{,}69}{1{,}08} = 0{,}639$

The transformed width of the flanges = (42 × 2,231) = 93,70 mm

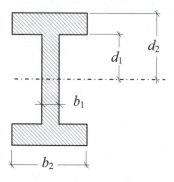

b_1 = 12 mm

b_2 = [(2 × 93,70) + 12] = 199,40 mm

d_1 = 212/2,0 = 106 mm

d_2 = (106 + 44) = 150,0 mm

$p = b_1/b_2 = (12/199{,}4) = 0{,}06$

$t = d_1/d_2 = (106{,}0/150{,}0) = 0{,}71$

Transformed area:

$A = (199{,}40 \times 300) - (2 \times 93{,}70 \times 212)$

$= 20{,}09 \times 10^3$ mm^2

$$k = \frac{4,5 \times \left[\frac{1}{p}(1-t)+t\right] \times \left\{\frac{1}{p^2}\left(\frac{t^5}{2}-t^3+\frac{t}{2}\right)+\frac{1}{p}\left[-t^5\left(\frac{3}{30\beta}+\frac{2}{3}\right)+t^3\left(\frac{1}{3\beta}+\frac{2}{3}\right)-\frac{t}{2\beta}+\frac{8}{30\beta}\right]+\frac{8t^5}{30}\right\}}{\left[\frac{1}{p}(1-t^3)+t^3\right]^2}$$

$$\left[\frac{1}{p}(1-t)+t\right] = \left[\frac{1}{0,06}\times(1-0,71)+0,71\right] = 5,543$$

$$\frac{1}{p^2}\left(\frac{t^5}{2}-t^3+\frac{t}{2}\right) = \frac{1}{0,06^2}\times\left(\frac{0,71^5}{2,0}-0,71^3+\frac{0,71}{2,0}\right) = 24,25$$

$$-t^5\left(\frac{3}{30\beta}+\frac{2}{3}\right) = -0,71^5\times\left(\frac{3,0}{30\times0,639}+\frac{2,0}{3,0}\right) = -0,149$$

$$t^3\left(\frac{1}{3\beta}+\frac{2}{3}\right) = 0,71^3\times\left(\frac{1}{3,0\times0,639}+\frac{2,0}{3,0}\right) = 0,425$$

$$\frac{t}{2\beta} = \frac{0,71}{2,0\times0,639} = 0,556 \qquad\qquad \frac{8}{30\beta} = \frac{8,0}{30,0\times0,639} = 0,417$$

$$\frac{8t^5}{30} = \frac{8,0\times0,71^5}{30,0} = 0,048$$

$$\left[\frac{1}{p}(1-t^3)+t^3\right]^2 = \left[\frac{1}{0,06}\times(1-0,71^3)+0,71^3\right]^2 = 122,31$$

$$k = \frac{(4,5\times5,543)\times\left[24,25+\frac{1}{0,06}\times(-0,149+0,425-0,556+0,417)+0,048\right]}{122,31} = \mathbf{5,42}$$

The shear deformation induced by a uniformly distributed load of W_{total} can de determined from the product of the k-value and the previously determined $\sum_{i=1}^{j} T_i v_i$ value.

$$w_v = k\times\frac{0,125W_{total}L}{AG} = 5,42\times\frac{0,125W_{total}L}{AG} = \frac{\mathbf{0,68W_{total}L}}{\mathbf{AG}}$$

The shear deformation can also be estimated based on the rectangular web area only with the value of k assumed to be 1,2, i.e.

$$w_v = k \times \frac{0,125 W_{total} L}{AG} = 1,2 \times \frac{0,125 W_{total} L}{A_w G} = \frac{0,15 W_{total} L}{A_w G}$$

In the case of the thin-webbed beam shown in Figure B.3:

$$w_v = \frac{0,15 W_{total} L}{A_w G} = \frac{0,15 W_{total} L}{(300 \times 12) G} = \frac{41,67 W_{total} L}{10^6 G}$$

Using the full transformed section value:

$$w_v = \frac{0,68 W_{total} L}{AG} = \frac{0,68 W_{total} L}{20,09 \times 10^3 G} = \frac{33,85 W_{total} L}{10^6 G}$$

In this case the deformation calculated using only the web is overestimated by $\approx 23\%$.

The value of k can be estimated using Charts 1 to 4 shown in Figures B.5 to B.8 for $\beta = 0,3, 0,5, 0,8$ and 1,0 respectively; for intermediate values of β interpolation is permitted.

Consider the ply-web beam in Example 5.4 in Chapter 5, ($k = 2,26$). In this case $\beta = 0,75$, $p = 0,19$ and $t = 0,72$. Chart 2 indicates a value of $k \approx 2,3$ for $\beta = 0,5$ and a value of $k \approx 2,25$ for $\beta = 0,8$ in Chart 3 as shown in Figures B6 and B7.

Using interpolation, the value corresponding to $\beta = 0,75$ can be determined as follows:

$$k = 2,3 + \frac{(0,75 - 0,50)}{(0,80 - 0,50)} \times (2,25 - 2,30) = 2,26$$

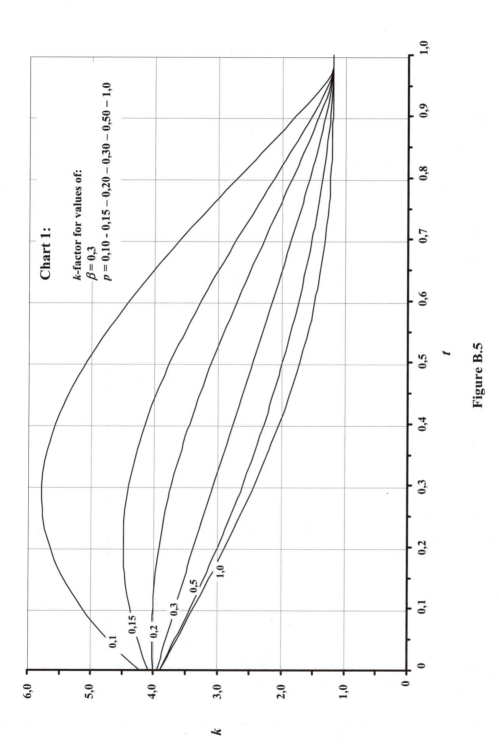

Chart 1:

k-factor for values of:
$\beta = 0{,}3$
$p = 0{,}10 - 0{,}15 - 0{,}20 - 0{,}30 - 0{,}50 - 1{,}0$

Figure B.5

Design of Structural Timber to EC5

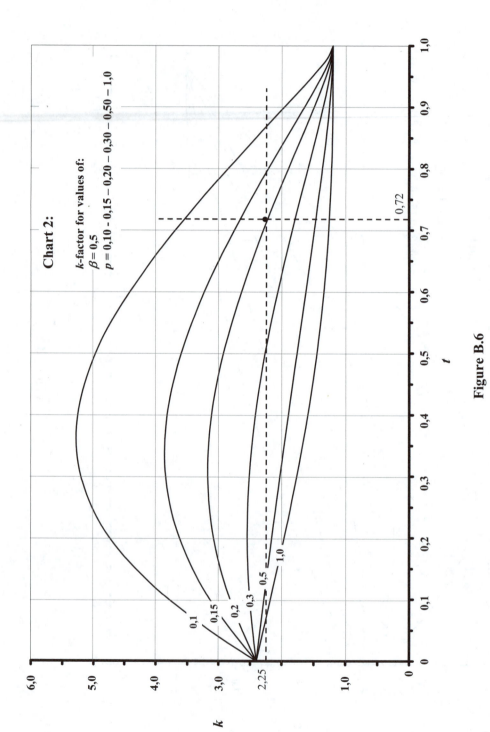

Figure B.6

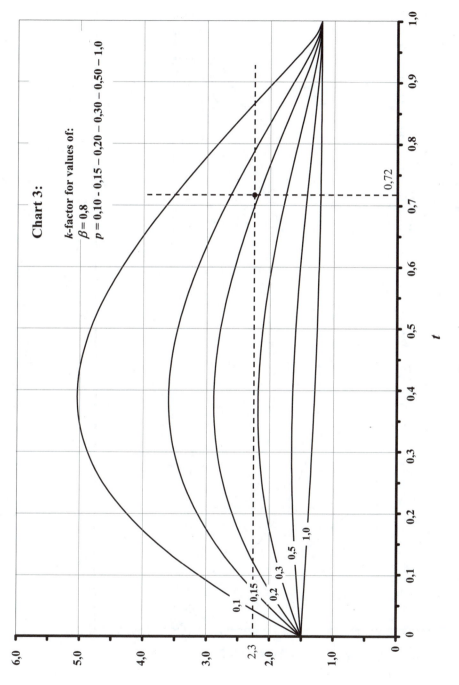

Chart 3:

k-factor for values of:

$\beta = 0,8$

$p = 0,10 - 0,15 - 0,20 - 0,30 - 0,50 - 1,0$

Figure B.7

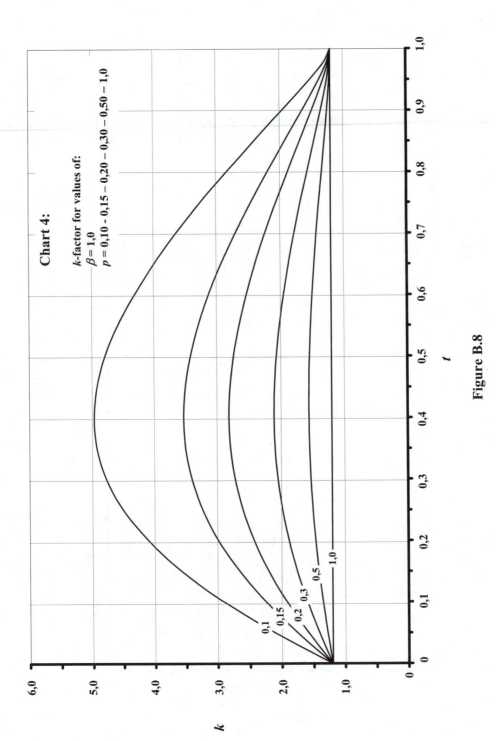

Figure B.8

Bibliography

1. **British Standards Institution (2002)**
   BS EN 1990: Eurocode - Basis of structural design

2. **British Standards Institution (2002)**
   NA to BS EN 1990: UK National Annex for Eurocode 0 — Basis of structural design

3. **British Standards Institution (2002)**
   BS EN 1991-1-1: Eurocode 1: Actions on structures —
   Part 1-1: General actions — Densities, self-weight, imposed loads for buildings

4. **British Standards Institution (2002)**
   NA to BS EN 1991-1-1: UK National Annex to Eurocode 1: Actions on structures —
   Part 1-1: General actions — Densities, self-weight, imposed loads for buildings

5. **British Standards Institution (2002)**
   BS EN 1991-1-2: Eurocode 1: Actions on structures —
   Part 1-2: General actions — Actions on structures exposed to fire

6. **British Standards Institution (2002)**
   NA to BS EN 1991-1-2: UK National Annex to Eurocode 1: Actions on structures –
   Part 1-2: General actions – Actions on structures exposed to fire

7. **British Standards Institution (2003)**
   BS EN 1991-1-3: Eurocode 1 — Actions on structures —
   Part 1-3: General actions — Snow loads

8. **British Standards Institution (2003)**
   NA to BS EN 1991-1-3: UK National Annex to Eurocode 1: Actions on structures —
   Part 1-3: General actions — Snow loads

9. **British Standards Institution (2005)**
   BS EN 1991-1-4: Eurocode 1: Actions on structures —
   Part 1-4: General actions — Wind actions

10. **British Standards Institution (2005)**
    NA to BS EN 1991-1-4: UK National Annex to Eurocode 1: Actions on structures —
    Part 1-4: General actions — Wind loads (Draft 4.13)

11. **British Standards Institution (2006)**
    BS EN 1991-1-7: Eurocode 1 — Actions on structures —
    Part 1-7: General actions — Accidental actions

12. **British Standards Institution (2003)**
    BS EN 1991-2: Eurocode 1: Actions on structures — Part 2: Traffic loads on bridges

13. **British Standards Institution (2004)**
    BS EN 1995-1-1: Eurocode 5: Design of timber structures —
    Part 1-1: General — Common rules and rules for buildings

14. **British Standards Institution (2004)**
    NA to BS EN 1995-1-1: UK National Annex to Eurocode 5: Design of timber
    structures – Part 1-1: General – Common rules and rules for buildings

15. **British Standards Institution (2004)**
    BS EN 1995-1-2: Eurocode 5: Design of timber structures —
    Part 1-2: General — Structural fire design

16. **British Standards Institution (2004)**
    NA to BS EN 1995-1-2: UK National Annex to Eurocode 5: Design of timber
    structures – Part 1-2: General – Structural fire design

17. **British Standards Institution (2004)**
    BS EN 1995-2: Eurocode 5: Design of timber structures — Part 2: Bridges

18. **British Standards Institution (2004)**
    NA to BS EN 1995-2: UK National Annex to Eurocode 5: Design of timber structures
    – Part 2: Bridges

19. **British Standards Institution (2006)**
    BS EN 300: Oriented Strand Boards (OSB) — Definitions, classification and
    specifications

20. **British Standards Institution (2006)**
    BS EN 301: Adhesives, phenolic and aminoplastic, for load-bearing timber structures
    — Classification and performance requirements

21. **British Standards Institution (2004)**
    BS EN 302-1: Adhesives for load-bearing timber structures — Test methods —
    Part 1: Determination of bond strength in longitudinal tensile shear strength

22. **British Standards Institution (2004)**
    BS EN 302-2: Adhesives for load-bearing timber structures — Test methods —
    Part 2: Determination of resistance to delamination

23. **British Standards Institution (2003)**
    BS EN 312: Particleboards — Specifications

24. **British Standards Institution (2006)**
    BS EN 335-1: Durability of wood and wood-based products — Definitions of use classes — Part 1: General

25. **British Standards Institution (2006)**
    BS EN 335-2: Durability of wood and wood-based products — Definition of use classes — Part 2: Application to solid wood

26. **British Standards Institution (2003)**
    BS EN 336: Structural timber — Sizes, permitted deviations

27. **British Standards Institution (2003)**
    BS EN 338: Structural timber — Strength classes

28. **British Standards Institution (1994)**
    BS EN 350-1: Durability of wood and wood-based products — Natural durability of solid wood — Part 1: Guide to the principles of testing and classification of the natural durability of wood

29. **British Standards Institution (1994)**
    BS EN 350-2: Durability of wood and wood-based products — Natural durability of solid wood — Part 2: Guide to natural durability and treatability of selected wood species of importance in Europe

30. **British Standards Institution (1996)**
    BS EN 351-1: Durability of wood and wood-based products — Preservative-treated solid wood — Part 1: Classification of preservative penetration and retention

31. **British Standards Institution (1996)**
    BS EN 351-2: Durability of wood and wood-based products — Preservative-treated solid wood — Part 2: Guidance on sampling for the analysis of preservative-treated wood

32. **British Standards Institution (2001)**
    BS EN 385: Finger jointed structural timber — Performance requirements and minimum production requirements

33. **British Standards Institution (2001)**
    BS EN 386: Glued laminated timber — Performance requirements and minimum production requirements

34. **British Standards Institution (2001)**
    BS EN 387: Glued laminated timber — Large finger joints — Performance requirements and minimum production requirements

35. **British Standards Institution (2001)**
    BS EN 460: Durability of wood and wood-based products — Natural durability of solid wood — Guide to the durability requirements for wood to be used in hazard classes

36. **British Standards Institution (2003)**
    BS EN 622-1: Fibreboards — Specifications — Part 1: General requirements

37. **British Standards Institution (2004)**
    BS EN 622-2: Fibreboards — Specifications — Part 2: Requirements for hardboards

38. **British Standards Institution (2004)**
    BS EN 622-3: Fibreboards — Specifications — Part 3: Requirements for medium boards

39. **British Standards Institution (1997)**
    BS EN 622-4: Fibreboards — Specifications — Part 4. Requirements for softboards

40. **British Standards Institution (2003)**
    BS EN 636: Plywood — Specifications

41. **British Standards Institution (2004)**
    BS EN 789: Timber structures — Test methods — Determination of mechanical properties of wood based panels

42. **British Standards Institution (2000)**
    BS EN 912: Timber fasteners — Specifications for connectors for timber

43. **British Standards Institution (1999)**
    BS EN 1194: Timber structures — Glued laminated timber — Strength classes and determination of characteristic values

44. **British Standards Institution (2004)**
    BS EN 1912: Structural timber — Strength classes — Assignment of visual grades and species

45. **British Standards Institution (2001)**
    BS EN 12369-1: Wood-based panels — Characteristic values for structural design — Part 1: OSB, particleboards and fibreboards

46. **British Standards Institution (2004)**
    BS EN 12369-2: Wood-based panels — Characteristic values for structural design — Part 2: Plywood

47. **British Standards Institution (2005)**
   BS EN 14080: Timber structures — Glued laminated timber — Requirements

48. **British Standards Institution (2005)**
   BS EN 14081-1: Timber structures — Strength graded structural timber with rectangular cross section — Part 1: General requirements

49. **British Standards Institution (2005)**
   BS EN 14081-2: Timber structures — Strength graded structural timber with rectangular cross section — Part 2: Machine grading; additional requirements for initial type testing

50. **British Standards Institution (2005)**
   BS EN 14081-3: Timber structures — Strength graded structural timber with rectangular cross section — Part 3: Machine grading; additional requirements for factory production control

51. **British Standards Institution (2005)**
   BS EN 14081-4: Timber structures — Strength graded structural timber with rectangular cross section — Part 4: Machine grading — Grading machine settings for machine controlled systems

52. **British Standards Institution (1968)**
   BS 4320: Metal washers for general engineering purpose metric series

53. **British Standards Institution (2007)**
   BS 4978: Visual strength grading of softwood – Specification

54. **British Standards Institution (2002)**
   BS 5268-2: Structural use of timber — Part 2: Code of practice for permissible stress design, materials and workmanship

55. **British Standards Institution (2006)**
   BS 5268-3: Structural use of timber — Part 3: Code of practice for trussed rafter roofs

56. **British Standards Institution (1996)**
   BS 6399-1: Loading for buildings — Part 1: Code of practice for dead and imposed loads

57. **British Standards Institution (2004)**
   PP 1990: Structural Eurocodes. Guide to the Structural Eurocodes for students of structural design

58. **Blass, H.J., Aune, P., Choo, B.S., Görlacher, R., Griffiths, D.R., Hilson, B.O., Racher, P. and Stek, G. (1995)**
Timber Engineering STEP 1: Basis of design, material properties, structural components and joints
1st edition, Centrum Hout, the Netherlands

59. **Blass, H.J., Aune, P., Choo, B.S., Görlacher, R., Griffiths, D.R., Hilson, B.O., Racher, P. and Stek, G. (1995)**
Timber Engineering STEP 2: Design – Details and structural systems
1st edition, Centrum Hout, the Netherlands

60. **Council of Forest Industries of British Columbia (COFI)**
Fir Plywood Web Beam Design

61. **Enjily, V. and Bregulla, J. (2008)**
Designers' Guide to EN 1995-1-1 Eurocode 5: Design of Timber Structures - Common Rules and Rules for Buildings
Thomas Telford

62. **Faherty, K.F. and Williamson, G.T. (1998)**
Wood Engineering and Construction Handbook
3rd edition, McGraw-Hill Professional

63. **Freudenthal, A.M. (1945)**
The Safety of Structures
Proceedings of the American Society of Civil Engineers

64. **Gulvanessian, H., Calgaro, J.-A. and Holicky, M. (2002)**
Designers' Guide to EN 1990 Eurocode: Basis of Structural Design
Thomas Telford

65. **Gulvanessian, H., Calgaro, J.A., Formichi, P. and Harding, G. (2008)**
Designers' Guide to EN 1991-1.1, 1991-1.3 and 1991-1.5 to 1.7 Eurocode 1: Actions on Structures: General Rules and Actions on Buildings
Thomas Telford

66. **HMSO (1963)**
The collapse of a precast concrete building under construction
Technical statement by the Building Research Station, London

67. **HMSO (1968)**
Report of the inquiry into the collapse of flats at Ronan Point, Canning Town, London

68. **Institution of Structural Engineers (1998)**
    Stability of Buildings

69. **Johansen, K.W. (1949)**
    Theory of Timber Connections
    International Association of Bridge and Structural Engineering
    Publication No. 9:249-262, Bern

70. **Kaminetzky, D. (1991)**
    Design and Construction Failures: Lessons from forensic investigations
    McGraw-Hill

71. **McKenzie, W.M.C. (2006)**
    Examples in Structural Analysis
    Taylor and Francis, London

72. **Orosz, I, (1970)**
    Simplified Method for Calculating Shear Deflection of Beams
    U.S.D.A. Forest Service Research Note FPL-0210

73. **Ozelton, E.C. and Baird J.A. (2002)**
    Timber Designers' Manual
    3rd edition, Blackwell Science Ltd

74. **Steer, P.J. (2002)**
    EN1995 Eurocode 5: Design of timber structures
    Institution of Civil Engineers

75. **TRADA Technology Ltd (2006)**
    Eurocode 5 - An introduction
    TRADA

76. **Veistinen, J. and Pennala, E. (1999)**
    Finnforest Plywood Handbook
    Finnforest

77. **Young, W.C. and Budynas, R.J. (2002)**
    Roark's Formulas for Stress and Strain
    7th edition, McGraw-Hill

Index